SYNCHRONY IN CELL DIVISION AND GROWTH

Synchrony in Cell Division and Growth

Edited by

ERIK ZEUTHEN

Biological Institute of the Carlsberg Foundation,
Copenhagen, Denmark

1964

INTERSCIENCE PUBLISHERS
a division of John Wiley & Sons Inc.
New York—London—Sydney

Library of Congress Catalog Card No. 64–15650

PRINTED IN THE UNITED STATES OF AMERICA

Contributors

IVAR AGRELL	Zoophysiological Institute, University of Lund, Sweden
VICTOR W. BURNS	Biophysics Laboratory, Stanford University, Stanford, California
ALLAN CAMPBELL	Department of Biology, University of Rochester, Rochester, New York
JAMES R. COOK	Laboratory of Nuclear Medicine and Radiation Biology, University of California, Los Angeles, California
JOSEPH ENGELBERG	Department of Physiology and Biophysics, Medical College of the University of Kentucky, Lexington, Kentucky
RALPH O. ERICKSON	Department of Biology, University of Pennsylvania, Philadelphia, Pennsylvania
J. WOODLAND HASTINGS	Division of Biochemistry, University of Illinois, Urbana, Illinois
THOMAS W. JAMES	Department of Zoology, University of California, Los Angeles, California
HARALD LORENZEN	Pflanzenphysiologisches Institut der Universität Göttingen, Göttingen, West Germany
YOICHI MARUYAMA	The Institute of Food Microbiology, Chiba University, Okubo, Narashino, Chiba-ken, Japan
DANIEL MAZIA	Department of Zoology, University of California, Berkeley, California
YUJI MORIMURA	The Tokugawa Institute for Biological Research, Tokyo, Japan
ROBERT J. NEFF	Vanderbilt University, Nashville, Tennessee
RUTH H. NEFF	Vanderbilt University, Nashville, Tennessee
ALISON A. NEWTON	Department of Biochemistry, University of Cambridge, England
GEORGE M. PADILLA	Biology Division, Oak Ridge National Laboratory, Oak Ridge, Tennessee
PAUL PLESNER	Biological Institute of the Carlsberg Foundation, Copenhagen, Denmark
D. M. PRESCOTT	Department of Anatomy, University of Colorado School of Medicine, Denver, Colorado

LEIF RASMUSSEN — Biological Institute of the Carlsberg Foundation, Copenhagen, Denmark

OTTO H. SCHERBAUM — Department of Zoology, University of California, Los Angeles, California

BEATRICE M. SWEENEY — Biology Department, Yale University, New Haven, Connecticut

B. SYLVÉN — The Cancer Research Division of Radiumhemmet, Karolinska Institute, Stockholm, Sweden

HIROSHI TAMIYA — The Tokugawa Institute for Biological Research, Tokyo, Japan

B. THORELL — The Department of Pathology, Karolinska Institute, Stockholm, Sweden

NORMAN E. WILLIAMS — Department of Zoology, State University of Iowa, Iowa City, Iowa

D. H. WILLIAMSON — The Cell Biology Department, John Innes Institute, Bayfordbury, Hertford, Hertfordshire, England

TOMOMICHI YANAGITA — Institute of Applied Microbiology, University of Tokyo, Tokyo, Japan

ERIK ZEUTHEN — Biological Institute of the Carlsberg Foundation, Copenhagen, Denmark

Preface

This volume for the first time collects results and views of workers who have been actively engaged in studies which aim at removing some of the many barriers which the minuteness of the single cell sets for the study of the events which bring it from division to division. If cells in a mass population can be induced to march together from division to division then the population represents the individual and the cell cycle will be open to direct studies on a macro-scale with modern chemical and physical methods.

The synchronous culture approach has now reached a sufficient degree of maturity to deserve application also to the more complex systems representing, at one end of the scale, the relation between cells and their viruses, and at the other end, the mixed societies of cells which make the higher plants and animals. Can we in part or fully phase or synchronize cells representing whole simple organisms, or whole tissues, normal or neoplastic? What would be the theoretical and, in the latter case, the therapeutic value? Can we induce synchrony in whole populations of parasites in a host organism, thus imitating the malaria situation?

The cyclic shift between growth and division is fundamental in Nature, and therefore at the root of all biological disciplines. This cycle deserves to be studied by all possible methods. This book presents one approach which we hope will be useful in many biological fields.

Biological Institute of the Carlsberg Foundation
Copenhagen, Denmark

ERIK ZEUTHEN

Contents

Introduction *Erik Zeuthen* 1

Section I

SYNCHRONY OF CELLULAR AND NUCLEAR DIVISION IN TISSUES

1. Synchronous cell and nuclear division in tissues of the higher plants *Ralph O. Erickson* 11
2. Natural division synchrony and mitotic gradients in metazoan tissues *Ivar Agrell* 39

Section II

SYNCHRONY OF CELL DIVISION IN MICRO-ORGANISMS

A. THE SINGLE CELL

3. The normal cell cycle *D. M. Prescott* 71

B. INDUCED SYNCHRONY IN PROTOZOA

4. The temperature-induced division synchrony in *Tetrahymena* *Erik Zeuthen* 99
5. Structural development in synchronously dividing *Tetrahymena pyriformis* *Norman E. Williams* 159
6. Biochemical studies on synchronized *Tetrahymena* *Otto H. Scherbaum* 177
7. Nucleotide metabolism during synchronized cell division in *Tetrahymena pyriformis* *Paul Plesner* 197
8. Induction of synchronous division in amoebae *Robert J. Neff* and *Ruth H. Neff* 213

C. INDUCED SYNCHRONY IN ALGAE

9. Growth and cell division of *Chlorella* *Hiroshi Tamiya* 247
10. Phased cell division in the marine dinoflagellates *J. Woodland Hastings* and *Beatrice M. Sweeney* 307
11. Induced division synchrony in the flagellates *Thomas W. James* 323

D. SYNCHRONY IN FUNGI

12. Division synchrony in yeasts *D. H. Williamson* 351

13. Enzymes in synchronized yeast populations *B. Sylvén* and *B. Thorell* 381

14. Germinating conidiospores of *Aspergillus niger* *Tomomichi Yanagita* 391

E. INDUCED SYNCHRONY IN BACTERIA

15. Synchrony of bacterial populations as established by filtration techniques *Yoichi Maruyama* 421

16. Synchronization of division in bacteria by nutritional changes *Victor W. Burns* 433

F. INDUCED SYNCHRONY IN MAMMALIAN CELLS

17. Synchronous division of animal cells in culture *Alison A. Newton* 441

Section III

GENERAL CONSIDERATIONS

A. MATHEMATICAL CONSIDERATIONS

18. The theoretical basis of synchronization by shifts in environmental conditions *Allan Campbell* 469

19. Age distribution of cells in logarithmically growing cell populations *James R. Cook* and *Thomas W. James* 485

20. Measurement of degrees of synchrony in cell populations *Joseph Engelberg* 497

B. DETECTION OF SYNCHRONY IN AMOEBAE

21. Detection of synchrony induction in amoebae; problems in cell counting *Robert J. Neff* 509

C. DEVELOPMENT OF TECHNIQUES FOR FLAGELLATES

22. The development of techniques for synchronizing flagellates *George M. Padilla* and *James R. Cook* 521

Outlook *Daniel Mazia* 537

Section IV

TECHNICAL PROCEDURES

A. PROTOZOA

23. Techniques used in the study of synchronous *Tetrahymena*
Paul Plesner, Leif Rasmussen and *Erik Zeuthen* 543

B. ALGAE

24. Synchronous cultures of *Chlorella* starting from a homogeneous population *Hiroshi Tamiya* and *Yuji Morimura* 565

25. Synchronization of *Chlorella* with light–dark changes and periodical dilution of the populations to a standard cell number
Harald Lorenzen 571

26. The use of the electronic cell counter in studies of synchronous cell division in dinoflagellates
Beatrice M. Sweeney and *J. Woodland Hastings* 579

C. FUNGI

27. Techniques for synchronizing yeast cells *D. H. Williamson* 589

D. BACTERIA

28. Technique of synchronizing bacterial culture by filtration
Yoichi Maruyama 593

29. Techniques involved in synchronization of *Lactobacillus acidophilus* R26 *Victor W. Burns* 599

E. TISSUE CELLS

30. Special methods relating to tissue cells *Alison A. Newton* 601

Author index 609

Subject index 621

Introduction*

Erik Zeuthen, *Biological Institute of the Carlsberg Foundation, Copenhagen, Denmark*

In the past 10 years there has been a considerable interest in developing and studying mass populations of micro-organisms in which cell divisions and interphase preparations for division are phased or synchronized. A synchronous population has unusual advantages because millions of cells do the same thing at the same time. In the first place representative samples can be withdrawn without disturbance of the cyclic chain of events that we want to study. In the second place large enough samples can be obtained to permit sophisticated structural studies and chemical analyses which we have no hope of performing with single or few cells, the only alternative when we want to know the age between divisions of the cells which we analyze.

Synchronous culture is thus an amplification method, aimed at the biological material, and there is no apparent theoretical limit. The limit for the amplification is in the engineering necessary for work with large populations. Today, many synchronized cell systems are available, permitting studies under defined environmental conditions of synchronized cell cycles in bacteria, yeast, algae, ciliates, amoebae, and tissue cells.

The word 'synchrony' must not be read to indicate absolute simultaneity. It is here used as a relative term. This reflects the limitation of the synchronous culture methods. There is an early limit to the precision of the phasing and to the degree of synchrony which can be hoped for. A certain degree of synchrony must always represent a balance between synchronizing factors imposed upon the system and a multitude of desynchronizing tendencies inherent in a population and expressing themselves in a spread of individual generation times. This spread (Powell, 1958, Kubitschek, 1962) appears to be much the same in a culture in the steady state of growth and in the state of well synchronized growth (chapter 4). For better phasing to be obtained, this spread must be controlled, by way of the environment, or by the choice of genetically suitable

* Quotations are limited to situations or points not covered by or made in the chapters of this book.

cells, and no good general proposal as to how this can be done is yet available.

For the reasons discussed, environmental changes and the resulting synchrony of the cellular cycles have not been fully separated in all systems. 'Continuous synchrony' is monitored by continued changes of the environment (chapters 8, 11, 21, 22, 25). In a constant environment desynchronizing tendencies will soon revert such a system to the asynchronous state. If, however, this return is gradual and occurs through several cell cycles, this should be accepted because in this case the separate phenomena of synchrony induction and of induced synchrony can be separately studied, the latter cycling phenomenon in a constant environment (chapters 4, 9, 10, 12, 14, 15, 16, 17).

In bacteria the induction phase has been separately studied in the simplest situation when the environment is shifted one time. Even though no good synchrony results there are many advantages to this approach. Reference is made to a series of reviews (Maaløe, 1961, 1962; Schaechter, 1961).

A stated degree of synchrony indicates that a defined fractional number of the population is engaged in a particular activity. The 'synchrony' must come out better for a longer, and poorer for a shorter, segment of the cell cycle in which the cell is engaged in the activity selected for consideration. This situation accounts, on the one hand for the high degrees of induced division synchrony in *Tetrahymena* (chapter 4), and of the excellent natural mitotic synchrony in batches of fertilized eggs (chapter 2); and on the other hand, it is responsible for the low degrees of mitotic or meiotic synchrony in most other well phased systems. The outstanding exception to the last part of the statement is the natural system represented by anthers of lily plants (chapter 1). The means by which Nature controls division synchrony in this and other less spectacular cases are unknown. Though tempting, no author seems yet to have viewed the topics of induced and natural synchrony from one angle. This book brings the two topics together and invites such attempts.

Continued biological mass increase is based on a cyclic shift between growth and division of the biological unit, the cell. In genetically different cells the processes of growth and division are balanced differently, so in Nature the sizes of free cells vary greatly. A moderate estimate for mono-nucleated cells is that size variation is by a factor 10^7; or from bacteria ($\sim 10^{-9}$ μl) to large amoebae ($\sim 10^{-2}$ μl). Metabolic data (Zeuthen, 1953) suggest that the rate of life varies more or less in proportion to the surface/volume ratio of cells; this ratio being perhaps one hundred times higher in small, fast-growing bacteria than in the large, slow-growing and multiplying amoebae. In Nature there must have been evolutionary selection for the balance between growth and division, probably because the setting of this balance has selective value. The parameters directly involved may be a multitude of rate constants for vital functions.

Within limits set by the genetic composition of a cell its size and chemical composition (DNA, RNA, protein) varies with environmental factors (Schaechter, Maaløe and Kjeldgaard, 1958). However, by rigid control of axenic cultures, a number of *balanced growth* states (Campbell, 1957) can be defined. For each set of conditions generation time and composition of the average cell stay constant for generations. When, during one or more generations, cells assume larger or smaller size, and change in average chemical composition, we have *unbalanced growth* (Cohen and Barner, 1954). In many cases such populations are engaged in regulating from one balanced state to another. Growth and cellular differentiation in mixed societies of cells, such as our body, must be largely unbalanced because in the course of not too many cell generations one cell (the egg) gives rise to cells which vary greatly in chemical composition and in size ($\sim 10^5$ times). In the extreme cases of unbalanced growth, we have cell division without growth (the cleaving egg), or growth without division (differentiating cell), thus complete dissociation between growth and division. As shown by new examples in this book such partial or full dissociation is experimentally possible with many cells. While the field here discussed is clearly apart from the general topic of cellular differentiation it should be borne in mind that it extrapolates in that direction.

Cells from populations in balanced growth are out of phase with respect to the discrete chemical events which take each cell from division to division or which accompany this progress. Cell divisions are randomly distributed in time because in their cycle-activities cells are independent units. By definition, any deviation from this fully *asynchronous state* indicates a degree of unbalanced growth and extrapolates towards the fully *synchronous state*.

It follows that apart from starting cells out from a natural resting phase by the process of fertilization (chapter 2), or by germination of spores (chapter 14), there are two possible general approaches, now to be dealt with, to the problem of establishing *mass synchrony* (Scherbaum and Zeuthen, 1954) in cell populations with the purpose of supplying unlimited amounts of biological material for a detailed chemical description of the cell cycle and for a causal analysis of the events which steer the cell from division to division.

The theoretically simplest, but apparently technically most difficult, approach is to start *synchronous culture* from cells which have been separated in quantity from a large culture in balanced exponential, but asynchronous, growth. The separated cells must represent a short segment of the balanced growth cycle. When isolated they go through several divisions in some degree of *unforced synchrony* (Abbo and Pardee, 1960). Selection is for small or large cells (chapters 15, 28). Abbo and Pardee using *Escherichia coli* suggest that when the separation for small cells is gentle enough not to disturb the balanced state—and this is the technical problem—then the synthesis of DNA, RNA and of protein show no discontinuity in relation to division. This would then be the normal state for bacterial cells in populations with balanced growth. For a discussion of this and other problems in synchronous bacterial growth and division, reference is made to a recent review by Maaløe (1962).

In the other general approach, a main topic of this book, the balanced state is purposefully, and in some cases perhaps maximally, disrupted by exposing populations with random division to environmental changes. In response the populations show stepwise syntheses of important cell components and stepwise multiplication, reflected sometimes in visibly synchronous cell divisions. This situation is referred to as forced or *induced synchrony*. For the induction of synchrony, temperature changes were independently proposed for bacteria by Hotchkiss (1954) and for ciliates by Scherbaum and Zeuthen (chapter 4). Dark–light changes were applied to *Chlorella* by Tamiya *et al.* (chapter 9). The hypothesis behind both temperature studies predicted that preparation for division, continuous between divisions, is by chemical reactions which display themselves sequentially during interphase. Some reactions and some developmental phases would be more influenced than others by the same environmental change, and this would give the cells an opportunity to come into phase with each other. Temperature changes were chosen because they are unspecific in the sense that they can be expected to interfere with the balance between a multitude of reactions, therefore including processes directly related to division. Temperature cycles (chapters 4, 8, 11, 17), dark–light cycles (chapters 9, 10, 11, 22) and general nutritional changes (chapters 12, 13) should all be listed as unspecific and therefore potentially useful inducers of synchrony. Other workers (chapter 16) have guessed that DNA synthesis must precede nuclear and cellular division and in bacteria they have induced synchrony by short-time starvation for DNA-precursors. In this case the purposefully established imbalance is by interference with a specified macromolecular synthesis. This should help us to understand the mechanism of the induction.

The now considerable number of studies on induced division synchrony have stimulated theoretical considerations dealt with in chapters 18, 19 and 20. These considerations are applicable to systems in which environmental changes interfere in a differential manner with *the rate of continued progress* of cells in the cycle.

When synchrony is induced by differential interference with cellular processes which unfold themselves sequentially in time it is fairly obvious that the synchronized system must be distorted relative to the starting situation. When many have labelled the systems with induced synchrony 'unnatural' or 'abnormal' they have only, and sometimes negatively, restated the experimental approach. Any distortion of a synchronized system is relative to a standard, most often represented by the situation of the cell suspension before the synchrony induction. This standard is quite arbitrary, although for the technical purpose of reproducibility it is desirable that it is a balanced growth situation, in itself more 'natural' in the laboratory than in Nature. Confusion can be avoided if users of the words 'normal' and 'abnormal' always state which is the 'normal standard' to which they refer. A suitable standard may be the single cell, and its progeny which for a few generations may show excellent unforced synchrony. For such a standard to be useful the single cell must be large enough to permit at least a few critical experiments to be done in parellel with studies on synchronous populations. Unfortunately clones more often than not refuse to show balanced growth. Their status must then be described as deviations from an average. Some of the chapters of this book include studies on single cells. Chapter 1 is meant as a general reference.

The synchronized cell systems display cellular control mechanisms in response to environmental changes or imposed stresses. In this demonstration of capacity for regulation—in itself in the definition of life—the cells cover the full range of 'normality' of which each of us seems to have a different definition. (To the present author a cell is normal in its whole viable range.) It seems at least permissible to suggest that, in their play with the environment, cells are more plastic than we perhaps usually think, and that phased biochemical activities and synchronous division induced into cultures with previously balanced growth are manifestations of back regulations from ill-defined stresses towards definable balanced states. The balanced growth situation and the 'unforced synchrony' show how elegantly, and probably economically, bacterial cells can fit their syntheses in time; we may expect with a minimum of enzymes lying idle. However, the chemical discontinuities demonstrated by unbalanced and synchronized growth systems in bacteria and higher cells are likely to tell us much more about dissociability of chemical and physiological events, and of control mechanisms in operation.

It must not be forgotten that at the organizational level of the developing multicellular organism balanced states can usually not be defined. On the contrary, the overall developmental course is across the whole scale from one maximally unbalanced situation to the opposite imbalance. It is from cell division without growth (cleavage) to growth without cell division (terminal differentiation). Control mechanisms studied in free cells, by the synchronous culture approach, may supply basic information, also for the study of development in the higher organism.

With regard to the systems with induced division synchrony, the basic difficulty may now be how to define the points of attack in the cells of the external factors which induce the synchrony. In other words: what is the specific stress produced by an unspecific agent? At the first glance the situation looks even more complicated than what has been discussed above, because the early ideas behind the temperature-induced synchrony have undergone gradual modification (chapter 4). In the ciliates the synchrony induction by mild heating or chilling is not achieved simply by differential interference with *the rate of progress* of the cells in the cycle as initially assumed. These treatments produce *differential set-back in time*; the older cell suffering more set-back than the younger. This is what makes cells accumulate within narrow segments of the cycle. There is now evidence from many angles that preparation for division is not along a single channel in which chemical processes are sequentially arranged but rather along several independent channels which perhaps may meet in the sense that products which have been independently synthesized must be assembled in complex structures, the performance of which are displayed, e.g. at division. Furthermore, a growing complex structure may become useless and preparation for division may therefore become reversed following slowed development along a particular channel, or interference with the correct synthesis. A note in chapter 12 (yeast) and two lines in chapter 17 (tissue cells) suggest that the set-back concept may soon spread to synchronized systems other than the ciliate system.

Obviously, in attempting to understand the mechanisms by which unspecific environmental changes induce synchrony we run the danger of putting over-specific questions. The important step may occur at an organizational level which is one or more steps above the one at which we interfere directly. Many unspecific interferences may have a common specific effect. For the same reason we may be too confident that we have understood the mechanism of induction when it is by interference with defined chemical steps.

Reflecting a general anxiety to bring the cell cycle into the domain of molecular genetics considerable efforts have been directed towards following the rates of synthesis of DNA, RNA and of protein through the cell cycle. At this time incorporation of labelled precursor molecules into macromolecules of single cells are being squeezed for the information they will yield (chapter 1). The amplification offered by synchronous cultures permit the same syntheses to be followed by direct chemical analyses and, most important, all sorts of both low—and high—molecular biochemistry can be studied on the same samples (especially chapters 6, 7, 9 and 13). Morphogenetic studies (chapter 5) can now be correlated with the biochemical approach.

The molecular geneticist studies primary genetic events at the molecular level. Because he studies the cell as it is here and now he is in the same boat as the molecular biologist and as the cellular physiologist in the sense that he is largely deprived of the tremendous biological amplifications offered by the field of classical genetics. Synchronous cultures offer, in a different way, the amplification needed. So, it can be suggested that, in the future, synchronous culture techniques will materially aid in fusing genetics and cell physiology. In future studies we should therefore make attempts to use cells which are manageable both for synchronization and for genetic experiments. The happy choice may not yet have been made.

The material covered in this book borders, but does not enter, the field of biological periodicity and clock mechanisms. Chapter 10 draws the demarkation line. It suggests that a biological clock induces a diurnal division periodicity in populations of *Gonyaulax* cells. However, only under the one exceptional, and in fact not yet established, condition that growth doubles the population every 24 hours is there hope that the clock will phase or synchronize the whole population. If the analogy of the biological clock is the physical oscillator, then the processes which steer the cell through the cycle are not clocks. They are relatively imprecise, temperature dependent, and probably chemical of Nature.

REFERENCES

ABBO, F. E. and A. B. PARDEE (1960). Synthesis of macromolecules in synchronously dividing bacteria. *Biochim. et Biophys. Acta* **39,** 478–485.

CAMPBELL, A. (1957). Synchronization of cell division. *Bacteriol. Rev.* **21**, 263.

COHEN, S. S. and H. D. BARNER (1954). Studies on unbalanced growth in *Escherichia coli. Proc. Natl. Acad. Sci. U.S.* **40**, 885.

HOTCHKISS, R. D. (1954). *Proc. Natl. Acad. Sci. U.S.* **40**, 46–55.

KUBITSCHEK, H. E. (1962). *Exptl. Cell Research* **26**, 439–450.

MAALØE, O. (1961). The control of normal DNA replication in bacteria. *Cold Spring Harbor Symp. on Quantitative Biology XXVI*, 45–52.

MAALØE, O. (1962). Synchronous growth. In *The Bacteria*, Volume 4, chapter 1, p. 1 (I. C. Gunsalus and R. Y. Stanier eds.). Academic Press, New York.

POWELL, E. O. (1958). An outline of the pattern of bacterial generation times. *J. Gen. Microbiol.* **18,** 382–417.

SCHAECHTER, M. (1961). Patterns of cellular control during unbalanced growth. *Cold Spring Harbor Symp. on Quantitative Biology XXVI*, 53–62.

SCHAECHTER, M., O. MAALØE and N. O. KJELDGAARD (1958). Dependency of medium and temperature on cell size and chemical composition during balanced growth of *Salmonella typhimurium. J. Gen. Microbiol.* **19,** 592–606.

SCHERBAUM, O. and E. ZEUTHEN (1954). *Exptl. Cell Research* **6,** 221–227.

ZEUTHEN, E. (1953). Oxygen uptake as related to body size in organisms. *Quart. Rev. Biol.* **28,** 1–12.

SECTION I

SYNCHRONY OF CELLULAR AND NUCLEAR DIVISION IN TISSUES

1

Synchronous Cell and Nuclear Division in Tissues of the Higher Plants *

RALPH O. ERICKSON, *Department of Biology, University of Pennsylvania, Philadelphia, Pennsylvania*

CONTENTS

I Introduction 11
II Meristems 13
III Syncytia 15
1. Heterosporous plants 15
2. Gymnosperms 15
3. Flowering plants 17
IV Antheridia 21
V Sporangia 22
VI Microsporangiate Strobili of the Gymnosperms 23
VII Anthers 24
1. Sensitivity to ionizing radiation 27
2. Permeability 27
3. Respiration 28
4. Sulfhydryl compounds 29
5. Nucleic acid synthesis 30
6. Soluble proteins and free amino acids 33
References 34

I. INTRODUCTION

By and large, the impression which one gains on examining vegetative tissues of higher plants, in which cell divisions are going on (i.e. meristems), is that the cells are dividing at random with no suggestion of synchrony. Diurnal fluctuations of the mitotic index have, to be sure, been demonstrated in the root meristem and a few others. Also, one finds, on examining a longitudinal section of a root meristem, for example, that pairs of adjacent cells belonging to the same file can be found which are in nearly the same mitotic stage, and sometimes this correlation extends over four or more cells of a file. Altogether though, the degree of synchrony found in this and other meristems is quite low. If the interest in synchrony of cell division lies in the fact that a population of synchronously dividing cells presents an opportunity for analytical and experi-

* This article was prepared with the assistance of Mrs. Jane W. Robie, employed under a grant to the author from the National Science Foundation.

mental studies of cell division, of a kind which cannot be done with single cells or with an asynchronous population, plant meristems are of little interest in the present context. The data on meristems will, however, be reviewed briefly.

In reproductive tissues of plants, on the other hand, there are striking examples of synchrony. In many plants the female gametophyte, embryo sac, proembryo or endosperm undergoes an early syncytial stage. Each of these structures originates from a single cell, and in early stages nuclei divide without the formation of cell walls, sometimes giving rise to several hundred free nuclei. In some cases, a strikingly high degree of synchrony prevails through a number of nuclear generations. In other cases, there appear to be waves of mitotic figures progressing through the syncytium. At a later stage of development, cell walls typically appear quite suddenly, converting the syncytium into a cellular tissue.

Another circumstance in which synchrony can be observed is in the development of antheridia in the bryophytes, as well as in the psilophytes, horsetails (*Equisetum*), some lycophytes and ferns. In the bryophytes particularly, an antheridium produces a large number of sperm cells (antherozoids). Frequently a section of an antheridium shows a striking pattern of identical mitotic figures extending over a large block of cells.

The cell divisions resulting in the formation of spores within sporangia, and of pollen grains within anthers, are also frequently synchronous. In all of the plants considered here the spores (meiospores) are the direct products of meiosis, and it is observed, not only that meiosis is sometimes synchronized, but that the premeiotic mitoses, and in the case of pollen, the divisions of the microspores also are.

In reviewing these cases of natural synchrony below, I can make no pretence of completeness, or of balance, but will only cite instances which have come to my attention. There have been a large number of life history studies of plants, and these are summarized in a number of text books and monographs. For the most part, authors have failed to comment on the matter of synchrony, and rarely publish illustrations of more than single cells or small groups. In any case, virtually nothing beyond the descriptive facts of histology is known about the majority of these cases.

Of the organs mentioned above, the anthers of some flowering plants are the only ones which have been exploited to any extent for other than morphological description. In the anthers of *Trillium erectum* and *Lilium longiflorum*, to mention two examples, both meiosis of the microsporocytes and mitosis of the microspores are synchronized. Studies of radiation sensitivity, respiratory metabolism, nucleic acid synthesis, etc. have shown striking correlations with the cytological events, and have thrown some light on the physiology of cell division. Certainly, much more work can be done with anthers, and there are undoubtedly many other organs among the plants discussed here which would be well worth similar investigation.

II. Meristems

There are a number of old investigations of mitotic periodicity in roots of several species of flowering plants (Friesner, 1920, and references cited there; Stålfelt, 1921). In these studies, cytological preparations have been made of roots collected around the clock, and these have been scored for the number of nuclei in stages other than interphase, per unit area of tissue, or for the proportion of cells in the designated stages (i.e. for the mitotic index). Such data are highly variable, and in the absence of statistical tests, the significance of the fluctuations in some cases seems to me to be open to question. Of the older studies, Laughlin's (1919) investigation of the onion (*Allium*) root appears quite critical and is a clear demonstration of the reality of diurnal fluctuations of the mitotic index in roots which were subject to diurnal changes in temperature and light, or light alone. In two experiments, he collected roots every 10 min through a mitotic maximum (10.00 to 12.00 a.m.), sectioned them and scored nuclei selected at random, for ten 'active' stages (early prophase to late telophase), and interphase. From these extensive data, he calculated for each root a mitotic index (number of 'active' stages/total), a stage index (number of nuclei of a particular stage/ number of 'active' nuclei), and a procession index which is the stage index corrected for the 'average relative duration' of the stage in question. The procession indices were entered in a two-way table under stage and sampling time, and the maxima of the index for each stage were traced through time. In this way he arrived at average durations for each of the ten stages and for the 'active' portion of the mitotic cycle. At a constant temperature of 20°C, for instance, he found that the average duration from early prophase through late telophase was 81 min. He estimated the temperature coefficient (Q_{10}) for these stages, between 10°C and 20°C, as 1·2, and between 20°C and 30°C, as 2·6.

Jensen and Kavaljian (1958) have plotted mitotic indices for four morphological regions of the *Allium* root against time, from 10.00 a.m. to 10.00 p.m., and in general find sharp peaks at noon for each of the regions. The rooted bulbs were 'allowed a daily fluctuation of light and temperature', as in Laughlin's preliminary experiment. Although Friesner maintained that fluctuations would occur under constant environmental conditions, I know of no convincing evidence that this is so, and Laughlin seems to have felt that mitotic fluctuations would not have occurred without temperature or light fluctuations.

These experiments surely demonstrate some degree of synchrony of division in the root. However, it is not high. As Laughlin's tables show (e.g. Table 9), nearly all stages are to be found at any time, and at any time the most frequent stage is interphase. In a sharply synchronized system, one would expect to find a time at which nearly all the cells would be, for instance, at metaphase. Jensen and Kavaljian expressed their belief that periodicity could be utilized in physiological experiments and biochemical determinations on roots. I would agree that it would be quite interesting if the cell divisions of a root could be highly synchronized, but I know of no procedure by which this can be done.

There is a limited amount of data on periodicity of mitosis in other meristems. In connection with the discussions by various morphologists of Buvat's (1952) suggestion that the most apical portion of the shoot meristem in flowering plants may be inactive during vegetative growth, fairly extensive mitotic counts have been made by some investigators. In some cases the possibility of periodicity has been considered. Lance (1952) has made mitotic counts of sections, prepared each 3 hr, of shoot apices of *Vicia faba* grown in a greenhouse during the winter, and out-of-doors during spring and summer. She has presented good evidence of a mitotic peak at 9 a.m., and a smaller peak at 9 p.m. Savelkoul (1957) has reported mitotic indices for shoot apices of *Elodea densa* collected at 3-hr intervals for 24 hr, from plants growing in a greenhouse pool. The index values are quite variable, and she concludes that the observed fluctuations are not statistically significant. Popham (1958) has drawn a similar conclusion from his study of the shoot apex of *Chrysanthemum morifolium*, in which he made hourly collections for 24 hr from plants maintained on a long-day regime in a greenhouse.

Speese (1939) has made mitotic counts at 2-hr intervals for 24 hr of young leaves of *Smilax bona-nox*, which were presumably growing out of doors. Samples were removed with a 1-mm diameter leaf punch, and the total number of 'mitoses' recorded for each sample. Although her data are somewhat variable, there appear to be two peaks, shortly after midnight and shortly after noon.

III. Syncytia

1. *Heterosporous Plants*

In heterosporous plants, the development of the female gametophyte within the megaspore walls is syncytial in early stages. In *Selaginella*, for instance, the haploid megaspore nucleus gives rise by a series of free nuclear divisions to a fairly large number of nuclei, which are suspended in a thin layer of cytoplasm surrounding a large central vacuole. Later, cell walls are formed. Most authors (e.g. those cited by Foster and Gifford, 1959) make no reference to synchrony in these divisions. Lyon (1901), however, has published a figure of *S. apus* in which twelve mitotic figures can be seen, all of which are telophases, and another later stage, in which the majority of the nuclei appear to be in metaphase or anaphase. Though she does not comment, these figures suggest that synchrony may hold at least to the thirty-two nucleus stage.

In *Isoetes*, development of the megagametophyte is similarly syncytial but we have found no information on synchrony or lack of synchrony in the nuclear divisions.

Of the remaining heterosporous plants, the ferns *Marsilea*, *Regnellidium* and *Pilularia* have a much more limited development of the female gametophyte. The course of events in *Salvinia* and *Azolla* does not appear to have been adequately described.

2. *Gymnosperms*

In most of the 'gymnosperms' there is a syncytial female gametophyte of considerable extent, and in some also a free-nuclear stage of the proembryo. It is interesting that these plants have had a student, Chamberlain, who, unlike most authors of life histories, made specific note of the occurrence of synchronous divisions when he saw them. In his books (1919, 1932, 1935) he has brought together data on the syncytial stages of many genera.

Among the cycads, *Dioon* appears to be one of the most striking in the extent of its syncytial stages (Chamberlain, 1935). The female gametophyte has about 1000 free nuclei (haploid) before wall formation sets in. Synchrony of the mitoses is apparently quite good for about eight generations, when the megaspore has given rise to 256 nuclei. Virtually the same statements apply to the proembryo, where, as Chamberlain (1935) states, the early divisions are highly synchronous. After the 256-nucleate stage, in *Dioon* the proembryo shows a polarity, the nuclei in the micropylar end dividing sooner than at the base. Free-nuclear divisions continue in the basal end of the proembryo after they have ceased in the larger part of it, and these are simultaneous for a fairly large number of nuclei. This course of events is quite general in the cycads. *Zamia floridana* for instance has 256 free nuclei in the proembryo, and *Bowenia serrulata* has the least extensive syncytial proembryo with sixty-four nuclei.

Ginkgo is quite similar to the cycads in these respects. The first five mitoses of the megagametophyte are reported to be synchronous, and more than 256 free nuclei are formed before walls are formed. In the proembryo, there are also said to be 256 free nuclei before wall formation starts, and the early mitoses are synchronous.

According to Chamberlain (1935) all of the conifers are characterized by a free-nuclear stage in the early development of the female gametophyte, just as are the cycads and *Ginkgo*. Ferguson (1904), whom he quotes, found the free-nuclear divisions to be synchronous for ten nuclear generations in *Pinus strobus*, and this does not appear to be a unique example. Another form which has been carefully studied is *Torreya taxifolia*, which has 256 free nuclei before walls are formed (Coulter and Land, 1905). These examples could be multiplied. Synchronous mitoses also occur in the early stages of proembryo development of the conifers, though the number of free nuclei formed is less than in cycads. As Chamberlain (1935) summarizes the number varies from four in *Torreya* to thirty-two or sixty-four in *Agathis*.

Free-nuclear megagametophytes also occur in the Gnetales. In *Ephedra trifurca*, according to Land's (1904, 1907) careful studies, divisions are synchronous at least to the sixty-four-nucleate stage. Pearson (1906) has apparently made the best studies of *Welwitschia*. He estimates that about 1024 nuclei are formed before wall formation begins. In *Gnetum* also an extensive free-nuclear gametophyte is formed.

Embryo development in the Gnetales is very imperfectly known, but there is apparently no extensive free-nuclear stage, in any of the three genera.

I know of no physiological work which has been done to characterize these syncytial gametophytes and proembryos, aside from the demonstration that the 'endosperm' of *Ginkgo* contains growth-promoting substances (see below). Virtually all that is known is the classical descriptive histology and this is quite incomplete. One or two suggestive morphological correlations can be mentioned. Chamberlain (1935) apparently felt that the large size of the megaspore or egg cell from which these structures develop accounted for the unusual cytological events. It is also of some interest to note that the female gametophyte of many of these plants is surrounded by a tapetum-like tissue, which develops from the inner layers of the nucellus. Nearly all authors ascribe a nutritive function to these cells, though there is, of course, no experimental justification for such statements.

The ovules of the cycads, *Ginkgo* and the conifers are relatively large, and apparently can be dissected rather easily. In the later stages of proembryo development, at least, the several proembryos of an ovule can be teased free of the gametophyte tissue easily with a needle (Chamberlain, 1932). Many histological preparations have been made in this way.

3. *Flowering Plants*

In the flowering plants also, the megagametophyte, or embryo sac, is a syncytial structure. There is great variability in the structure of embryo sacs. The cytological details in a wide variety of plants have been studied, and well reviewed by Maheshwari (1950). In the 'normal', or, as Maheshwari classifies it, the *Polygonum* type of embryo-sac development, the megaspore nucleus gives rise, by three cycles of synchronous mitoses, to an eight-nucleate embryo sac. Some cell walls then appear and fertilization occurs at the eight-nucleate stage, one of the cells at the micropylar end of the embryo sac serving as the egg. In many plants only four nuclei and in others sixteen nuclei are formed. Examples of the latter are tetrasporic, in that the megasporocyte undergoes meiosis to form four haploid nuclei, and each of these divides twice to form sixteen nuclei. In all the cases with which I am familiar the mitoses which occur before fertilization are highly synchronous, although later divisions, which sometimes occur among the antipodal cells, may not be so.

Another structure in the angiosperm life cycle in which syncytial development occurs is the endosperm. The megagametophyte of the gymnosperms referred to above is sometimes called an endosperm, although it originates from a megaspore and is haploid. The endosperm of the angiosperms on the other hand originates by a second fertilization of one or more nuclei of the embryo sac, by a second sperm nucleus. It is usually triploid, or sometimes more highly polyploid. The extent and nature of endosperm development is highly variable (Maheshwari, 1950). In some cases walls are formed with each nuclear division; in other cases there is an extended syncytial stage, followed by wall formation; and in a third class, the helobial type, a transverse wall is formed at the first division of the primary endosperm nucleus. Of the two resulting cells, one develops syncytially, and the other by the usual cellular pattern. As an example of the free-nuclear type of development, Johri and Kapil's (1953) study of the endosperm of *Acalypha* might

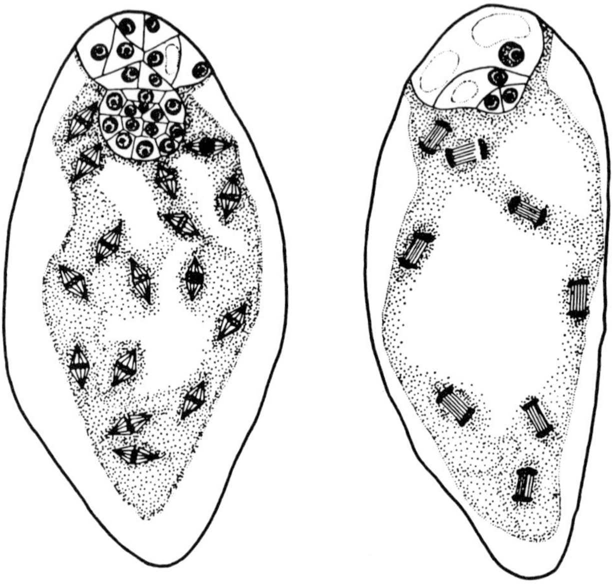

Fig. 1. Synchronous mitoses in the endosperm of *Acalypha indica* (Johri and Kapil, 1953).

be cited. They illustrate two ovules (Fig. 1) from which it can be concluded that close synchrony of nuclear division extends to at least the thirty-two-nucleate stage. Another example, the endosperm of *Fritillaria* (Fig. 2), is from Tischler (1922). In order to obtain a sample of the extensive morphological literature we have looked at each of the papers in the journal *Phytomorphology*, from 1951 to the present, which deals with endosperm development in angiosperms, about sixty-five in all. Of these, forty-five deal with the free-nuclear or helobial type of development, and about a dozen make some comment about synchrony of division, or illustrate synchronous divisions. The majority, however, make no mention of synchrony, although I suspect that it occurs quite generally.

A small amount of experimental and analytical work has been done with the endosperm of certain flowering plants, a part of it inspired by the discovery of Van Overbeek, Conklin and Blakeslee (1941) that coconut milk, which is the syncytial part of the endosperm of *Cocos nucifera*, contains a substance which promotes the growth of excised embryos of *Datura* and many other plant embryos and tissues in culture. Cutter and others have made a cytological study of nuclei isolated from coconuts, and describe various nuclear aberrations and multinucleate cells (Cutter, Wilson and Freeman, 1955; Cutter and Freeman, 1955). There have also been studies of

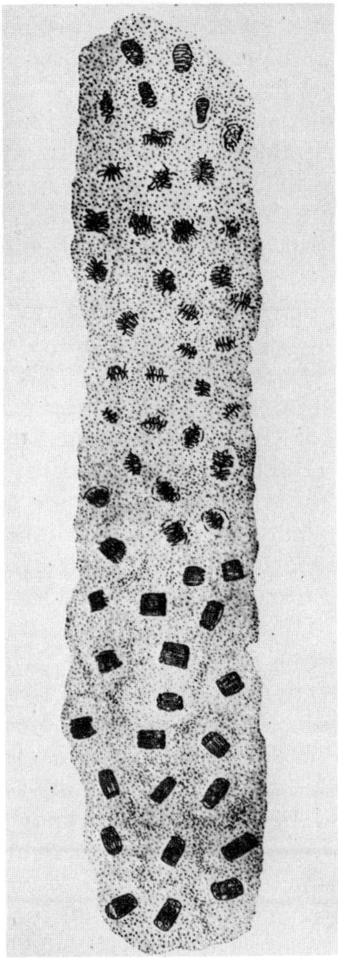

FIG. 2.
Endosperm of *Fritillaria imperialis* showing a gradient of mitotic stages (Tischler, 1922).

the respiration and acid phosphatase activity of this endosperm (Cutter, Wilson and Dubé, 1952; Wilson and Cutter, 1955). Nothing is said about synchrony of nuclear division in this syncytium, but the stages studied were quite advanced, and the early stages are apparently quite imperfectly known.

Steward *et al.* (1961) have studied the chemical nature of substances contained in coconut milk, and have also investigated the 'milk' of immature fruits of horse chestnut (*Aesculus*), walnut (*Juglans*), corn (*Zea*), banana (*Musa*) and *Ginkgo*. In their fractionation studies of the liquid endosperm of coconut, they have distinguished an active fraction and a neutral fraction, both of which are necessary to promote rapid cell division. Pollard *et al.* (1961) have identified three hexitols as active ingredients of the neutral fraction.

Bajer and Molè-Bajer (1954, and other refs.) have investigated the suitability of the endosperm of a number of species of flowering plants for study of mitosis in living nuclei. They point out that the endosperm has the advantage of lacking cell walls, and in early stages of being relatively free of cytoplasmic inclusions. They have worked out techniques for cutting open ovules, pressing out the endosperm and making hanging drop mounts in which the cells remain alive for many hours and can be flattened for study under high magnification. They give details of the appropriate stage at which to obtain fruit, and comment on the suitability of many species. Many genera of the Liliaceae, Amaryllidaceae and Iridaceae are suitable for these studies, because of their large chromosomes and the ease with which the endosperm may be obtained. Several dicotyledons are also suitable. In particular, Bajer has used *Haemanthus katharinae* extensively. He comments on the synchrony of nuclear division in a few cases.

It is interesting to note that Bajer speaks of the endosperm as consisting of separate cells, which are somewhat ameboid, and his illustrations bear this out. Despite their lack of a cellulose wall, a typical cell plate and phragmoplast are formed in cell division, as in other plant cells. It is therefore doubtful if one should refer to the endosperm as a syncytium, as I have done in this discussion. It would be interesting to know if the megagametophyte (or endosperm), of gymnosperms, which also appears as a syncytium in fixed material, shares these characteristics.

Bajer has made striking time-lapse motion pictures of mitosis in this endosperm material, and made detailed analyses of the movement of chromosomes during mitosis in normal cells (Bajer, 1958a, and other refs.), and of cells which have been subjected to β-irradiation (1958b), and to colchicine treatment (Molè-Bajer, 1958). Changes of mass of the chromosomes and other parts of the cell during mitosis have been measured with an interference microscope (Richards and Bajer, 1961), and studies of the birefringence of the spindle have been made with a polarizing microscope (Inoué and Bajer, 1961).

A number of authors have speculated about the existence of substances in the endosperm and other syncytia which are responsible for the synchrony of division. Tischler (1922) for instance hypothesized the existence of a diffusible *Reizstoffe* in embryo sacs and endosperms. Dixon (1946) suggested that there may be a mitotic hormone in the endosperm, on the basis of his observations of *Fritillaria imperialis* (Fig. 2). He fixed young seeds, and after slitting them open was able to dissect out the endosperm lining the inner surface of the ovule, and to mount it in balsam. He observed transverse bands of nuclei which were in successively more advanced stages of mitosis, from the antipodal to the micropylar end of the ovule, and suggested that a hormone initiating mitosis diffused from the micropylar end basally. At later stages, when cell walls have appeared in the endosperm, the synchrony of mitosis is not so apparent, and he suggests that the free diffusion of the hormone which was possible in the free-nuclear endosperm is now hindered by the walls. Dixon supposes that accidental blows, or local temperature fluctuations, on the walls of the fruit might alter the diffusion pattern, and hence mitotic pattern, but he does not report any experiments.

In contrast to the gymnosperms, the embryos of flowering plants are cellular from the first division of the zygote, and the cell divisions are not synchronous. One exception to this statement is *Paeonia* (Yakovlev and Yoffe, 1957) where the embryo appears to be free-nuclear until the sixteen-cell stage.

In summary, the syncytial stages in the life history of a number of plants would appear to present attractive material for experimental cytology. However, little experimental work has been done with them, and there are many gaps in our knowledge of their descriptive cytology.

IV. Antheridia

Throughout the bryophytes, rather massive antheridia are formed, containing several thousand sperm cells at maturity. It is a common observation that, in later stages of development, the androgonia, which give rise mitotically to the sperm cells, are arranged in rectangular blocks, within which divisions are synchronous. Each of these blocks of cells is apparently descended from a single cell of an earlier stage.

From an incomplete review of publications on the life history of liverworts (Hepaticae), it appears that most attention has been given to the sequence and position of laying down of the cell walls in the earliest stages, and of the cytological details of spermiogenesis in the final stage of sperm development. We have found references to simultaneous division of blocks of androgonia in the following genera: *Sphaerocarpos* (Nevins, 1932), *Plagiochila* (Johnson, 1929), *Riccia* (Black, 1913), and *Fossombronia* (Humphrey, 1906).

From the similarity in appearance of the antheridia of the Anthocerotae and the mosses, I expect that synchrony occurs in sperm development in these groups also, but have not felt it worthwhile to search for references.

Antheridia also occur in the psilophytes, horsetails, lycopods and ferns. In general they are considerably smaller than in the bryophytes, and are imbedded in the tissue of the gametophyte. It is my impression that synchrony of division of the androgonial cells is the rule in these groups also. A single instance which might be cited is *Osmunda* (Campbell, 1892).

It might be mentioned in passing that Döpp (1950) and Näf (1958) have demonstrated the existence of a specific substance, produced by fern prothalli, which initiates the formation of antheridia. This substance has been highly concentrated, and its physiology has been worked out in some detail. However, I do not believe that it is known whether this substance plays a role in the continued development of the antheridia, once initiated.

In heterosporous forms, and in all of the seed plants, the development of the microgametophytes is quite limited, and only a small number of sperm cells are produced from each microspore.

V. Sporangia

In all of the bryophytes, the sporophyte phase of the life cycle is limited in extent, and largely devoted to spore production. Often the number of spores produced in a single capsule is very large, particularly in the liverworts. In *Marchantia*, for instance, there is O'Hanlon's (1926) often quoted estimate of 300,000. After some search, we have found no specific information about synchrony or the lack of it in the development of spores in the liverworts.

In the moss, *Mnium hornum*, Wilson (1909) states that the premeiotic mitoses are simultaneous for the groups of sixteen to thirty-two cells into which the archesporium is divided. The meiotic divisions, however, are simultaneous for all the sporocytes within the capsule. Other examples could no doubt be found.

Of the lower vascular plants, *Psilotum*, *Tmesipteris* and *Equisetum* are characterized by a peculiar disintegration of much of the sporogenous tissue into a plasmodial mass, which is supposed to nourish the remaining sporocytes and resulting spores. We have found no information as to synchrony in these processes. Synchrony of the premeiotic mitoses and of meiosis has been described for *Lycopodium* (Sykes, 1908). *Isotes* is said to have the largest sporangia among living plants, being 5 mm or so long and in the case of the microsporangia, containing of the order of 10^6 microspores (Eames, 1936). According to Smith (1900) the premeiotic divisions leading to microsporocytes are not synchronous, but the meiotic divisions are quite well synchronized.

Among the ferns, the more primitive groups have relatively large sporangia, sometimes containing a few thousand spores each. In *Botrychium*, for instance, Cardiff (1905) states that divisions up to the sporocyte stage are simultaneous throughout the sporogenous mass. At about the time that the sporocytes differentiate, they separate into blocks, and the synchrony of meiosis which follows is poorer, although it is maintained within each block of cells. Since there appear to be at least 2000 sporocytes formed, this would imply at least eleven cycles of synchronous mitosis. The situation in *Ophioglossum* and *Helminthostachys* appears to be similar (Burlingame, 1907). In these genera, the sporangia are restricted to a portion of a frond, and there is a suggestion that different sporangia may be at nearly the same stage of development.

Other instances of synchronous spore development may be cited among the many other genera of ferns. In the majority, which are leptosporangiate, the sporangia are smaller, often containing thirty-two or sixty-four spores. Genera such as *Osmunda*, *Onoclea*, *Aneimia*, etc. in which separate sporophylls are produced might be of some interest in physiological studies.

VI. Microsporangiate Strobili of the Gymnosperms

In the cycads, *Ginkgo*, the conifers and the Gnetales, the microspores develop into pollen grains within the sporangia. The microsporophylls, which bear various numbers of sporangia, are in turn clustered into strobili or male cones. Though we have made no survey, I suspect that some degree of synchrony in these developmental processes is the rule. One example will suffice. Chamberlain (1935) has illustrated a portion of a microsporangium of *Pinus laricio* with about twenty microsporocytes in meiotic stages ranging from first telophase to quartets of microspores. He and other authors regularly cite specific dates at which particular stages were collected, implying perhaps that there may be a considerable amount of correlation of developmental stages of pollen throughout a tree. As a matter of fact, one is reminded of the reports of great clouds of pollen shed simultaneously by numerous pines in a forest, and is tempted to speculate about the possible adaptive value of synchrony of the developmental processes leading to pollen formation.

VII. Anthers

In the anthers of flowering plants, a succession of mitoses of the sporogenous tissue of the young anther gives rise to microsporocytes. These undergo meiosis, and the resulting microspores undergo a further mitotic division to produce a tube cell and a generative cell. In some plants the generative cell divides again to form the two sperm nuclei before the pollen is released; in others this last division occurs during pollen germination. These stages have been examined in a very large number of plants, especially in connection with studies of chromosome number and morphology in application to taxonomy (Darlington and Wylie, 1955). It is a commonplace observation in such studies that an anther of the right stage will contain a high proportion of cells at the same stage of meiosis or mitosis. In some plants, such as those of Ranales, with numerous stamens, there may be stamens of various stages of development within a single flower. In families with a small and definite number of stamens in a flower, these may all be in very nearly the same stage. As in the instances of synchronous division referred to above, the majority of descriptive cytological papers make no explicit mention of synchrony, and therefore we have made no attempt to review this literature. Instead, the remainder of this article will be largely devoted to work of a more analytical and experimental sort, principally with the liliaceous genera, *Trillium* and *Lilium*, in which the synchrony of division has been clearly recognized and made use of.

In *Trillium*, and presumably in other spring-flowering plants with underground perennating organs, flower primordia are differentiated in autumn, and grow slowly during winter. In *Trillium erectum*, Huskins and Smith (1935) reported that early prophase stages of meiosis of the microsporocytes were found in rhizomes collected in November in Quebec, and that meiosis is completed in late winter. Sparrow and Sparrow (1949) have made recommendations about obtaining rhizomes, storing them at low temperature, and making smear preparations from the anthers. They report approximate durations in days for the various stages of microsporogenesis, at 4–6°C: more than thirty days for leptotene and zygotene, 40–60 days for the remaining meiotic stages, 28–40 days for the microspore interphase and 8–14 days for microspore mitosis.

It is argued by several authors that this slow rate of division, together with the natural synchrony and the low number of large chromosomes, present advantages for many kinds of investigations. Sparrow, Moses and Steele (1952) have published a striking figure to demonstrate the degree of synchrony found at meiosis (Fig. 3).

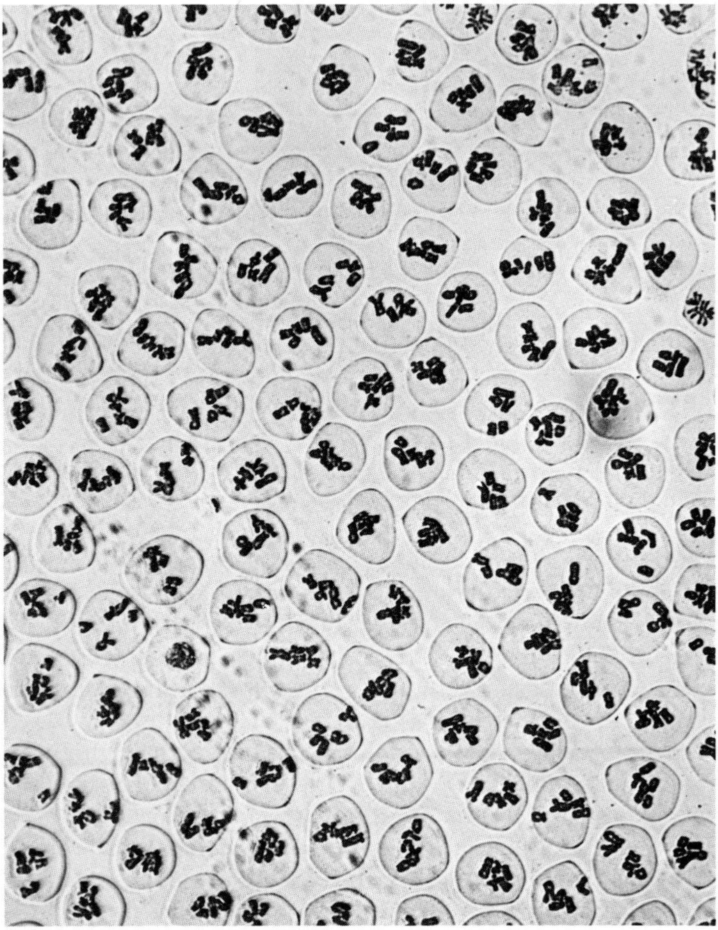

Fig. 3. Smear of microsporocytes from an anther of *Trillium erectum* in first meiotic division. (Courtesy of A. H. Sparrow.)

In *Lilium longiflorum*, Erickson (1948) has made a detailed study of the growth of the flower buds and organs contained within them.

In *Lilium* the flower primordia are not formed within the bulb, but appear at the apex of the shoot after it has emerged from the bulb. The flower buds are accessible for measurement without injury to the plant from the time they are about 10 mm in length. From this stage until a day or two before anthesis, when the flower is about 150 mm long, growth in length is exponential, i.e., the logarithm of bud length is a linear function of time, and hence serves as a con-

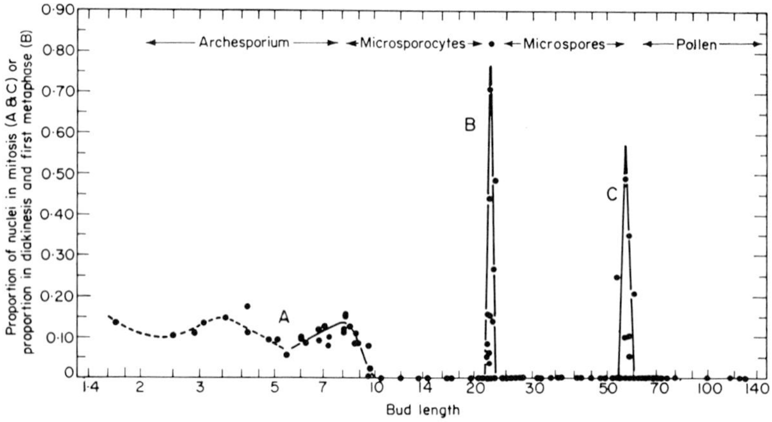

FIG. 4. Proportions of cells in premeiotic mitosis (A), diakinesis and first metaphase of meiosis (B), and microspore mitosis in anthers of *Lilium longiflorum* related to flower bud length (Erickson, 1948).

venient and precise 'developmental index'. Using this index, growth curves for the anthers, filaments, styles and ovaries were worked out by dissecting out these organs, and plotting measurements of various kinds against log bud length. Cytological studies were made of transverse sections of the youngest anthers, and of smear preparations of the contents of older ones (Fig. 4).

The degree of synchrony was found to be quite high for the meiotic stages, and fairly high at the microspore mitosis. The earliest stages of meiotic prophase were recognized at about 13·5 mm bud length, the first metaphase at about 22·5 mm, and meiosis was completed at 25 mm. The microspore mitosis occurs at 60 mm more or less, depending on the variety and cultural conditions. Some cytological data on the stages of embryo-sac development were also presented. Information about the mitotic divisions in the tapetum has been presented by other authors (Taylor and McMaster, 1954). These cytological processes are rather slow in *Lilium* also, the interval from first metaphase of meiosis to the microspore mitosis, for example, is nearly 2 weeks under normal greenhouse conditions.

The anthers of *Trillium* and *Lilium* have been used for a variety of studies in which the synchrony of meiosis and mitosis have played a role. Topics which have been investigated are sensitivity of the chromosomes to breakage by ionizing radiation, permeability of the cell membrane at various stages of division, the respiratory metabolism of the dividing cells, and nucleic acid synthesis of the cells of the anther.

1. *Sensitivity to Ionizing Radiation*

Sparrow and other investigators have carried out a series of studies to determine at which stage of the meiotic coiling cycle chromosomes are most susceptible to breakage by X-radiation. The work on *Trillium* is summarized by Sparrow (1951) and Sparrow, Moses and Dubow (1952). Anthers containing microsporocytes or microspores at known stages were irradiated, and scored at later stages, such as the first meiotic anaphase or the microspore metaphase, for the number of chromosome fragments found. It was found that the number of fragments varied not only with the stage irradiated, but also depended upon the stage at which the scoring was done. When cells were scored at anaphase I of meiosis, the early prophase stages of meiosis appeared to be most sensitive to radiation damage, whereas scoring at the microspore metaphase indicated that the sensitivity was greatest at diplotene, diakinesis and metaphase I, when the chromonemata are most contracted. The total number of fragments seen at both anaphase I and microspore metaphase also indicated that the chromosomes are most sensitive to damage at their most contracted stages. Conclusions by other investigators working with other organisms and tissues appear to be at variance with Sparrow's. Clearly the phenomenon of chromosome breakage is a complex one, involving among other factors the rejoining of broken ends, and I am not able to evaluate the issues in this controversy critically. Sparrow's position appears to be that the anthers which he has used offer advantages over other materials, in that a large number of cells can be irradiated at nearly the same stage, and that these cells can be examined at subsequent divisions so that the total number of aberrations can be estimated.

2. *Permeability*

The permeability of microsporocytes, microspores and pollen of *Trillium erectum* was investigated by Stern (1946a, b, 1948). His technique was to plasmolyze the cells in solutions of sucrose and other solutes, and to record the time required for deplasmolysis. The reciprocal of deplasmolysis time is a rough measure of the permeability of the cells to the solute. He found that in the earliest stages of meiosis, the microsporocytes are so highly permeable to sucrose (1·5 *M*) that they cannot be plasmolyzed. As meiosis progresses, the permeability decreases, and the resulting microspores are highly impermeable to sucrose, as are other plant cells, with deplasmolysis times of the order of 6 hr. At prophase of the microspore mitosis the deplasmolysis time decreases again, and at metaphase becomes only a few minutes. With completion of this mitosis, the cells again become highly impermeable to sucrose. Stern has presented evidence that the cells are similarly permeable to other solutes, when they are dividing, and that there are other protoplasmic changes correlated with division processes.

3. *Respiration*

The respiration of *Lilium* anthers was studied by Erickson (1947), by determining the rate of oxygen uptake manometrically for anthers from buds of known length. When these rates ($\mu l\ O_2.hr^{-1}.g^{-1}$ fresh weight) were plotted against log bud length, a fairly complex curve resulted, showing, in addition to a general decline as the anther grows from an early stage to maturity, two minima at bud lengths slightly over 20 mm, and between 50 and 60 mm. These are the bud lengths at which the first metaphase of meiosis, and the microspore mitosis, respectively, were found in the cytological study (Erickson, 1948). In an earlier paper Guilcher (1937) presented similar respiratory data, but no cytological information, for anthers of *Lilium croceum*. Stern and Kirk (1948) have carried out a respiratory study of anthers of *Trillium erectum*, using a specially designed microrespirometer. The stages of development of the anthers were judged from cytological examination of each, and embraced the premitotic interphase, microspore mitosis, and postmitotic stages. Their data indicate a respiratory minimum at mitosis, just as in *Lilium*, but in addition a maximum just preceding prophase of mitosis. It was thought by both Erickson and Stern that these fluctuations in respiratory rate were damped by the presence of wall tissue, and that they might be more pronounced if microspores were isolated. Accordingly Nasatir (1958) devised methods for preparing microspore suspensions from anthers of *Lilium*, counting them and determining their rate of oxygen consumption and carbon dioxide evolution. Nasatir and Stern (1959) have also studied the activity of the glycolytic enzymes, aldolase and triose phosphate dehydrogenase, in isolated microspores. The respiratory data are broadly consistent with those on whole anthers. As anticipated they show greater fluctuations in rate, with a high rate preceding mitosis, an exceedingly low rate during mitosis, and a subsequent return to a high rate. Although some authors have suggested that glycolysis provides the energy for mitotic processes, the rate of carbon dioxide production and the aldolase and dehydrogenase activities are at a low level during active mitosis, just as the rate of oxygen consumption is. These striking metabolic findings certainly deserve further study.

4. *Sulfhydryl Compounds*

Studies of sulfhydryl compounds in the anthers of *Lilium* and *Trillium* made by Stern (1956, 1958) have shown marked fluctuations which correlate with the cytological processes. In *Lilium*, there are two peaks in the concentration of soluble sulfhydryl compounds such as glutathione, corresponding with the zygotene stage of meiotic prophase, and with microspore mitosis, respectively. Protein-SH determinations were also made from about 40 to 100 mm bud length to include stages before, during and after the microspore

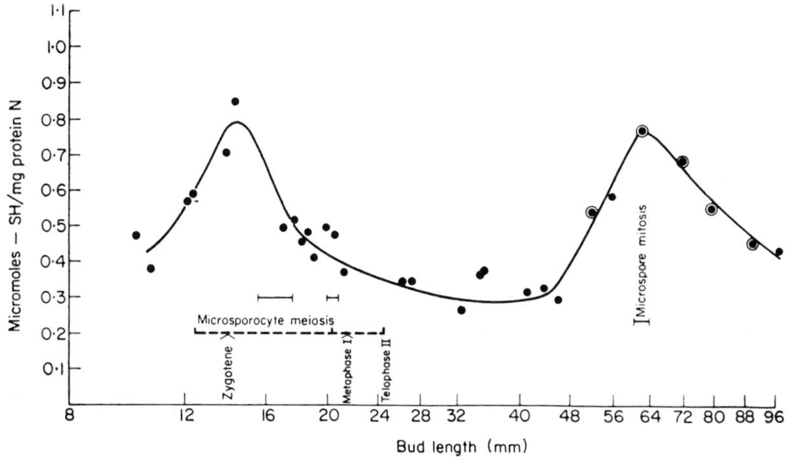

FIG. 5. Concentration of sulfhydryl compounds in anthers of *Lilium longiflorum* related to bud length (Stern, 1958).

mitosis, and showed little change in concentration (Fig. 5). Somewhat similar results were obtained with the *Trillium* anthers. Stern appears to favor the view that the high concentrations of sulfhydryl result by *de novo* synthesis, and that they have a metabolic role, rather than a purely structural role in spindle formation. Perhaps sharper cytological correlations would be found if analyses were made on isolated microspores and pollen.

5. *Nucleic Acid Synthesis*

Lilium anthers have been used by several investigators for studies of nucleic acid synthesis in relation to meiosis and mitosis. Ogur *et al.* (1951) prepared suspensions of microspores and pollen from buds of known length, estimated the number of cells, and analyzed them for deoxyribose and ribose nucleic acid (DNA and RNA) content. The amounts of DNA and RNA per cell were plotted against log bud length to show the time course of deposition of the nucleic acids. These data indicated a rapid doubling of the DNA content of the microspores preceding visible mitosis. In addition there appeared to be a continual slow increase during the microspore interphase which precedes mitosis, and a further increase during the interphase of the maturing pollen grains, such that a mature pollen grain contains about seven times as much DNA as the microspore from which it develops. There now appears to be some doubt of the validity of the extraction techniques used for these analyses, and they have not been directly verified. Foster and Stern (1959; also Stern 1960a, b) have made analyses of whole anthers by similar methods and find a broad peak for DNA content at 30–40 mm bud length (Fig. 7). Since their data pertain to the anther wall and tapetum, as well as the microspores, it is not surprising that they do not correspond to the data of Ogur *et al.* Linskens' (1958) data on *Lilium henryi* cannot be reconciled with these, since they are expressed on a dry weight basis, rather than as DNA per cell or anther.

The periods of DNA synthesis in microspore development have also been investigated by microphotometric analyses of the cells, and by autoradiographic detection of the incorporation of labelled phosphate and nucleotides. In these studies, cuttings of *Lilium* and of *Tradescantia* were placed in solutions of phosphate containing ^{32}P, and at various intervals after this treatment smear preparations and radioautographs were made (Taylor, 1953, Taylor and McMaster, 1954). It was found that incorporation of the ^{32}P into the chromosomes occurred only at three specific short periods (Fig. 6). The first of these was in late premeiotic interphase, at 10–11 mm bud length (*Lilium*). Another period of incorporation was during the latter part of the microspore interphase, at 50–55 mm, and the third during the interphase following the microspore mitosis, associated with the differentiation of the generative nucleus of the pollen grain, at 70–75 mm. The microphotometric determinations of the DNA content of the cells were plotted against time of development, and confirmed the radioautographic conclusions. The premeiotic period of DNA synthesis has also been identified by autoradiographic detection of ^{14}C and ^{3}H in nuclei of microsporocytes after treatment with labelled orotic acid and cytidine (Taylor, 1959), and the same techniques showed that rapid protein synthesis also occurs in the

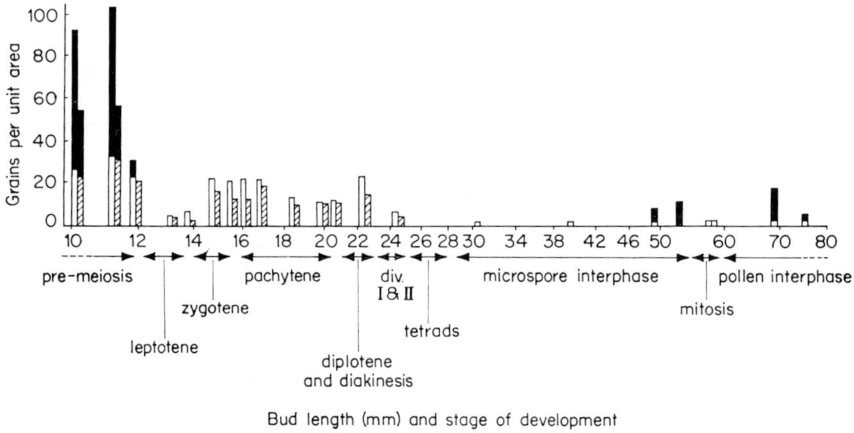

FIG. 6. Periods of incorporation of ^{32}P into nuclei of cells in the anther of *Lilium longiflorum* (black bars), related to flower bud length (Taylor and McMaster, 1954).

premeiotic interphase. Taylor has discussed a basis on which the data of Ogur *et al.* can be reconciled with the results of his tracer studies.

In *Trillium* Hotta and Stern (1961) have presented DNA analyses of microspores and pollen which clearly show a doubling during the microspore interphase, well before mitotic prophase, and this is supported by their (1963) data on incorporation of thymine and cytosine into the microsporocytes and microspores.

There are also correlations of the cytological events and of nucleic acid synthesis with transient peaks in concentration of presumed nucleic acid precursors. In *Lilium* anthers Foster and Stern (1958, 1959) have found transient peaks of concentration of deoxyribosides some twenty-five times as high as the base level (Fig. 7). The earliest of these peaks, which may also consist of nucleotides, occurs in anthers from 10 mm buds, and appears to precede the cycle of DNA synthesis which occurs in the premeiotic interphase of the microsporocytes. Between 50 and 60 mm bud length, two sharp peaks occur which are associated with the DNA doubling in late microspore interphase; and again between 75 and 90 mm bud length, two peaks occur at the time when DNA is synthesized in the maturing pollen grains. The materials extracted have not been completely characterized, but they appear not to be entirely mononucleotides, and there are differences in solubility of the premeiotic and the microspore pools. The presumption is that they are breakdown products of the DNA of the tapetum and wall tissue,

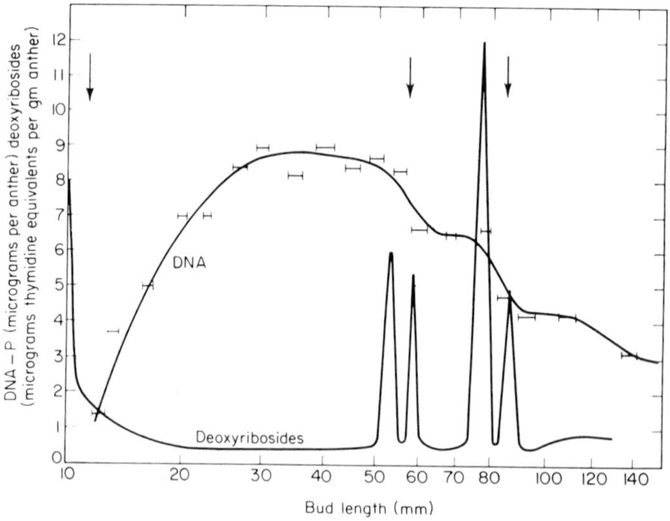

FIG. 7. Concentration of deoxyribosides and of DNA in anthers of *Lilium longiflorum*. Vertical arrows indicate periods of DNA synthesis in microspores, etc. (Stern, 1960).

and that they form a pool of precursors for the synthesis of DNA of the microsporocytes, microspores or pollen. Stern (1961) has also found remarkable transient appearances of deoxyribonuclease in anthers of particular stages, and relates the activity of this enzyme to the appearance of the high concentrations of deoxyribonucleotides which he previously found. The anthers also exhibit marked activity in deamination of deoxycytidine and 5-methyldeoxycytidine at 50–55 mm and again at 75–90 mm bud length with lower activities at other stages (Hotta and Stern, 1961a). It is remarkable in this case that it is the anther walls which are active, the microspores and pollen having virtually no activity. The isolated microspores of *Trillium*, on the other hand, show a transient high activity in phosphorylation of thymidine (Hotta and Stern, 1961b) and of incorporation of ^{14}C-leucine (Hotta and Stern, 1963) during the interphase preceding division of the microspore nuclei.

These results, indicating a localization of catabolic activities in the anther wall, and of synthetic activities in the microspores, are reminiscent of the classical cytological interpretation that the tapetum, which lines the anther locules, serves a nutritive function for the developing microspores and pollen. To investigate this question more directly, Takats (1962) undertook to determine by means of tritium-labelled thymidine whether the label incorporated into tapetal nuclei would later appear in the nuclei of microspores or pollen. There are two periods of tapetal DNA synthesis at which the label can be introduced. At the second of these, at 22 mm bud length, when the microsporocytes were in pachytene, Takats was able to label tapetal nuclei and to get continued normal growth of the anthers. After 18 days, radioautographs of the sectioned anthers showed that the tapetum had disintegrated and that the microspore walls, but not microspore nuclei, were heavily labelled. This confirms the classical cytological interpretation that the tapetum plays a role in formation of the outer walls of spores, and indicates that tapetal DNA does not contribute to the synthesis of microspore DNA.

6. *Soluble Proteins and Free Amino Acids*

In addition to these nucleic acid data, some preliminary information has been obtained about soluble proteins and free amino acids in *Lilium* anthers. Nasatir *et al.* (1960) made phosphate-buffer extracts of anthers from 40, 50, 60 and 80 mm flower buds, and analyzed them by electrophoresis in the Tiselius apparatus. Four components were identified, of which one was most prominent at 40 and 50 mm bud length, preceding the microspore mitosis. Two other components were prominent at 60 mm, the time of microspore mitosis, and at lower concentrations at other bud lengths. The fourth component remains at fairly uniform concentration throughout. In a study of free amino acids, by column chromatography, Nasatir *et al.* (1961) found changes in the concentrations of several amino acids, with anther development. The most striking were the high concentrations of histidine and arginine at 52·5 mm coincident with the peak in deoxyribosidic compounds reported by Foster and Stern (1958), and of lysine and histidine at 60 mm when mitosis occurs.

REFERENCES

BAJER, A. (1958a). Cine-micrographic studies on mitosis in endosperm. IV. The mitotic contraction stage. *Exptl. Cell Research* **14,** 245–256.

BAJER, A. (1958b). Cine-micrographic studies on chromosome movements in β-irradiated cells. *Chromosoma* **9,** 319–331.

BAJER, A. and J. MOLÈ-BAJER (1954). Endosperm: Material for study of the physiology of cell division. *Acta Soc. Botan. Polon.* **23,** 69–110.

BLACK, C. A. (1913). The morphology of *Riccia frostii* Aust. *Ann. Botany* (*London*) **27,** 511–532.

BURLINGAME, L. L. (1907). The sporangium of the Ophioglossales. *Botan. Gaz.* **44,** 34–56.

BUVAT, R. (1952). Structure, évolution et fonctionnement du point végétatif de *Vicia faba* L. *Ann. sci. nat., Botan. et biol. végétale* ser. 11, **13,** 301–339.

CAMPBELL, D. H. (1892). On the prothallium and embryo of *Osmunda claytoniana* L. and *O. cinnamomea* L. *Ann. Botany* (*London*) **6,** 49–92.

CARDIFF, I. D. (1905). The development of the sporangium in *Botrychium*. *Botan. Gaz.* **39,** 340–347.

CHAMBERLAIN, C. J. (1919). *The Living Cycads.* University of Chicago Press, Chicago.

CHAMBERLAIN, C. J. (1932). *Methods in Plant Histology.* University of Chicago Press, Chicago.

CHAMBERLAIN, C. J. (1935). *Gymnosperms, Structure and Evolution.* University of Chicago Press, Chicago.

COULTER, J. M. and W. J. G. LAND (1905). Gametophytes and embryo of *Torreya taxifolia*. *Botan. Gaz.* **39,** 161–178.

CUTTER, V. M., Jr., and B. FREEMAN (1955). Nuclear aberrations in the syncytial endosperm of *Cocos nucifera*. *J. Elisha Mitchell Sci. Soc.* **71,** 49–58.

CUTTER, V. M., Jr., K. S. WILSON and L. F. DUBÉ (1952). The endogenous oxygen uptake of tissues in the developing fruit of *Cocos nucifera*. *Am. J. Botany* **39,** 51–56.

CUTTER, V. M., Jr., K. S. WILSON and B. FREEMAN (1955). Nuclear behavior and cell formation in the developing endosperm of *Cocos nucifera*. *Am. J. Botany* **42,** 109–115.

DARLINGTON, C. D. and A. P. WYLIE (1955). *Chromosome Atlas of Flowering Plants.* (2nd ed.) Allen & Unwin, London.

DIXON, H. H. (1946). Evidence for a mitotic hormone: observations on the mitosis of the embryo-sac of *Fritillaria imperialis*. *Sci. Proc. Roy. Dublin Soc.* **24,** 119–124.

DÖPP, W. (1950). Eine dei Antheridienbildung bei Farnen förderndes Substanz in die Prothallien von *Pteridium aquilinum* (L.) Kühn. *Ber. Deut. Botan. Ges.* **63,** 139–147.

EAMES, A. J. (1936). *Morphology of Vascular Plants, Lower Groups.* McGraw-Hill, New York.

ERICKSON, R. O. (1947). Respiration of developing anthers. *Nature* **159,** 275.

ERICKSON, R. O. (1948). Cytological and growth correlations in the flower bud and anther of *Lilium longiflorum*. *Am. J. Botany* **35,** 729–739.

FERGUSON, M. C. (1904). Contributions to the life-history of *Pinus*, with special reference to sporogenesis, the development of the gametophytes, and fertilization. *Proc. Wash. Acad. Sci.* **6,** 1–202.

FOSTER, A. and E. GIFFORD (1959). *Comparative Morphology of Vascular Plants.* W. H. Freeman, San Francisco.

FOSTER, T. S. and H. STERN (1958). Soluble deoxyribosidic compounds in relation to duplication of deoxyribonucleic acid. *Science* **128,** 653–654.

FOSTER, T. S. and H. STERN (1959). The accumulation of soluble deoxyribosidic compounds in relation to nuclear division in anthers of *Lilium longiflorum. J. Biophys. Biochem. Cytol.* **5,** 178–192.

FRIESNER, R. C. (1920). Daily rhythms of elongation and cell division in some roots. *Am. J. Botany* **7,** 380–407.

GUILCHER, J. M. (1937). Intensité respiratoire comparée de l'androcée et du gynécée. *Compt. Rend. Acad. Sci., Paris* **204,** 190–192.

HOTTA, Y. and H. STERN (1961a). Deamination of deoxycytidine and 5-methyldeoxycytidine in developing anthers of *Lilium longiflorum* (var. Croft). *J. Biophys. Biochem. Cytol.* **9,** 279–284.

HOTTA, Y. and H. STERN (1961b). Transient phosphorylation of deoxyribosides and regulation of deoxyribonucleic acid synthesis. *J. Biophys. Biochem. Cytol.* **11,** 311–319.

HOTTA, Y. and H. STERN (1963). Inhibition of protein syntheses during meiosis and its bearing on intracellular regulation. *J. Cell Biol.* **16,** 259–279.

HUMPHREY, H. B. (1906). The development of *Fossombronia longiseta* Aust. *Ann. Botany* (*London*) **20,** 83–108.

HUSKINS, C. L. and S. G. SMITH (1935). Meiotic chromosome structure in *Trillium erectum* L. *Ann. Botany* (*London*) **49,** 119–150.

INOUÉ, S. and A. BAJER (1961). Birefringence in endosperm mitosis. *Chromosoma* **12,** 48–63.

JENSEN, W. A. and L. G. KAVALJIAN (1958). An analysis of cell morphology and the periodicity of division in the root tip of *Allium cepa. Am. J. Botany* **45,** 365–372.

JOHNSON, D. S. (1929). Development of antheridium and spermatozoid in *Plagiochila adantioides* Lindb. (Swartz). *Botan. Gaz.* **88,** 38–62.

JOHRI, B. M. and R. N. KAPIL (1953). Contribution to the morphology and life history of *Acalypha indica* L. *Phytomorphology* **3,** 137–151.

LANCE, A. (1952). Sur la structure et le fonctionnement du point végétatif de *Vicia faba* L. *Ann. sci. nat., Botan. et biol. végétale* ser. 11, **13,** 301–339.

LAND, W. J. G. (1904). Spermatogenesis and oogenesis in *Ephedra trifurca. Botan. Gaz.* **38,** 1–18.

LAND, W. J. G. (1907). Fertilization and embryogeny in *Ephedra trifurca. Botan. Gaz.* **44,** 273–292.

LAUGHLIN, H. H. (1919). Duration of the several mitotic stages in the dividing root tip cells of the common onion. *Carnegie Inst. Wash. Publ.*, 265.

LINSKENS, H. F. (1958). Physiologische Untersuchungen zur Reifeteilung. II. Mitteilung. Über die Änderung des Nukleinsäurengehaltes während der Pollenmeiose und Pollenentwicklung von *Lilium henryi. Acta Botan. Neerl.* **7,** 61–68.

LYON, F. M. (1901). A study of the sporangia and gametophytes of *Selaginella apus* and *Selaginella rupestris. Botan. Gaz.* **32,** 124–141, 170–194.

MAHESHWARI, P. (1950). *An Introduction to the Embryology of Angiosperms.* McGraw-Hill, New York.

MOLÉ-BAJER, J. (1958). Cine-micrographic analysis of C-mitosis in endosperm. *Chromosoma* **9,** 332–358.

NÄF, U. (1958). Physiology of antheridium formation in the bracken fern, *Pteridium aquilinum* (L.) Kühn. *Physiol. Plantarum* **11,** 728–746.

NASATIR, M. (1958). Changes in respiration and fermentation associated with microspore mitosis in *Lilium longiflorum* Thunb., Ph.D. dissertation Univ. of Pennsylvania.

NASATIR, M., A. M. BRYAN and A. RAKE (1961). Simultaneous appearance of free arginine and deoxyribosidic compounds during mitosis. *Science* **134,** 666.

NASATIR, M., A. M. BRYAN and S. D. RODENBERG (1960). Changes in soluble proteins of developing lily anther. *Science* **132,** 897–898.

NASATIR, M. and H. STERN (1959). Changes in the activities of aldolase and of D-glyceraldehyde-3-phosphate dehydrogenase during the mitotic cycle in microspores of *Lilium longiflorum. J. Biophys. Biochem. Cytol.* **6,** 189–192.

NEVINS, B. I. (1932). Cytological studies of the antheridia of *Sphaerocarpos donnellii. Cellule rec. cytol. histol.* **41,** 293–334.

OGUR, M., R. O. ERICKSON, G. U. ROSEN, K. B. SAX and C. HOLDEN (1951). Nucleic acids in relation to cell division in *Lilium longiflorum. Exptl. Cell Research* **2,** 73–89.

O'HANLON, S. M. E. (1926). Germination of spores and early stages in the development of the gametophytes of *Marchantia polymorpha. Botan. Gaz.* **82,** 215–222.

PEARSON, H. H. W. (1906). Some observations on *Welwitschia mirabilis* Hooker. *Phil. Trans. Roy. Soc. London Ser. B* **198,** 265–304.

PEARSON, H. H. W. (1906). Further observations on *Welwitschia. Ibid.* **200,** 331–402.

POLLARD, J. K., E. M. SHANTZ and F. C. STEWARD (1961). Hexitols in coconut milk: their role in nurture of dividing cells. *Plant Physiol.* **36,** 492–501.

POPHAM, R. A. (1958). Cytogenesis and zonation in the shoot apex of *Chrysanthemum morifolium. Am. J. Botany* **45,** 198–206.

REYNOLDS, L. G. (1924). The female gametophyte of *Microcycas. Botan. Gaz.* **77,** 297–300.

RICHARDS, B. M. and A. BAJER (1961). Mitosis in endosperm. Changes in nuclear and chromosome mass during mitosis. *Exptl. Cell Research* **22,** 503–508.

SAVELKOUL, R. M. H. (1957). Distribution of mitotic activity within the shoot apex of *Elodea densa. Am. J. Botany* **44,** 311–317.

SMITH, R. W. (1900). Structure and development of the sporophylls and sporangia of *Isoetes. Botan. Gaz.* **29,** 225–259, 323–346.

SPARROW, A. H. (1951). Radiation sensitivity of cells during mitotic and meiotic cycles with emphasis on possible cytochemical changes. *Ann. N. Y. Acad. Sci.* **51,** 1508–1540.

SPARROW, A. H., M. J. MOSES and R. DUBOW (1952). Relationships between ionizing radiation, chromosome breakage and certain other nuclear disturbances. *Exptl. Cell Research, Suppl.* **2,** 245–267.

SPARROW, A. H., M. J. MOSES and R. STEELE (1952). A cytological and cytochemical approach to an understanding of radiation damage in dividing cells. *Brit. J. Radiol.* **25,** 182–188.

SPARROW, A. H. and R. C. SPARROW (1949). Treatment of *Trillium erectum* prior to and during mass production of permanent smear preparations. *Stain Technol.* **24,** 47–55.

SPEESE, B. M. (1939). Mitosis in leaves of *Smilax*. *Am. J. Botany* **26,** 852–855.

STÅLFELT, M. (1921). Studien über die Periodizität der Zellteilung und sich daran ausschliessend Erscheinungen. *Kgl. Svenska Vetenskapsakad. Handl.* **62,** 36–48.

STERN, H. (1946a). The formation of polynucleated pollen mother cells. *J. Heredity* **37,** 46–50.

STERN, H. (1946b). The permeability of cells undergoing nuclear division. *Trans. Roy. Soc. Can.* **40,** 141–148.

STERN, H. (1948). Protoplasmic changes during nuclear division in pollen mother cells and microspores of *Trillium erectum*. *Plant Physiol.* **23,** 388–396.

STERN, H. (1956). Sulfhydryl groups and cell division. *Science* **124,** 1292–1293.

STERN, H. (1958). Variations in sulfhydryl concentration during microsporocyte meiosis in the anthers of *Lilium* and *Trillium*. *J. Biophys. Biochem. Cytol.* **4,** 157–161.

STERN, H. (1960a). Aspects of deoxyriboside metabolism in relation to the mitotic cycle. *Ann. N. Y. Acad. Sci.* **90,** 440–454.

STERN, H. (1960b). Biochemical sequences in mitosis. In *Developing Cell Systems and Their Control. 18th Growth Symposium*. Ronald Press, New York.

STERN, H. (1961). Periodic induction of deoxyribonuclease activity in relation to the mitotic cycle. *J. Biophys. Biochem. Cytol.* **9,** 271–277.

STERN, H. and P. L. KIRK (1948). The oxygen consumption of the microspores of *Trillium* in relation to the mitotic cycle. *J. Gen. Physiol.* **31,** 243–248.

STEWARD, F. C., E. M. SHANTZ, J. K. POLLARD, M. O. MAPES and J. MITRA (1961). Growth induction in explanted cells and tissues: metabolic and morphogenetic manifestations. In *Molecular and Cellular Structures, 19th Growth Symposium* (D. Rudnick, ed.). Ronald Press, New York, pp. 193–246.

SYKES, E. (1908). Notes on the morphology of the sporangium bearing organs of the Lycopodiaceae. *New Phytologist* **7,** 41–60.

TAKATS, T. (1962). An attempt to detect utilization of DNA breakdown products from the tapetum for DNA synthesis in the microspores of *Lilium longiflorum*. *Am. J. Botany* **49,** 748–758.

TAYLOR, J. H. (1953). Autoradiographic detection of incorporation of P^{32} into chromosomes during meiosis and mitosis. *Exptl. Cell Research* **4,** 164–173.

TAYLOR, J. H. (1959). Autoradiographic studies of nucleic acids and proteins during meiosis in *Lilium longiflorum*. *Am. J. Botany* **46,** 477–484.

TAYLOR, J. H. and R. D. MCMASTER (1954). Autoradiographic and microphotometric studies of desoxyribose nucleic acid during microgametogenesis in *Lilium longiflorum*. *Chromosoma* **6,** 489–521.

TISCHLER, G. (1922). Allgemeine Pflanzenkaryologie. In *Handbuch der Pflanzenanatomie*. Abt. I, Teil 1. Borntraeger, Berlin.

VAN OVERBEEK, J., M. E. CONKLIN and A. F. BLAKESLEE (1941). Factors in coconut milk essential for growth and development of very young *Datura* embryos. *Science* **94,** 350–351.

WILSON, K. S. and V. M. CUTTER, Jr. (1955). Localization of acid phosphatase in the embryo sac and endosperm of *Cocos nucifera*. *Am. J. Botany* **42,** 116–119.

WILSON, M. (1909). On spore-formation and nuclear division in *Mnium hornum*. *Ann. Botany* **23,** 141–157.

YAKOVLEV, M. S. and M. D. YOFFE (1957). On some peculiar features of the embryogeny of *Paeonia* L. *Phytomorphology* **7,** 74–81.

2

Natural Division Synchrony and Mitotic Gradients in Metazoan Tissues

IVAR AGRELL, *Zoophysiological Institute, University of Lund, Sweden*

CONTENTS

I Introduction 39
II Synchronous Division in the Initial Phase of Embryogenesis . 40
III The Transition from Synchronous to Asynchronous Division during Embryogenesis. Mitotic Gradients 44
IV Reinforced Mitotic Activity during the Later Embryogenesis and in Adults 50
V Morphological and Functional Characteristics of the Synchronously Dividing Cell 51
1. The synchronous cleavage cycle in fertilized eggs . . . 52
2. Cytochemical characteristics of eggs dividing synchronously . 53
3. Syncytial contacts within a synchronously dividing tissue . 55
4. Biochemical storage as a factor for division synchrony . . 57
VI Graded Mitotic Activity and Tissue Differentiation . . . 60
References 63

I. INTRODUCTION

In an adult human being the total number of cells may be estimated at 2×10^{14}. These many cells result from about forty-nine cellular duplications. An adult rotifer results from less than eleven divisions and the animal has less than a thousand cells. In the rotifers the number of cells is said to be strictly constant for a species.

By and large, the difference between a big and a small metazoan reflects a difference in cell numbers rather than in average cell size, cf. Conklin (1912) and Rensch (1948). The metabolic laws are such that an animal above a certain size could not exist were it not subdivided into small cellular units (Zeuthen, 1953a).

Contrariwise, within the protozoa, the difference in cell size is enormous. As an example, the biggest uninuclear amoeba may have a million times larger mass than the smallest one.

In the individual life of a metazoan, the rate of mitosis is most intense during the very early embryogenesis. Mitosis may be accompanied by cell division right from the beginning or from a somewhat advanced stage of development. In the latter case a syncytium is temporarily formed. Syncytial conditions may persist locally for long periods. In any case the big egg cell splits into an increasing number of smaller and smaller cells. There is no growth of the embryo as a whole until it begins to take food from the outside, enterally or parenterally. Only a kind of 'inner growth' occurs in the early embryo. This is a transformation of the stored components of the egg cell into active protoplasm. This transformation is reflected in a steadily increasing respiration rate. Sooner or later the intense cellular multiplication slows down, differently in different species, and more or less reflecting increasing cellular differentiation. There is a degree of incompatibility between cellular division and cellular differentiation.

The very ability to divide is intimately linked with the destiny of the different cells. At their developmental terminus some cells find themselves incapable of any further division be they ever so provoked, e.g., the specialized nerve cells. Others stop their division in the course of the normal embryonic or postembryonic development, but they can be forced to divide again, e.g., liver cells during the process of regeneration. Still other cells divide throughout the life of the animal (e.g., bone marrow and epithelia). Some cells pass through the labyrinth of embryonic development and settle down as quiet ordinary members of the cell community. Later, for some reason they may tear themselves loose from controlling influences exerted by the community and start new cellular division, in this case in an irregular and uncontrolled manner, producing neoplasms.

The early span of the embryonic life, cleavage, and the period of transition between predominating mitotic activity and beginning cellular differentiation, has attracted very little interest in comparison to the later embryogenesis. If we except the studies of the cell lineage in a number of mosaic eggs, little is known about the initial period of the embryonic development.

II. Synchronous Division in the Initial Phase of Embryogenesis

The first mitotic divisions in the embryo are usually synchronous. This division synchrony is very precise. Not only are all the nuclei of the blastomeres either in mitosis or in interphase at the same time, but all the nuclei of the embryo pass separate phases of the mitosis in beautiful synchrony. Synchrony may persist for a different length of time in embryos of different animal species. The classic example of a long sustained synchrony is the dividing egg of the holothurian *Synapta digitata*. Cell divisions are completely synchronous for nine cell cycles in which 512 cells are produced. Then gastrulation is initiated (Selenka, 1883). Among the coelenterates, the embryo of the hydrozoan *Aequorea forskålea* remains synchronous for at least six cell cycles (Häcker, 1892). The amphibian embryo shows synchrony for not less than fifteen cell cycles. However, the later divisions are somewhat out of phase, in *Bufo* spreading from the animal pole and proceeding to the vegetal pole (Dan, 1960; Sirakami, 1958). The *Amphioxus* egg shows an exacting synchrony for only three cycles, and a relative synchrony for four more cycles. After division 3 the cleavage of the blastomeres in the animal half are somewhat ahead of those in the vegetal half (Cerfontaine, 1907). Each ring of the blastomeres seems to divide simultaneously (Hatschek, 1882). In the sea urchin embryo there is also a complete synchrony but for only three cycles. At the fourth division the micromeres are formed at the vegetal pole and these cells exhibit a slow cleavage rhythm (Zeuthen, 1951). However, the segmentation of the macro- and mesomeres proceeds in absolute synchrony for many more cycles. In *Echinus esculentus* this synchrony lasts for eight cycles, in *Psammechinus miliaris* for seven cycles and in *Paracentrotus lividus* for six cycles (Agrell, 1956a). Also in the *Spatangus purpureus* egg the synchrony will last for six cycles, but in *Arbacia lixula* the synchrony within the embryo is poor already after five cycles (Agrell, unpublished data). After the breakdown of complete synchrony of the macro- and mesomere divisions there follows a period of relative synchrony. In this period mitotic gradients can be demonstrated.

These reports of a prolonged cleavage synchrony all concern eggs of the regulative type of development. In eggs with a mosaic development the period of synchronous cleavage of the blastomeres is more restricted. Among eggs with a spiral cleavage the gastropods and lamellibranchs usually show a strict synchrony only for three cell cycles (Robert, 1902) and the annelids for four cell cycles (Mead, 1897; Robert, 1902). After this the mosaic development soon makes itself apparent in a mosaic pattern of cleavage. The ascidian embryo, with a marked type of bilateral cleavage, displays complete synchrony only for three cycles. A period of relative division synchrony follows, and after about seven divisions a mosaic pattern of cleavage is established (Conklin, 1905).

All the eggs mentioned thus far show total cleavage. In eggs with partial cleavage, we may again observe prolonged synchrony of early divisions. This is valid for the fish egg showing discoidal cleavage. The events in the embryo of *Belone acus* are elucidating (Kopsch, 1901). Complete synchrony continues within the blastoderm disc for eight division cycles, but declines during the following two cycles.

After this, synchrony of nuclear division continues in the marginal cells for at least three more cycles. Mitoses are followed by incomplete cell divisions, so a peripheral syncytium results (Fig. 1). Similar observations have been made in the egg of the squids. Also in this case the germinal disc shows prolonged cleavage synchrony, continuing for some time in the syncytial blastocones at the margin adjacent to the yolk (Watase, 1891; Vialleton, 1888).

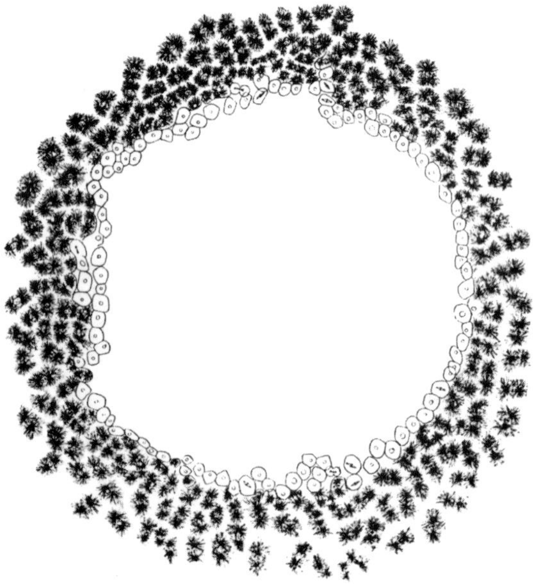

FIG. 1. Complete division synchrony in the syncytial marginal segments of the *Belone* embryo during the thirteenth cell cycle. The inner part of the blastoderm is not drawn. (From Kopsch, 1901.)

Extended division synchrony is not general for all discoblastic eggs. In the *Pyrosoma* egg there is division synchrony for not more than five cycles, or for as long as the embryo remains one-layered (Korotneff, 1906). The early cleavages of the hen's egg may be synchronous only for three cycles. Thereafter the central cells of the blastodisc begin to multiply more rapidly than those of the margin (Patterson, 1910).

Also among the eggs with peripheral cleavage a pronounced synchrony is often found. In the insect egg the synchrony concerns mitosis more than the cell division. The early divisions are limited to the nuclei and their surrounding plasma islands. It is only when the nuclei surrounded by their islands have migrated to the surface of the egg, that one multiple cell division produces a continuous blastoderm. Generally, the nuclear divisions are synchronous for many cycles. For instance, the first five mitoses are synchronous in the embryo of the midge *Sciara* (DuBois, 1932). The first nine mitoses are synchronous in the damsel fly *Platychnemis* (Seidel, 1929), and in the moth *Ephestia* (Sehl, 1931). In the fruit fly *Drosophila* there may be twelve synchronous mitoses (Sonnenblick, 1950), but in another fly, *Calliphora*, I have found (Agrell, 1962) that the ninth to the twelfth cleavage cycles proceed in waves. In *Gryllus* there was no apparent synchrony of the widely-scattered nuclei after five or six nuclear divisions (Agrell, unpublished). Tiegs and Murray (1938) report that initially there is no synchrony in the egg of the beetle *Calandra*. However, their samples were sparse, and no batch of eggs was followed during continued development. They report pronounced synchrony at the time when the blastoderm forms. In the moth egg the synchrony disappears when the nuclei reach the egg surface (Huie, 1918). The spider embryo shows similar development. When the cleavage nuclei move into the egg cortex after six cycles of division synchrony, the synchrony is lost (Holm, 1952).

An extremely bewildering type of division synchrony can occur among turbellarian worms. The egg cocoon of the alloeocoelan *Plagiostomum girardi* contains ten to twelve eggs, which lie enclosed in a syncytial mass of yolk. All these separate eggs start to divide in complete synchrony, in spite of the fact that the continued cleavage into blastomeres, within each single embryo, is asynchronous (Bresslau, 1904). These blastomeres of the single embryos separate completely and move into the yolk mass. Only later, after many divisions, do the blastomeres fuse again to form the definite embryos. This phenomenon also exists in the polyembryonic capsules of *Triclades*, in which, however, no division synchrony has been found. Also, no synchrononous cleavage seems to occur in the polyembryonic eggs of parasitic wasps (Martin, 1914).

This section would not be complete without mentioning cases in which division synchrony does not extend beyond the first cleavage of the egg. Already at the two-cell stage the two blastomeres divide at different times. With further development the asynchrony becomes more and more manifest. This is so in the eggs of the trematode *Polystomum* (Goldschmidt, 1902), of the nematode *Ascaris* (Boveri, 1887) and of the rotifers (Nachtwey, 1925). Also the mammalian embryo shows no division synchrony (Dalcq, 1957), however with the exception of the meroblastic egg of the *Mono-*

tremata. In *Echidna* division synchrony continues for five cycles (Flynn and Hill, 1947).

III. The Transition from Synchronous to Asynchronous Division during Embryogenesis. Mitotic Gradients

In the mosaic eggs synchrony is soon replaced by asynchrony, but the asynchrony develops in a regular and orderly manner. The mosaic development primarily reveals itself in a strict divisional pattern. The general rule seems to be that corresponding cells in each quadrant divide synchronously. A typical example is given in Fig. 2.

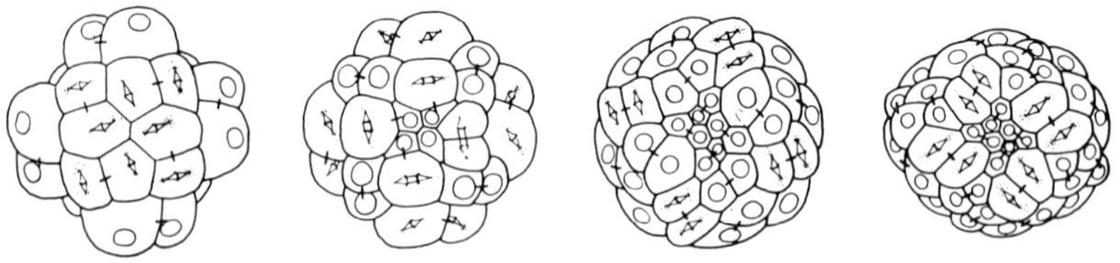

Fig. 2. Mosaic cleavage pattern in subsequent stages of the *Trochus* embryo. (Redrawn from Robert, 1902.)

The fact also that in the more regulative eggs the transition from a synchronous to an asynchronous mode of division takes place in an orderly manner has been mainly unnoticed. As an example, I will use the sea urchin egg with which I have worked myself (Agrell, 1955a). In the embryo of *Paracentrotus lividus* (Fig. 3) synchronous division goes on for six cell cycles, except for the micromeres which after they are formed in the fourth cleavage exhibit a mitotic rhythm of their own. When the blastomeres of the embryo start the mitosis for the seventh cell cycle the new prophases appear first in the macromeres situated on top of the micromeres. From the macromeres the mitotic activity spreads to the mesomeres; gradually a complete mitotic gradient develops, reaching from telo–anaphases in the macromeres at the vegetal poles over metaphases to prophases at the animal pole. Then all the macro- and mesomeres gradually pass into interphases which last for some time. Then a new mitotic wave, the eighth, starts at the vegetal pole, but to begin with it concerns solely the macromeres. Only when the macromeres have reached telo–interphase do the mesomeres start their mitotic activity. Then again all blastomeres stay for some time in interphase. During the next cleavage cycle, the ninth, the same phenomenon is repeated. Mitosis begins in the macromeres and only when almost all macromeres have entered telo–interphase is mitosis initiated in the

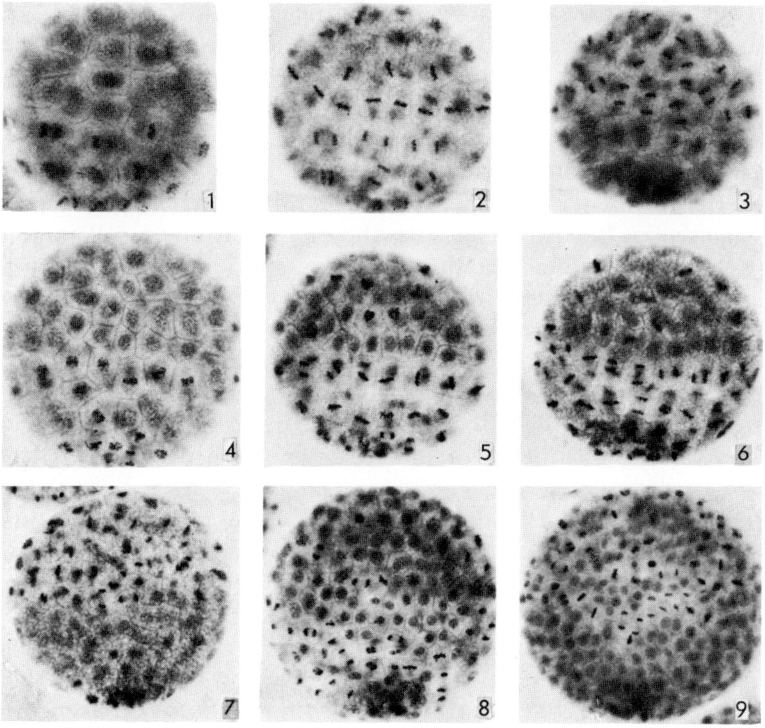

Fig. 3. Mitotic gradients during the development of *Paracentrotus lividus* following the division synchrony. Micromeres and vegetal pole pointing downwards and animal pole upwards. Consecutive mitotic stages during the seventh division step, 1–3, during the eighth division, 4–7, and during the ninth division cycle, 8–9. (From Agrell, 1956a.)

mesomeres. When the embryo enters the next stage of division this graded mitotic activity is no longer observable. However, the tendency for a vegetal–animal mitotic gradient is still there and can be demonstrated if the mitoses are accumulated by the action of a mitotic inhibitor, e.g., oestradiol (Agrell, 1960).

Also in other sea urchin species there is graded mitotic activity. In *Psammechinus miliaris* a complete gradient is developed during the eighth cleavage cycle followed by the alternating mitotic activity in the vegetal and animal half during the ninth cycle. In *Echinus esculentus*, where the synchrony goes on for eight cycles, a corresponding graded mitotic activity was seen (Agrell, 1956a). *Spatangus purpureus* and *Arbacia lixula* behave in this respect about the same as *Paracentrotus lividus*. The *Arbacia* embryo which has a less pronounced cleavage synchrony also shows less distinct mitotic gradients (Agrell, unpublished). The same seems to occur to some extent in all the species mentioned. When in a batch of embryos the cleavage synchrony is not distinct, the later mitotic gradients are less conspicuous too.

Using microphotometric measurements on interphase nuclei made by McMaster (1955), the graded mitotic activity (sea urchin *Lytechnius*) can be traced back to the eight or sixteen cell stage. At this stage the Feulgen reaction reaches its maximum intensity first in the macromeres, later in the mesomeres, latest in the micromeres. Markman (1961) observed strongest incorporation of ^{14}C-adenine in the nuclei of the animal region of the early blastula (sea urchin *Paracentrotus*). It remains undecided whether incorporation is in RNA or in DNA or in both. RNAase removes RNA only with difficulty from the nuclei of the early sea urchin embryo (Agrell, 1961).

Between themselves the micromeres in the sea urchin embryo show a graded mitotic activity. When the micromeres, which arise as four cells, have divided into eight they form an outer ring of larger cells and an inner ring of smaller cells. The cleavage of the outer micromeres is always ahead of the inner ones (Agrell, 1956a; Zeuthen, 1951).

Mitotic gradients at the transition from synchrony to asynchrony may occur also in other metazoan eggs with holoblastic cleavage. They may often have been overlooked because they can only be observed in suitable preparations of whole eggs (Agrell, 1958d). The detection of a mitotic gradient in a sectioned material must be extremely difficult.

Recently there has also been a report of mitotic gradients in the anuran egg (Dan, 1960; Sirakami, 1958). In *Bufo vulgaris* each cleavage step occurs in a wave of divisions, in this case passing from the animal to the vegetal pole. This should concern the greater part of the first fifteen cleavage steps. More detailed reports are to be hoped for. As was mentioned in the preceding section, a corresponding mitotic wave operates in the *Amphioxus* embryo during the fourth to seventh cleavage cycles, likewise moving from the animal to the vegetal half (Cerfontaine, 1906).

Also in the mosaic eggs there is a tendency for the mitoses to develop in waves from pole to pole. In the ascidian embryo for three cell cycles the cells of the dorsal hemisphere divide in advance of those in the ventral half. Furthermore, the anterior part of the embryo tends to divide earlier than the posterior part (Conklin,

1905). In the molluscan egg a vegetal–animal mitotic gradient can be seen at the time of the fourth and fifth division cycles. During later cleavage this gradient may be reversed. Animal cells tend to divide earlier than vegetal cells (Robert, 1902). The same seems to hold for the annelid embryo in which there is a tendency for the mitotic gradient to reverse around the fifth or the sixth cleavage cycle.

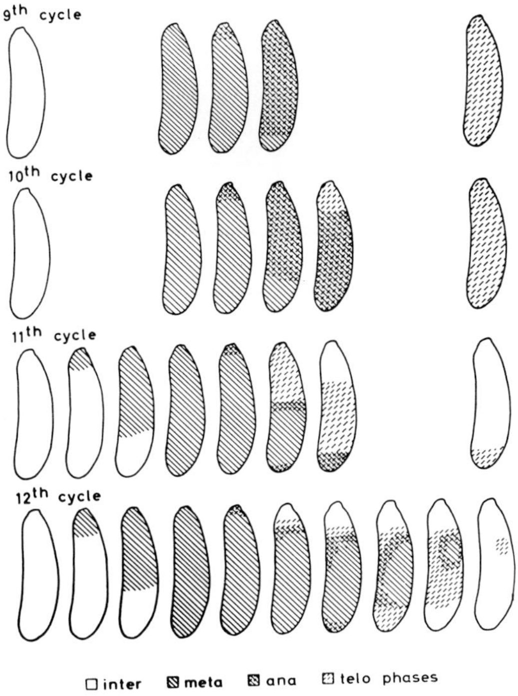

FIG. 4. Scheme of the cleavage waves in the blastoderm of the *Calliphora* embryo during the four cell cycles which show a graded mitotic activity. (From Agrell, 1962.)

In the meroblastic egg the period of synchronous cellular division may be followed by a period in which the mitotic phases travel as waves over the embryo. At any one time a regular succession of mitotic stages can be demonstrated along the axis of the embryo. As an example (Fig. 4), I shall take the egg of the fly *Calliphora erytrocephala* which shows the superficial type of cleavage (Agrell, 1962). At the eighth division cycle the nuclei have reached the periplasm at the outer margin of the egg and the blastoderm has

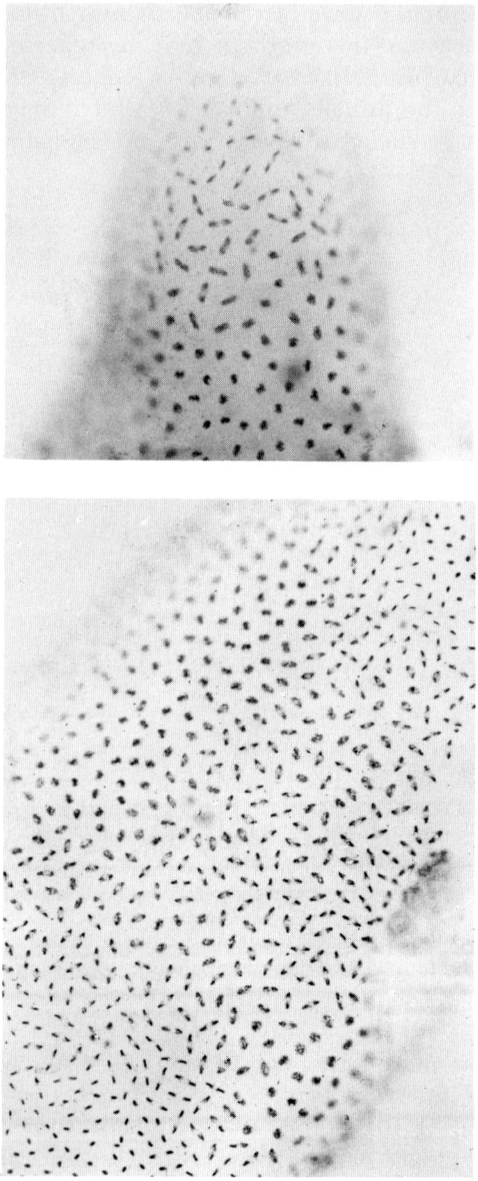

(a)

(b)

Fig. 5. Cleavage wave in the blastoderm of the *Calliphora* embryo. (a) Tenth cell cycle. The cleavage wave proceeds from the anterior pole. (b) Twelfth cycle. Two cleavage waves meet at the presumptive prothoracic region. (From Agrell, 1962.)

formed. From now on cleavage represents not only nuclear but also cellular segmentation. Some nuclei remain as yolk nuclei in the interior of the egg. During the ninth cell cycle the cleavage synchrony in the blastoderm is lost. Instead a wave of divisions passes from the anterior to the posterior pole of the egg. During the tenth cycle a corresponding but even more pronounced cleavage wave passes over the embryo (Fig. 5). During this cycle the germ cells bud off at the posterior pole and immediately establish a mitotic rhythm of their own. During the eleventh cycle a narrow girdle of anaphases travels from the anterior pole towards the posterior pole but before reaching the pole another wave makes its appearance at the posterior pole. During the twelfth cycle a double mitotic gradient is firmly established. The two cleavage waves flow into each other dorsally at the middle of the embryo and surround an area of more slowly dividing cells (Fig. 5). After the twelfth cycle all the nuclei of the blastoderm, except for the pole cells, show a very long interphase. The new mitotic activity is initiated and the foldings and invaginations of the whole embryo occur almost simultaneously. From now on there is neither mitotic synchrony nor obvious graded mitotic activity. However, resumed cell division seems first to be in the invaginations and only later in the non-invaginated areas of the blastoderm.

This very regular graded mitotic activity may not be special for the *Calliphora* embryo. It may be present in some form in many other insect eggs, and most likely in eggs in which the cleavage nuclei enter the periplasm at the same time over the whole egg surface. In fact a mitotic gradient has been noted in the *Drosophila* embryo (Sonnenblick, 1950) but it was explained as resulting from the diffusion of the fixative through the egg. This artifact was excluded in the case of the *Calliphora* egg. Also there is an observation (Tiegs and Murray, 1938) of a single mitotic band in the late blastoderm of the embryo of the beetle *Calandra*. In the *Gryllus* egg, which shows poor division synchrony, there is definitely no graded mitotic activity (Agrell, unpublished data). The closely studied embryo of the odonate *Platychnemis*, which has the same primitive type of embryogenesis as *Gryllus*, shows no mitotic waves after the series of synchronous mitoses (Seidel, 1929).

It is not known whether or not there are mitotic gradients in the blastoderm of the fish embryo. However, in the squid egg, which also shows partial, discoidal cleavage there may be some polarity of the mitotic activity in the blastodermal disc. In the *Loligo* embryo, which shows fairly good division synchrony for eight cycles, a group of six inner cells and a pair of marginal cells along the first cleavage furrow divide the slowest in the whole embryo. The anterior part of the disc divides more rapidly than the posterior part, and the division of the left side may be ahead of that of the right side (Watase, 1891).

IV. Reinforced Mitotic Activity during the Later Embryogenesis and in Adults

With continued embryonic development the rate of the mitotic activity slows down. The division synchrony disappears, mitotic gradients fade off and gradually the cell divisions are rather evenly distributed within the embryo. Later, renewed mitotic activity can be observed in restricted parts of the embryo. This is for instance the case in the sea urchin embryo at the time for gastrulation. Increased mitotic activity sets in at the vegetal pole, the site of the first invagination (Agrell, 1954; Kinnander and Gustafson, 1960). A vegetal–animal mitotic gradient thus reappears. Generally, the different foldings of the embryo which occur in embryogenesis tend to be accompanied by local mitotic activity. Cases in point are the fly egg (Agrell, 1962), the mammalian brain (Frank, 1925; Källén, 1956), the bone primordia in the eyeball of the chick embryo (Hale, 1956), and the primitive streak of the chick embryo (Emanuelsson, 1961b).

As was mentioned in the introduction, some mitotic activity may persist in some organs during the whole individual life of the metazoan. This mitotic activity often shows a diurnal period. The control may be humoral. Cells of an ascites tumour cell, when transplanted into the body cavity of the mouse, show the diurnal mitotic period of the host organs (Agrell and Welin-Berger, 1957). These temporary and most often only very slight variations in cellular proliferation scarcely deserve the designation 'division synchrony', and I will not enter here into further details. I may refer to the comprehensive articles of Bullough (1952) and Halberg (1959).

However, there is one adult organ in which a high degree of division synchrony exists, or can be said to have reappeared during ontogony from egg to adult. This is the male gonad. The spermiogenesis includes steps of divisions which are highly synchronous. This is seen most clearly in the follicular tubes of the insect testis, where the maturation from spermatogonium to sperm cell proceeds longitudinally in a series of clusters, or cysts. For instance, in the grasshopper testis it has been stated that all the spermatogonia in each cyst are the direct descendants of a single primary spermatogonium. At every time point they are in the same phase (Weber, 1933) of the division cycle (Davis, 1908). In the vertebrate testis the maturation of the sperm proceeds in radial direction towards the cavity of the tubule. Through a blocking at metaphase with colchicine it can be seen that the spermatogonia divide synchronously over five cycles. Each mitotic maximum results in a doubling of the number of spermatogonia. The first three peaks give birth to type A spermatogonia, the fourth to type A and type B spermatogonia and the fifth to spermatocytes (Leblond and Clermont, 1952). External and internal factors may influence the mitotic activity. They serve as pacemakers, and act simultaneously at several stages of the sperm development (Roosen-Runge and Giesel, 1950). See also the report by Humphrey (1922) for the urodeles.

One type of somatic cells has been reported to exhibit division synchrony similar to that which is found during spermiogenesis. The cnidoblasts of *Hydra* give rise to the nematocysts, and they appear in clusters of up to sixteen cells. They result from a series of synchronous cellular divisions (Fawcett *et al.*, 1959).

In the female gonad of the mouse meiotic prophases of the ovary are synchronized. During the sixteenth to eighteenth day *post coitum* (the fourteenth day of fetal life) almost exclusively pachytene nuclei are seen. After this stage meioses never again appear in the ovary. This is in contrast to the male gonad, in which the germ cells enter meiosis throughout mature life (Borum, 1961). In animals which release a great number of eggs at the same time, all at the same stage and ready for fertilization, we may expect that there is synchrony at least of certain steps in ovogenesis. However, I have not been able to find relevant information in the literature.

V. Morphological and Functional Characteristics of the Synchronously Dividing Cell

We may guess that division synchrony persists in a tissue, embryonic or adult, because the dividing cells are inherently alike with regard to nucleus and/or cytoplasm, furthermore because the cells live under closely similar extraneous conditions. It therefore needs to be investigated to what extent the synchronously dividing system is uniform and to what extent and how the internal and external environment differentiates when the division synchrony transforms into a mosaic cleavage or develops into a mitotic gradient. Therefore it is necessary to give an account of some known cytological and cytochemical features of synchronously dividing cells.

1. *The Synchronous Cleavage Cycle in Fertilized Eggs*

The synchronous division cycle in the early embryo is very rapid, usually there is less than an hour between divisions of ordinary temperature. In the fly egg the nuclear division cycle takes less than 10 min (Sonnenblick, 1950). The division rate of the fertilized egg can not easily be accelerated and may be nearly maximal at each temperature. An attempt to affect the synchrony of sea urchin embryos during the period of division synchrony by applying a series of moderate thermal shocks failed completely. Regardless of when the shocks were applied, the subsequent cleavage rate was unaffected (Agrell, 1957a). There are a few reports of substances that may be mitotic stimulators when tested upon early cleavage. Thus the sea urchin egg has been reported to contain two cleavage factors, viz., an accelerating dinucleotide, possibly uracil dinucleotide, and a retarding factor, probably a polynucleotide (Menkin, 1959). Furthermore, the cleavage rate is claimed to be affected by single nucleotides. I myself tested a series of normal purine and pyrimidine compounds but found acceleration only in batches of sea urchin eggs, which for some reason cleaved abnormally slowly. However, if the eggs are sensitized using oestradiol, which seems to split DNA from protein, making DNA accessible to the action of DNAase (Agrell and Persson, 1956a), then the early cleavage rate becomes extremely sensitive in both positive and negative direction depending on the proportion of bases used.

During the early and synchronous phase of egg cleavage the division rate and also the time relation between interphase and mitosis is constant (Zeuthen, 1953b; Dan, 1960). This is so even though the nucleo-cytoplasmic ratio increases steeply when the egg segments into smaller and smaller cells (Godlewski, 1908; Zeuthen, 1951). After the period of division synchrony the cell cycle becomes longer. As mentioned before, the rate of division may slow differently in different parts of the embryo, the pattern depending on the organism. The interphase rather than the mitosis extend in time. The mitotic time may even shorten, and the relative time occupied by the various mitotic phases may change, as seen from Fig. 6 (Agrell, 1957a). This may in part be a reflection of the increased nucleo-cytoplasmic ratio.

During the division synchrony a respiratory rhythm has been demonstrated to match the cell cycles (Holter and Zeuthen, 1957). The rate of oxygen consumption increases during interphase and decreases during mitosis. Also the rate of uptake of ^{32}P (Zeuthen, 1951) is lower during mitosis than during interphase. These findings led Zeuthen (1951) to the important suggestion that the nucleus may have the alternative choice of reorganizing itself or of controlling growth of the whole cell. Recently this rhythm in energy metabolism has been extended by the demonstration that in sea urchin eggs the amount of acid-labile, energy-rich phosphates increases during interphases and decreases during mitosis (Nilsson, 1961). The inorganic phosphorus varies inversely. The number of mitochondria per egg changes in correspondence to the respiratory cycles (Agrell, 1955c). In the dividing sea urchin egg the huge mitotic apparatus engages more than 10 per cent of the total protein. This indicates that the mitosis must be energy requiring (Mazia, 1957).

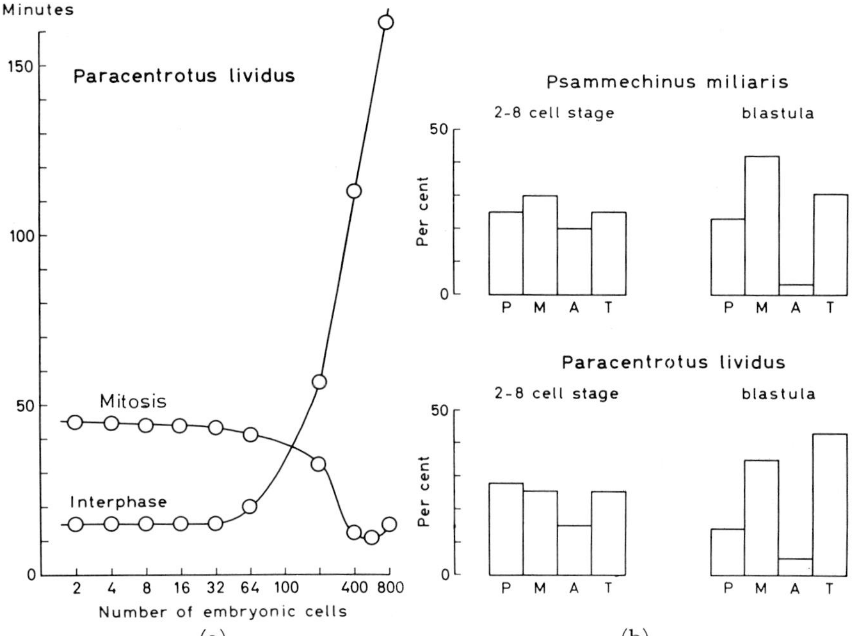

FIG. 6. Changes in the cell cycles during the early development of the sea urchin embryo. (a) The change in the average duration of interphase and mitosis (prophase–telophase). (b) The changes in the mitotic spectrum at two stages of development, early cleavage and blastula stage. P, M, A, T = pro-, meta-, ana-, telo-phase. (Redrawn from Agrell, 1957a.)

2. *Cytochemical Characteristics of Eggs Dividing Synchronously*

The cytochemistry of the rapidly dividing embryonic cell shows some peculiarities. Recently (Agrell, 1961) it was found that RNAase or acid hydrolysis attacks RNA of the nuclei of the early sea urchin embryo but does not remove it. Pre-treatment with DNAase depletes the cell of DNA, and later treatment with RNAase substantially removes the rest of the basophilic substances. Thus, RNA seems to be linked to DNA. Neither DNAase nor RNAase, applied separately, demonstrably expose free basic groups in the nuclear structure. These groups appear only if first DNA and then RNA is enzymatically removed. It is concluded that only RNA is fixed to basic protein, and that RNA is probably linked between DNA and protein. This peculiar RNA may be a precursor of the nucleolar RNA. It is remembered that one characteristic of the early embryonic cell is that a visible nucleolus does not appear at interphase. In the early blastomeres also the cytoplasmic RNA has strange features. It combines with methyl green and is directly linked to a basic protein. The early blastomeres share the characteristics which concern the nucleoproteins with other rapidly dividing cells, for instance with *Tetrahymena* (Agrell, 1961). The peculiarities gradually disappear as embryonic development progresses. For the mouse embryo it has been demonstrated by Alfert (1959) that the normal histone reaction gradually appears during embryogenesis.

It is concluded that the autosynthetic nuclei of the early embryo are different from the nuclei of the more advanced embryo and of the adult. All the nuclei of the early embryo may be identical as can be postulated from the now classic experiments made by Briggs and King (1957) on frog embryos. They are equipotential and totipotential as well. Each single division nucleus may serve as a zygote nucleus and support completely normal development if implanted into a fertilized egg deprived of its zygote nucleus. After a certain developmental stage, soon after the period of relative division synchrony, the nuclei have changed irreversibly because of interactions with a subdivided and differentiated cytoplasm.

With regard to nuclear differentiation the insect embryo displays some similarities to the frog embryo. When at the multi-nucleated stage the nuclei move to the surface of the egg and some of them invade and interact with the cytoplasm, rich in RNA, at the posterior pole. The germ cells are derived from this interaction (Huettner, 1923). Chance is the deciding factor as to which nuclei are incorporated into the germ plasm; so the resulting nuclear differentiation is to be referred to the influence of the local cytoplasm. In certain groups of midges this nuclear differentiation is later to be directly observed as a chromatine elimination in the somatic nuclei. Only the nuclei of the germ plasm escape this process of elimination (Beermann, 1956; Dubois, 1932). Similar visible nuclear differentiation is well known in the *Ascaris* embryo where nuclear segregation appears at the second division—this early division being consistently asynchronous. Very likely early, visible nuclear differentiation is much more frequent than would appear from the few cases known.

As has been pointed out above, there is a good correlation between long division synchrony and late differentiation. The division synchrony simply reflects a persistently high degree of equality within the dividing tissue, both with regard to the cytoplasms and to the nuclei. Eggs which show early differentiation (determinate cleavage) have only a short early period of division synchrony. However, there is at least one striking exception. This is the mammalian egg (Seidel, 1952) which shows an asynchronous division from the beginning, yet it is conspicuously regulative, as shown, for example, by the fact that pronounced polyembryony occurs in certain armadillos where one egg may give rise to as many as eight identical twins. The special case of the mammalian egg should be related to the fact that the division cycle is extremely extended in the early mammalian embryo. There may be 24 hr between divisions (Dalcq, 1954; Lewis and Hartman, 1933). During the long interphase the nuclei develop complicated nucleoli (Dalcq, 1957; Ludwig, 1954), and thus demonstrate hetero-synthetic activity. Such activity might produce differences or add to initial differences between the blastomeres, which for this reason develop separate division rhythms.

Because of its early syncytial development the insect embryo represents a very special case with respect to the relation between cleavage and differentiation. In this case the real problem seems to be why a relative division synchrony, in the form of mitotic gradients, persists for some time, also after the omnipotent nuclei have moved into the periplasm with its mosaic pattern of potentialities. It may be relevant here to mention that division synchrony over four early cycles is to be observed also in the mosaic egg of the annelid worms. It is further significant that in those insect eggs where a periplasm is missing there is no mosaic determinate pattern, nor are division waves to be seen.

3. *Syncytial Contacts within a Synchronously Dividing Tissue*

One way to approach the problem of uniformity within the dividing system seems to be to establish to what extent the cells in a synchronously dividing tissue represent a syncytium. This question is relevant because well defined syncytia may show pronounced mitotic synchrony. The classic examples are the plasmodia of the myxomycetes. The plasmodium of *Physarum* covers many square centimetres, and in the whole plasmodium the mitoses occur almost simultaneously as shown by Howard (1932). If the plasmodium is shaken into pieces, the mitoses within each fragment remain synchronous, but there is no correlation from piece to piece (Rusch cited from Swann, 1958). In one case in the multinucleate *Amoeba carolinensis* twenty-nine of a total of thirty-three nuclei were in late anaphase and four in early telophase at the same time, thus displaying a pronounced division synchrony (Short, 1946).

Syncytial conditions may not necessarily lead to mitotic synchrony. An elucidating example is represented by certain multi-micronucleate ciliates (Seshachar and Padmavathi, 1956). The micronuclei in *Frontonia leucas* divide in complete synchrony, in *Oxytricha bifaria* the situation is conspicuous and in *Spirostomum ambiguum* only 20–30 per cent of the micronuclei divide together. In the *Calliphora* egg, examined at the stage of blastoderm formation, the yolk nuclei show no division synchrony (Agrell, 1962). A special case can be seen during fertilization. If the pronuclei in the fertilized sea urchin egg are prevented from syngamy, through the action of oestradiol (Agrell, 1957b), or by low temperature (Agrell, 1959), they pass through the mitotic cycle independently of each other – the female pronucleus always being somewhat ahead of the male pronucleus. Even within one nucleus the structural changes of the different chromosomes may not occur synchronously. A case in point is the endomitosis of the highly ploid cell nuclei of the wings of butterfly pupae (Lipp, 1955).

Thus, even though a syncytial communication between dividing nuclei is not the only factor which evokes and maintains a division synchrony, it may definitely be regarded as a promoting factor. There are likewise statements that cells in synchronously dividing metazoan tissues are in syncytial junction. Electron-microscopic observations have demonstrated this for spermatogenic tissues at the time of meiosis, and for the cnidoblasts in *Hydra* (Fawcett, 1961). For embryonic tissues similar reports are less definite. The synchronously dividing yolk nuclei in the fish egg are obviously joined in a syncytium (Henneguy, 1896), and intercellular bridges between the dividing blastomeres are reported for the trout. Filamentous contacts between blastomeres are observed in early cleavage stages of echinoderms (Andrews, 1898; Gray, 1924), annelids, molluscs, nemertins, and Acrania (Andrews, 1898). Possibly these cell contacts persist longer in eggs (echinoderms) which show the persistent division synchrony. However, whether or not filamentous contacts join blastomeres to form syncytial units is unknown. The early insect embryo shows pronounced synchronous mitosis of the free nuclei, each of which is surrounded by an island of cytoplasm. It remains obscure to what extent cytoplasmic connections between the islands is responsible for the synchrony of the nuclear divisions.

In the course of development the syncytial contacts between blastomeres may at some stage be broken. Cells, thereby separated, may enter a mitotic rhythm of their own. This may be so in the case of the micromeres of the sea urchin embryo and of the pole cells of the insect egg. The latter, perhaps also the former, may give rise to the germ cells. It seems to me likely that the prospective germ cells should on the earliest possible occasion evade the influence of the cell continuum.

For the cleaving frog egg no protoplasmic contacts between the blastomeres are reported. However, if the blastomeres are separated the differences in cleavage time between animal and vegetal cells, normally responsible for division waves, disappear (Sirakami, 1958). Possibly, when the blastomeres are separated the syncytial contacts, necessary for the propagation of the wave, are disrupted. A similar phenomenon can be produced in sea urchin embryos. If the first two blastomeres are separated, the division synchrony within each of them continues, but from blastomere to blastomere there is asynchrony (Harnden, cited from Swann, 1958). These results are reminiscent of the effect of cutting the *Physarum* plasmodium.

In the sea urchin embryo it has frequently been observed that at stages when division synchrony and mitotic gradients have disappeared there appear in the blastoderm areas of adjacent cells, which all enter mitosis at the same time (Fig. 4 – cycle 9). Possibly these areas represent former syncytial cellular communities.

4. *Biochemical Storage as a Factor for Division Synchrony*

In the foregoing we have discussed the possible continuity from cell to cell and its significance in the control of synchronous cell division. We shall now consider whether from cell to cell in a tissue there may be biochemical homogeneities which are favourable for synchronous division in that population. The early embryo is charged with chemical entities which become utilized during embryogenesis. First of all we shall consider the nucleic acids. It seems indisputable that the frog egg (ovocytes, unfertilized and fertilized eggs) contains large amounts of cytoplasmic DNA and of deoxyribosides (Zeuthen, 1951; Hoff-Jørgensen, 1954; Hoff-Jørgensen and Zeuthen, 1952; Grant, 1958; and Løvtrup, 1959).

The DNA content of the embryo remains constant until the early blastula stage; after that it increases. The DNA which is stored in the frog egg is estimated to be sufficient for about twelve to thirteen division cycles. It was mentioned earlier that in the toad a relative cleavage synchrony can be traced over about fifteen division cycles. It seems not unlikely that in amphibia a contributing factor in the maintenance of the long division synchrony is the presence of preformed DNA and of deoxyribosides in the cytoplasm.

Also in insect eggs there is DNA stored in the cytoplasm. In this case DNA is clearly supplied by the nurse cells. The egg of the cricket *Gryllus* may contain DNA sufficient for about ten division cycles (Durand, 1955). However, it has been found that the mitotic synchrony in the *Gryllus* egg is less pronounced than in other insect eggs investigated. Also the *Drosophila* egg seems to contain cytoplasmic DNA (Levenbook *et al.*, 1953; Nigon and Daille, 1958). During the first thirteen hours of development the DNA content in the embryo increases five times. At the same time the nuclear multiplication must be more than a thousand-fold (Nigon and Daille, 1958). Thus, in the insect embryo also DNA storage may contribute to the division synchrony.

The hen's egg taken as a whole (white, yolk and embryo) contains DNA equal to 5×10^7 times the contents of a diploid cell from the adult (Hoff-Jørgensen, 1954). However, almost all the DNA is outside the asynchronously cleaving blastoderm, the embryo proper. Each blastomere of the unincubated chick embryo contains twice, or less than this, the amount of DNA of a cell from a 26-hr embryo or from an adult (Emanuelsson, 1961a). In the unfertilized egg of the sea urchin *Paracentrotus lividus*, the amount of DNA is about twenty-five times that of the sperm as demonstrated by Elson and Chargaff (1952), and by Hoff-Jørgensen (1954). The contents of DNA stays constant for the first two to three divisions and then rises (Hoff-Jørgensen, 1954). In *Psammechinus miliaris* the DNA content appears to be about ten times higher and it keeps constant for seven division cycles (Agrell, 1955b). In *Paracentrotus lividus* this DNA level was found constant for the first six divisions, and in *Echinus esculentus* even for the first eight divisions (Agrell and Persson, 1956b). Strikingly, a sudden net synthesis of DNA always seems to be initiated at exactly the stage—different from species to species—when the division synchrony of the macro- and mesomeres ceases. The early mitotic cycles are always unaffected by testosterone, applied immediately after fertilization. However, in the three species this compound blocks further development at the time of the first asynchronous division (Agrell, 1955b; Agrell and Persson, 1956b). These findings pinpoint the termination of the division synchrony as a critical phase in embryonic life, or as an early stage of differentiation. Using varied and refined preparation techniques, we have later been able to demonstrate that the 'stored DNA' in the sea urchin egg is not DNA but a nucleotide-containing glucoprotein, stable to moderate acid and alkaline hydrolysis (Agrell and Persson, unpublished). This complex is relatively rich in guanylic acid but it also contains some thymine and deoxyribosides. During early development it varies inversely to RNA (Agrell and Persson, 1956b) with which there is little obvious chemical relationship. This nucleotide complex is present in consistently larger amounts in those embryos which have a longer period of division synchrony than in those which have only a short period of synchrony.

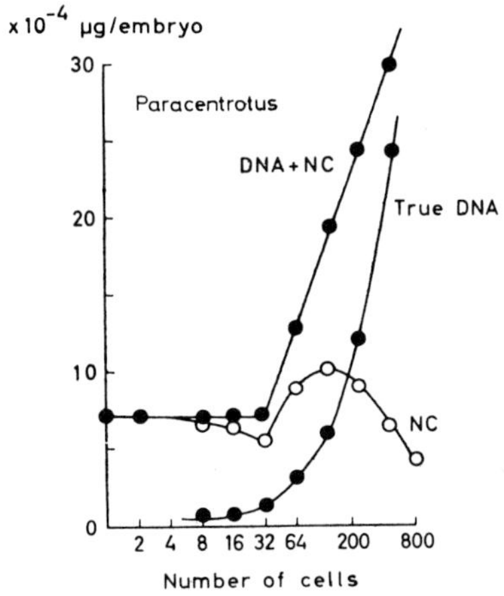

FIG. 7. Changes in the fractions of 'DNA' during the early embryogenesis of the sea urchin embryo—from one and the same batch—all fractions acid and alkali stable. NC = nucleotide-containing glucoprotein extractable with 0·1 *M* NaCl. True DNA not extractable with 0·1 *M* NaCl. DNA + the nucleotide complex together extracted with 1·7 *M* NaCl.

After an initial decrease during early development the complex increases in amount to show a peak around the time when the division synchrony comes to an end. At the same time it changes in composition, becoming poorer in guanine and richer in other bases. It may be composed of two fractions, one which from the beginning is stored in the cytoplasm and another which is synthesized during embryogenesis (Fig. 7). On the other hand, the amount of true DNA in the sea urchin embryo was found by us to double with each cleavage, at least from the four or eight-cell stage. In Hoff-Jørgensen's biological assay of DNA, used in the cited references (Hoff-Jørgensen, 1954; Hoff-Jørgensen and Zeuthen, 1952; Løvtrup, 1959), an analysis is made of the deoxyribosidic linkage present in compounds liberated from the cell structure by the action of DNAase. All these results can, to a certain extent, be harmonized if we assume that polydeoxyribosides combined to glucoproteins represent the postulated 'stored DNA' which is a useful precursor for DNA, although different from true DNA. Discussing this matter from a biological viewpoint Zeuthen (1951) has previously suggested that the 'cytoplasmic DNA' could not carry genetic information. It is therefore concluded that in the sea urchin embryo not cytoplasmic DNA, but special stores of combined bases, may be important for maintaining the division synchrony. Because there are such low amounts of true DNA in the unfertilized sea urchin egg some authors (Marshak and Marshak, 1953) have suggested that in this cell there is no DNA at all, not even in the nucleus. However, in the pronucleus I have demonstrated Feulgen-positive material and true chromosomes attached to the nuclear membrane (Agrell, 1958b, c).

VI. Graded Mitotic Activity and Tissue Differentiation

It may now be time to raise the question as to how mitotic gradients arise. Are their premises inherent in the egg, or are they induced? From the beginning all eggs have polarity. In the amphibian egg an animal–vegetal RNA gradient has been demonstrated (cf. Brachet, 1942, 1960). This gradient could very well represent the chemical basis for a later division gradient. In the fly egg, the nuclei with their plasma islands reach, and are absorbed into, a superficial continuous sheet of periplasm, which from the beginning must be differentiated to be a substrate on which mitotic gradients can later develop. In the primitive insect egg (e.g., the odonates) no mitotic gradients develop and there is no periplasm to represent the necessary continuum.

In some eggs we can pinpoint developmental steps in which mitotic gradients may be induced. In the sea urchin embryo this step seems to be when the micromeres bud off from the vegetative pole. The micromeres exert strong 'vegetalizing' effects. It seems possible that they do so by inducing a mitotic gradient. The gradient is initiated from the vegetative pole of the embryo, and the cytoplasm of the micromeres have a much more intense nucleic acid turnover measured by changes in intensity of RNA staining than have the macromeres and mesomeres (Agrell, 1958a). Unfortunately it is not yet known how the mitotic gradient develops after removal of the micromeres, but lithium which is a vegetalizing agent interferes with the division synchrony and with the mitotic gradient in the sea urchin embryo (Agrell, 1956b). Normally, within the single *Paracentrotus* embryo there is division synchrony for six cycles and within the batch for five cycles. Lithium extends the batch-synchrony for two more cycles. This may indicate that the sequence of cell divisions within the embryo has some relation to the process of vegetalization and thus to morphogenesis. Animalizing agents destroy the synchrony in single sea urchin embryos (Agrell, unpublished) but this may not be significant because many other agents (narcotics; Polowzow, 1924: oestradiol; Agrell, 1955a) are desynchronizing.

Also in the fly embryo the mitotic gradient may be polarly controlled. The first mitotic wave starts at the anterior end so that the cells at the posterior pole of the embryo cleave the latest. An inhibiting influence may spread from the posteriorly situated germ plasm (since the pole plasm has a delaying action upon the mitotic activity of the nuclei which enter this specific cytoplasm). When the germ plasm has budded off as the pole cells, the inhibition may be cancelled and this may release the second mitotic gradient which is initiated at the posterior pole.

Finally, this discussion must touch upon the question of the possible interrelation between cellular division and cellular differentiation. If the nucleus has the alternative choice (Zeuthen, 1951) of reorganizing itself or of controlling growth and differentiation in the cell, then the degree of differentiation may be related to the duration of the interphase. When there is a rapid sequence of synchronous divisions the cells may have limited time for differentiation, and when division rate slows differentially to produce a mitotic gradient, then graded cellular differentiation may be demonstrable along the mitotic gradient.

There are (cf. Waymouth, in Hughes, 1952) few direct observations of antagonisms between cellular division and cellular differentiation. One case has recently been demonstrated. When muscle cells differentiate by showing synthesis of myosin and of actin they do not concurrently synthesize DNA. On the other hand, presumptive myoblasts which synthesize DNA do not concurrently synthesize contractile protein (Stockdale and Holtzer, 1961).

I myself have observed (unpublished) that myoblasts which are taken from the heart of 4–5 day old chicken embryos and which are cultivated on glass may divide, and myofibrils previously present disappear. When during continued growth the medium no longer supports division, the cells grow bigger and myofibrils form, gradually developing cross-striation. When fresh growth medium is added the myofibrils again disappear and the cells divide. This sequence can be repeated many times with the same culture. The experiment clearly demonstrates an antagonism between cell division and visible differentiation. In the sea urchin embryo a second mitotic gradient arises in the mesenchyme blastula (Agrell, 1954). This gradient is inverse to a mitochondrial animal–vegetal gradient (Gustafson and Lenique, 1952). If the formation of new mitochondria is regarded as a manifestation of cellular differentiation we have another example demonstrating antagonism between mitosis and differentiation.

A graded mitotic activity within an embryo may tentatively be regarded as an actual cause of differentiation, because the polarity of the embryo is increased. When in a fast and synchronous mitotic system a slower division rhythm sets in, spreading from a centre a mitotic gradient arises. If lower mitotic activity signals extended interphase times, then the mitotic wave would pass regions with progressively longer interphase times. Furthermore, a clear tendency was reported that with developing asynchrony the interphases continue to extend in time. Therefore the differences in interphase time within the embryo may not be restricted to the one cell cycle which corresponds to the first mitotic wave, but it may persist during more cycles. Thus the polarity in the embryo signalled by the first mitotic wave will become more and more amplified. If, as we have discussed in this account, the protein synthesis and, thereby, the cellular differentiation takes place during interphase rather than during mitosis, then we may expect that differentiation gradients run parallel to the direction taken by the mitotic waves and opposite to the slope of the mitotic gradients. Consequently, graded mitotic activity may assist in directing embryonic differentiation. An early centre of low mitotic activity may be a centre of early differentiation. An immediate result of the differentiation process itself is that initial small differences within the tissue become amplified. Increased polarity produced by graded mitotic activity may be involved in normal embryogenesis both in regulatory and determinate eggs, the two types in this account being mostly represented by the sea urchin egg and by the fly egg.

REFERENCES

AGRELL, I. (1954). A mitotic gradient in the sea urchin embryo during gastrulation. *Arkiv Zool.* **6,** 213.

AGRELL, I. (1955a). L'action de l'oestradiol sur les premières phases du développement embryonnaire de l'Oursin *Psammechinus miliaris. Compt. Rend. Soc. Biol.* **149,** 1322.

AGRELL, I. (1955b). L'action de la testostérone sur les premières phases du développement embryonnaire de l'Oursin *Psammechinus miliaris. Compt. Rend. Soc. Biol.* **149,** 1754.

AGRELL, I. (1955c). A mitotic rhythm in the appearance of mitochondria during the early cleavages of the sea urchin embryo. *Exptl. Cell Research* **8,** 323.

AGRELL, I. (1956a). A mitotic gradient as the cause of the early differentiation in sea urchin embryo. In *Bertil Hanström, Zoological Papers* (K. G. Wingstrand, ed.), Zoological Institute, Lund, Sweden, p. 27.

AGRELL, I. (1956b). The synchronous mitotic rhythm and the effect of lithium during the early development of the sea urchin embryo. *Acta Zool.* (*Stockholm*) **37,** 53.

AGRELL, I. (1957a). The cellular division during the early embryonic development of the sea urchin embryo as influenced by temperature changes. *Kgl. Fysiograf. Sällskap. Lund, Förh.* **27,** 61.

AGRELL, I. (1957b). Partial fertilization of the sea urchin egg by the action of oestradiol. *Arkiv Zool.* **10,** 491.

AGRELL, I. (1958a). A cytoplasmic production of ribonucleic acid during the cell cycle of the micromeres in the sea urchin embryo. *Arkiv Zool.* **11,** 435.

AGRELL, I. (1958b). Cytochemical indication of deoxyribonucleic acid in the pronucleus of the mature sea urchin egg. *Arkiv Zool.* **11,** 457.

AGRELL, I. (1958c). The structure of the chromatine in the pronucleus and the syncarion of the sea urchin egg. *Arkiv Zool.* **12,** 95.

AGRELL, I. (1958d). Whole amounts of small embryos attached directly to glass slides. *Stain Technol.* **33,** 265.

AGRELL, I. (1959). The influence of the temperature upon the nuclear changes during syngamy in the sea urchin egg. *Arkiv Zool.* **12,** 291.

AGRELL, I. (1960). Detection of a graded mitotic activity within the sea urchin embryo through the use of oestradiol. *Arkiv Zool.* **12,** 411.

AGRELL, I. (1961) Cytochemical demonstration of a varied linkage of nuclear RNA during growth and division. *Pathol.Biol. Semaine Hop.* **9,** 775.

AGRELL, I. (1962). Mitotic gradients in the early insect embryo. *Arkiv Zool.* ser. 2, **15,** 143.

AGRELL, I. and H. PERSSON (1956a). The action of oestradiol on desoxyribonucleoprotein as the cause of its inhibitory effect upon mitosis in the sea urchin embryo. *Biochim. et Biophys. Acta* **20,** 543.

AGRELL, I. and H. PERSSON (1956b). Changes in the amount of nucleic acids and free nucleotides during early embryonic development of sea urchins. *Nature* **178,** 1398.

AGRELL, I. and E. WELIN-BERGER (1957). A diurnal cycle in the mitotic activity of ascites tumour cells in mice. *Nature* **180,** 705.

ALFERT, M. (1959). Cytochemische Untersuchungen an basischen Kernproteinen während der Gametenbildung, Befruchtung und Entwicklung. *Chem. Gene. Berlin,* **9,** 73.

ANDREWS, E. A. (1898). Filose activity in metazoan eggs. *Zool. Bull.* **2,** 1.

BEERMANN, W. (1956). Nuclear differentiation and functional morphology of chromosomes. *Cold Spring Harbor Symp. Quant. Biol.* **21,** 217.

BORUM, K. (1961). Oogenesis in the Mouse. A study of the meiotic prophase. *Exptl. Cell Research* **24,** 495.

BOVERI, T. (1887). Über Differenzierung der Zellkerne während der Furchung des Eies von *Ascaris megalocephala. Anat. Anz.* **2,** 688.

BRACHET, J. (1942). La localisation des acides pentosenucléiques dans les tissus animaux et les œufs d'Amphibiens en voie de développement. *Arch. Biol. (Liège)* **53,** 207.

BRACHET, J. (1960). *The Biochemistry of Development.* Pergamon Press, Oxford.

BRESSLAU, E. (1904). Beiträge zur Entwicklungsgeschichte der Turbellarien. I. Die Entwicklung der Rhabdocölen und Alloiocölen. *Z. Wiss. Zool.* **76,** 213.

BRIGGS, R. and T. J. KING (1957). Changes in the nuclei of differentiating endoderm cells as revealed by nuclear transplantation. *J. Morphol.* **100,** 269.

BULLOUGH, W. S. (1952). The energy relations of mitotic activity. *Biol. Rev. Cambridge Phil. Soc.* **27,** 133.

CERFONTAINE, P. (1906–1907). Recherches sur le développement de l'Amphioxus. *Arch. Biol. (Liège)* **22,** 229.

CONKLIN, E. G. (1905). The organization and cell-lineage of the Ascidian egg. *J. Acad. Nat. Sci. Philad.* **13,** 1.

CONKLIN, E. G. (1912). Body size and cell size. *J. Morphol.* **23,** 159.

DALCQ, A. (1954). Fonctions cellulaires et cytochimie structurale dans l'œuf de quelques Rongeurs. *Compt. Rend. Soc. Biol.* **148,** 1332.

DALCQ, A. (1957). *Introduction to General Embryology.* Clarendon Press, Oxford.

DAN, K. (1960). Cyto-embryology of Echinoderms and Amphibia. *Intern. Rev. Cytol.* **9,** 321.

DAVIS, H. S. (1908). Spermatogenesis in *Acrididae* and *Locustidae. Bull. Museum Comp. Zool. Harvard, Coll.* **53,** 59.

DUBOIS, A. M. (1932). A contribution to the embryology of *Sciara* (*Diptera*). *J. Morphol.* **54,** 161.

DURAND, M. C. (1955). L'acide desoxyribonucléique des gametes de *Gryllus domesticus. Compt. Rend. Acad. Sci., Paris* **241,** 1340.

ELSON, D. and E. CHARGAFF (1952). On the desoxyribonucleic acid content of sea urchin gametes. *Experientia* **8,** 143.

EMANUELSSON, H. (1961a). Photometric estimation of the relative amount of DNA in early chick embryos. *Acta Physiol. Scand.* **52,** 197.

EMANUELSSON, H. (1961b). Mitotic activity in chick embryos at the primitive streak stage. *Acta Physiol. Scand.* **52,** 211.

FAWCETT, D. W. (1961). Intercellular bridges. *Exptl. Cell Research Suppl.* **8,** 174.

FAWCETT, D. W., S. ITO and D. SLAUTTERBACK (1959). The occurrence of intracellular bridges in groups of cells exhibiting synchronous differentiation. *J. Biophys. Biochem. Cytol.* **5,** 453.

FLYNN, T. T. and J. P. HILL (1947). The development of the *Monotremata.* VI. The later stages of cleavage and the formation of the primary germ-layers. *Trans. Zool. Soc., London* **26,** 1.

FRANK, G. M. (1925). Über Gesetzmässigkeiten in der Mitosenverteilung in den Gehirnblasen im Zusammenhange mit Formbildungsprozessen. *Arch. Mikroskop. Anat.* **104,** 262.

GODLEWSKI, E., Jr. (1908). Plasma und Kernsubstanz in der normalen und der durch aussere Faktoren veränderten Entwicklung der Echiniden. *Arch. Entwicklungsmech. Organ.* **26,** 278.

GOLDSCHMIDT, R. (1902). Bemerkungen zur Entwicklungsgeschichte des Polystomum integerrimum Rud. *Z. Wiss. Zool.* **72,** 180.

GRANT, P. (1958). The synthesis of deoxyribonucleic acid during early embryonic development of *Rana pipiens. J. Cellular Comp. Physiol.* **52,** 227.

GRAY, J. (1924). The mechanism of cell-division. I. The forces which control the form and cleavage of the eggs of *Echinus esculentus. Proc. Cambridge Phil. Soc. Biol. Sci.* **1,** 164.

GUSTAFSON, T. and P. LENIQUE (1952). Studies on mitochondria in the developing sea urchin egg. *Exptl. Cell Research* **3,** 251.

HÄCKER, V. (1892). Die Furchung des Eis von *Aequorea Forskålea. Arch. Mikroskop. Anat.* **40,** 243.

HALBERG, F. (1959/60). Physiologic 24-hour periodicity; general and procedural considerations with reference to the adrenal cycle. *Intern. Z. Vitaminforsch. Beih.* **10,** 225.

HALE, L. J. (1956). Mitotic activity during the early differentiation of the scleral bones in the chick. *Quart. J. Microscop. Sci.* **97,** 333.

HATSCHEK, B. (1882). Studien über Entwicklung des *Amphioxus. Arb. Zool. Inst. Univ. Wien* **4,** 1.

HENNEGUY, L. (1896). *Leçons sur la Cellule.* Georges Carré, Paris.

HOFF-JØRGENSEN, E. (1954). Deoxynucleic acid in some gametes and embryos. In *Recent Developments in Cell Physiology* (J. A. Kitching, ed.). Butterworths, London, p. 79.

HOFF-JØRGENSEN, E. and E. ZEUTHEN (1952). Evidence of cytoplasmic deoxyribosides in the frog's egg. *Nature* **169,** 245.

HOLM, Å. (1952). Experimentelle Untersuchungen über die Entwicklung und Entwicklungsphysiologie des Spinnenembryos. *Zool. Bidr. Uppsala* **29,** 293.

HOLTER, H. and E. ZEUTHEN (1957). Dynamics of early echinoderm development, as observed by phase contrast microscopy and correlated with respiratory measurements. *Pubbl. Staz. Zool. Napoli* **29,** 285.

HOWARD, F. L. (1932). Nuclear division in plasmodia of *Physarum. Ann. Botany* **46,** 461.

HUETTNER, A. F. (1923). The origin of the germ cells in *Drosophila melanogaster. J. Morphol.* **37,** 385.

HUGHES, A. (1952). *The Mitotic Cycle.* Butterworths, London, p. 167.

HUIE, L. H. (1918). The formation of the germ-band in the egg of the Holly Tortrix Moth, *Eudemis nævana* (*Hb.*) *Proc. Roy. Soc. Edinburgh* **38,** 154.

HUMPHREY, R. R. (1922). The multiple testis in *Urodeles. Biol. Bull.* **43,** 45.

KÄLLÉN, B. (1956). Contributions to the knowledge of the regulation of the proliferation processes in the vertebrate brain during ontogenesis. *Acta Anat.* **27,** 351.

KINNANDER, H. and T. GUSTAFSON (1960). Further studies on the cellular basis of gastrulation in the sea urchin larva. *Exptl. Cell Research* **19,** 278.

KOPSCH, F. (1901). Die Entstehung des Dottersackentoblasts und die Furchung bei *Belone acus. Intern. Monatsschr. Anat. Physiol.* **18,** 43.

KOROTNEFF, A. (1906). Zur Embryologie von *Pyrosoma. Mitt. Zool. Sta. Neapel* **17,** 295.

LEBLOND, C. D. and I. CLERMONT (1952). Definition of the stages of the cycle of the seminiferous epithelium in the rat. *Ann. N. Y. Acad. Sci.* **55,** 548.

LEVENBOOK, L., E. TRAVAGLINI and J. SCHULTZ (1953). Nucleic acids and free polynucleotide fragments in the egg of *Drosophila. Anat. Record* **117,** 585.

LEWIS, W. H. and C. G. HARTMAN (1933). Early cleavage stages of the egg of the monkey (*Macacus Rhesus*). *Contrib. Embryol.* **24,** 189.

LIPP, Ch. (1955). Beitrag zur somatischen Cytologie der Schmetterlinge. *Chromosoma* **7,** 1.

LUDWIG, K. S. (1954). Das Verhalten der Thymonukleinsäure (DNA) während der Befruchtung und den ersten Segmentationsstadien bei der Ratte und dem Goldhamster. *Arch. Biol.* (*Liège*) **65,** 135.

LØVTRUP, S. (1959). Synthesis of desoxyribonucleic acid during amphibian embryogenesis at different temperatures. *J. Exptl. Zool.* **141,** 545.

MARKMAN, B. (1961). Regional differences in isotopic labelling of nucleic acid and protein in early sea urchin development. An autoradiographic study. *Exptl. Cell Research* **23,** 118.

MARSHAK, A. and C. MARSHAK (1953). Desoxyribonucleic acid in *Arbacia* eggs. *Exptl. Cell Research* **5,** 288.

MARTIN, F. (1914). Zur Entwicklungsgeschichte des polyembryonalen *Chalcidiers Ageniaspis* (*Encyrtus*) *fuscicollis Dalm. Z. Wiss. Zool.* **110,** 419.

MAZIA, D. (1957). *Chemical Basis of Heredity*. Baltimore, p. 169.

MCMASTER, R. D. (1955). Desoxyribose nucleic acid in cleavage and larval stages of the sea urchin. *J. Exptl. Zool.* **130,** 1.

MEAD, A. D. (1897). The early development of marine annelids. *J. Morphol.* **13,** 227.

MENKIN, V. (1959). Nature of factors regulating the rate of cell division. *J. Exptl. Zool.* **140,** 441.

NACHTWEY, R. (1925). Untersuchungen über die Keimbahn, Organogenese und Anatomie von *Asplanchna priodonta Gosse. Z. Wiss. Zool.* **126,** 239.

NIGON, V. and J. DAILLE (1958). La synthèse de l'acide désoxyribonucléique au cours du développement de la drosophile. *Biochim. et Biophys. Acta* **29,** 246.

NILSSON, R. (1961). Acid-soluble nucleotides during early embryonic development of the sea urchin *Paracentrotus lividus. Acta Chem. Scand.* **15,** 583.

PATTERSON, J. T. (1910). Studies on the early development of the hen's egg. I. History of the early cleavage and of the accessory cleavage. *J. Morphol.* **21,** 101.

POLOWZOW, W. (1924). Über die Wirking der Alkoholnarkose auf die Entwicklung der Seeigeleier. II. Teil. Polysyncytien. *Arch. Mikroskop. Anat.* **103,** 1.

RENSCH, B. (1948). Histological changes correlated with evolutionary changes of body size. *Evolution* **2,** 218.

ROBERT, A. (1902). Recherches sur le Développement des Troques. *Arch. Zool. Exptl. Gen.* **10,** 269.

ROOSEN-RUNGE, E. C. and L. O. GIESEL, Jr. (1950). Quantitative studies on spermatogenesis in the albino rat. *Am. J. Anat.* **87,** 1.

SEHL, A. (1931). Furchung und Bildung der Keimanlage bei der Mehlmotte *Ephestia Kuehniella* Zell. nebst einer allgemeinen Übersicht über den Verlauf der Embryonalentwicklung. *Z. Wiss. Biol., Abt. A* **20,** 533.

SEIDEL, F. (1929). Untersuchungen über das Bildungsprinzip der Keimanlage im Ei der Libelle *Platycnemis pemmipes* I–V. *Arch. Entwicklungsmech. Organ.* **119,** 322.

SEIDEL, F. (1952). Die Entwicklungspotenzen einer isolierten Blastomere des Zwelzellenstadiums im Säugetierei. *Naturwiss.* **39,** 355.

SELENKA, E. (1883). *Studien über Entwicklungsgeschichte der Thiere. Die Keimblätter der Echinodermen.* Kreidel's Verlag, Wiesbaden.

SESHACHAR, B. R. and P. B. PADMAVATHI (1956). Synchronous micronuclear divisions in multi-micronucleate ciliates. *Current Sci.* (*India*) **25,** 281.

SHORT, R. B. (1946). Observations on the giant amoeba *Amoeba carolinensis. Biol. Bull.* **90,** 8.

SIRAKAMI, K. (1958). Cyto–embryological studies. I. On the cleavage of isolated blastomeres from morulae of *Bufo vulgaris.* (A preliminary note.) *Mem. Fac. Liberal Arts Educ.* **9,** 181.

SONNENBLICK, B. P. (1950) The early embryology of *Drosophila melanogaster.* In *Biology of Drosophila* (M. Demerec, ed.). John Wiley, New York, p. 62.

STOCKDALE, F. E. and H. HOLTZER (1961). DNA synthesis and myogenesis. *Exptl. Cell Research* **24,** 508,

SWANN, M. M. (1957). The control of cell division: A review. I. General mechanisms. *Cancer Research* **17,** 727.

SWANN, M. M. (1958). The control of cell division: A review. II. Special mechanisms. *Cancer Research* **18,** 1118.

TIEGS, O. W. and F. MURRAY (1938). The embryonic development of *Calandra oryza. Quart. J. Microscop. Sci.* **80,** 159.

VIALLETON, M. L. (1888). Premières phases du développement de la seiche. (*Sepia officinalis*). *Ann. Sci. Nat., Zool.* **6,** 165.

WATASE, S. (1891). Studies on Cephalopods. I. Cleavage of the ovum. *J. Morphol.* **4,** 247.

WEBER, H. (1933). *Lehrbuch der Entomologie.* Verlag Gustav Fischer, Jena, p. 492.

ZEUTHEN, E. (1951). Segmentation, nuclear growth and cytoplasmic storage in eggs of Echinoderms and Amphibia. *Pubbl. Staz. Zool. Napoli* **23,** suppl. 47.

ZEUTHEN, E. (1953a). Oxygen uptake as related to body size in organisms. *Quart. Rev. Biol.* **28,** 1.

ZEUTHEN, E. (1953b). Biochemistry and metabolism of cleavage in the sea urchin egg, as resolved into its mitotic steps. *Arch. Neerl. Zool.* **10,** *Suppl.* 1, 31.

SECTION II

SYNCHRONY OF CELL DIVISION IN MICRO-ORGANISMS

A.	The single cell	Chap. 3
B.	Induced synchrony in protozoa	Chaps. 4–8
C.	Induced synchrony in algae	Chaps. 9–11
D.	Induced synchrony in fungi	Chaps. 12–14
E.	Induced synchrony in bacteria	Chaps. 15–16
F.	Induced synchrony in mammalian cells	Chap. 17

3

The Normal Cell Cycle

D. M. Prescott, *Biology Division, Oak Ridge National Laboratory,* Oak Ridge, Tennessee*

CONTENTS

I	Introduction	71
II	Growth Patterns of Single Cells	74
III	DNA and Histone Syntheses	80
IV	Cycle Variations in RNA Synthesis	86
V	Protein Synthesis and Levels of Enzyme Activity during the Cycle	88
VI	Respiration and Energy Metabolism	89
VII	Concluding Remarks	90
	References	91

I. Introduction

The cell life cycle includes one complete interphase and one cell division, and ordinarily the completion of cell division is the most convenient means of marking the end of one cycle and the beginning of the next. Succinctly summarized, the general events of the cycle are increase in cell mass, replication of the genetic code, and the division of the cell mass and duplicated code into two independent functional units. These components are under detailed study from many points of view and in a wide variety of cell types, and although the cycle of each cell type has characteristic features of its own, the interpretation of information is usually based to some degree on the premise—assumed but seldom explicitly stated—that the cycle operates according to a set of biochemical principles of replication and reproduction that is common to all cells.

The two key events of cell reproduction are DNA synthesis and nuclear division, and progress of the cell in interphase is now commonly related temporally to one or both of these events. The terminology introduced by Howard and Pelc (1953) is convenient for designating the cycle status of a cell and has been widely adopted. The interval between the end of cell division and the beginning of DNA synthesis is designated as the G_1 period, the interval occupied by DNA synthesis is called the S period, and the G_2 period covers the

* Operated by Union Carbide Corporation for the United States Atomic Energy Commission.

interval between the end of DNA synthesis and the first visible sign of cell division. The letter D is appropriate for the division period and is intended to include the full course of cytoplasmic and nuclear divisions. This terminology may not be completely satisfactory for bacteria in which nuclear and cytoplasmic divisions do not occupy the same time interval.

Subdivision of the complete cycle into these four successive intervals, G_1, S, G_2, and D, carries recognition that cell reproduction is a continuous process, the continuity of which is maintained by an uninterrupted stream of causally related events. Knowledge of the S and D periods is far from complete, but at least these two intervals have positive definitions. G_1 and G_2, on the other hand, represent periods to which it is not yet possible to ascribe any biochemical events that are specific parts of cycle progress or continuity, and the positive basis for the existence of these two intervals is not known. RNA and protein syntheses and general mass increase of the cell occur during G_1 and G_2 (and S), but the role of these activities in advancing the cell within the cycle is unknown. The co-ordination of DNA replication and nuclear division must be mediated across the G_1 and G_2 intervals, but for the present the linking events can be little more than guessed at. In some bacteria DNA synthesis may be almost continuous, so that G_1 and G_2 periods may be virtually absent. This apparent simplicity of the cycle composition does not, however, diminish the complexity of the basic problems. The causal relations that co-ordinate DNA synthesis and nuclear division still remain to be dealt with and are unfortunately still largely a mystery.

At the cytological level, mechanisms for accomplishing various steps can differ markedly among cell types. Most cells, for example, achieve an equal distribution of DNA between the two daughter cells at division by mitosis, but in ciliated protozoa the DNA of the macronucleus is divided by a nonmitotic or amitotic mechanism. In amicronucleated strains of *Tetrahymena*, mitosis is absent completely. (Nuclear fragmentation is occasionally observed in nuclei that normally reproduce by mitosis and is often incorrectly referred to as amitosis.) Evidence of mitosis in bacteria is not convincing, and the separation of DNA into two complete sets is apparently accomplished by a nonmitotic process. Cytokinesis in animal cells is ordinarily achieved by constriction initiated at the cell surface or cortex and in plants by a deposition of a septum between the newly separated nuclei. A complete list of such cytological modifications would be very long. Finally, the various cell types follow a diversity of time scales and schedules. The absolute time or the time span relative to the complete cycle length required to accomplish a particular step may differ widely among cell types. Variations in the period of DNA synthesis are a good example. All of these numerous modifications increase the difficulties of arriving at a unitary hypothesis of the continuity of the cell life cycle.

Normally cells double in size between divisions, and a balance between cell growth and cell reproduction is maintained. Growth, however, is a measure of the total net increase of all the accumulatory and synthetic processes of a cell, and many of these are of no direct importance in the preparations for cell division. Cell growth and reproduction are connected by a flexible relationship, but the balance between the two processes can be upset temporarily without affecting cell viability. The rates of growth and reproduction in *Paramecium*, for example, can be altered to a large degree independently by manipulating nutrient conditions (Kimball *et al.*, 1959). In the heat-shock synchronization of *Tetrahymena*, cell division is inhibited by the induction treatment and abnormally large cells are produced by the continuation of growth (Scherbaum and Zeuthen, 1954). In the synchronized divisions that occur following release from the treatment, the cells do not grow rapidly enough to double in size between successive divisions, and the average cell size is reduced markedly after a few divisions. These and other observations prove that cell mass or the rate of cell-mass increase have no direct role in the control of the cell cycle or in the initiation of cell division.

The partial, mutual independence between cell growth (or cell size) and progress of the reproductive cycle is evidence that the metabolism of the cell is divisible into two very general categories: those activities that immediately advance the cycle and those that are only remotely or indirectly concerned with maintaining cycle continuity (Burns, 1961; Prescott, 1961b). This recently achieved awareness of the existence of the two categories has led to the beginning of attempts at sorting out the activities of the growth-duplication cycle.

Such central problems of the cell cycle as the links between replication (DNA synthesis) and reproduction (cell division), maintenance of the balance between growth and reproduction, and co-ordination of nuclear and cytoplasmic divisions are still difficult to approach experimentally in a direct way, but at least the questions have become much more clearly defined during the past few years. For additional discussions of these and other aspects of the cell life cycle see Anderson (1956a, b), Burns (1961), Campbell (1957), Hughes (1952), Mazia (1958), Prescott (1961b), Stern (1956), Swann (1957, 1958) and Zeuthen (1958).

II. Growth Patterns of Single Cells

Methods now available for measuring growth in single cells are not completely satisfactory, and obtaining an accurate record of cell growth over the complete cycle is still a difficult, tedious task. In an ideal system, accurate measurements of mass, protein, or other constituents would be recorded continuously on individual cells without in any way disturbing progress of the normal cycle. A certain amount of compromise is necessary with current methods either in the accuracy or frequency of measurements, by the necessity of killing the cell at the time of measurement, in accuracy of cycle timing, by disturbance of cell activities, increases in normal generation time, etc. Volume or mass changes (by interference microscopy, X-ray absorption, or the Cartesian diver balance) over the full cycle have been recorded with reasonably good success for a number of different cells.

Determinations of cell growth patterns from one division to the next have given clear answers to two questions. First, increase in cell size or attainment of a 'critical' size have no immediate role either in maintaining progress of the cycle or in the initiation of cell division. Second, division-to-division growth of the cell is not an autocatalytic phenomenon. Hypotheses encompassing the idea that cell growth or size in some manner acts as a stimulus of cell division have a long history, and in spite of a broad range of circumstantial and direct evidence against these hypotheses, modern cell biology has been slow to abandon them. Cell growth and reproduction ordinarily maintain a balance such that cell mass approximately doubles between successive divisions, a fact that has undoubtedly contributed to the impression that cell mass might have a direct bearing on the initiation of division. In amoeba, however, cell mass ceases to increase several hours before any recognizable sign of mitosis and cannot be immediately involved in the stimulation of cell division (Prescott, 1955). The mass increase of an amoeba can be stopped in early interphase by withdrawing nutrients from the medium, but even these cells divide anyway (Prescott, 1956). The more severely growth is limited in such experiments, the longer is the resulting interphase. Apparently the necessary events leading to division cannot be accomplished as rapidly using the internal resources of the cell. The variability among cell types with respect to the strength of this reserve potential for continuing the reproductive cycle is probably dependent largely upon the extent of reserves in the form of precursor pools, storage materials, etc. In this respect, cell mass is indirectly important for cell division because it is a measure, to some degree, of both the amount of reserve or precursor materials available and the amount of metabolic machinery upon which advancement of the cell in the cycle is either directly or indirectly dependent. Other examples illustrating the lack of a critical relationship between cell size or growth and cell division are discussed elsewhere (Prescott, 1961b). Two other illustrations on this point—with *Tetrahymena* and *Paramecium*—were mentioned in the Introduction.

With one exception the individual cell growth curves clearly demonstrate that division-to-division growth cannot be autocatalytic. The absolute growth rate does not increase as cell size increases and is obviously regulated by a mechanism that is relatively independent of cell size. Growth rate in mass of single amoebae, determined with the Cartesian diver balance, is most rapid immediately after division, declines slightly through interphase and falls to zero several hours before division (Prescott, 1955). The rate of protein synthesis during interphase in *Tetrahymena*, measured by the incorporation of ^{14}C-methionine, does not change (Prescott, 1960b). The volume of individual *Tetrahymena* also increases at a constant absolute rate through most of interphase but increases very rapidly during division (Cameron and Prescott, 1961). In Zeuthen's (1951, 1953) remarkable studies on respiration in single *Tetrahymena*, increases in respiration rate were considered to reflect corresponding increases in the respiratory machinery and hence to be a measure of growth. The increase in respiration rate of single *Tetrahymena* between divisions was linear, but during division itself the respiration rate remained constant or declined slightly, suggesting a possible cessation of growth during division. These three criteria of growth, i.e. increases in volume, respiratory rate and ^{14}C-methionine incorporation, show the same linear course. Finally, both protein and RNA syntheses, measured as ^{3}H-histidine and ^{3}H-uridine incorporation, continue during cell division in *Tetrahymena* without any marked change from the rate prevailing during interphase (author's unpublished experiments). Thus, during division, cell volume increases rapidly, respiration ceases to increase at all, and protein and RNA syntheses continue without any pronounced change in rate. These different results with the same cell are not necessarily in conflict but simply illustrate that the various quantities used to indicate growth may at times vary independently of one another.

The capacity of *Tetrahymena* to continue protein synthesis and of the *Tetrahymena* nucleus to continue RNA synthesis during division indicates that macronuclear function is probably not appreciably changed during amitotic splitting. In mammalian tissue culture cells, RNA synthesis is completely blocked when the chromosomes are heavily condensed during mitosis (mid prophase to late telophase) (Feinendegen, Bond, Shreeve and Painter, 1960; Prescott and Bender, 1962; Taylor, 1960a), and during this absence of RNA synthesis, protein synthesis in the whole cell decreases sharply (Prescott and Bender, 1962). Blockage of RNA synthesis during mitosis has also been described in root cells of *Vicia faba* (Davidson, 1963). This behavior with respect to RNA synthesis emphasizes the fundamental difference in the functional state of mitotically and amitotically-dividing nuclei.

In the earlier studies by Bayne-Jones and Adolph (1932) volume increase in the budding yeast, *Saccharomyces cerevisiae*, took place at a constant rate during most of the cycle, but a sharp reduction in growth rate occurred during the last 15 min before formation of a new cell bud. Lindegren and Haddad (1954), and Mitchison (1958) confirmed these volume measurements. Mitchison has also determined mass increase in *S. cerevisiae*. Growth was approximately linear through the whole cycle but, unlike volume increase, no reduction in the rate of mass growth was apparent. Mitchison (1957) has reported very similar results for volume and mass increase in the fission yeast, *Schizosaccharomyces pombe*. The absence of increase in volume of this yeast just before and during division has also been reported by Knaysi (1940). Mitchison's careful measurements show that cell density rises steadily for a time just before completion of cytokinesis and falls slowly thereafter. The significance of these density changes in yeast is unknown. Mitchison has suggested that limitations on the rate of cell wall synthesis may prevent volume increase at the time of septum formation between the two daughter yeast cells. An increased density in HeLa cells during division has also been reported and is apparently due to water loss from both nucleus and cytoplasm (Sandritter, Schiemer, Kraus and Dörrien, 1960).

In *Bacillus megatherium* (Adolph and Bayne-Jones, 1932) the rate of volume growth increases during the interdivision interval, but the rate increase is too low to justify description of an exponential curve. In *Streptococcus faecalis*, however, the rate of volume growth decreases over the cycle, and for a distinct period before the appearance of a cytokinetic furrow, volume increase ceases altogether (Mitchison, 1961). Mitchison's dry mass measurements of this bacterium are an important contribution to the study of cell growth. Evidently in bacteria generally, the cell and nuclear division cycles are out of phase with one another, and in *S. faecalis*, this phase relationship apparently varies with incubation temperature. By following cell increase at two temperatures, Mitchison found evidence that changes in the rate of volume growth occur cyclically in relation to cytokinetic events, but that changes in the rate of increase in dry mass are consistently related to nuclear fission. Dry mass increases most rapidly immediately after nuclear division, and the rate of growth gradually decreases and reaches a minimum just before the next nuclear fission. This correlation between nuclear behavior and mass growth reinforces the argument that the control of growth is dependent upon nuclear activities.

Sandritter *et al.* (1960) undertook the difficult task of measuring mass growth by interference microscopy in a mammalian tissue culture cell (HeLa) during the cycle, and although mass changes on the same cell over the complete cycle could not be followed, the large amount of data collected on many cells over long segments of the cycle have permitted construction of an averaged growth curve that is probably close to the true picture. Total dry weight of the HeLa cell begins to rise at a roughly linear rate after the nucleolus reappears in late telophase and reaches double the telophase value long before the next mitosis. As in amoeba (Prescott, 1955), therefore, a long plateau of no mass growth separates the end of the growth period from the visible beginning of mitosis. The volume of both the cytoplasm and nucleus increased at a gradually diminishing rate to three times the telophase amount. Apparently much of the volume increase is due to water intake. Water appeared to be lost from both nucleus and cytoplasm during preprophase. Maximum cell density was therefore reached during mitosis and minimum cell density during the middle of interphase.

The only cell thus far studied that follows an interdivision growth course compatible with the notion of autocatalytic increase is *Paramecium*. Kimball *et al.* (1959) constructed a mass increase curve from determinations by both interference microscopy and X-ray absorption methods on a series of cells fixed at known postdivision ages. Clearly, mass increase in this cell follows an exponential-like rise. Kaudewitz (1957–1958) has found that the rate of $^{32}PO_4$ uptake by *P. aurelia* accelerates through the interkinetic period but not sufficiently to yield the exponential-like curves for mass increase obtained in the careful studies of Kimball *et al.* (1959). It is reasonable to expect that mass and protein increases should follow parallel courses, but Woodard, Gelber and Swift (1961) measured protein over the cycle in *P. aurelia* by a microspectrophotometric method and found a greater than exponential increase. Although some lack of agreement among these studies is evident, there is no doubt that *Paramecium* has an accelerating growth rate over the cycle and in this respect is distinctly different from the other cell types examined.

Curves showing an accelerating growth rate, and even closely following an exponential type course, certainly do not necessitate the interpretation of autocatalytic growth [for further discussion of this see Prescott (1961b)]. The growth curves for bacteria, yeast, amoeba, HeLa cells, and *Tetrahymena*, on the other hand, suggest that the growth rate is dependent upon some activity that proceeds at a relatively unchanging rate. Although the control of growth is a complex phenomenon involving many interacting processes, the nucleus probably exerts primary control. Mitchison's (1961) experiments, suggesting a correlation between growth rate and the nuclear reproductive cycle, supports such a conclusion. Amoeba and *Tetrahymena* (Prescott, 1961a) from which the nucleus has been removed and *Paramecium* that failed to receive a macronucleus during division all discontinue growth very soon, if not immediately, although some amino acid incorporation into protein may continue for a time (Prescott and Kimball, unpublished experiments).

The interphase nucleus is the major if not the exclusive site of RNA synthesis—at least in cells that do not contain DNA in the cytoplasm (Goldstein *et al.*, 1960; Perry, 1960; Prescott, 1959a, 1960a, 1961a; Zalokar, 1960)—and RNA is released to the cytoplasm by the nucleus continuously during interphase (Amano and Leblonde, 1960; Feinendegen *et al.*, 1960; Goldstein and Micou, 1959a, b; Perry, 1960; Perry, Errera, Hell and Dürwald, 1961; Perry, Hell and Errera, 1961; Prescott, 1961a; Woods, 1959; Zalokar, 1960). These conclusions, combined with the knowledge of the role of the various fractions of RNA in protein synthesis, are sufficient reason to give serious consideration to the hypothesis that nuclear control over growth is mediated through RNA synthesis and transfer. In this hypothesis the control of RNA synthesis then becomes a question of central importance. Currently, *in vitro* studies of RNA synthesis in cell-free, semipurified systems and *in vivo* and *in vitro* studies of the status of RNA synthesis during enzyme induction are perhaps the more promising means of gaining specific information on this point. Kimball and Perdue (1962) have, however, reported some evidence that the rate of RNA synthesis in the macronucleus of *Paramecium* may be proportional to the amount of DNA. During the DNA synthesis period, the rate of RNA synthesis rises more or less in parallel with macronuclear volume, indicating indirectly a proportionality between the rate of RNA synthesis and the amount of DNA.

Although there is ordinarily a general correspondence between the amount of RNA in a cell and the rate of protein synthesis, the presence of RNA is not enough to insure protein synthesis. In enucleated *Tetrahymena*, protein synthesis stops completely when less than 50 per cent of the anucleates' RNA has been lost (Prescott, 1962). The experiment suggests that some fraction of RNA (Volkin and Astrachan, 1957) essential for all protein synthesis, produced in the nucleus and transported to the cytoplasm, would be a means for mediating a relatively direct control of growth by the nucleus. Finally, the speculations that the cell growth rate must increase as cell size increases or that growth is an autocatalytic phenomenon are based on a superficial examination of the problem and in view of single cell growth studies, in particular, should be dropped from further serious consideration.

The single cell growth studies demonstrate one additional important point. Except in *Paramecium* (Kimball *et al.*, 1959), the combined growth rate for the two daughter cells increases abruptly at cell division. This event coincides with division of nuclear material into two independent functional units but does not appear to be related to cytokinesis. In *S. faecalis*, the rate of mass growth rises sharply after nuclear division whether or not cytokinesis follows shortly after or is delayed (Mitchison, 1961). In amoeba, cytokinesis can be prevented by immersing dividing cells in a protein solution (Prescott, 1959b). Mitosis proceeds normally, and the resulting binucleate amoeba begins to increase in size at a rate that is roughly double the absolute growth rate for uninucleates. Completion of cytokinesis, therefore, has no role in accelerating the growth rate. Aside from nuclear division, there is no known change that can be correlated with this increased rate of growth, and possibly separation of the chromatids into independent units is the responsible event. The rise in growth rate after division cannot, for example, be accounted for by any synthetic activity during mitosis. Synthesis of DNA-histone is completed well before nuclear division, RNA synthesis stops completely during mitosis (Feinendegen *et al.*, 1960; Prescott and Bender, 1962; Taylor, 1960a) [but not during amitosis in ciliates (author's unpublished experiments)], protein synthesis declines sharply during mitosis (Prescott and Bender, 1962), and respiration drops during the same period (Frydenberg and Zeuthen, 1960; Zeuthen, 1950 a, b, 1951, 1953, 1955, 1958, 1960).

Some efforts have been made to measure nuclear growth during the cycle. In general, there is an increase in nuclear volume or mass in most cells during the *S* period (see, for example, Carrière *et al.*, 1961; Kimball *et al.*, 1960; Tschermak-Woess *et al.*, 1957; Woodard, Rasch and Swift, 1961). Except for this correlation, no consistently repeated patterns of nuclear-size changes have emerged from measurements on the nuclei of various cell types.

III. DNA and Histone Syntheses

The DNA synthesis interval has been accurately determined for many cell types, usually by microspectrophotometry or autoradiographic detection of incorporated labeled thymidine. With the exceptions of bacteria (Schaechter *et al.*, 1959) and the grasshopper neuroblast (Gaulden, 1956) DNA synthesis occupies only part of the interdivision interval. The cycle position of synthesis is similar in all vertebrate cells so far studied (newt, chicken, rat, mouse, hamster, guinea pig, man, cat). Typically, a G_1 period of several hours separates telophase from the beginning of the *S* period (Bender and Prescott, 1962; Edwards *et al.*, 1960; Firket and Verly, 1957; Harris, 1959; Mendelsohn *et al.*, 1960; Painter and Drew, 1959; Painter *et al.*, 1960; Richards *et al.*, 1956; Siskin and Kinosita, 1961; Stanners and Till, 1960; Taylor, 1960b; Walker and Yates, 1952). Synthesis then proceeds for several hours and is completed several hours before the first sign of prophase. The pattern in plant root cells is similar except that the DNA synthesis period is positioned a little earlier in interphase (Deeley *et al.*, 1957; Howard and Pelc, 1951; Siskin, 1959; Wimber, 1960). In the generative nucleus of the pollen grain, however, DNA synthesis occurs in early interphase (Taylor, 1958a; Taylor and McMaster, 1954; Woodard, 1958). In the slime mold, synthesis is limited to a relatively short space of interphase immediately after mitosis (Nygaard *et al.*, 1960), but in grasshopper neuroblasts (Gaulden, 1956) the *S* period fills all of interphase (from middle telophase to very early prophase).

In *Tetrahymena*, the *S* period for the macronucleus begins shortly after cytokinesis and ends about mid interphase (McDonald, 1958; Prescott, 1960b). As in the slime mold (Nygaard *et al.*, 1960) the G_1 period is almost nonexistent and G_2 is long. The mitotically dividing micronucleus of *Tetrahymena* doubles its DNA immediately after telophase (McDonald, 1960) and before cytokinesis is completed. There is therefore a very clear, temporal separation of DNA synthesis in these two nuclei that share the same cytoplasm. In *Paramecium*, macronuclear DNA synthesis occurs during the last half of the interdivision interval (Kimball and Barka, 1959; Kimball and Perdue, 1962) although there is some evidence that DNA is being synthesized very slowly during the first half of the interdivision interval (Kimball and Perdue, 1962). In contrast to the situation in *Tetrahymena*, the micronucleus synthesizes DNA just before the bulk of macronuclear synthesis (Kimball and Perdue, 1962). DNA synthesis in the macronucleus of Stentor goes on during 90 per cent of interphase (Guttes and Guttes, 1960).

The replication of the macronucleus in the ciliate *Euplotes* shows a particularly unique and interesting pattern. The macronucleus of this ciliate has the shape of a long narrow rod or ribbon (in *E. eurystomus*, 140 by 8μ). Under good culture conditions the generation time is about 12 hr. From 2½ to 3 hr after cytokinesis a band-like structure, referred to as the reorganization band in protozoological literature, originates at each tip of the elongated nucleus, and during the remaining 8 to 9 hr of interphase the two bands travel through the nucleus, meet at the center, fuse, and disappear just before the amitotic splitting of the nucleus during cell division. DNA synthesis is restricted to the bands (Gall, 1959; Prescott and Kimball, 1961) and therefore passes through the nucleus as two waves during the last three-quarters of interphase. DNA replication in the micronucleus occurs just after telophase (Kimball, 1961) and, similar to that in *Tetrahymena* (McDonald, 1958, 1960), is completely separated temporally from macronuclear synthesis.

Studies on synchronized populations of bacteria indicated that DNA doubles immediately preceding a doubling in the number of nuclear bodies (Barner and Cohen, 1956; Lark and Maaløe, 1956). Schaechter *et al.* (1959), however, have reported that in a population of bacteria dividing with complete asynchrony every cell is labeled after a very short pulse of ^{3}H-thymidine. The authors conclude that DNA synthesis occurs during 80 per cent or more of the cycle and suggests that results of studies on cells subjected to treatments inducing synchrony may not represent the 'normal' situation. It is an interesting speculation that such treatments might bring about synchronized synthesis of DNA molecules and thereby shorten the total time and percentage of cycle time required for a complete DNA doubling.

Obviously, the period of DNA synthesis occupies very different proportions of the cycle and shows a wide variety of temporal relations to nuclear division in different cell types. Except in grasshopper neuroblasts (Gaulden, 1956) and possibly in bacteria (Schaechter *et al.*, 1959), DNA synthesis takes up only part of the interdivision interval and never occurs during fission of the nucleus. In many cases, DNA synthesis is completed long before any recognizable sign of nuclear division. Therefore, although DNA synthesis is an absolute requirement for the successful cell life cycle, completion of such synthesis does not immediately bring the cell to the stage recognized as the beginning of nuclear division. Current definitions of the beginning of nuclear division are based on some recognizable cytological change at the light microscope level. In most cases, visible condensation of chromosomes is the first recognizable sign of division, but since condensation proceeds for some time, possibly a long time, before it reaches a state detectable by light microscopy, the real beginning of mitosis cannot be marked. Electron microscopy may eventually give some information on the very early stages of mitosis that are beyond light microscope study. In any case, it is certain that the true D period is longer, and the true G_2 period is correspondingly shorter than accorded by present criteria. Such accuracy in defining the transition from G_2 to D perhaps seems superfluous at the present time, but G_2 contains the link between the end of DNA synthesis and the initiation of nuclear division, and the G_2–D transition is a key part of this link.

The role of histone in the structure and function of the genetic material is still unknown, and the question of whether histone molecules are conserved from cycle to cycle like DNA has not been answered satisfactorily. There is no doubt, however, that new histone synthesis occurs in close association with new DNA synthesis (Alfert, 1959; Bloch and Godman, 1955; Gall, 1959; Prescott and Kimball, 1961). In the reorganization band of *Euplotes*, ^{3}H-histidine (Prescott and Kimball, 1961) and ^{3}H-arginine (author's unpublished experiments) incorporation into protein is very heavy in the very narrow region of DNA synthesis. This is a strong indication that DNA and histone synthesis are virtually simultaneous. In the region of these syntheses, RNA is apparently absent, suggesting that histone synthesis is directly supported by DNA.

One of the central questions about the maintenance of cycle continuity and progress is the initiation and control of DNA synthesis. It has been proposed that DNA synthesis may be controlled through precursor production with polymerization being set off when the accumulation of deoxyribonucleoside triphosphates reaches a critical level. Foster and Stern (1959) have demonstrated the accumulation of soluble deoxyribosidic compounds in a short interval just before DNA synthesis in anthers of *Lilium* (cf. Erickson's chapter in this book). The thesis shifts attention to the question of how the rapid accumulation of DNA precursors is induced. Initiation and control of synthesis of DNA appear to be too complex, however, to be explained by the build-up of precursor pools. Evidence against the idea is varied. For example, addition of all four deoxyribonucleosides in various concentrations does not initiate precocious synthesis of DNA in *Tetrahymena* and does not increase the rate of DNA synthesis once it has started (author's unpublished experiments). In *Tetrahymena* and *Euplotes* DNA synthesis in the macronucleus and micronucleus are completely separated temporally from one another, although the two nuclei are enclosed in the same cytoplasm, are separated from one another by no more than 1 or 2 μ, and are probably exposed to the same precursor pools. In the cell-free, semipurified DNA polymerase system, the appropriate deoxynucleoside triphosphates are present but the initiating event resides in the changes that accompany the transition from native DNA to a denatured form (Bollum, 1959).

The cytoplasm does appear to have a role in synchronizing the initiation of DNA synthesis at distantly separated points within a cell. In *Euplotes* DNA synthesis begins at each tip of the elongated macronucleus at the same time, and the tips are ordinarily separated by 100 or more μ of intervening cytoplasm or nucleus (Gall, 1959; Kimball and Prescott, 1962; Prescott and Kimball, 1961). That this synchrony of initiation is mediated through the cytoplasm and not through the nucleus itself is proven by the fact that in double cells containing two macronuclei, which are well separated from each other by cytoplasm, synthesis starts synchronously in both macronuclei. The initiation of synthesis of the micro- and macronuclei in the same cytoplasm are well separated from each other temporally, and synthesis in these two nuclei must be dependent upon different sets of conditions, at least part of which must be fulfilled by the cytoplasm.

In the macronucleus of *Euplotes* it is clear that only a small fraction of DNA is engaged in replication at any one moment—certainly less than 5 per cent and probably less than 1 per cent (Prescott and Kimball, 1961). Asynchrony in the synthesis of the DNA molecules is apparent in other types of nuclei. Tritiated thymidine labeling of chromosomes of *Crepis* indicates that DNA synthesis begins at the chromosome ends and advances toward the kinetochore (Taylor, 1958b). In Chinese hamster tissue culture cells, asynchrony is apparent within individual chromosomes and between them (Taylor, 1960b). A short pulse of ^{3}H-thymidine results in labeling of only part of the complement and frequently only a limited length of individual chromosomes. Taylor (1960b) has reported evidence that chromosomes enter DNA synthesis in a definite sequence in Chinese hamster cells, but whether chromosomes consistently replicate according to a prescribed order in cells generally remains to be determined. In the plant *Bellevalia* (Taylor, 1958b), in human leucocytes in culture (Bender and Prescott, 1962) and in HeLa cells (Painter *et al.*, 1960) virtually no asynchrony of DNA synthesis between chromosomes is apparent but a very short pulse of ^{3}H-thymidine labeling would be necessary to detect a small degree of asynchrony. When a 30-min pulse of ^{3}H-thymidine is used, frequently only a limited length of human leucocyte chromosomes are labeled. After similarly exposing leucocyte interphase to ^{3}H-thymidine, labeling is often restricted to several separate points and sometimes evenly dispersed within a single nucleus. The patterns of labeling have not been studied in relation to the position of the nuclei in the *S* period, but the findings suggest that DNA synthesis may start at a few points within the leucocyte nucleus and spreads from there. Harris (1959) has described asynchrony of DNA synthesis in rat heart cells with ^{3}H-thymidine beginning at the nucleolus and proceeding into the chromosome.

The problem of how DNA synthesis is initiated has begun to yield to experiment. *In vitro* studies on DNA synthesis by the polymerase system, particularly the work of Bollum (1959, 1960) show that native DNA (presumably independent and separate double helices) is inadequate to support DNA synthesis. DNA denatured by treatments that should break hydrogen bonds does support synthesis, although there is no direct evidence that denaturing represents a change from double helix to single-chain polymer. The naturally occurring single-chain DNA from ΦX-174 coliphage (Sinsheimer, 1959) is an efficient substitute for denatured DNA (Bollum, 1960). This does not necessarily mean that DNA synthesis *in vivo* proceeds by unwinding of double helix DNA into single chains. A local disruption of hydrogen bonds with a few bases rotating out to provide template, such as proposed in the conservative DNA replication model of Bloch (1955), is still a contending hypothesis. As pointed out by Cavalieri and Rosenberg (1961a, b) semiconservative replication of DNA has not been *proven* by any experimental evidence. Cavalieri *et al.* (1959) have presented some evidence that the molecular weight of DNA is halved just before synthesis in *Escherichia coli*. According to these authors, however, the reduction does not represent the formation of single chains but rather a change in the organization of DNA in which laterally linked DNA double helices become disassociated from one another. There seems little doubt that DNA engaged in information dissemination (presumably supporting RNA synthesis) cannot replicate without first markedly reorganizing. The *Euplotes* macronucleus offers a striking example of this thesis. As a wave of DNA synthesis traverses this elongated nucleus the granule-like structures containing the DNA disappear just ahead of the synthesis zone. Electron microscope examination shows the DNA dispersed into coarse fibrils (Fauré-Fremiet *et al.*, 1957; Kluss, 1962). As the region of synthesis is reached these coarse fibrils disappear. Sufficiently high electron microscope resolution has not been achieved to determine what structure exists in this region, but any DNA fibrils in the synthesis zone would certainly have to be less than 100 Å in diameter to have escaped detection in studies completed so far (Kluss, 1962). After synthesis the DNA returns to the organization of granule-like bodies. RNA synthesis, moreover, is present only in nonreplicating regions of the nucleus (Prescott and Kimball, 1961). Thus, in this case, DNA undergoes two distinct changes in organization and these are accompanied by changes in nuclear function. In regions with granularly organized DNA (more than 90 per cent of the nucleus), RNA synthesis is continuous. In the transition zone of dispersed, coarse DNA fibrils, neither DNA nor RNA synthesis is present. In the zone of DNA synthesis, RNA synthesis is absent and DNA is much more dilute and dispersed into a unit unresolved as yet by electron microscopy of sectioned material. How these changes leading to DNA replication are produced and controlled, of course, remains to be determined.

The measurement of DNA polymerase levels over the cycle is one possible approach to the problem of control of DNA synthesis. Polymerase may persist through non-DNA synthesis periods but at some stage, conceivably at the beginning of or during the *S* period new polymerase must be produced. In regenerating liver the level of DNA polymerase increases several-fold sometime within the first 24 hr after partial hepatectomy (Bollum and Potter, 1958). The rate of DNA synthesis begins to rise at about 18 hr after partial hepatectomy and reaches a high level by 24 hr (Hecht and Potter, 1956).

Finally, any hypothesis designed to explain the initiation and control of DNA synthesis must take into account the fact that DNA is exactly doubled during each *S* period, and apparently each DNA molecule replicates only once in the course of each normal cell cycle. Synthesis of DNA molecules in an individual nucleus during a single *S* period is far from synchronous, suggesting that replication produces some changes in the property of DNA that normally prevents further replication of the same molecule until the next cycle. Periods of DNA replication normally alternate with nuclear division, and reorganization during division may return the genetic material to a state allowing another replication of DNA. There is still no satisfactory explanation of why the replication of viral DNA is not subjected to the same control.

The length of the preceding consideration of DNA synthesis is warranted in a discussion of the cell cycle because the key events of cycle continuity are centered around problems of the control of DNA synthesis and function and its relation to cell division.

IV. Cycle Variations in RNA Synthesis

On the basis of the DNA–RNA-protein hypothesis for flow of genetic information, a depression in RNA synthesis would be predicted for two periods in the cell cycle—during DNA synthesis and during mitosis (Bender and Prescott, 1962). A decreased rate of RNA synthesis has been detected in a number of cell types during the *S* period (Moses and Taylor, 1955; Nygaard *et al.*, 1960; Siskin, 1959; Taylor, 1958a; Taylor and McMaster, 1954). The assumption is that DNA engaged in replication is unavailable to serve as a template, or in any other manner, in support of RNA synthesis. Because only a small fraction of the DNA molecules in an individual nucleus is undergoing replication at any one time, presumably a major fraction of DNA is always functioning in RNA synthesis. This is a possible explanation for the limited decrease in RNA synthesis during the *S* interval in many cells.

A more detailed examination of the relation between DNA and RNA syntheses has been possible with *Euplotes* because of the unique features of its macronuclear cytology (Prescott and Kimball, 1961). DNA synthesis moves in two waves through the macronucleus and at any moment is restricted to a very narrow region in each wave. RNA synthesis occurs throughout the nucleus but is undetectable in the region of DNA synthesis. These observations support the hypothesis of a complete mutual exclusiveness of the syntheses of the two nucleic acids. They are also in agreement with the reported findings that two different forms of DNA are necessary in cell-free

RNA and DNA synthesis systems. Native DNA is required for cell-free RNA synthesis in a number of systems (Burma *et al.*, 1961; Furth *et al.*, 1961a, b; Huang *et al.*, 1960; Hurwitz *et al.*, 1960; Ochoa *et al.*, 1961; Stevens, 1961; Weiss, 1960; Weiss and Nakamoto, 1961 a, b), and denatured DNA is required to achieve cell-free synthesis of DNA (Bollum, 1959, 1960).

Concerning RNA production during mitosis, the rate of RNA synthesis decreases during early prophase in mammalian tissue culture cells. Synthesis ceases completely at about mid prophase and is not resumed again until later telophase (Feinendegen *et al.*, 1960; Prescott and Bender, 1962; Taylor, 1960a). Similarly during meiosis in the mouse testis, RNA synthesis is completely absent in cells with advanced chromosome condensation (Welshons *et al.*, unpublished). These findings have been interpreted as an inability of DNA to serve as a template, or otherwise, when tightly condensed into a mitotic or meiotic chromosome. During the absence of RNA synthesis in hamster cells engaged in mitosis, the rate of protein synthesis declines rapidly, reaching a minimum in telophase and rising quickly thereafter (Prescott and Bender, 1962). The situation suggests that some fraction of RNA may be 'used up' during protein synthesis, and that the rate of protein synthesis falls off during mitosis because the critical fraction of RNA is no longer being replenished.

Enucleation experiments on a number of cells support the idea that the nucleus is the exclusive site of RNA synthesis (Goldstein and Micou, 1959a; Prescott, 1959a, 1960a, 1961a). Nuclear transplantation experiments with amoeba (Goldstein and Plaut, 1955) and time-course studies of incorporation of radioactive precursors into RNA have given considerable evidence that RNA is transferred from the nucleus to the cytoplasm continuously during interphase (Amano and Leblond, 1960; Feinendegen *et al.*, 1960; Goldstein and Micou, 1959a, b; Perry, 1960; Perry, Errera, Hell and Dürwald, 1961; Perry, Hell, and Errera, 1961; Prescott, 1961a; Woods, 1959, Zalokar, 1960). In *Euplotes*, nuclear RNA appears to be completely replaced by new RNA during each DNA synthesis period, possibly by transfer of RNA to the cytoplasm. The dissemination of genetic information to the cytoplasm, therefore, is assumed to be continuous during the interphase interval but to cease during mitosis. Also, an abrupt release of a large amount of RNA to the cytoplasm during mitosis in amoeba (Rabinovitch and Plaut, 1956) and at the point of nuclear membrane breakdown in prophase in HeLa cells (Prescott and Bender, 1962) has been demonstrated.

There is no evidence of an equivalent inhibition of RNA synthesis in cells with nuclei that do not divide by mitosis. In the macronucleus of *Euplotes* and *Tetrahymena*, for example, RNA synthesis continues during amitotic splitting (author's unpublished experiments). In *Tetrahymena*, net synthesis of RNA can be prevented by eliminating essential pyrimidines from the medium, but cells go through several divisions before this treatment takes effect (Prescott, 1960b). Net synthesis of DNA apparently continues during these several cycles either through utilization of precursor pools or breakdown products of RNA (author's unpublished experiments). Proliferating *Tetrahymena* are rich in RNA and probably contain enough to fulfill the pyrimidine requirements for DNA synthesis and RNA turnover for several cycles. During such pyrimidine starvation the amount of RNA per cell decreases as the cells divide, but the amount of protein per cell is much less affected, indicating synthesis of protein in the absence of net increase of RNA (Lederberg and Mazia, 1960).

V. Protein Synthesis and Levels of Enzyme Activity during the Cycle

Increase in total protein during the cycle has been used in several instances as a measure of growth and is discussed in Section II. During mitosis, protein synthesis—measured by ^{3}H-histidine incorporation—decreases markedly, possibly as a consequence of the complete inhibition of RNA production during the same period. During amitotic splitting of the macronucleus of *Tetrahymena* there is no such clearly apparent diminution in protein synthesis and no cessation of RNA synthesis. With respect to specific proteins, the evidence for synthesis of histone simultaneously with DNA is discussed in Section III. Synthesis of DNA polymerase was also discussed in Section III. Very little is known about the level of various enzymes from one division to the next. Nasatir and Stern (1959) found decreases in aldolase and D-glyceraldehyde-3-phosphate dehydrogenase during mitosis in lily anther, possibly reflecting a decrease in glycolysis during this period. Peptidase and proteinase activities were found to rise markedly in interphase and decrease during cell budding in yeast (Sylven *et al.*, 1959) but the significance of these changes in relation to cycle progress is unknown. Certainly, detailed knowledge of enzyme activity levels during the cycle is scant. Accurate methods for measuring activities of enzymes in single cells or small populations of cells have not been developed for more than a few enzymes. Moreover, except for DNA polymerase, there is little or no information to guide the choice of enzymes to be measured as a function of the cell reproductive cycle.

Acting as enzymes or structural units, proteins contribute to virtually every phase of cell growth and cycle continuity and, as expected, interference with protein synthesis has relatively immediate effects on progress of the cell in the reproductive cycle. Treatment with chloramphenicol or ribonuclease brings a sharp decrease in protein synthesis and prevents cell division. Ribonuclease produces these effects in *Tetrahymena* through digestion of one or more of the cell's RNA fractions (Price and Prescott, unpublished). In *Tetrahymena*, deprivation of an essential amino acid prevents cells in the first 70 to 80 per cent of the interdivision interval from entering the division (Prescott, 1960b). Withdrawal of the amino acid from the medium during the last 20 per cent of the interdivision interval does not block cell division, either because the cell has sufficient reserves to reach division or because some or all of the protein synthesis of this period is not indispensable for division. For a further discussion of protein synthesis in relation to division see Zeuthen in this volume.

VI. Respiration and Energy Metabolism

Zeuthen's extensive work (1950a, b, 1951, 1953, 1955, 1958, 1960; also Frydenberg and Zeuthen, 1960) with the Cartesian diver microrespirometer has consistently shown a drop in respiration during cell division. Erickson (1947) has observed a depression in respiration during meiotic and mitotic divisions of microspores in lily anthers. According to Stern and Kirk (1948) oxygen consumption in anthers rises during premeiotic stages and drops immediately in the period preceding mitosis and during mitosis. In amoeba (Mazia and Prescott, 1954) and in sea urchin eggs (Zeuthen, 1958), phosphate uptake is reduced during mitosis, possibly reflecting a decline in the generation of high energy phosphate esters, although at least part of the decrease in phosphate utilization is probably due to the absence of RNA synthesis during mitosis. These data on respiration and phosphate uptake and the evidence for lowered rate of glycolysis during mitosis in lily anthers suggest that the energy demands of cells in mitosis and meiosis are reduced. Since there is good reason to believe that the complete cessation of RNA synthesis and the reduction in protein synthesis during mitosis are a consequence of condensation of DNA into a tightly packed nonfunctional state (Prescott and Bender, 1962) the observed decline in energy metabolism at this part of the cycle is more likely a result than a cause of the decreased synthetic activity.

Bullough (1952, 1955) has proposed that a build-up of energy in preprophase allows the cell to begin division. Since energy production is the final requirement in this scheme, carbohydrates such as glucose are raised to the status of stimulators of division. The hypothesis is based largely on the stimulation of mitosis in explanted fragments of mouse ear. However, similar experiments by Gelfant (1960) on the same material have indicated that Bullough's interpretations are at least partly incorrect and that energy metabolism has no direct role in initiating mitosis. Based principally on experiments demonstrating inhibition of division by respiratory poisons, Swann (1954, 1957) has suggested that the build-up of an energy pool during interphase acts as a control on the initiation of division. The production of biologically useful energy is, of course, essential for the synthesis activities that constitute cycle progress toward division, but there is as yet little or no evidence to indicate that energy metabolism *regulates* the initiation of division or any other part of the cycle. Production of biologically useful energy is probably geared to the demands of the cell, fluctuating as the intensity or type of cell activities shift.

The work of Plesner (1958; see also chapter 7 this book), showing an accumulation of ATP in *Tetrahymena* just before division, is not interpreted by him as supporting evidence for the energy pool hypothesis.

VII. Concluding Remarks

The essence of the cell cycle is the duplication of the genetic code and its separation into independent functional units. Successful completion of the cell cycle is certainly dependent, however, upon a number of other activities proceeding concurrently and more or less closely integrated with the genetic code cycle. Centriolar reproduction and separation, production of mitochondria, synthesis of ribosomes, etc., all contribute in some way to the progress of the cycle, but how these events are timed and accomplished and how they are interrelated with the genetic code cycle is completely unknown. Presumably the same principles of integration that maintain the balanced condition of cell metabolism during any instant of time are also responsible for the co-ordinated and orderly increases in cell structures and functions that accompany each cell cycle. Primary propulsion of the cell cycle, however, is provided by the cyclic duplication and division of the genetic code.

REFERENCES

ADOLPH, E. F. and S. BAYNE-JONES (1932). Growth in size of micro-organisms measured from motion pictures. II. *Bacillus megatherium. J. Cellular Comp. Physiol.* **1,** 409.

ALFERT, M. (1959). Variations in cytochemical properties of cell nuclei. *Exptl. Cell Research Suppl.* **6,** 227.

AMANO, M. and C. P. LEBLOND (1960). Comparison of the specific activity time curves of ribonucleic acid in chromatin, nucleolus and cytoplasm. *Exptl. Cell Research* **20,** 250.

ANDERSON, N. G. (1956a). Cell division. I. A theoretical approach to the primeval mechanism, the initiation of cell division, and chromosomal condensation. *Quart. Rev. Biol.* **31,** 169.

ANDERSON, N. G. (1956b). Cell division. II. A theoretical approach to chromosomal movements and the division of the cell. *Quart. Rev. Biol.* **31**, 243.

BARNER, H. D. and S. S. COHEN (1956). Synchronization of division of a thymineless mutant of *Escherichia coli. J. Bacteriol.* **72,** 115.

BAYNE-JONES, S. and E. F. ADOLPH (1932). Growth in size of micro-organisms measured from motion pictures. I. Yeast, *Saccharomyces cerevisiae. J. Cellular Comp. Physiol.* **1,** 387.

BENDER, M. A. and D. M. PRESCOTT (1962). DNA synthesis and mitosis in cultures of human peripheral leukocytes. *Exptl. Cell Research* **27**, 221.

BLOCH, D. P. (1955). A possible mechanism for the replication of the helical structure of desoxyribonucleic acid. *Proc. Natl. Acad. Sci., U.S.* **41,** 1058.

BLOCH, D. P. and G. C. GODMAN (1955). A microphotometric study of the syntheses of desoxyribonucleic acid and nuclear histone. *J. Biophys. Biochem. Cytol.* **1,** 17.

BOLLUM, F. J. (1959). Thermal conversion of nonpriming deoxyribonucleic acid to primer. *J. Biol. Chem.* **234.**

BOLLUM, F. J. (1960). Deoxyribonucleic acid synthesis by mammalian enzymes: Some observations on primers. In *The Cell Nucleus.* Butterworths, London, p. 60.

BOLLUM, F. J. and V. R. POTTER (1958). Incorporation of thymidine into deoxyribonucleic acid by enzymes from rat tissues. *J. Biol. Chem.* **233,** 478.

BULLOUGH, W. S. (1952). The energy relations of mitotic activity. *Biol. Rev. Cambridge Phil. Soc.* **27,** 133.

BULLOUGH, W. S. (1955). Hormones and mitotic activity. *Vitamins and Hormones* **13,** 261.

BURMA, D. P., H. KRÖGER, S. OCHOA, R. C. WARNER and J. D. WEILL (1961). Further studies on deoxyribonucleic acid-dependent enzymatic synthesis of ribonucleic acid. *Proc. Natl. Acad. Sci., U.S.* **47,** 749.

BURNS, V. W. (1961). Relations among DNA and RNA synthesis and synchronized cell division in *L. Acidophilus. Exptl. Cell Research* **23,** 582.

CAMERON, I. L. and D. M. PRESCOTT (1961). Relations between cell growth and cell division. V. Cell and macronuclear volumes of *Tetrahymena pyriformis* HSM during the cell life cycle. *Exptl. Cell Research* **23,** 354.

CAMPBELL, A. (1957). Synchronization of cell division. *Bacteriol. Rev.* **21,** 263.

CARRIÈRE, R., C. P. LEBLOND and B. MESSIER (1961). Increase in the size of liver cell nuclei before mitosis. *Exptl. Cell Research* **23,** 625.

Cavalieri, L. F. and B. H. Rosenberg (1961a). The replication of DNA. I. Two molecular classes of DNA. *Biophys. J.* **1,** 317.

Cavalieri, L. F. and B. H. Rosenberg (1961b). The replication of DNA. II. The number of polynucleotide strands in the conserved unit of DNA. *Biophys. J.* **1,** 323.

Cavalieri, L. F., B. H. Rosenberg and J. F. Deutsch (1959). The subunit of deoxyribonucleic acid. *Biochem. Biophys. Res. Commun.* **1,** 124.

Davidson, D. (1963). In press.

Deeley, E. M., H. G. Davies and J. Chayen (1957). The DNA content of cells in the root of *Vicia faba. Exptl. Cell Research* **12,** 582.

Edwards, J. L., A. L. Koch, P. Youcis, H. L. Freese, M. B. Laite and J. T. Donalson (1960). Some characteristics of DNA synthesis and the mitotic cycle in Ehrlich ascites tumor cells. *J. Biophys. Biochem. Cytol.* **7,** 273.

Erickson, R. O. (1947). Respiration of developing anthers. *Nature* **159,** 275.

Fauré-Fremiet, E., Ch. Rouiller and M. Gauchery (1957). La réorganisation macronucléaire chez les *Euplotes*. Étude au microscope électronique. *Exptl. Cell Research* **12,** 135.

Feinendegen, L. E., V. P. Bond, W. W. Shreeve and R. B. Painter (1960). RNA and DNA metabolism in human tissue culture cells studied with tritiated cytidine. *Exptl. Cell Research* **19,** 443.

Firket, H. and W. G. Verly (1957). Autoradiographic visualization of syntheses of deoxyribonucleic acid in tissue culture with tritium-labelled thymidine. *Nature* **181,** 274.

Foster, T. S. and H. Stern (1959). The accumulation of soluble deoxyribosidic compounds in relation to nuclear division in anthers of *Lilium longiflorum. J. Biophys. Biochem. Cytol.* **5,** 187.

Furth, J. J., J. Hurwitz and M. Goldmann (1961a). The directing role of DNA in RNA synthesis. *Biochem. Biophys. Res. Commun.* **4,** 362.

Furth, J. J., J. Hurwitz and M. Goldmann (1961b). The directing role of DNA in RNA synthesis specificity of the deoxyadenylate deoxythymidylate copolymer as a primer. *Biochem. Biophys. Res. Commun.* **4,** 431.

Frydenberg, O. and E. Zeuthen (1960). Oxygen uptake and carbon dioxide output related to the mitotic rhythm in the cleaving eggs of *Dendraster excentricus* and *Urechis caupo. Compt. Rend. Trav. Lab. Carlsberg* **31,** 423.

Gall, J. G. (1959). Macronuclear duplication in the ciliated protozoan *Euplotes. J. Biophys. Biochem. Cytol.* **5,** 295.

Gaulden, M. E. (1956). DNA synthesis and X-ray effects at different mitotic stages in grasshopper neuroblasts. *Genetics* **41,** 645.

Gelfant, S. (1960). The energy requirements for mitosis. *Ann. N. Y. Acad. Sci.* **90,** 536.

Goldstein, L. and J. Micou (1959a). On the primary site of nuclear RNA synthesis. *J. Biophys. Biochem. Cytol.* **6,** 301.

Goldstein, L. and J. Micou (1959b). Nuclear-cytoplasmic relationships in human cells in tissue culture. III. Autoradiographic study of interrelation of nuclear and cytoplasmic ribonucleic acid. *J. Biophys. Biochem. Cytol.* **6,** 1.

Goldstein, L., J. Micou and T. T. Crocker (1960). Nuclear-cytoplasmic relationships in human cells in tissue culture. IV. A study of some aspects of nucleic acid and protein metabolism in enucleate cells. *Biochim. et Biophys. Acta* **45,** 82.

GOLDSTEIN, L. and W. PLAUT (1955). Direct evidence for nuclear synthesis of cytoplasmic ribose nucleic acid. *Proc. Natl. Acad. Sci., U.S.* **41,** 874.

GUTTES, E. and S. GUTTES (1960). Incorporation of tritium-labelled thymidine into the macronucleus of *Stentor coeruleus. Exptl. Cell Research* **19,** 626.

HARRIS, H. (1959). The initiation of deoxyribonucleic acid synthesis in the connective-tissue cell, with some observations on the function of the nucleolus. *Biochem. J.* **72,** 54.

HECHT, L. I. and V. R. POTTER (1956). Nucleic acid metabolism in regenerating rat liver. I. The rate of deoxyribonucleic acid synthesis *in vivo. Cancer Research* **16**, 988.

HOWARD, A. and S. R. PELC (1951). Nuclear incorporation of P^{32} as demonstrated by autoradiographs. *Exptl. Cell Research* **2,** 178.

HOWARD, A. and S. R. PELC (1953). Synthesis of desoxyribonucleic acid in normal and irradiated cells and its relation to chromosome breakage. In *Symposium on Chromosome Breakage* (Supplement to *Heredity,* Vol. VI). Charles C. Thomas, Springfield, p. 261.

HUANG, R. C., N. MAHESHWARI and J. BONNER (1960). Enzymatic synthesis of RNA. *Biochem. Biophys. Res. Commun.* **3,** 689.

HUGHES, A. F. W. (1952). *The Mitotic Cycle.* Butterworths, London.

HURWITZ, J., A. BRESLER and R. DIRINGER (1960). The enzymatic incorporation of ribonucleotides into polyribonucleotides and the effect of DNA. *Biochem. Biophys. Res. Commun.* **3,** 15.

KAUDEWITZ, F. (1957–1958). Versuche mit radioaktivem Phosphor an einzelnen Zellen verschiedener Protozoenarten. *Arch. Protistenk.* **102,** 321.

KIMBALL, R. F. (1961). Postirradiation processes in the induction of recessive lethals by ionizing radiation. *J. Cellular Comp. Physiol.* **58,** suppl. 1, 163.

KIMBALL, R. F. and T. BARKA (1959). Quantitative cytochemical studies on *Paramecium aurelia.* II. Feulgen microspectrophotometry of the macronucleus during exponential growth. *Exptl. Cell Research* **17,** 173.

KIMBALL, R. F., T. O. CASPERSSON, G. SVENSSON and L. CARLSON (1959). Quantitative cytochemical studies on *Paramecium aurelia.* I. Growth in total dry weight measured by the scanning interference microscope and X-ray absorption methods. *Exptl. Cell Research* **17,** 160.

KIMBALL, R. F. and S. W. PERDUE (1962). Quantitative cytochemical studies on *Paramecium aurelia.* V. Autoradiographic studies on nucleic acid syntheses. *Exptl. Cell Research* **27,** 405.

KIMBALL, R. F. and D. M. PRESCOTT (1962). DNA synthesis and distribution during growth and amitosis of the macronucleus of *Euplotes. J. Protozool.* **9,** 88.

KIMBALL, R. F., L. VOGT-KÖHNE and T. O. CASPERSSON (1960). Quantitative cyto-chemical studies on *Paramecium aurelia.* III. Dry weight and ultraviolet absorption of isolated macronuclei during various stages of the interdivision interval. *Exptl. Cell Research* **20,** 368.

KLUSS, B. (1962). Electron microscopy of the macronucleus of *Euplotes eurystomus. J. Cell Biol.* **13,** 462.

KNAYSI, G. (1940). A photomicrographic study of the rate of growth of some yeasts and bacteria. *J. Bacteriol.* **40,** 247.

LARK, K. G. and O. MAALØE (1956). Nucleic acid synthesis and the division cycle of *Salmonella typhimurium. Biochim. et Biophys. Acta* **21,** 448.

LEDERBERG, S. and D. MAZIA (1960). Protein synthesis and cell division in a pyrimidine-starved protozoan. *Exptl. Cell Research* **21,** 590.

LINDEGREN, C. C. and S. A. HADDAD (1954). Growth rates of individual yeasts cells. *Genetica* **27,** 45.

MAZIA, D. (1957–1958). Cell division. In *The Harvey Lectures.* Academic Press, New York, p. 130.

MAZIA, D. and D. M. PRESCOTT (1954). Nuclear function and mitosis. *Science* **120,** 120.

McDONALD, B. B. (1958). Quantitative aspects of deoxyribose nucleic acid (DNA) metabolism in an amicronucleate strain of tetrahymena. *Biol. Bull.* **114,** 71.

McDONALD, B. B. (1960). Time of DNA synthesis in micro- and macronuclei of *Tetrahymena pyriformis. J. Protozool. Suppl.* **7,** 10.

MENDELSOHN, M. L., F. C. DOHAN, Jr., and H. A. MOORE, Jr. (1960). Autoradiographic analysis of cell proliferation in spontaneous breast cancer of C^3H mouse. I. Typical cell cycle and timing of DNA synthesis. *J. Natl. Cancer Inst.* **25,** 477.

MITCHISON, J. M. (1957). The growth of single cells. I. *Schizosaccharomyces pombe. Exptl. Cell Research* **13,** 244.

MITCHISON, J. M. (1958). The growth of single cells. II. *Saccharomyces cerevisiae. Exptl. Cell Research* **15,** 214.

MITCHISON, J. M. (1961). The growth of single cells. III. *Streptococcus faecalis. Exptl. Cell Research* **22,** 208.

MOSES, M. J. and J. H. TAYLOR (1955). Desoxypentose nucleic acid syntheses during microsporogenesis in *Tradescantia. Exptl. Cell Research* **9,** 474.

NASATIR, M. and H. STERN (1959). Changes in the activities of aldolase and of D-glyceraldehyde-3-phosphate dehydrogenase during the mitotic cycle in microspores of *Lilium longiflorum. J. Biophys. Biochem. Cytol.* **6,** 189.

NYGAARD, O. F., S. GÜTTES and H. P. RUSCH (1960). Nucleic acid metabolism in a slime mold with synchronous mitosis. *Biochim. et Biophys. Acta* **38,** 298.

OCHOA, S., D. P. BURMA, H. KRÖGER and J. D. WEILL (1961). Deoxyribonucleic acid-dependent incorporation of nucleotides from nucleoside triphosphates into ribonucleic acid. *Proc. Natl. Acad. Sci., U.S.* **47,** 670.

PAINTER, R. B. and R. M. DREW (1959). Studies on deoxyribonucleic acid metabolism in human cancer cell cultures (HeLa). I. The temporal relationships of deoxyribonucleic acid synthesis to mitosis and turnover time. *Lab. Invest.* **8,** 278.

PAINTER, R. B., R. M. DREW and B. G. GIAQUE (1960). Further studies on deoxyribonucleic acid metabolism in mammalian cell cultures. *Exptl. Cell Research* **21,** 98.

PERRY, R. P. (1960). On the nucleolar and nuclear dependence of cytoplasmic RNA synthesis in HeLa cells. *Exptl. Cell Research* **20,** 216.

PERRY, R. P., M. ERRERA, A. HELL and H. DÜRWALD (1961). Kinetics of nucleoside incorporation into nuclear and cytoplasmic RNA. *J. Biophys. Biochem. Cytol.* **11,** 1.

PERRY, R. P., A. HELL and M. ERRERA (1961). The role of the nucleolus in ribonucleic acid and protein synthesis. I. Incorporation of Cytidine into normal and nucleolar inactivated HeLa cells. *Biochim. et Biophys. Acta* **49,** 47.

PLESNER, P. E. (1958). The nucleoside triphosphate content of *Tetrahymena pyriformis* during the division cycle in synchronously dividing mass cultures. *Biochim. et Biophys. Acta* **29,** 462.

PRESCOTT, D. M. (1955). Relation between cell growth and cell division. I. Reduced weight, cell volume, protein content, and nuclear volume of *Amoeba proteus* from division to division. *Exptl. Cell Research* **9,** 328.

PRESCOTT, D. M. (1956). Relation between cell growth and cell division. II. The effect of cell size on cell growth rate and generation time in *Amoeba proteus. Exptl. Cell Research* **11,** 86.

PRESCOTT, D. M. (1959a). Nuclear synthesis of cytoplasmic ribonucleic acid in *Amoeba proteus. J. Biophys. Biochem. Cytol.* **6,** 203.

PRESCOTT, D. M. (1959b). Microtechniques in amoeba studies. *Ann. N. Y. Acad. Sci.* **78,** 655.

PRESCOTT, D. M. (1960a). The nuclear dependence of RNA synthesis in *Acanthamoeba sp. Exptl. Cell Research* **19,** 29.

PRESCOTT, D. M. (1960b). Relation between cell growth and cell division. IV. The synthesis of DNA, RNA, and protein from division to division in *Tetrahymena: Exptl. Cell Research* **19,** 228.

PRESCOTT, D. M. (1961a). RNA synthesis in the nucleus and RNA transfer to the cytoplasm in *Tetrahymena pyriformis*. In *Biological Structure and Function.* (T. W. Goodwin, ed.). Academic Press, New York, pp. 527–536.

PRESCOTT, D. M. (1961b). The growth-duplication cycle of the cell. In *International Review of Cytology,* Vol. XI. Academic Press, New York, p. 255.

PRESCOTT, D. M. (1962). Nucleic acid and protein metabolism in the macronuclei of two ciliated protozoa. *J. Histochem. Cytochem.* **10,** 145.

PRESCOTT, D. M. and M. A. BENDER (1962). Synthesis of RNA and protein during mitosis in mammalian tissue culture cells. *Exptl. Cell Research*; in press.

PRESCOTT, D. M. and R. F. KIMBALL (1961). Relation between RNA, DNA, and protein syntheses in the replicating nucleus of euplotes. *Proc. Natl. Acad. Sci., U.S.* **47,** 686.

PRESCOTT, D. M., R. F. KIMBALL and R. F. CARRIER (1962). Comparison between the timing of micronuclear and macronuclear DNA synthesis in *Euplotes eurystomus. J. Cell Biol.* **13,** 175.

RABINOVITCH, M. and W. PLAUT (1956). Cytochemical and autoradiographic observations in nuclear ribonucleic acid in *Amoeba proteus. Exptl. Cell Research* **10,** 120.

RICHARDS, B. M., P. M. B. WALKER and E. M. DEELEY (1956). Changes in nuclear DNA in normal and ascites tumor cells. *Ann. N. Y. Acad. Sci.* **63,** 831.

SANDRITTER, W., H. G. SCHIEMER, H. KRAUS and U. DÖRRIEN (1960). Interferenzmikroskopische Untersuchungen über das Wachstum von Einzelzellen (HeLa-Zellen) in der Gewebekultur. *Frankfurt. Z. Pathol.* **70,** 271.

SCHAECHTER, M., M. W. BENTZON and O. MAALØE (1959). Synthesis of deoxyribonucleic acid during the division cycle of bacteria. *Nature* **183,** 1207.

SCHERBAUM, O. H. and E. ZEUTHEN (1954). Induction of synchronous cell division in mass cultures of *Tetrahymena pyriformis. Exptl. Cell Research* **6,** 221.

SINSHEIMER, R. L. (1959). A single-stranded deoxyribonucleic acid from bacteriophage Φ X174. *J. Molec. Biol.* **1,** 43.

SISKIN, J. E. (1959). The synthesis of nucleic acids and proteins in the nuclei of *Tradescantia* root tips. *Exptl. Cell Research* **16,** 602.

SISKIN, J. E. and R. KINOSITA (1961). Timing of DNA synthesis in the mitotic cycle *in vitro*. *J. Biophys. Biochem. Cytol.* **9,** 509.

STANNERS, C. P. and J. E. TILL (1960). DNA synthesis in individual L-strain mouse cells. *Biochim. et Biophys. Acta* **37,** 406.

STERN, H. (1956). The physiology of cell division. *Ann. Rev. Plant Physiol.* **7,** 91.

STERN, H. and P. L. KIRK (1948). The oxygen consumption of the microspores of trillium in relation to the mitotic cycle. *J. Gen. Physiol.* **31,** 243.

STEVENS, A. (1961). Net formation of polyribonucleotides with base compositions analogous to deoxyribonucleic acid. *J. Biol. Chem.* **236,** PC 43.

SWANN, M. M. (1954). The control of cell division. In *Colston Papers*, Vol. VII. Butterworths, London, p. 185.

SWANN, M. M. (1957). The control of cell division: A review. I. General mechanisms. *Cancer Research* **17,** 727.

SWANN, M. M. (1958). The control of cell division: A review. II. Special mechanisms. *Cancer Research* **18,** 1118.

SYLVEN, B., C. A. TOBIAS, H. MALMGREN, R. OTTOSON and B. THORELL (1959). Cyclic variations in the peptidase and catheptic activities of yeast cultures synchronized with respect to cell multiplication. *Exptl. Cell Research* **16,** 75.

TAYLOR, J. H. (1958a). Incorporation of phosphorus-32 into nucleic acids and proteins during microgametogenesis of *Tulbaghia*. *Am. J. Botany* **45,** 123.

TAYLOR, J. H. (1958b). The organization and duplication of genetic material. *Proc. 10th Intern. Congr. Genet.* Aug. 20–27, 1958, Montreal, **1,** 63.

TAYLOR, J. H. (1960a). Nucleic acid synthesis in relation to the cell division cycle. *Ann. N. Y. Acad. Sci.* **90,** 409.

TAYLOR, J. H. (1960b). Asynchronous duplication of chromosomes in cultured cells of chinese hamster. *J. Biophys. Biochem. Cytol.* **7,** 455.

TAYLOR, J. H. and R. D. MCMASTER (1954). Autoradiographic and microphotometric studies of desoxyribose nucleic acid during microgametogenesis in *Lilium longiflorum*. *Chromosoma* **6,** 489.

TSCHERMAK-WOESS, E. and R. DOLEZAL-JANISCH (1957). Rhythmic nuclear growth during the mitotic interphase of *Vicia faba*. *Österr. Botan. Z.* **103,** 588.

VOLKIN, E. and L. ASTRACHAN (1957). RNA metabolism in T2-infected *Escherichia coli*. In *The Chemical Basis of Heredity* (W. D. McElroy and B. Glass, eds.). The Johns Hopkins Press, Baltimore, p. 686.

WALKER, P. M. B. and H. B. YATES (1952). Nuclear components of dividing cells. *Proc. Roy. Soc. Edinburgh* **B140,** 274.

WEISS, S. B. (1960) Enzymatic incorporation of ribonucleoside triphosphates into the interpolynucleotide linkages of ribonucleic acid. *Proc. Natl. Acad. Sci., U.S.* **46,** 1020.

WEISS, S. B. and T. NAKAMOTO (1961a). Net synthesis of ribonucleic acid with a microbial enzyme requiring deoxyribonucleic acid and four ribonucleoside triphosphates. *J. Biol. Chem.* **236,** PC 18.

WEISS, S. B. and T. NAKAMOTO (1961b). On the participation of DNA in RNA biosynthesis. *Proc. Natl. Acad. Sci., U. S.* **47,** 694.

WIMBER, D. E. (1960). Duration of the nuclear cycle in *Tradescantia paludosa* root tips as measured with H-3 thymidine. *Am. J. Botany* **47,** 828.

WOODARD, J. W. (1958). Intracellular amounts of nucleic acids and protein during pollen grain growth in *Tradescantia*. *J. Biophys. Biochem. Cytol.* **4,** 383.

WOODARD, J. W., B. GELBER and H. SWIFT (1961). Nucleoprotein changes during the mitotic cycle in *Paramecium aurelia*. *Exptl. Cell Research* **23,** 258.

WOODARD, J. W., E. RASCH and H. SWIFT (1961). Nucleic acid and protein metabolism during the mitotic cycle in *Vicia faba*. *J. Biophys. Biochem. Cytol.* **9,** 445.

WOODS, P. S. (1959). RNA in nuclear-cytoplasmic interaction. *Brookhaven Symposia in Biology No.* 12, 153.

ZALOKAR, M. (1960). Sites of protein and ribonucleic acid synthesis in the cell. *Exptl. Cell Research* **19,** 559.

ZEUTHEN, E. (1950a). Respiration during cell division in the egg of the sea urchin *Psammechinus miliaris*. *Biol. Bull.* **98,** 144.

ZEUTHEN, E. (1950b). Respiration and cell division in the egg of *Urechis caupo*. *Biol. Bull.* **98,** 152.

ZEUTHEN, E. (1951). Segmentation, nuclear growth and cytoplasmic storage in eggs of Echinoderms and Amphibia. *Pubbl. Staz. Zool. Napoli, Suppl.* **23,** 47.

ZEUTHEN, E. (1953). Growth as Related to the Cell Cycle in Single-cell Cultures of *Tetrahymena piriformis*. *J. Embryol. Exptl. Morphol.* **1,** 239.

ZEUTHEN, E. (1955). Mitotic respiratory rhythms in single eggs of *Psammechinus miliaris* and of *Ciona intestinalis*. *Biol. Bull.* **108,** 366.

ZEUTHEN, E. (1958). Artificial and induced periodicity in living cells. *Advances in Biol. and Med. Phys.* (C. A. Tobias and J. H. Lawrence, eds.), Vol. VI. Academic Press, New York, p. 37.

ZEUTHEN, E. (1960). Cycling in oxygen consumption in cleaving eggs. *Exptl. Cell Research* **19,** 1.

4

The Temperature-Induced Division Synchrony in *Tetrahymena*

ERIK ZEUTHEN, *Biological Institute of the Carlsberg Foundation, Copenhagen, Denmark*

CONTENTS

I Introduction 99
II Description of the Synchronized *Tetrahymena* Systems . . . 100
1. Multiplication rate as a function of temperature . . . 100
2. Responses of the population to single temperature shocks . . 101
3. Induction of division synchrony by a series of temperature shocks 104
4. Comparison of synchronous and logarithmic systems . . 111
A. Biochemical parameters 111
B. Sensitivity to chemical agents 116
C. Sensitivity to temperature shocks 118
D. Generation times 118
E. Desynchronization of synchronized systems 118
5. Synchronization of microscopically visible events . . . 121
6. Synchronization of biochemical events 123
A. Growth in nitrogen and in dry matter 123
B. Protein, enzymes 124
C. Rate of uptake of ^{14}C-histidine and of ^{32}P-phosphate . . 125
7. Radioautographic studies 125
A. Protein, RNA 125
B. Asynchrony of DNA synthesis 127
III Mechanisms in the Induction of the Division Synchrony . . 131
1. Syntheses involving amino acids as a condition for division . 131
2. Synchronization with *p*-fluorophenylalanine and with gaseous nitrogen 139
3. The time point of physiological transition 139
4. Kinetosomes and cell division 144
5. 'Protein synthesis', 'set-back' and 'physiological transition' . 147
6. Nuclear function and synchronous division 150
IV Summarizing Discussion 151
V Conclusions 154
References 156

I. INTRODUCTION

In his earlier years the author enjoyed many visits to marine biological laboratories in Europe and in the United States. He took an interest in studying biochemical periods accompanying the naturally synchronized division period in batches of dividing eggs from single marine animals. In order to be able to continue this type of work in his home laboratory, he looked for other types of naturally synchronized material. Finding none very useful, he decided to attempt to induce synchrony in mass populations of a suitable micro-organism which could be readily grown in any laboratory.

The ciliate protozoon *Tetrahymena pyriformis* is a good-sized cell ($40 \times 60\ \mu$) well suited both for population and for single-cell studies. It was the first animal cell to be grown axenically on a fully defined medium (Kidder and Dewey, 1951). The amino acid requirements of this organism closely resemble those of mammals. For these reasons *Tetrahymena*, with which the author got acquainted during a stay in Dr. George W. Kidder's laboratory, was selected (1950) for the series of studies which are to be reported in this chapter and in the chapters by Williams, Plesner and Scherbaum.

The present account reports (section II) how, using temperature changes, division synchrony can be induced into logarithmically growing populations previously random for division. On the cellular level the mechanism of induction is one of differential delay of cell division, the delay being directly related to the age of the cells since the previous division.

The induction of the synchrony and the recovery, by division, from the induction are treated as closely related, but in their effects oppositely directed, biochemical phenomena (section III). For this reason studies aimed at elucidating biochemical mechanisms in the preparation of synchronous division at the same time shed light on the biochemical mechanisms by which the synchrony was induced, and vice versa.

It will be seen that synchronous *Tetrahymena*, and especially when division is studied in an inorganic medium, is a useful substitute for the marine eggs which are so inaccessible to many.

II. Description of the Synchronized Tetrahymena Systems

1. *Multiplication Rate as a Function of Temperature*

Strain GL (Lwoff, 1923) used by us shows the relationship (Thormar, 1956; 1962) of Fig. 1 between temperature and rate of multiplication in the first (X, curve (a)), the second (O) and sixth or seventh generations (●) after the transfer from room temperature to any of the temperatures shown. Curve (a) shows a maximum at 27·5°C. The division rate decreases sharply at the higher temperatures. The curve is fairly linear ($\mu = 14{,}700$ cal) from around 26·0°C to around

17·5°C, but it changes into a steeper slope (μ = 32,000 cal) in the temperature range 17·5–11°C. The curve for the second and later generations (curve (b)) is significantly different from curve a but shows related discontinuities. The maximum multiplication rate is at 29°C. We may accept this curve as linear from 28°C to 16°C (μ = 14,700 cal).

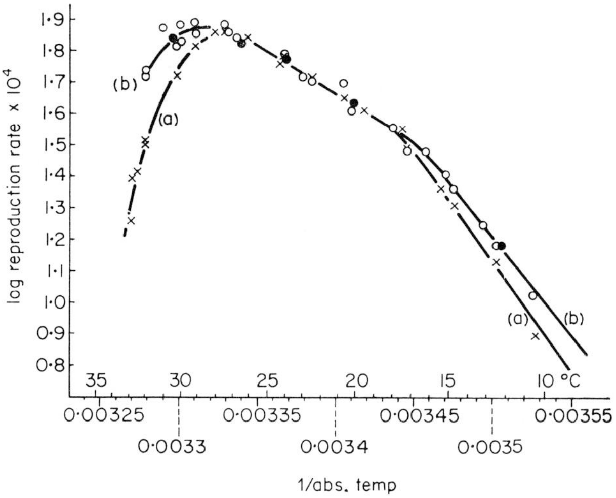

FIG. 1. Relation between multiplication rate and temperature (*Tetrahymena pyriformis* GL) in the first (X), the second (O) and the sixth or seventh (●) generations after transfer from room temperature to the temperatures shown (Thormar, 1962). (By courtesy of Academic Press.)

Cell sizes, measured (Thormar, 1956) just before and after division after four to five generations at each temperature, are shown in Table I. Cell size at division is smallest around the optimum temperature for growth and division and increases at both sides of the temperature optimum. This has also been observed for strain HS (Prescott, 1957).

2. *Responses of the Population to Single Temperature Shocks*

A temperature shock is defined as a shift to a temperature different from the selected growth temperature, later followed by a shift back to the growth temperature. The duration of a shock is defined as the time interval between the shift (as signalled by the controlling clock) away from and back to the growth temperature. For closer description full temperature curves are necessary. This is because it takes some time for the cultural temperature to go from one

TABLE I. Size of *Tetrahymena pyriformis* GL grown at 10–32°C for 4–5 generations. Fixation before division (a) and after division (b). (From Thormar, 1956.)[a]

°C	GT (hr)	A (μ)	B (μ)	B/A	Volume (μ^3)	Relative volume	Number of cells measured
			(a)				
32	3·1	58·0±0·8	39·0±0·7	0·67	46300	152	18
29·2	2·3	60·1±0·5	32·5±0·4	0·54	33300	109	22
26·4	2·4	61·3±0·6	30·8±0·2	0·50	30500	100	20
22·7	3·0	61·3±0·5	30·4±0·3	0·50	29800	98	25
19·3	4·1	64·4±0·6	31·6±0·3	0·49	33700	110	23
15·4	6·5	66·0±0·6	32·2±0·3	0·49	35900	118	22
10·6	14·4	67·3±0·9	34·1±0·4	0·51	41000	132	25
			(b)				
32		32·0±0·6	38·9±0·7	1·20	25300	83	22
29·2		31·9±0·3	31·6±0·3	0·99	16700	54	20
26·4		33·5±0·5	29·9±0·3	0·89	15700	51	18
22·7		32·9±0·5	29·4±0·4	0·89	14900	49	20
19·3		34·2±0·3	29·6±0·3	0·87	15700	51	18
15·4		36·5±0·5	31·9±0·4	0·88	19500	64	22
10·6		35·7±0·5	31·6±0·3	0·89	18700	61	22

GT: generation time A: length of cell B: width of cell

[a] Published as curves by H. Thormar (1962).

level to another. When cultures are temporarily cooled below the selected growth temperature, we speak of 'cold shocks'; when they are temporarily heated, we speak of 'heat shocks'.

When logarithmic cultures grown at 28°C are shocked for periods of 1 or 2 hr (Zeuthen, unpublished data) at the various temperatures indicated in Fig. 2, then the population responses, observed at 28°C and shown by the curves, are obtained. In all cases zero time is when the population is returned to 28°C. The division indices (fraction of all cells which show cleavage furrows) are followed in time by visual estimates (cf. Plesner, Rasmussen and Zeuthen, chapter 23). While cultures shocked in the range 18–31°C show a rather undisturbed balance between the rate of entry of cells into division and exit of cells from division, and thus a fairly constant division index, this is not so for cultures which have been heated to above 31°C or chilled to 16°C or below this temperature. In this case the shock is followed first by reduced, then by increased division rate. Later, the system comes back to, or close to, normal.

Results essentially similar to those shown in Fig. 2 are obtained when the growth and observation temperature is anywhere between 22 and 28°C. According to Thormar (1962) 32°C is the highest temperature at which the strain GL is capable of continued growth and multiplication. At 33°C maximally one division can be performed. So, when the multiplication rate is concerned, there is a dramatic discontinuity between 32 and 33°C. Nachtwey (personal communication) observes that when the growth temperature is 32°C, then temperature shocks to 33°C are quite, though not maximally, effective in producing the type of imbalance shown in Fig. 2.

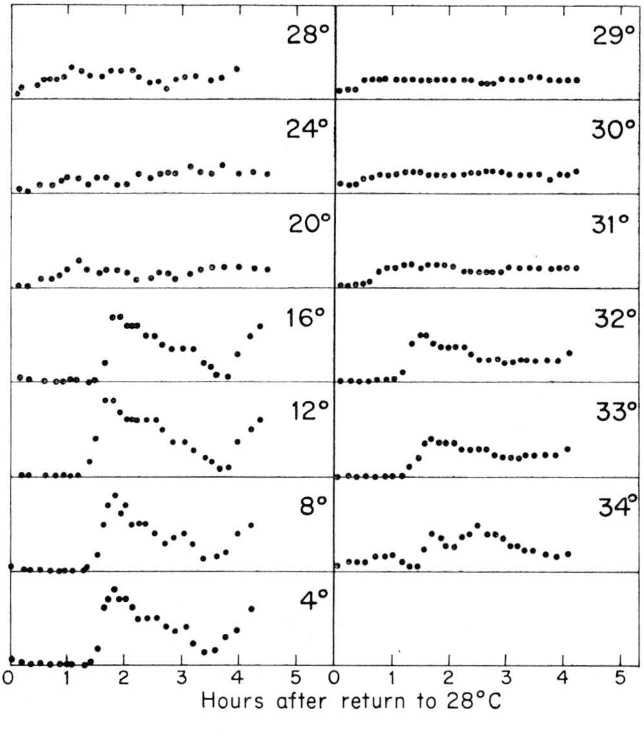

FIG. 2. Division indices (*Tetrahymena pyriformis* GL), visually estimated, after exposure of logarithmic cultures (2 ml in upright test tubes, density 2–3 $\times$ 10^4/ml) for 2 hr to the temperatures indicated. Previous growth in 150 ml shaker flasks at 28°C. Observations in 1 ml salt cellars at 28°C (Zeuthen, unpublished data). In each curve the ordinate is linear and goes from 0·0 to 0·5 (0 per cent to 50 per cent of all cells dividing). The highest division index observed in this series is 40 per cent.

As seen from Fig. 2, one temperature shock may—after a lag—temporarily cause a very definite increase in the division index above the logarithmic value. This, however, is not so if the shocks are within a temperature interval (18–28°C) over which the value of μ (Fig. 1) is constant. With proper temperature changes a partial synchronization of the cell divisions is thus obtained. (A single shift in temperature may do the same as does a shock. Prescott (1957) found induction of partial synchrony in strain HS when the temperature was shifted from constant 18·4°C to constant 27·7°C. In strain HS there is a break in the value of μ at 23·5°C.) If after a shock our populations are permitted to perform one partially synchronous division, and are then given another temperature shock, allowed to divide, shocked again, and so forth, we obtain a system which keeps cycling for cell division in an environment which itself changes. Essentially the same result is obtained whether (Zeuthen, 1958; Zeuthen and Scherbaum, 1954) we cool to 7°C (Fig. 3a) or whether (Zeuthen, 1958) we heat to 33°C (Fig. 3b). We then have a 'one-shock-per-division system' like those later reported for the flagellates (chapters 11 and 22) and for the amoebae (chapter 8). We may at that early time (Zeuthen and Scherbaum, 1954) have taken a too pessimistic stand when we felt that in their uses these systems are limited by the fact that periodic cellular functions (division) and environmental changes are not fully separated.

3. *Induction of Division Synchrony by a Series of Temperature Shocks*

When a series of temperature shocks are applied at shorter intervals than mentioned in the above paragraph a 'multi-shock-system' is obtained. The cells do not divide between the shocks which together act as a block for cell division. At the same time the shocks permit continued growth so that greatly over-sized cells are produced. [This is not the case in the one-shock-per-division systems (Fig. 3) in which size increase after one shock is largely compensated by size decrease due to later cellular fission.]

After return to constant growth temperature there is a lag with no division, followed by a burst of division. This first division maximum may be followed by one or two more maxima. Fig. 4 demonstrates one experiment (Zeuthen and Scherbaum, 1954). Fig. 5 shows temperature regimens which induce synchrony in *Tetrahymena pyriformis* GL. We use the following code to describe the systems, e.g. (Fig. 5a) 8 × 30′34/30′28 (briefly 30′/30′). This code describes the system published by Scherbaum and Zeuthen (1954).

It says that eight heat shocks, each of 30 min at 34 °C were applied. Between the shocks there was recovery for periods of 30 min at 28°C. Normally, the recovery temperature equals the growth temperature before and after the shocks. In Scherbaum and Zeuthen (1954) volumes of 10 ml 2 per cent proteose peptone

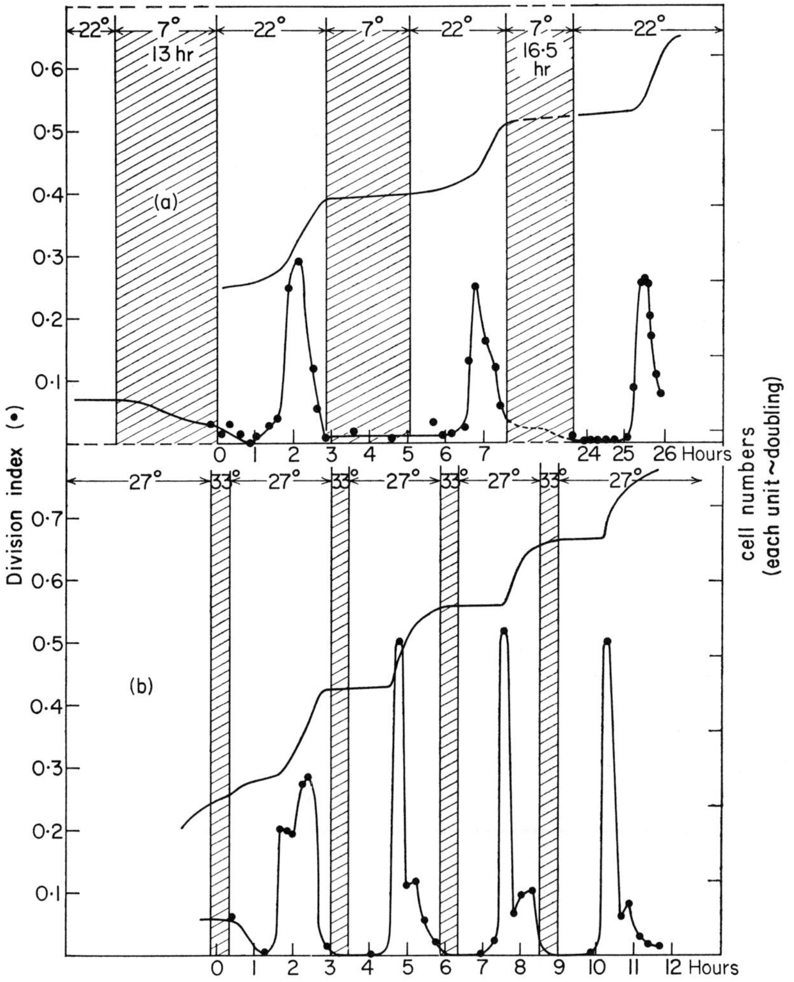

FIG. 3. Synchronization of *Tetrahymena pyriformis* by 'one-shock-per-division' methods: (a) using cold shocks (7°C) of various duration (Zeuthen and Scherbaum, 1954); (b) using 30-min heat shocks to 33°C (Zeuthen, 1958). The calculated multiplications have since been confirmed by actual counts. (By courtesy of Academic Press.)

plus 0·1 per cent liver fraction were used and observation after the shocks was at room temperature (24°C). The culture occupied a shallow layer of 1·75 mm to secure rapid heat transfer and good aeration without shaking. The two next reports (Scherbaum and

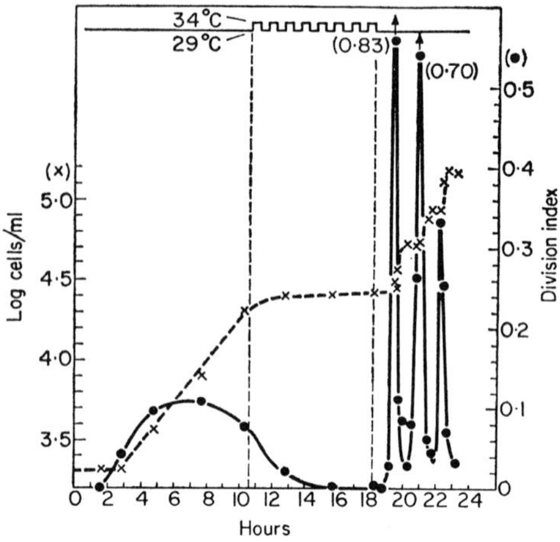

FIG. 4. Synchronization of cell divisions in *Tetrahymena pyriformis* GL. The fully drawn curve shows the division index during lag and log phases of logarithmic growth (0–11 hr), during the period of changing temperature (11–18 hr) and during synchronous divisions at constant temperature (18–23 hr). The dotted curve represents the cell counts (from Zeuthen and Scherbaum, 1954).

Zeuthen, 1955; Zeuthen and Scherbaum, 1954) describe how the system was scaled to 150-ml volumes in shaker flasks. Observation was at 28°C. Synchrony could also be induced (Scherbaum and Zeuthen, 1955) when growth was in Kidder's and Dewey's synthetic medium (1951). The largest volumes routinely used (30′/30′ system) in our laboratory are 700-ml shaker flasks (cf. chapter 23). Different techniques are used for still larger cultural volumes (cf. Plesner, chapters 7 and 23).

A slightly twisted heat synchronized system (Rasmussen and Zeuthen, 1962) is represented (Fig. 5b) as 6 (or 7) × 20′34/40′28 (for brevity 20′/40′). In this system fewer (6 or 7) and shorter shocks (20′) give as good synchrony as is observed in the 30′/30′ system.

Cold synchronization (Zeuthen, unpublished data) can be obtained by the exposure of logarithmic cells (2 per cent proteose peptone plus 0·4 per cent liver fraction, 28°C) to a series (usually 5) of chillings to 7–8°C (Fig. 5c) for periods of 2 hr with recovery for 40 min at 28 °C. This system, still new, will be referred to as 5 × 120′, 8/40′28 (for brevity 120′/40′).

In the experiments of Figs. 6 and 7d the cold shocks terminate when the culture is returned to 28°C for development of synchronous division. If we apply shocks (2 hr) to 10°C and omit the transfer to 28 °C, thus leaving the cells at 10 °C, and observe for synchrony at this temperature we find excellent synchronous division, in this case 17 hr after the temperature was shifted for the last time from 28°C to 10°C (Fig. 5d). While this system is of limited practical value it has considerable theoretical interest.

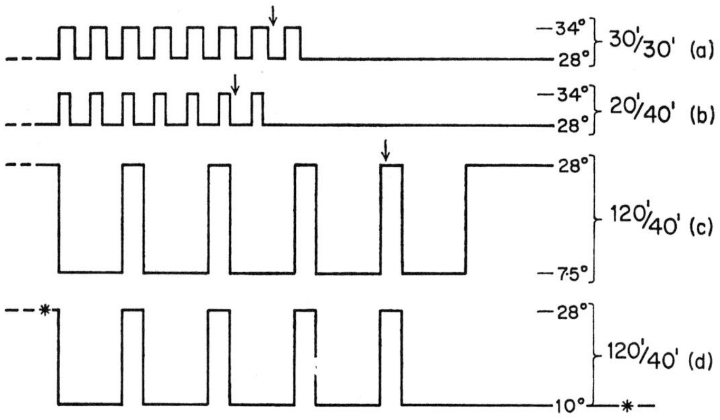

FIG. 5. Temperature regimens (switches of controlling clock) which induce synchrony in *Tetrahymena* populations (*T. pyriformis* GL). (*Arrow:* time when the organic medium may be replaced by an inorganic medium. *Asterisks:* see text, section II-4-A.)

In the multi-shock systems the height of the first division maximum after the shocks builds up as a function of the number of shocks applied. With heat at least, there is the tendency also for the later division maxima to become higher as more and more synchronizing shocks are applied. Results obtained with heat (20′/40′) and with cold (120′/40′) are shown in Fig. 6. In the 20′/40′ system results in 2 per cent proteose peptone (a) are compared with results obtained when 0·1 per cent (b) and 0·4 per cent (c, d) liver extract is added. Above five and below ten shocks Fig. 6 suggests about equal synchrony, regardless of whether heat or cold shocks are used, and (with heat shocks) regardless of whether liver extract is present or not. If more than ten heat shocks are used the synchrony breaks down the sooner the richer is the medium. In the 120′/40′ system (d, medium as in c) maximal effects are obtained after three shocks. There is no change if the number of shocks increases to 10.

In Fig. 7 the same experiments that have been plotted in Fig. 6 are analysed for time between the termination of the last heat or cold shock and division maximum 1 (first synchronous generation) and for time intervals between division maxima 1 and 2, sometimes also 2 and 3 (second and third synchronous generation). The generation times first shorten then tend to become constant. With many heat shocks in the rich media (b, c) the generations again extend, more or less in correspondence with the lowering of the division maxima (Fig. 6).

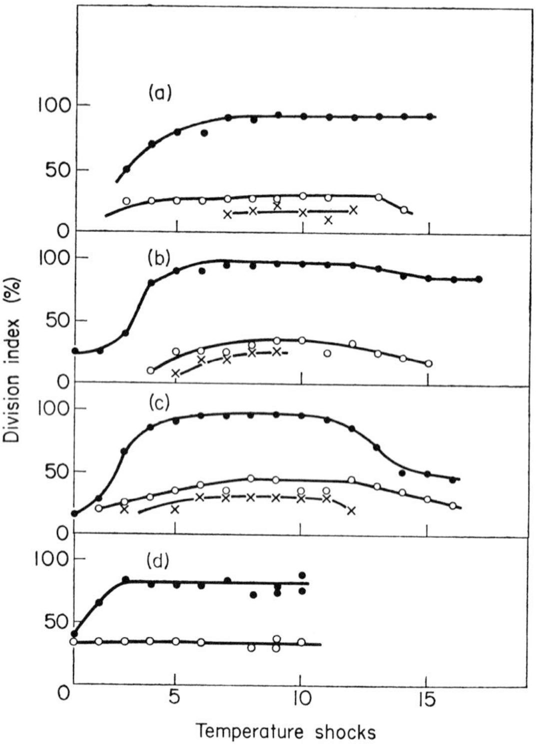

Fig. 6. *Tetrahymena pyriformis* GL. Height of the division index at the time of maximal division 1 (●), 2 (O) and 3 (X) as a function of the number of heat shocks (a, b, c, 20′/40′ system) or cold shocks (d, 120′/40′ system). The growth medium is 2 per cent proteose peptone (a), same plus 0·1 per cent liver extract (b), same plus 0·4 per cent liver extract (c, d). [a, b, c (Zeuthen, 1961a), d (Zeuthen, unpublished data).] (By courtesy of Basic Books, Inc.)

The synchronized *Tetrahymena* systems demonstrate striking cases of dissociation between growth and division. During the synchronization with heat (30′/30′, also 20′/40′) there is continued though reduced rate of growth, but no division. During the syn-

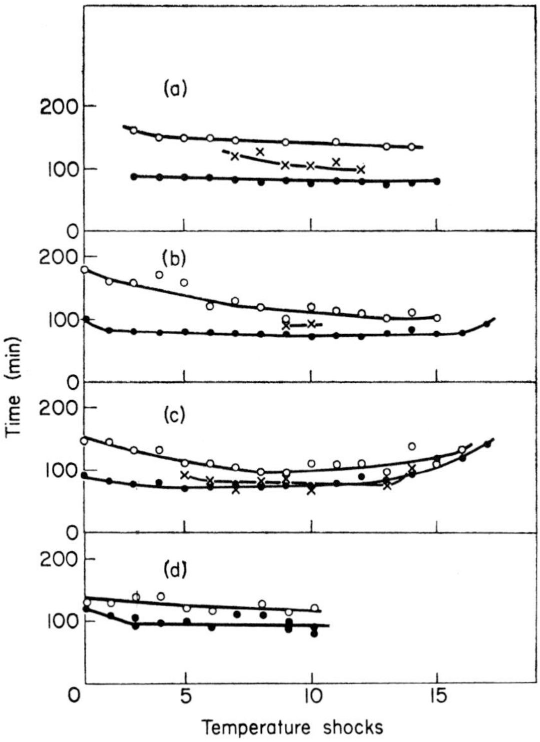

FIG. 7. Duration of synchronous generations (1st (●), 2nd (O) and 3rd (X)) as a function of the number of temperature shocks applied. a, b, c: 20′/40′—d: 120′/40′. Growth media as described in Fig. 6. [a, b, c (Zeuthen, 1961a), d (Zeuthen, unpublished data).] (By courtesy of Basic Books, Inc.)

chronous divisions there is reduced rate of growth and increased rate of division. Comparison is with logarithmic growth and division before the heat treatment, and after the system, by synchronous division, has reverted to logarithmic growth and multiplication. The reduced rate of growth after the heat may find an explanation in the fact that for several generations the synchronized cell is oversized but has only one functional mouth like the smaller logarithmic cell has. It may also (Zeuthen and Scherbaum, 1954) reflect competition within the cell between general growth mechanisms and specific mechanisms that prepare the cell for division.

We may arrange the experiment so that after the temperature shocks (30′/30′, 20′/40′, 120′/40′), we eliminate growth and retain one or two useful divisions (Fig. 8). After, or before, the last tem-

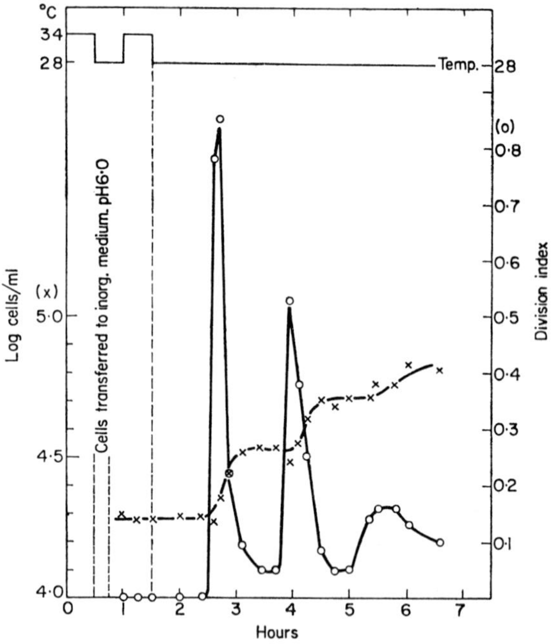

FIG. 8. *Tetrahymena pyriformis* GL synchronized with seven heat shocks (30′/30′) in 2 per cent proteose peptone, 0·1 per cent liver fraction followed by one shock in the inorganic medium (from Hamburger and Zeuthen, 1957). (By courtesy of Academic Press.)

perature shock the growth medium is replaced (Fig. 5, a, b, c, arrow) with a simple inorganic medium (Hamburger and Zeuthen, 1957) (cf. chapter 23). This works in both heat and cold synchronized systems. The resulting system of division without growth is as close as we have got to 'laying an egg' (cf. Introduction). In the inorganic medium each new shock redefines the time EH after which a full first division maximum, and also a second smaller maximum, may develop in due time. In the 30′/30′ system a maximum of eight additional shocks have been applied in the inorganic medium after seven synchronizing shocks in the organic medium (0·1 per cent liver fraction), but sometimes only two or three additional shocks can be given before the synchrony breaks down. After a shock in the inorganic medium the population division maximum (30′/30′, 20′/40′) usually comes 5–10 min earlier than in the control which is in the organic medium. The average cell also acquires earlier insensitivity (cf. section II-4-C) to heat, cold and chemical agents, so that after a shock only part of the population may react to the next shock. The chances of a breakdown of the synchrony with continued shocking in the inorganic medium is therefore higher in the 20′/40′ than in the 30′/30′ system. Division synchrony as studied in the non-nutrient medium is a very useful tool as shown in this and in Plesner's chapter.

The inverse relation between generation time and cell-size observed for synchronous cells which have not yet been over-shocked (the early part of the curves in Figs. 6 and 7) is not unique. It is also observed for cells which represent logarithmic cultures and which for unknown reasons vary considerably in size. Løvlie (personal communication) related size (measured as reduced weight) at division with the duration of the preceding generation. The larger the cell the shorter the generation time. Løvlie's logarithmic cells and the synchronized cells appeared similar in the sense that in both cases the cells regulate towards smaller cell size by increased fission rate.

4. *Comparison of Synchronous and Logarithmic Systems*

A. Biochemical parameters

Available information (30′/30′ system) has been collected in Table II. The populations were sampled at a number of selected growth points: (1) logarithmic growth prior to the initiation of the heat shocks (cf. discussion in chapter 23); (2) EH = termination of the seven or eight synchronizing temperature shock; (3) prior to division 1; (4) after this division; (5) prior to division 2; (6) after this division. During the heat shocks there is a considerable (2–4 ×) and more or less proportional increase in important cell parameters, such as dry weight, cell volume, total protein, and nucleic acid contents. This increase continues to division 1. During the synchronous divisions growth continues, but the cell parameters decrease because growth is overcompensated by the faster rate of splitting into new cells. At no time are there any obvious changes in the composition of the pool of free amino acids, or in the protein amino acids (Christensson, 1959; Scherbaum, James and Jahn, 1959). The base ratios of both nucleic acids remain constant (Scherbaum, 1957b). However, 'structural proteins' (insoluble

TABLE II. Biochemical parameters (*Tetrahymena pyriformis*)

	1 Logarithmic	2 At EH	3 Prior to (division 1)	4 After (division 1)	5 Prior to (division 2)	6 After (division 2)	7 Number of heat shocks	8 Ref.
Dry weight	4·1	8·2	9·8	5·1			7	Scherbaum (1957b)
(mg × 10^{-6}/cell)		8·6	10·0		5·9		8	Hamburger and Zeuthen (1960)
Cell volume[k]	2·14	6·98			2·75		8	Zeuthen and Scherbaum (1954)
(μl × 10^{-5}/cell)	1·6	4·4	5·1	3·0			7	Scherbaum and Zeuthen (1954)
	2·5	9·3	9·4	6·1			7	Williams and Scherbaum (1959)
Macronuclear vol. (μl × 10^{-7}/cell)	2·65		10·6	4·3			7	Scherbaum, Louderback and Jahn (1958)[a]
SNA/macronuclear vol.[a]	0·72%			2·8%			7	
Protein								
insoluble	11	28		15			8	Christensson (1959)[b,c]
soluble	1	11		6				
total	15	40		22				
(mg × 10^{-7}/cell)								
Free amino acids (mg × 10^{-7}/cell	2·56	6·7	8·4	4·9			8	Christensson (1959)[d,e]
Total P (mg × 10^{-7}/cell)		1·75	2·12		1·28		8	Hamburger and Zeuthen (1960)
RNA	1·28	1·71		1·43		1·2	8	Iverson and Giese (1957)[f]
(mg × 10^{-7}/cell)	2·46	5·3	6·8		3·39		7	Scherbaum (1957b)

In RNA (%)								
Adenine	30·9	29·8	30·4		30·8		7	Scherbaum (1957b)
Guanine	19·8	20·4	21·9		19·1		7	
Cytosine	19·3	21·4	19·8		20·6		7	
Uracil	30·0	28·4	27·9		29·5		7	
DNA								
(mg × 10^{-8}/cell)	1·36	3·02	3·84		1·97		7	Scherbaum (1957b)[g]
	3·03	5·53		2·83		3·05	8	Iverson and Giese (1957)[h]
	1·5	6·0				3·4	8	Zeuthen and Scherbaum (1954)[i]
IN DNA (%)								
Adenine	36·6	35·3	33·5		36·9		7	Scherbaum (1957b)
Guanine	11·5	11·6	12·6		10·4		7	
Cytosine	15·0	15·3	15·8		15·4		7	
Thymine	36·9	37·8	38·1		37·3		7	
Desoxyribotides (mg × 10^{-9}/cell)	1·86	3·36		3·19		1·67	8	Iverson and Giese (1957)[j]

All data are for strain GL per cell, except those of Iverson and Giese (1957) which are for strain W.

[a] SNA; subnuclear aggregates dispersed in cytoplasm.

[b] Data read from curves of Christensson (1959).

[c] Soluble–insoluble in 0·1 *M* KCl–0·011 *M* HAc.

[d] Sum of 16, separately determined.

[e] More data in Scherbaum, James and Jahn (1958).

[f] Combined Schmidt–Thanhauser and Schneider procedure. More data in Christensson (1962).

[g] Schmidt–Thanhauser.

[h] Indole.

[i] Microbiological, with DNAase.

[j] Microbiological, without DNAase.

[k] More data in Scherbaum (1956).

TABLE III. Effects of purine, pyrimidine and amino acid analogues and inhibitors of protein synthesis on synchronized cell division and population growth of *T. pyriformis* GL (Holz *et al.*, 1963)

Compound and source	Division delays (min)[a] Conc. (m*M*) 5	1/2	1/20	1/200	Growth Conc.[b] m*M*	Normal metabolite[c]
PURINES						
8-Azaguanine	> 40	10	0		0·0025	Na- guanylate $4H_2O$
8-Azaxanthine	0				0·01	,,
8-Azahypoxanthine	0[d]				>1·0	,,
8-Azaadenine	> 35	0			>1·0	,,
6-Methylpurine	>300	200	15	0	0·005	,,
6-Mercaptopurine	15	0			>1·0	,,
2-Mercaptopurine	0				>1·0	,,
2, 6-Diaminopurine	10	0			>1·0	,,
Benzimidazole	90	10	0		>1·0	,,
Purine	10	0			>1·0	,,
PYRIMIDINES						
5-Fluoro-2-deoxyuridine	0				0·0005	uracil
5-Fluorouracil	0	0			0·005	,,
6-Azauracil	10	0			>1·0	,,
6-Azathymine	0				>1·0	,,
5-Bromouracil	0				>1·0	,,
5-Iodouracil	0				>1·0	,,
5-Nitrouracil	0				>1·0	,,
4-Hydroxy-2-thiopyrimidine	0				>1·0	,,
4-Hydroxy-5-methyl-2-thiopyrimidine	0				>1·0	,,
5-Amino-2, 4-dioxypyrimidine	0				>1·0	,,
Isobarbituric acid	0				>1·0	,,
Orotic acid	0				>1·0	,,

5-Mercaptouracil	0				> 1·0	,,
2, 4-Dihydroxy-6-methylpyrimidine	0				> 1·0	,,
AMINO ACIDS						
DL-*p*-Fluorophenylalanine	>300	>300	>150	25	0·15	DL-phenylalanine
DL-β-2-Thienylalanine	>200	>100	50	0	0·5	,,
DL-5-Methyltryptophane	0				2·0	L-tryptophane
L-Ethionine	110	75	30	20	0·3	DL-methionine
D-Ethionine	20	15	0		0·5	,,
DL-Methionine-DL-sulfoximine	0				>2·0	,,
DL-Norvaline	0				>2·0	DL-valine
DL-Isovaline	0				>2·0	,,
DL-β-2-Thienylserine	0				>2·0	DL-serine
L-2-Thiolhistidine	0				>2·0	L-histidine·HCl·H_2O
DL-α-Methyl glutamic acid	0				>2·0	L-glutamic acid [e]
L-Canavanine	>250	>250	120	25	0·12	L-arginine
SPECIFIC INHIBITORS OF PROTEIN SYNTHESIS						
Puromycin [f]	>300 [g]	>300	>200	10	0·03 [h]	
Chloramphenicol [f]	25	0			0·3 [i]	

[a] Delays (min) of the synchronous divisions (inorganic medium) when exposed to the indicated concentrations of antimetabolites.
[b] Concentration (m*M*) required to completely inhibit growth in the presence of an amount of the normal metabolite which would support half maximal growth; 60 hr, 24°C.
[c] Component of the medium; absolute requirement.
[d] The cells tend to get stuck in division.
[e] Stimulatory to growth; not an absolute nutritional requirement.
[f] Full complement of amino acids present in the growth medium.
[g] All cells killed after 15 hr.
[h] Half maximal growth at 0·01 m*M*.
[i] Half maximal growth at 0·1 m*M*.

in 0·1 *M* KCl–0·11 *M* HAc) increase relatively more during the heat and perhaps less after the heat than do the soluble proteins (Christensson, 1959). During the heat glycogen becomes deposited faster than other compounds (Scherbaum and Levy, 1961). The most striking variations in biochemical parameters yet observed are at the level of the free nucleotides and of certain nucleotide–amino acid complexes. Reference is made to Plesner's chapter in this book.

After nine heat shocks in the 20′/40′ system the same degree of synchrony is obtained in 2 per cent proteose peptone plus 0, plus 0·1, or plus 0·4 per cent liver fraction L. Taking the logarithmic cell size as unity, the cell sizes relate as 1·5–2 to 2–4 to 6–10 in the three media (Zeuthen, 1961a). When synchronization is obtained with cold shocks there is very little size increase of the cells. In curve d (Fig. 5) two asterisks mark two time points at which dry weights were measured. Before synchronization the average cell weighed $3{\cdot}25 \times 10^{-6}$ mg, after synchronization it weighed $4{\cdot}6 \times 10^{-6}$ mg (Zeuthen, unpublished data). Synchronous division took place in the growth medium (2 per cent proteose peptone, 0·4 per cent liver fraction L) after 17 hr at 10°C. As a whole there is poor correlation between division synchrony and growth of the cells during the synchronization procedures. While we cannot yet exclude the possibility that some preliminary mass growth is necessary for synchrony to be induced, it seems likely that the mechanisms involved in the induction of synchrony are more specifically related to the cell's preparation for division.

B. Sensitivity to chemical agents

Holz, Rasmussen and Zeuthen (1963) made a comparison of the logarithmic growth system (chemically defined medium) with the synchronous division system (20′/40′, inorganic medium). Assuming that the main difference between the two systems would be that growth is eliminated in the synchornous system, a number of purine and pyrimidine analogues, several amino acid analogues and potential inhibitors of protein synthesis were tested (Table III). The main difference traced was in the response to two purines (8-azaguanine, 8-azaxanthine) and to two fluorinated pyrimidines, 5-fluoro-2-desoxyuridine and 5-fluorouracil, all presumed inhibitors of nucleic acid synthesis. The growth system is highly sensitive to all four compounds, the synchronous system is insensitive to the three latter compounds, and sometimes sensitive, sometimes not (Zeuthen, 1961a) to 8-azaguanine. The results are interpreted to mean that the main difference between the two systems is not

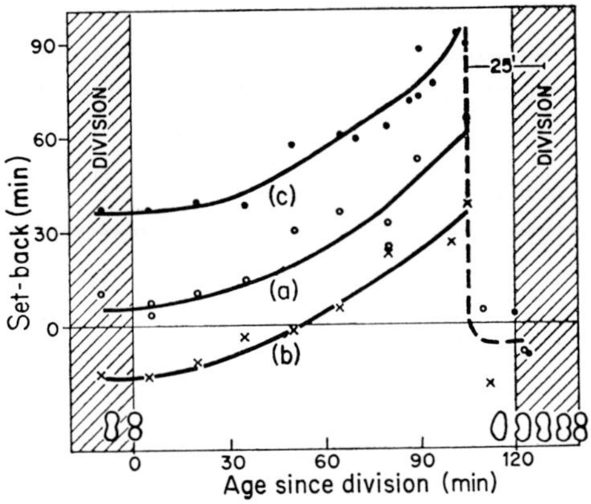

FIG. 9. 'Set-back' (delay minus time for which exposed to a new temperature) of a logarithmic division in response to a heat or cold shock initiated at various times between two divisions. (a) 9·3°C for 30 min. (b) 31·1°C for 20 min. (c) 34·1°C for 15 min. Cells were grown and observed at 28·5°C. (Compiled and redrawn from Thormar, 1959.) (Figures 9, 10, 12, 15–27 are reproduced by courtesy of the editor of *Compt. Rend. Trav. Lab. Carlsberg.*)

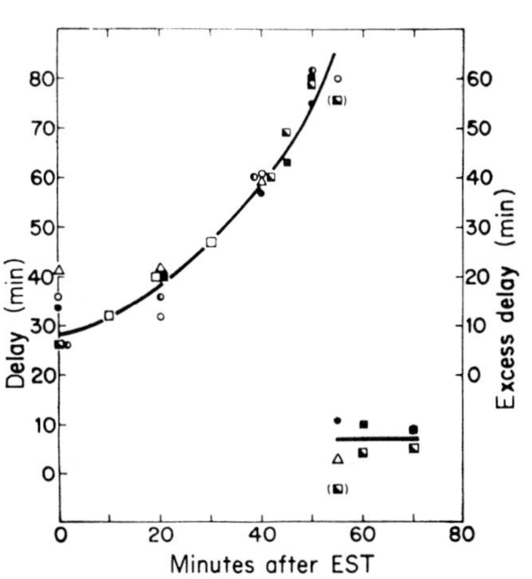

FIG. 10. Delay (left ordinate) and 'excess delay' (='set-back', right ordinate) of the first synchronous division, when the cells are exposed to 20-min shocks to 34°C at various times after the termination of the last heat shock (EST = EH as used elsewhere in this paper). Six experiments are combined. When points are in parenthesis the population corresponds heterogeneously (cf. Fig. 11). Maximal division is at 84–88 min after EST. The medium is 2 per cent proteose peptone plus 0·4 per cent liver extract. Synchronization 20′/40′. (From Frankel, 1962.)

growth or lack of growth, as first assumed, but rather the presence of excess stores of nucleic acids in the synchronized cells, not existing in the logarithmic cells (Table II). We shall return to this question in later sections.

C. *Sensitivity to temperature shocks*

The older a cell is—as measured in time after EH or after a normal or synchronous division—the more can it be delayed in its preparation of the subsequent division when exposed to a temperature shock.

Figures 9–11 demonstrate the sensitivity to heat of logarithmic [Fig. 9 (Thormar, 1959)] and of heat synchronized cells (20′/40′) in the first (Fig. 10; Frankel, 1962) and in the second (Fig. 11; Zeuthen, unpublished data) synchronous generation. The sensitivity to cold shocks are shown in Figs. 9 (Thormar, 1959) and 12 (Frankel, 1962). Thormar's data (1959) have been reinterpreted as described in section III-3. In all cases the sensitivity is in terms of delays of the subsequent division. The sensitivity to temperature shocks follows the same pattern in all cases. The cells cease to be sensitive some time before division. However, the heat synchronization makes the cells less sensitive to heat, evidence of an adaptation, which ultimately (Zeuthen, 1961a) permits some cells to break through the barrier for division set up by the heat shocks. This is seen best when we compare the excess delays obtained when the cells are shocked immediately after a logarithmic division (Fig. 9), after the last heat shock (Fig. 10) or after a synchronous division (Fig. 11). A 15-min shock to 34·0°C excess-delays the logarithmic cell 40 min, but a 20-min exposure to 34°C excess-delays the first and the second synchronous division <10 min. The sensitivity to cold shocks seems not to be influenced by the heat synchronization (cf. Figs. 9 and 12).

D. *Generation times*

Table IV compares generation times of logarithmic and of synchronized cells studied under identical conditions. The latter cells always show shorter generation times.

E. *Desynchronization of synchronized systems*

As shown by the curves for division indices and cell counts the division synchrony in the *Tetrahymena* populations is never complete. While the division maxima are clearly set off from the minima there is never an abrupt shift, from all to none dividing. The maxima only approach 100 and the minima only get close to 0.

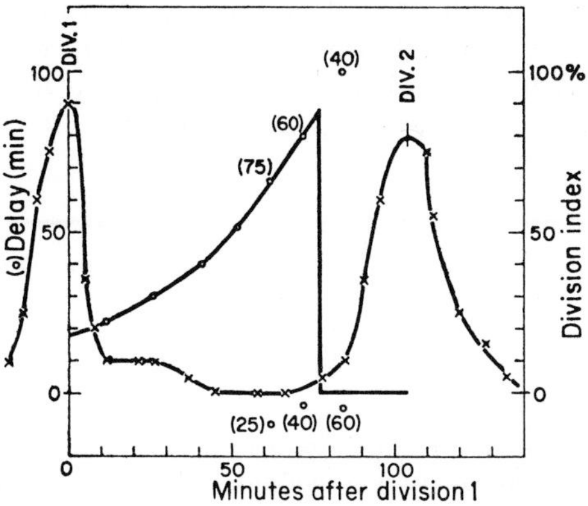

FIG. 11. *Open circles:* Delay (left ordinate) of the second synchronous division, when the cells are exposed to 20-min shocks to 34°C at various times between synchronous divisions 1 and 2. When points are in parenthesis the population responded heterogeneously, some cells being delayed (upper point of a pair) others not (lower point). The attached figures show which percentage fractions behave one or the other way. *Crosses:* Division indices. Medium and synchronization as in Fig. 10. (Zeuthen, unpublished data.)

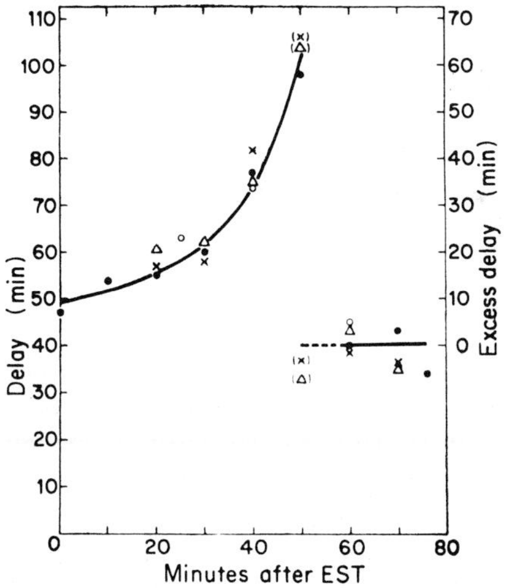

FIG. 12. Delay (left ordinate) and 'excess delay' (= 'set-back', right ordinate) of the first synchronous division when the cells are exposed to cold shocks (4–12°C) for 40 min at various times after EST (= EH). Four experiments are combined. Points in parentheses indicate heterogeneous responses (cf. Fig. 11). Synchronous division occurred at 85–90 min after EST. The medium and synchronization are as in Figs. 10 and 11. (From Frankel, 1962.)

The heat and cold synchronized systems quite often show less than 100 per cent increase in cell counts as a result of divisions 1 and 2. In Fig. 13 (a and b) the cell counts increase by 84–94 per cent in the first and second synchronous divisions, so 84–94 per cent of the cells divide.

TABLE IV. Generation times in *Tetrahymena pyriformis* GL

System	Medium	Temp.	Generation time (min)	Reference
logarithmic	2% PP+0·1% L	28·5	140	Scherbaum and Zeuthen (1955)
30′/30′ 1st gen.	2% PP+0·1% L	28·5	75[a]	
30′/30′ 2nd gen.	2% PP+0·1% L	28·5	100	
30′/30′ 3rd gen.	2% PP+0·1% L	28·5	108	
30′/30′ 1st gen.	inorganic	28	69	Hamburger and Zeuthen (1957)
30′/30′ 2nd gen.	inorganic	28	77	
30′/30′ 3rd gen.	inorganic	28	100	
20′/40′ 1st gen.	2% PP+0·4% L	22	99±12·6	Zeuthen (1961a)
20′/40′ 2nd gen.	2% PP+0·4% L	22	127±14·5	
20′/40′ 3rd gen.	2% PP+0·4% L	22	110±20	
20′/40′ 4th gen.	2% PP+0·4% L	22	135± 8·8	
logarithmic	2% PP+0·4% L	28	150	Rasmussen and Zeuthen (1962)
20′/40′ 1st gen.	inorganic	28	80	
20′/40′ 2nd gen.	inorganic	28	108	
120′/40′ 1st gen.	2% PP+0·4% L	28	88	[b]
120′/40′ 2nd gen.	2% PP+0·4% L	28	115	[b]

[a] Generally 75–85 (Rasmussen and Zeuthen, 1962). [b] Zeuthen, unpublished.

Comparison of these curves (Fig. 13) with curves (Fig. 14) representing the multiplication of one cell (GL) into eight or sixteen (third and fourth generations) suggests that the induced synchrony is inferior to the theoretically best, represented by the progeny from one cell. On the other hand, comparison with Prescott's results for strain HS (Fig. 13c) invites the view that the induced synchrony is close to a theoretical limit. Prescott (1959) followed the multiplication in fifty parallel clones identically grown from single cells picked from a large logarithmic culture. The common zero time was the time when each cell divided for the first time. The curve Fig. 13c is constructed by summing the cells in all clones at a number of common clonal time points. The desynchronization due to spread in individual generation times of this 'constructed synchronous population' runs about as fast as in the actually synchronized mass populations. The question remains whether the variation between Prescott's subcultures matches a variation in the mother culture or whether environmental differences between the subcultures play a part (see also chapter 18, this book).

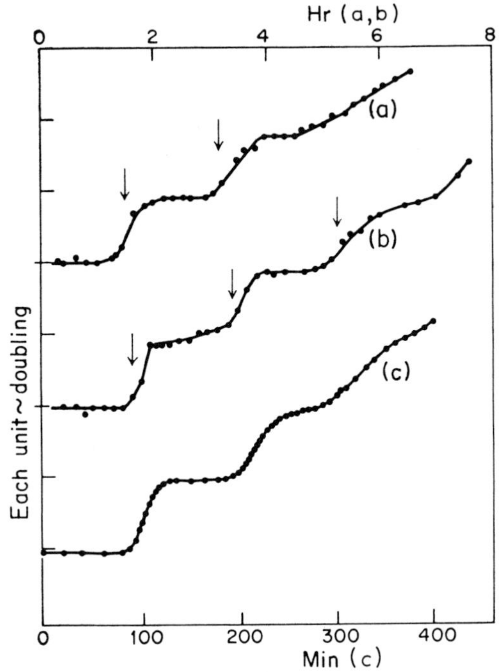

FIG. 13. (a) Cold synchronization (5 × 120′/40′, 2 per cent proteose peptone, 0·4 per cent liver extract). Division maxima at arrows. (b) Heat synchronization (7 × 20′/40′, medium as in a). Division maxima at arrows. (c) Constructed synchronous (logarithmic) population. (a, b) strain GL, 28°C (Zeuthen, unpublished data). (c) Strain HS, 32·5°C (Prescott, 1959).

5. *Synchronization of Microscopically Visible Events*

After EH cell divisions, macronuclear divisions (Holz, 1960; Holz *et al.*, 1957; Scherbaum, Louderback and Jahn, 1958; Scherbaum and Zeuthen, 1954, 1955; Williams and Scherbaum, 1959; Zeuthen and Scherbaum, 1954) and micronuclear divisions (Holz *et al.*, 1957, strain WH-1, mating type 1, variety 1) are synchronized. A synchronized nucleolar period (strain WH-6) has been demonstrated using the electron microscope (Elliott *et al.*,

1962).* Cortical morphogenesis at the level of the second mouth is synchronized by the first five temperature shocks in the 30′/30′ (Holz *et al.*, 1957; Williams and Scherbaum, 1959) and in the 20′/40′ treatment (Frankel, 1962).

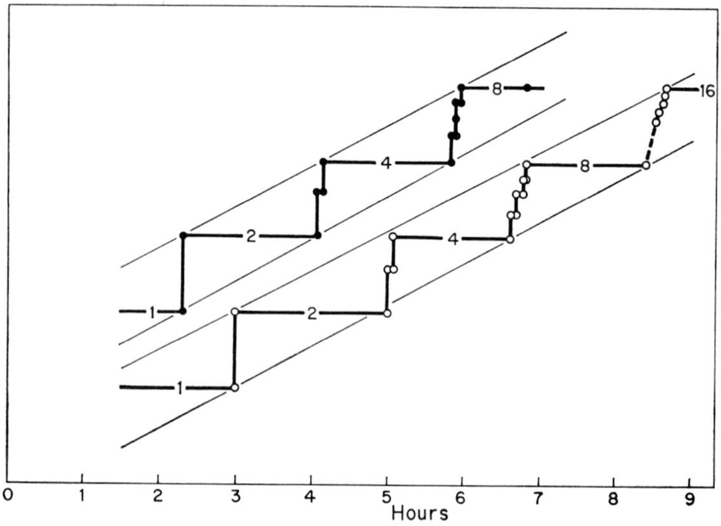

Fig. 14. *Tetrahymena pyriformis* GL. Stepwise multiplication of one logarithmic cell into eight or sixteen at 27°C. 2 per cent proteose peptone, 0·1 per cent liver extract (Zeuthen, 1953). (By courtesy of the Clarendon Press.)

Holz (1960) studied the interval from the completion of the first to the completion of the second synchronous division in synchronized strain WH-1 (mating type 1, variety 1) and made a comparison with results obtained by other workers using strains T–P and GL. The first strain carries a micronucleus, the two last strains do not. Holz reached the conclusion that except for the absence of a micronucleus in the last two strains no differences in the relationships between nuclear and morphogenetic events were noted when (1) normal cells of strain GL and mating type 1, variety 1, were compared, (2) when normal and synchronized cells of the latter strain were compared, and (3) when normal T–P and normal mating type 1, variety 1, were compared.

The ciliary apparatus, the endoplasmic reticulum (strain GL,

* In this study the initiation of the temperature shocking and the inoculation were simultaneous events. The nucleolar period may well have been induced by the latter event.

cf. Williams, chapter 5, this book) and the mitochondria (strain W, Sato, 1960) show synchronized changes during the first synchronous generation.

The present evidence suggests that *Tetrahymena* GL and other strains which have been synchronized for cell division are synchronized to a comparable extent with regard to all structural changes which have been carefully examined and which normally change visibly through the division cycle.

6. *Synchronization of Biochemical Events*

A. Growth in nitrogen and in dry matter

In Fig. 15a are shown curves for growth of total nitrogen (Zeuthen, 1958) and, in a combined experiment, of total dry weight and of total phosphorus (Hamburger and Zeuthen, 1960). All curves show growth between EH and synchronous division 1 (first generation) and between synchronous divisions 1 and 2 (second generation). Four experiments (Hamburger and Zeuthen, 1960) demonstrate

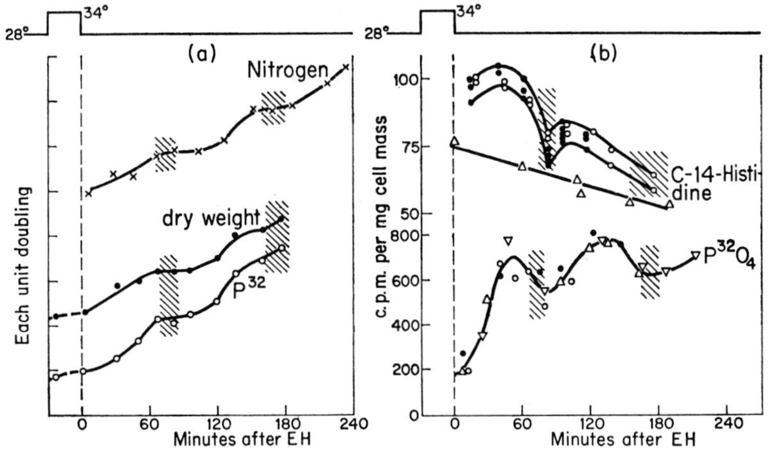

FIG. 15. (a) Cumulative growth in total nitrogen, dry matter, and phosphorus in synchronized (30′/30′) *Tetrahymena pyriformis* growing in 2 per cent proteose peptone plus 0·1 per cent liver extract. (b) Rates of uptake of ^{14}C-histidine and of $^{32}PO_4$, same medium. *Nitrogen:* Hamburger and Zeuthen (from Zeuthen, 1958). (*Other data:* same authors 1960.)

22 per cent (16–26) increase in dry weight in the first and 57 per cent (36–70) increase in the first plus the second synchronous generations. For total phosphorus the corresponding figures are 28 per cent (21–34) and 79 per cent (46–92). For total nitrogen

the increase (curve of Fig. 15a) is 16 per cent in the first and 42 per cent in the two first synchronous generations. So phosphorus is accumulated faster than total cell matter as measured by nitrogen and by dry weight. This reflects restoration of an effect of the heat shocks which act to depress more strongly (section III-5) the accumulation of P than of dry matter (Hamburger and Zeuthen, 1960).

B. Protein, enzymes

For growth in total protein Christensson (1959) describes a period resembling the one for total nitrogen and for dry matter. Hamburger (unpublished data), for cells in 2 per cent proteose peptone, 0·1 per cent liver extract, has followed the activity of several enzymes through the synchronous division cycles (Fig. 16). There is maximal activity some time before division. The reduced activity for succinic dehydrogenase at division (Fig. 16a, O) has also been found by Sullivan and Sparks (1961). At this stage ultraviolet (u.v.) radiation

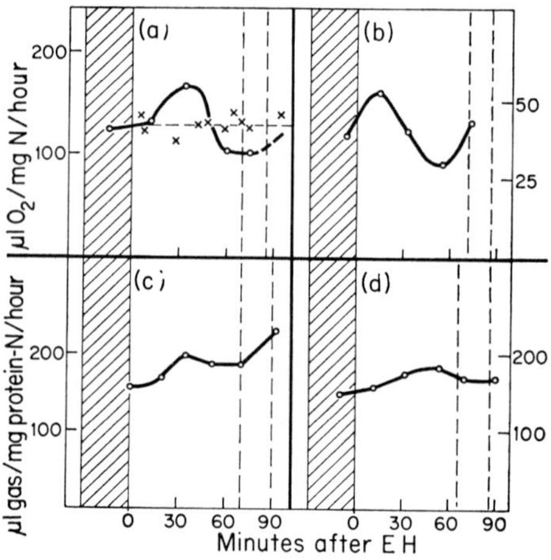

FIG. 16. Changes in enzyme activity measured during the first generation of synchronized *Tetrahymena pyriformis* GL. The cells were kept in proteose peptone (O) or in inorganic medium (X). Cell-free preparations were made by removing the proteose peptone in two washings and concentrating the cells. From the inorganic medium the cells were only concentrated. The packed cells were frozen twice at −20°C. Periods of 34°C are shaded, otherwise 28°C. The time of division is between the dashed lines. The abscissa shows the time elapsed since the end of heat treatment, the ordinate shows the enzyme activity (μl. gas exchange/mg total protein-N/hour). (From Hamburger, unpublished data.) (a) Succinic dehydrogenase. (b) Lactic oxidase. (c) β-Hydroxybutyric dehydrogenase. (d) Aldolase.

increased the activity significantly. Hamburger (unpublished data) repeated her determinations of succinic dehydrogenase, this time using the cells in the non-nutrient medium (Fig. 16a, ×). No period in enzyme activity could be observed, so when present this period is likely to reflect the growth, not the division cycle.

C. Rate of uptake of ^{14}C-histidine and of ^{32}P-phosphate

In Fig. 15b two separate experiments demonstrate discontinuous rate of uptake of ^{14}C-histidine (Hamburger and Zeuthen, 1960) into synchronized cells when studied in the growth medium (2 per cent proteose peptone, 0·1 per cent liver extract L, 30′/30′ system). The straight line represents two experiments with logarithmic cells. The downwards general slope towards the right of all curves probably indicates that the specific activity obtained by adding a standard dose of ^{14}C-histidine to the medium becomes lower the later the dose is added. This may be a reflection of the known fact that *Tetrahymena* excretes proteolytic enzymes to the medium (Elliott, 1959). Histidine may be liberated from proteose peptone so that the label becomes the more diluted the later it is added to an aliquot. On top of the general slope there is a distinct period in the rate of uptake of radioactive histidine into the synchronous cell system. The rate increases from EH to about 40 or 50 min. It then decreases to reach a sharp minimum during division 1. The cycle repeats itself between divisions 1 and 2. The maxima coincide fairly well with periods of maximum rates of mass growth of the population as measured by the cumulative curves for total nitrogen and for dry weight (Fig. 15a).

The rate of uptake of $^{32}PO_4$ (Fig. 15b) is also periodic and shows maximum and minimum values at about the times when nitrogen and dry weight accumulation (Fig. 15a) and the rate of histidine uptake (Fig. 15b) are fastest.

7. *Radioautographic Studies*

Special caution must be exerted when interpreting data obtained with, mostly biochemical, methods in which many pooled cells represent a sample. The smaller the biochemical fluctuations, the more justified is the suspicion that they might involve perhaps only a minor part of the population.

A. Protein, RNA

The autoradiographic methods are largely limited to the study of the synthesis of macromolecules in single cells. The result for the population is the sum of the results for the single cells,

TABLE V. Percentage inhibition of macromolecular syntheses (Cerroni and Zeuthen, 1962a)

	6-MP $1{\cdot}6 \times 10^{-3}M$	5-FUDR $5 \times 10^{-3}M$	dG $2 \times 10^{-4}M$	dA $2 \times 10^{-4}M$	Puromycin $1 \times 10^{-3}M$	DL-*p*-FPhe $1{\cdot}8 \times 10^{-4}M$
1. Cell division	no division	normal	normal	normal	no division	no division
2. DNA	—	88 v.s.	88 v.s.	90 v.s.	—	—
3. RNA	80 v.s.	86 v.s.	45 v.s.	45 v.s.	75 v.s.	–2 n.s.
4. Protein	45 v.s.	—	5 n.s.	1 n.s.	93 v.s.	40 v.s.

6-MP = 6-methyl purine
5-FUDR = 5-fluoro-2′-deoxyuridine
dG = deoxyguanosine
dA = deoxyadenosine
v.s. = very significant
n.s. = non-significant
The experiments with 6-MP and 5-FUDR were repeated by addition at 50 min. The results checked those reported in the table.

individually observed. When the sum curves are compared with curves obtained by biochemical analyses of aliquots removed at intervals and each representing high numbers of cells, statements about degrees of synchrony are possible. Preliminary results (Cerroni and Zeuthen, unpublished data) show that all cells in the synchronized populations (inorganic medium) incorporate ^{14}C-alanine, ^{14}C-adenine and ^{3}H-uridine at all times between EH and division. The labels were uniformly distributed over the single cell. Incorporation times were 4 or 5 min. The radioautographic data do not refute the view that the synthesis of protein and of RNA is synchronized with cell division. Our data are, as yet, insufficient to allow us to present rate curves comparable to those discussed by Prescott for the individual logarithmic cell (this book).

RNA is stored (Christensson, 1962; Hamburger and Zeuthen, 1960; Iverson and Giese, 1957; Scherbaum, 1957a) in excess amounts in the synchronized compared to the logarithmic cell. 5-Fluoro-2′-deoxyuridine (FUDR) and other compounds (Table V) inhibit the incorporation of ^{3}H-uridine (Cerroni and Zeuthen, 1962a). Still, in the presence of FUDR two synchronous divisions are performed at the same time as in the controls. Stores of RNA are distributed to three or four daughter cells at the same time as the RNA synthesis is reduced to less than 15 per cent of the control value.

B. *Asynchrony of DNA synthesis*

When the washed cells are incubated with ^{3}H-thymidine the silver grains are largely restricted to above the macronucleus, evidence of incorporation into newly-formed DNA. In the experiments to be reported (Cerroni and Zeuthen, 1962b) high specific activities (2·5 μC/0·5 ml ~ 'standard dose') but low chemical amounts of thymidine were used (~ 10 per cent of what was present in the DNA of about 50,000 cells/0·5 ml, usually used per sample). The label is not fully removed from the medium within 20–30 min of exposure (Fig. 17). Those cells which take the label seem to continue to do so for the indicated length of time. However, only a fraction of the population takes the label in the period between EH and the first synchronous division. In one experiment (Fig. 18, curve b) separate aliquots were incubated with ^{3}H-thymidine at various times after EH, and always for 4 min. Regardless of when incubated within 50 min after EH, 20–25 per cent of the population took the label. In the second experiment (Fig. 19, ●), incubations were made also at later times and through division 1. This time labelling declined with time. During division no cells took the label. After division there was new labelling. This experiment was repeated (Fig. 19, ○) in the growth medium (0·4 per cent liver extract). In Fig. 18, curve a, one aliquot was given six standard doses at 10-min intervals. Five minutes after it was added, each dose was reduced to one-thirtieth. In view of the high specific activities used this should not be taken to indicate removal of the label. The procedure served the purpose of keeping the specific activity in a range (< 2·8μ C/0·5 ml) which permits comparison with the other experiments. Linear extrapolation of curve a to the time of division 1 suggests that a maximum representing 50 per cent or so of the population synthesizes DNA in the period from

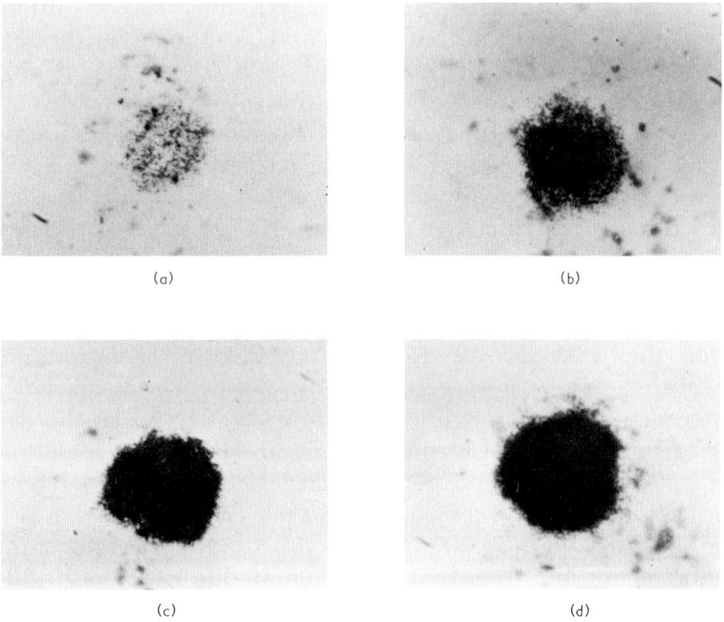

FIG. 17. The photographs (a–d) demonstrate increasing labelling in the course of the first 30 min of exposure to the label of those nuclei which take the label. In this experiment four aliquots of washed cells were taken at the time EH and incubated with the standard dose of ^{3}H-thymidine for the following times: (a) 4 min (27 per cent); (b) 10 min (30 per cent); (c) 20 min (31 per cent); (d) 30 min (27 per cent). The fractions which show macronuclear incorporation are indicated in parentheses. 95 per cent division at 85 min after EH. Even after long labelling most cells were free of label. All were Feulgen positive. (From Cerroni and Zeuthen, 1962b.)

EH to division 1. All this together represents evidence of an asynchrony of DNA synthesis. Because no cells synthesize DNA around division, the population as a whole is likely to show periodicity, linked to the division synchrony, with respect to the synthesis of DNA. This periodicity should not be confused with true synchrony.*

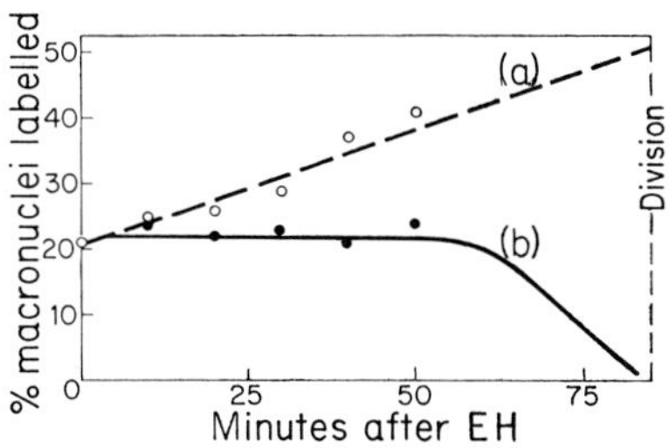

Fig. 18. Curve a: To one aliquot of cells were given standard doses of ^{3}H-thymidine at 0, 10, 20, 30, 40, and 50 min. Four min after addition each dose was largely removed and a sample was taken for radioautography. While the results are plotted at the times when the doses were given, they include a further incorporation time of 4 min. Curve b: control to curve a. Separate standard doses were given to separate aliquots at 10, 20, 30, 40, and 50 min. The results are plotted at the times indicated but include a further incorporation time of 4 min.

Extrapolation of curve I is linear to give an extrapolated maximum (50 per cent). (From Cerroni and Zeuthen, 1962b.)

The microbiological assay (Zeuthen and Scherbaum, 1954) and the chemical analysis (Scherbaum, 1957b) of the population show that the average amount of DNA per cell has increased very considerably during the synchronizing (30′/30′) treatment. Most results (Iverson and Giese, 1957; Scherbaum, 1957b) suggest more than a doubling, some slightly less (Scherbaum *et al.*, 1959). Whereas the absolute values are meaningful the relative values may not always be because in most cases (cf. chapter 23) we can question the value of the logarithmic control which serves as

* It is generally held (cf. Mazia, 1961) that chromosomal coiling during mitosis is incompatible with DNA synthesis which, for this reason we can add, becomes restricted to the interdivision periods as observed in many cells (Mazia, 1961, Table 1) and also in logarithmic *Tetrahymena* (McDonald, 1958; Prescott, 1960). The present observations suggest direct suppression of DNA synthesis during division in synchronized *Tetrahymena*. This supplies indirect evidence of the equivalent, in a ciliate, of chromosomal coiling during division.

a basis for the comparisons. As used in the comparisons the logarithmic cell is likely to be too big. The results for the average cell are supported by spectro-photometric analyses (Scherbaum *et al.*, 1959) of enough Feulgen-stained, separate cells to represent the population at the various stages. So, in the present writer's view, all data, including those in Scherbaum *et al.*'s report (1959), support the sweeping statement that at EH a synchronized cell has more than double the amount of DNA of the logarithmic, newly-divided cell. Exactly how important this is for a ciliate is difficult to say. But we may consider that when DNA is concerned, at EH the synchronized cell is mature, or greatly over mature, for division.

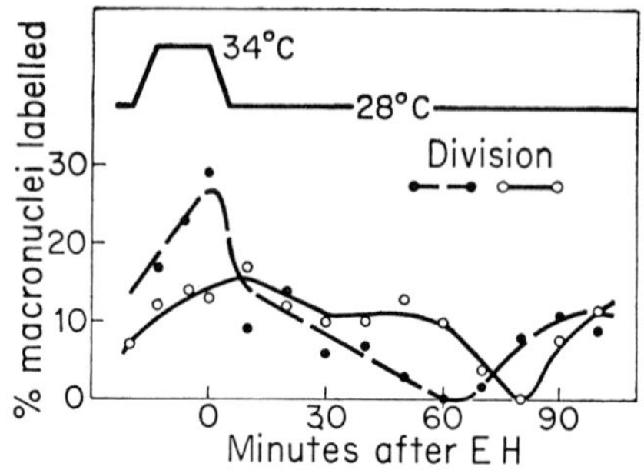

Fig. 19. Per cent macronuclei taking the labelled thymidine which is given in standard doses to separate aliquots at different time points before synchronous division. In each case the incorporation time is 4 min. ●——●: Washed cells. O——O: Cells in proteose peptone. (From Cerroni and Zeuthen, 1962b.)

In the logarithmic cell the synthesis of DNA seems to be an immediate condition for division; in the oversized synchronous cell the synthesis of this compound may be related to division in a more distant way. As previously observed by Iverson and Giese (1957) analysing the synchronized system (in proteose peptone) for sensitivity to u.v. (265 mμ): 'the u.v. sensitivity is inversely proportional to the amount of DNA present per cell and shows a direct relationship with the amount of DNA which the cell must synthesize before division'. As shown in Table V (Cerroni and Zeuthen, 1962a), DNA synthesis in actively synthesizing synchronized cells can be suppressed by 80–90 per cent, e.g. by the use of 5 m*M* 5-fluoro-2′-desoxyuridine (FUDR). There is no obvious consequence for divisions 1 and 2 (Cerroni and Zeuthen, 1962a, b). By contrast, the logarithmic cell reacts with blocked first division to FUDR added to the growth medium in a concentration of 10^{-3} m*M* (Cerroni and Zeuthen, 1962b).

Some of the inhibitors included in Table V were tested for effects on the incorporation of amino acids into proteins. The table suggests that cell division is inhibited whenever the synthesis of protein is inhibited, within wide limits regardless of inhibition of the synthesis of one or of both nucleic acids, both present in vast excess in the synchronized cells. This suggests that the channel which limits division in synchronized *Tetrahymena* is different from the one in which the nucleic acids are synthesized. It is one along which proteins, most likely specific for division, are produced. In section III we shall discuss this aspect in more detail.

III. MECHANISMS IN THE INDUCTION OF THE DIVISION SYNCHRONY

1. *Syntheses Involving Amino Acids as a Condition for Division*

For some time after the termination of the heat shocks the synchronized cells are highly sensitive to a number of inhibitors of protein synthesis. Table VI (from Rasmussen and Zeuthen, 1962) shows the delays of the first synchronous division in the 'washed cells' (cells in the inorganic medium) resulting from the addition of the various compounds, in the concentrations and at the times indicated. No agent is active when added later than 55 min after EH. Some (among them *p*-fluorophenylalanine—to be dealt with in detail) must be added considerably earlier to have an effect, but this may only mean that because of differences in permeability barriers, intracellular pools of normal metabolites, preliminary metabolic handling, etc., some compounds need longer time to take effect than others. For a given compound to block or delay division, addition must be the earlier the lower the concentration used. We have concluded (Rasmussen and Zeuthen, 1962) that 55 min is the latest time before division for an agent, interfering with protein synthesis, to be added and still delay division. It represents a 'physiological transition' after which blockage of protein synthesis can no longer interfere with the forthcoming division, maximal in the population at 80–85 min, thus 25–30 min later. A later section deals with the physiological transition.

While Table VI shows the effects of the inhibitors, added to stay at the times indicated, the following deals with experiments in which the cells are exposed to an inhibitor at defined times and for defined lengths of time. By analogy with the 'temperature shocks' we can speak of 'shock-exposures to antimetabolites'. *p*-Fluorophenylalanine (*p*-FPhe) proved particularly useful because it could be effectively removed by four washings with 1 mM phenylalanine, competitively antagonizing *p*-FPhe (Fig. 20), but itself having no

TABLE VI. Delays (in min) of synchronous division in *Tetrahymena pyriformis* GL. Continuous exposures (Rasmussen and Zeuthen, 1962)

Compound	Conc. mM	Minutes after EH for addition of compound 5	15	25	35	45	55	65	75	EH → 1. div.
1. Azaserine	0·05	270	270	270	270	—	0	0	0	85
2. D-Ethionine	38	125	125	8	5	0	0	0	0	85
3. DL-Ethionine	3	>240	>240	>240	8	0	0	0	0	85
4. DL-Ethionine	10	>240	>240	>240	8	0	0	0	0	85
5. DL-*p*-ClPhe	0·8	35	20	15	10	0	0	0	0	75
6. DL-*m*-FPhe	0·8	230	230	50	25	10	0	0	0	85
7. DL-*p*-FPhe	0·16	>180	>180	>180	20	15	0	0	0	80
8. DL-*p*-FPhe	0·8	>230	>230	>230	>230	20	0	0	0	85
9. DL-*p*-FPhe	4·0	>230	>230	>230	>230	25	0	0	0	80
10. β-2-Thienylalanine	1·0	40	40	50	10	5	0	0	0	85
11. Chloramphenicol	3·0	30	16	16	16	12	0	0	0	85
12. L-Canavanine	0·1	100	100	100	80	ill def.	10	0	0	85
13. L-Canavanine	0·5	200	200	200	200	div.	20	0	0	85
14. L-Canavanine	1·0	>200	>200	>200	200	max.	20	0	0	85
15. Puromycin	0·1	>200	>200	>200	>200	20	5	0	0	80
16. Puromycin	0·4	>200	>200	>200	>200	>200	ill def. div. max.	0	0	85

DL-*p*-ClPhe = DL-*p*-chlorophenylalanine
DL-*m*-FPhe = DL-*m*-fluorophenylalanine

effect. Fig. 21 shows the relation between response of the washed cells (20′/40′ system) and concentration of *p*-FPhe added at 5 min (a), at 25 min (b), and at 55 min (c) after EH. The exposure is for a standard time of 30 min. The response is measured by the delay of the division relative to the untreated control. Curves a and b relate to division 1, curve c to division 2. For both divisions maximal effects are obtained with 0·16 m*M* *p*-FPhe and with concentrations above this. Fig. 22 shows the relation between exposure time (abscissa) and 'set-back' of washed cells dumped into the inhibitor

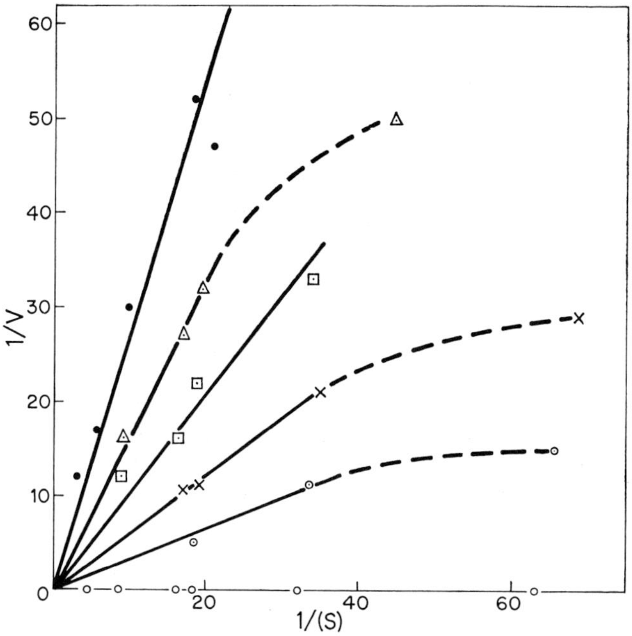

FIG. 20. Competitive antagonism between DL-*p*-fluorophenylalanine and L-phenylalanine in their effect on division in *Tetrahymena*. Lineweaver–Burk plot. The fully-drawn curves demonstrate competitive release of DL-*p*-FPhe inhibition (1/V) by L-phenylalanine (1/(S)). 1/V represents the delay of division 1 produced by five different concentrations of DL-*p*-FPhe. (1/S) represents the reciprocal of the millimolar concentration of L-phenylalanine. The controls contain L-phenylalanine, but are without DL-*p*-FPhe. Over a wide concentration range L-phenylalanine alone has no effect on division (curve represented by the abscissa line).

○ controls
⊙ 0·025 m*M* DL-*p*-FPhe
× 0·050 m*M* DL-*p*-FPhe
⊡ 0·100 m*M* DL-*p*-FPhe
△ 0·200 m*M* DL-*p*-FPhe
● 0·360 m*M* DL-*p*-FPhe

(From Rasmussen and Zeuthen, 1962.)

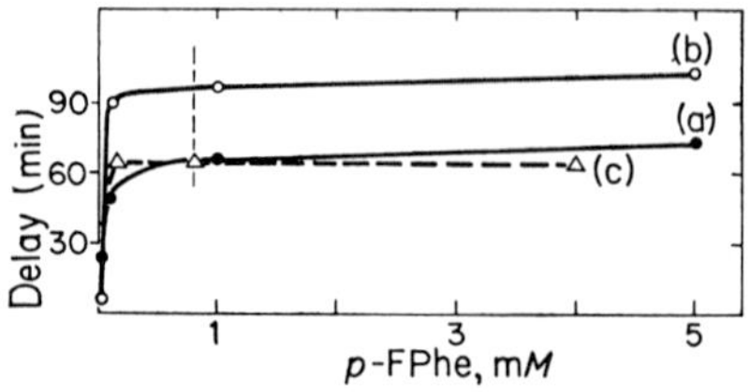

FIG. 21. Exposures to various concentrations of DL-*p*-FPhe for 30 min. The ordinate indicates the delays of the division maxima caused by the treatment. Curve a: exposure from 5 to 35 min after EH. Delay of division 1. Curve b: exposure from 30 to 60 min after EH. Delay of division 1. Curve c: the delay of division 2 when the cells were exposed from 55 to 85 min after EH. (From Rasmussen and Zeuthen, 1962.)

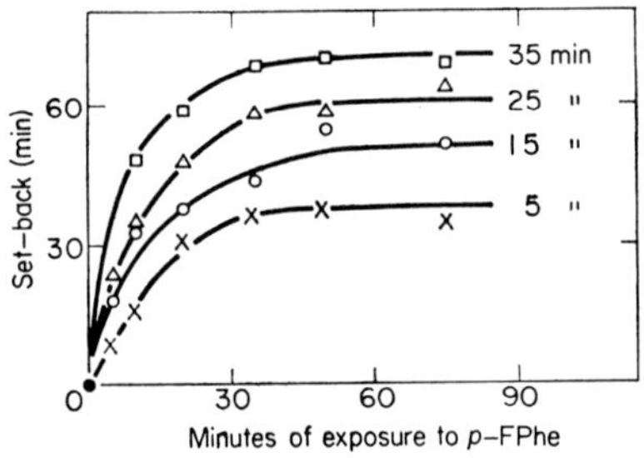

FIG. 22. Short-time exposures to 0·8 m*M* DL-*p*-FPhe (exposure times varied). The set-back of division 1 is related to the duration of exposure to a standard concentration of DL-*p*-FPhe. The treatments were initiated at the four different time points indicated. (From Rasmussen and Zeuthen, 1962.)

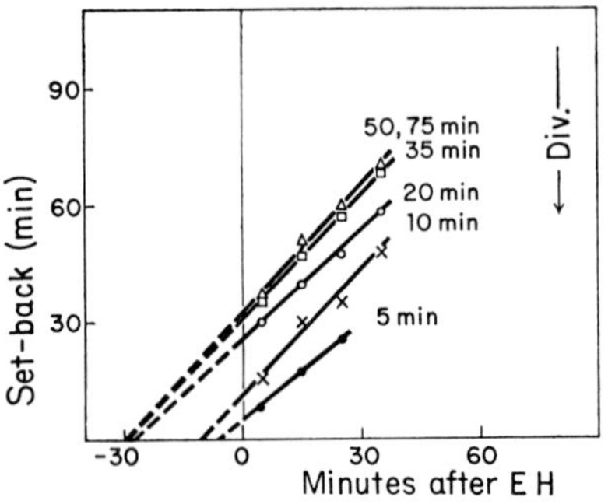

FIG. 23. Short-time exposures to 0·8 m*M* DL-*p*-FPhe. The set-back of division 1 is related to the age (EH = 0) of the cells when exposed to a standard concentration of DL-*p*-FPhe. The curves show the effects of varying the duration of exposures. The curves have been extrapolated towards the left to intersect with the abscissa. (Same data as in Fig. 22.) (From Rasmussen and Zeuthen, 1962.)

(0·8 m*M* *p*-FPhe) at defined times after EH. The 'set-back' is the delay of the division minus the time for which the cells were in contact with the inhibitors. Separate curves are shown for cells immersed at 5, 15, 25, and 35 min after EH, thus at intervals of 10 min. The maximal set-back takes time to develop, but it is nicely defined in each case. It becomes roughly 10 min longer for every 10 min by which we postpone the addition of the analogue. Fig. 23 is based on the same data as shown in Fig. 22. It relates 'set-back' and the time of beginning exposure to 0·8 m*M* *p*-FPhe. Separate observations are made for the six exposure times of 5, 10, 20, 35, 50, and 75 min. The two latter curves coincide. All curves tend to be linear. Proper extrapolation defines intersection points with the abscissa at −6, −11, −27, −29, −30, and −30 min. From Figs. 21, 22, and 23 it is concluded that shocks of maximal strengths (0·8 m*M* for more than 35 min) yield a curve which in the coordinate system chosen (two equal time scales) is linear and slopes at 45°. This means that for washed cells which have not yet matured to become insensitive to *p*-FPhe, the time necessary for recovery from treatment with the amino acid analogue is linearly related to the age of the cells when they are exposed to the agent. Furthermore, when age is defined from an absolute zero at 30 min *before* EH the recovery time equals the age of the cells when the treatment is initiated. The situation is simple: When *Tetrahymena* cells are exposed to *p*-FPhe in a high enough concentration and for a sufficient length of time, all time previously spent in preparation for division is lost. This preparation is in terms of synthesis of proteins or at least of compounds in which amino acids are part. Our method says nothing about rates at which the special proteins are synthesized because we only measure times. Also it says nothing of chemical amounts involved. They may be quite small in view of the fact that we study division in cells which do not grow. But, because we measure displacement in time of cell division, the postulated synthesis may have a special relation to the process of cell division. For convenience, and pending further studies, we shall refer to the postulated proteins as 'division proteins' (Zeuthen, 1961b). If we make the assumption (cf. Fig. 21) that the washings quickly and effectively remove the intracellular pool of *p*-FPhe, and that for this reason it is justified to speak of an absolute zero time at 30 min before EH, then it follows that the heat has not set the cells back to the absolute zero time. At time EH they may therefore still contain small, but—being synchronized—standard, amounts of 'division proteins'. What is there would correspond to what is synthesized in the course of the first 30 min after the *p*-FPhe treated cells have been washed free of the amino acid analogue. The other possibility is that it takes 30 min for the treated and washed cells to free themselves of intracellular *p*-FPhe, and only after this do they synthesize in preparation for division. In this case the cells are fully set back in time by heat.

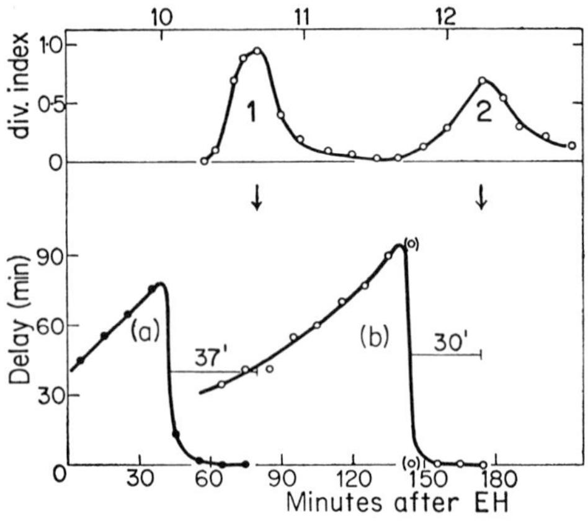

FIG. 24. Short-time exposures to 0·8 m*M* DL-*p*-FPhe for 20 min. Upper curve: The changing division index in the culture flask (control). Two division maxima are followed. Lower curve a: Delays of division 1 are plotted against the age of the cells when treatment with DL-*p*-FPhe was initiated. Lower curve b: Delays of division 2. Two points in parentheses indicate that at a particular time the treatment splits the cell population into two groups. (From Rasmussen and Zeuthen, 1962.)

When before division 1 the cells cease to be sensitive to *p*-FPhe (Fig. 24, curve a), sensitivity with respect to division 2 can be demonstrated (curve b). Chemosynthetic preparation for division is thus continuous. This, by the way, is our best, but not our only (cf. section III-3), evidence that the striking discontinuities demonstrated in this chapter are not the results of trivial variations in permeability and other barriers between the agent and its ultimate effect. The chemosynthetic phase in preparation for division 2 runs before and through division 1, and it culminates in a physiological transition before division 2. Results closely related to those obtained for the washed cells can also be obtained with synchronized cells which remain in the growth medium. For logarithmic cells studied in the growth medium transition before division (Rasmussen and Zeuthen, 1962) can also be demonstrated. In other words: preparation for division by synthesis of 'division protein' is not limited to the most suitable test system: the washed cells. It may be continuous through the synchronous and normal division cycles by which the growing synchronized system reverts to and continues normal logarithmic growth and division.

We can now characterize what we may call a chemosynthetic phase of the division cycle. In this phase the synchronized cells prepare for cell division by syntheses involving amino acids. In the heat synchronized cells this lasts 55 min, at which time the 'physiological transition' sets in. However, when the system is maximally set back in time by treatment with *p*-FPhe the chemosynthetic phase can (possibly: see above) be extended by another 30 min, thus to 85 min. It is followed by a phase of no sensitivity of division 1 to agents which interfere with protein synthesis. This latter phase may largely be a mechano-phase of cell division (cleavage of the cell and of the nucleus). It lasts 25–30 min as measured from the 'physiological transition' at 55 min after EH to maximal synchronous division at 80–85 min (cf. Table VI). The mechano-phase superimposes the chemosynthetic phase of division 2.

In the above discussion of the results obtained with *p*-fluorophenylalanine, the possibility that the effect of this compound is mediated via false proteins containing the analogue has been neglected. This is because the effects of *p*-FPhe, where studied, are mimicked by such specific inhibitors of protein synthesis as puromycin and (in high concentrations) chloramphenicol. These inhibitors cannot be expected to give rise to false proteins. Furthermore, related results (Fig. 25) are obtained with metabolic inhibitors (Hamburger, 1962; Hamburger and Zeuthen, 1957), and under defined conditions curves similar to those of Figs. 22 and 24 can be obtained when the synchronized cells are given shocks of anaerobiosis (Rasmussen, 1963). Inhibition of the respiratory metabolism is likely to inhibit protein synthesis.

A sharp drop in sensitivity to metabolic inhibition prior to division has been demonstrated for many cells (cf. Swann, 1954; Mazia, 1961). The pattern is essentially the one demonstrated in our Figs. 24 and 25 with the exception that *Tetrahymena* reacts excessively with 'set-backs' where most other cells seem only to be blocked. Following Swann (1954) we (Hamburger and Zeuthen, 1957) have previously argued that a labile energy depot is used at a critical time point before division. Failing to demonstrate variations in ATP in the sea urchin egg, Swann has suggested that the energy may be tied up in other energy-rich compounds, e.g. thiolesters.

Following the indicated line of argument for *Tetrahymena*, labile energy stores, but also labile stores into which are built amino acids, disappear at a common time before division. The reasonable suggestion is therefore that the two stores are identical and that

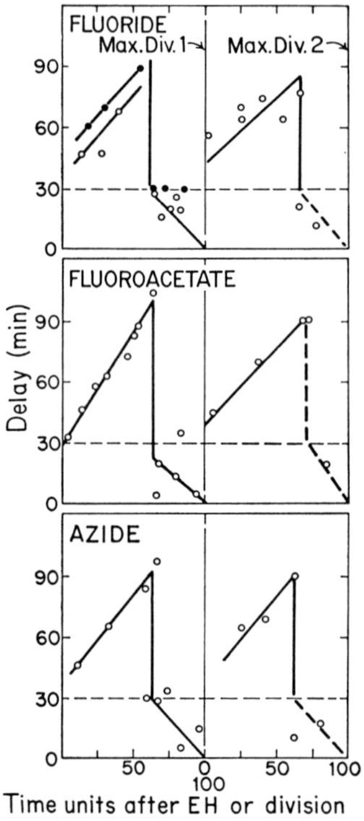

FIG. 25. *Tetrahymena pyriformis* GL and G1-R. Synchronized cells in proteose peptone treated for 30 min with 25 m*M* (O) and 34 m*M* (X) fluoride (upper curve), 78 m*M* fluoroacetate (middle curve) and 1 m*M* azide (lower curve). (From Hamburger, 1962.)

the energy is deposited in labile proteins or amino acid containing substances which at the 'physiological transition' do not disappear but cease to be labile. Mazia (1961), working on the sea urchin egg, points at the mitotic apparatus as a labile visible structure which is made by the expenditure of energy. In *Tetrahymena* we point at unidentified cortical protein structures which orient kinetosomes and only thereby become marked (cf. sections III-3 and III-4-2).

2. *Synchronization with p-Fluorophenylalanine and with Gaseous Nitrogen*

Populations of *Tetrahymena* cells in logarithmic growth (2 per cent proteose peptone, 0·4 per cent liver fraction) can be partially synchronized by one shock exposure to either *p*-fluorophenylalanine (Jensen, Rasmussen and Zeuthen, unpublished data) or gaseous nitrogen (Rasmussen, 1963). The amino acid analogue is applied in a concentration of 16 m*M* and it is removed after 70 min by three washings with fresh medium. As a result a division maximum of 70 per cent is observed 140 min after the end of the exposure. When a culture is subjected to anaerobiosis for 40 min a division maximum of 40–50 per cent is seen 170 min after the treatment.

These results are close to what could have been expected, judging from the results of studies of the effects of *p*-FPhe (Rasmussen and Zeuthen, 1962) and of anaerobiosis (Rasmussen, 1963) on single logarithmic cells: The asynchronous logarithmic populations contain cells (the majority) which are set back to their absolute zero-time point without dividing in the drug or under anaerobiosis. Other cells (the minority) go through a division in *p*-FPhe or they divide immediately after the exposure to nitrogen has ended. Continued studies with *Tetrahymena* as a model may help us to develop rational chemical approaches towards inducing synchrony in systems (e.g. tissue cells *in vivo*) which are not readily accessible by physical means.

3. *The Time Point of Physiological Transition*

In previous papers (Hamburger and Zeuthen, 1957; Zeuthen, 1958) we have referred to the time when before the first synchronous division the cells cease to be sensitive to the addition of an inhibitory agent, as the ‘critical point’ with respect to that agent. Table IV showed that even for a series of agents which can be expected to have a basically similar effect, the ‘critical point’ varies with the inhibitor used. The terminology ‘critical point’ is in accord with common usage (Swann, 1954), but it may be misleading because it is only the experimenter who faces a crisis at the time when the cell goes through a ‘critical point’: he runs the danger of adding his agent too late. The cell is in no crisis but it undergoes a transition from sensitivity to non-sensitivity to the agent studied. The transition point is differently defined for every agent, and for a given agent it is somewhat dependent on the concentration in which it is applied. When evaluating the significance of the differences in the transition points in Table VI—all dealing with agents which interfere with protein synthesis—it is important to remember (Rasmussen and Zeuthen, 1962) that the rate of metabolic handling, including transport across the membrane, may be quite different for the different compounds. There may be different lags between the time of addition and the time when a common effect is exerted, so the different ‘transition points’ in Table VI may signal only one ‘physiological transition’ of the synchronized population, and this transition cannot be earlier than the time at which the most powerful agents of the series take effect (Rasmussen and Zeuthen, 1962). The most powerful agents listed in Table VI are puromycin and L-canavanine, and these agents show transition points at or around 55 min after EH.

Table VII shows the analysis of various types of inhibition of division in synchronized *Tetrahymena pyriformis*. The inhibitions are sufficiently different for us to suppose that together they have more than one, and presumably several different chemical points of attack. In most cases 'shock exposures' were made. The pattern of effects on cell division is basically similar with all agents. All agents induce delays which are greater the later the time of exposure. The delay is always longer than the time for which exposure to the agent is made, hence it is 'excess delay' also referred to as 'set-back'. It reaches a maximum value immediately before transition to no excess delay.

The various compounds listed in Table VII were chosen for presumed variation in primary biochemical targets: protein synthesis (group I), energy metabolism (II), RNA and/or cofactor synthesis (III), sterol synthesis (IV). For comparison (V) the results of analysis with heat and with cold are included.

All inhibitors of groups I–III may be considered as one in as far as one way or another all substances may interfere with the synthesis of the postulated 'division proteins'. Groups I, II and III have been dealt with elsewhere in this paper. Group IV deserves some remarks. MER/29 citrate is one of a series of presumed inhibitors of lipid and sterol synthesis tested (Holz, 1961; Holz *et al.*, 1963) on the washed, synchronized cells. A wide variety of sterols, sterol precursors, fatty acids, glycerides, phospholipids, and their precursors and components were tested for releasing effects. Only oleic acid and other oleate-containing compounds were effective. Considered together with the other results of Table VII the results suggest that fatty acid metabolism has a function either in the postulated synthesis of the division proteins or in controlling proper function of such proteins.

In our laboratory, studies by Frankel (1962) of the morphogenetic effects of shocks by heat, cold and *p*-FPhe have added greatly to our understanding of the nature of the 'physiological transition'. Heat, cold and *p*-FPhe all have the same morphogenetic effect. Exposure before a morphogenetic 'stabilization point' results in the halting of further development and, at certain stages, also in the initiation of a resorption of the developing oral structures. When some time after the shock the regression is completed, a new set of

TABLE VII. The effects of various inhibitors on division in synchronized *T. pyriformis* GL (Modified from Frankel, 1962.)

		Molar conc.	Medium	Minutes exposure	Maximum excess delay	Approximate transition point	Reference
V	heat (34°)		PPL, IM	20	60	−25	Frankel (1962)
	cold (11°)		PPL	40	60	−25	
I	*p*-FPhe	4×10^{-3}	IM	25	80	−30	Frankel (1962)
	puromycin	4×10^{-4}	IM	—[a]	—	−30	Rasmussen and Zeuthen (1962)
	L-canavanine	5×10^{-4}	IM	—[a]	—	−40	
	DL-ethionine	3×10^{-3}	IM	—[a]	—	−50	
III	8-azaguanine	$3{\cdot}5 \times 10^{-4}$	IM	—[a]	—	−30	Zeuthen (1961a)
II	nitrogen	(>99·9%)	PPL, IM	40	60	−30	Rasmussen (1963)
	dinitrophenol	5×10^{-5}	PPL, IM	30	45	−30	Hamburger and Zeuthen (1957)
	azide	1×10^{-3}	PPL	30	60	−30	Hamburger (1962)
	fluoride	$3{\cdot}4 \times 10^{-2}$	PPL	30	60	−30	
	fluoroacetate	$7{\cdot}8 \times 10^{-2}$	PPL	30	70	−30	
IV	MER/29 citrate	$1{\cdot}5 \times 10^{-5}$	IM	—[a]	—	−30	Holz *et al.* (1963)

[a] = continuous exposure

When several concentrations of the inhibitor are employed, the lowest concentration which gives maximal effects is tabulated. The maximum excess delay is in all cases observed just prior to the time of sudden transition to no excess delay. The approximate transition point is given as a negative number which represents the amount of time prior to the division maximum at which the transition takes place. The times indicated are in most cases accurate to ± 5 min.

PPL = Proteose peptone, liver extract
IM = Inorganic medium

oral structures are formed and division takes place. Shocking the cell after a morphogenetic stabilization point neither induces delay of division nor regression of the developing oral structures. In most cases both the final modelling of the oral structures and the development of the division furrow (at the site indicated by the fission line) take place normally even in the presence of the interfering agent.

TABLE VIII. A summary of observations on *Tetrahymena* GL cultures subjected to temperature or *p*-FPhe treatments (Frankel, 1962)

Agent:	Temperature			*p*-Fluorophenylalanine	
Intensity:	34°		11°	$4 \times 10^{-3}M$	$4 \times 10^{-5}M$
Medium:	PPL	IM	PPL	IM	IM
1. transition point	−25	−25	(−25)[a]	−30	−42
2. stabilization point	−20	−20	−20	−20	−20
3. (2) minus (1)	5	5	(5)[a]	10	22
4. time to halt development	~7	5–7	0	10–15	20–30
5. time to elicit damage	~5		~20	(≦25)[a]	
6. primordium regression	slow	rapid	slow	rapid	rapid
7. abortive division	yes	no	yes	no	no
8. no. new oral primordia	1 or 2	1	1 or 2	1	1

The results, in lines 1–5, are expressed in minutes. The numbers in lines 1 and 2 are negative because they represent the amount of time prior to the division maximum at which transition or stabilization takes place.
[a] Not precisely determined.

Obviously, when a developing structure must first fully regress and then develop *de novo* there will be a relation between the degree of development before the damage and the amount of repair that must be made before the previous state of development is recovered. Because amounts of repair must be a function of time there is good harmony between the more chemical data shown in Figs. 21–25 inclusive and Tables VI, VII, and the morphogenetic observations reported.* Table VIII represents a summary of Frankel's observations. In his study the transition point measured with the agents, applied in the media indicated, came 25 to 42 min (as specified in line 1) before the division maximum in the synchronous population. The morphogenetic stabilization point came 20 min before division (as listed in line 2), thus 5 to 22 min later than the corresponding

* The physiological experiments suggest set-back to a considerably earlier time point than that (~40 after EH) at which Williams (this book) finds a 'morphogenetic arrest'. This might only reflect the fact that dynamics within an 'anarchic field' cannot be observed.

transition point (as specified in line 3). With heat shocks applied in proteose peptone–liver (PPL) and in the inorganic medium (IM) the time necessary at the high temperature to halt development (line 4) corresponds closely with the observed time difference between the transition point and the stabilization point, suggesting correspondence in time between the 'physiological transition' and the 'morphogenetic stabilization'. The same holds true when the agent used is *p*-FPhe dissolved in two concentrations in the inorganic medium. With cold (PPL) the morphogenetic development is halted immediately, but it takes (line 5) about 20 min for the morphogenetic damage to develop. So halt of development and morphogenetic damage seem not to be causally related.

The rate of primordium resorption is a function, not of the shocks, but of the medium (line 6). In proteose peptone, 0·1 per cent liver fraction, the rate of respiratory metabolism is three times higher than in the inorganic medium (Hamburger and Zeuthen, 1957). Whether or not a fraction of the population shows abortive division (line 7) during the shock (persisting for some time after EH) or develops supernumerary oral primordia (line 8) is correlated with the medium, with the rate of metabolism and with the rate of primordium regression.

Frankel's results lend strong support for the suggestion that differences between transition points recorded with various agents and with different concentrations of one agent show differences only between the agents, and between effects of various intensities of one agent (Rasmussen and Zeuthen, 1962). The studies reported thus far demonstrate only one point of 'physiological transition'. The 'physiological transition point' is also a 'morphogenetic stabilization point'. It occurs prior to the formation of the buccal cavity, fibre bundle, and cytostome. It comes 5–10 min before the division furrow, but after the fission line has appeared.

We can speak of a 'transition interval' (Frankel, 1962) within which the population changes from a state when all cells are sensitive to a state when almost all cells are insensitive to an agent. The 'transition point' for the population is that point where 50 per cent of the cells are insensitive. At the time when the cleavage furrow is making its first unilateral appearance (the 'notch' stage) a cell may be caught in division by a 15 or 20 min temperature shock to 34°C. This observation has been made by Thormar (1959) for the logarithmic cell, and by Frankel (1962) for the synchronized cell studied in PPL (20′/40′ system). During and after the shock the cell makes progress towards division but the division is abortive. After a long lag a new division develops and occasionally two oral primordia develop (cf. Table VIII). In any event final division is excessively delayed. This accounts for the maximal delay (many observations) in Thormar's Fig. 3 (1959).

We have reason to consider (Thormar, 1959) that cells which react to the temperature shocks in the atypical manner described are present in fair numbers in the logarithmic population and that they may account for the fact that quite often after synchronization cell numbers increase by less than 100 per cent in the synchronous division steps. Cells reacting this way represent a nuisance in synchronization work. If the last temperature shock sets in too late, quite a few may react by abortive division, which sets in immediately after the last shock. The later synchronous division begins in time but it is not well developed because those cells which earlier showed abortive division divide late.

When proper correction is made (Frankel, 1962) for the abortive dividers the 'physiological transition point' is moved by 5 min from 55 (p. 137) to 60 min after EH. The abortive dividers, scored as dividers when the synchronous system is studied by the exclusive observation of the division maxima in time, and correctly as non-dividers in the single-cell study of Thormar (1959), have helped to create a confusing statement for which the present writer is responsible (Zeuthen, 1958). Contrary to what we then believed, there is no obvious difference in the time of physiological transition between synchronized cells and logarithmic cells when time is reckoned from the transition to a defined stage of division. Also another factor has helped to support the earlier view. In Thormar's study the time of earliest 'notch' formation was taken as the beginning of division, the time of completed furrowing as the end of division. Thus, end of division in Thormar's study (1959) is close to what is scored as division maximum in the synchronous system. Translating Thormar's results for the single cells representing logarithmic (1959) and synchronous (Zeuthen, 1958) populations into what he would have observed had he studied synchronous populations by following the division indices, it appears that maximal response to 34°C (applied for 15 min) is 25 min or slightly more before 'maximal division' in the mass populations. This is very close to what is now agreed by workers using the synchronous system. Thus, there is no evidence of a substantial difference between logarithmic cells and synchronized cells (first and second synchronized generation, *Tetrahymena pyriformis* GL) when the time interval between the 'physiological transition' and the division is considered.

The discussion presented in this section modifies some of the views expressed in section V, 3e–g in Zeuthen (1958).

4. *Kinetosomes and Cell Division*

Synchronous division represents a special case of delayed division induced into a logarithmic culture. Till now we have emphasized the delay. We shall now focus attention on the fact that for several generations after the induction by a delay mechanism, synchronous division runs at a faster rate than the normal logarithmic (Table IV). So, using logarithmic growth as a basis for comparison, synchronous division develops prematurely. When for other cells we discuss cell division we usually consider that for this process to be initiated, units (chromosomes, centrioles, or, specially for the ciliates, cytostome, and contractile vacuole pores) must double their numbers since last division. Synchronous Tetrahymenas are abnormally large (Table II) but the populations are engaged in a regulation towards smaller average cell size. This regulation occurs in the synchronous division steps. The cell body, the macronucleus, the DNA of the macronucleus, and the RNA, all less than double between divisions. When we starve the cell (Hamburger and Zeuthen, 1957) and, in addition, block the synthesis of the two nucleic acids by 90 per cent (Cerroni and Zeuthen, 1962a) the chemical mass of important organelles is simply halved in the unaffected first, and again in the second, synchronous division. If we insist on relating the abnormally fast division rate of the synchronized *Tetrahymena* to large body size, and at the same time observe the basic fact that divisions are separate in time, then we need postulate that entry of the cell into division is dependent on biochemical or structural units which increase or multiply between synchronous divisions. Such units must be related to some cell parameter, but not simply to mass.

Taking the somatic kinety n-2 as a model Williams and Scherbaum (1959) counted an average of 57·6 longitudinally arranged kinetosomes at time EH and 59·9 just prior to division. In the balanced growth situation the logarithmic cell must double its kinetosomes from division to division, so it immediately appears that most of the longitudinal kinetosomal multiplication before synchronous division must have taken place during the synchronizing heat treatment. At EH somatic kinetomes are thus stored. After a comparable stage has been reached in the formation of an anarchic mouth field, by lateral multiplication of the kinetosomes of kinety 1, the development of the normal and of the synchronized cell can be confidently compared. Same developmental stage is 80 min after division in the logarithmic cell and 23 min after EH in the synchronized cell (read from Frankel's Fig 2, 1962). In Frankel's study (1962) only longitudinal multiplication of the kinetosomes of kinety 1 took place for $\sim$ 65 min after a logarithmic division. It is this phase which is largely skipped in the synchronized cell. So, in spite of small differences in later morphogenetic rate of the two cells the difference in generation time ultimately observed ($150 - 90 = 60$ min) is accumulated early.

When synchronization with heat shocks is initiated and continued with more and more shocks, the cells get bigger, kinety n-2 holds more kinetosomes (Williams and Scherbaum, 1959) and the first synchronous generations (Zeuthen, 1961a) shorten (Fig. 7). These are gradual developments. If too many shocks are applied, the cells may grow to excessive sizes, the kineties may coil badly, and division of the resulting monsters may be extremely delayed (Zeuthen, 1961a). The point to be made here is that both the gradual shortening, and the later gradual (Fig. 7) or excessive lengthening, of the first synchronous generation may be related to the state of the kinetosomal apparatus: kinetosomes packed in straight, or fairly straight, kineties may allow more rapid preparation for division. Kinetosomes structurally misplaced in coiled kineties correlate with delayed division. In both cases kinetosomes, or associated structures, seem to have some influence over division. The first case has been discussed in detail for the *Gymnodinoides* by Lwoff (1950), the second may be related to the fact that random fusion of more than five Stentors result in a giant which is unable to divide, except after cutting into smaller pieces (Tartar, 1961).

One would now like to proceed to the synchronous generations 2, 3, and 4 (Table IV). We may suggest that these generations, too, are shorter than the average logarithmic generation because after a preceding division there is an excess of kinetosomes so that the necessary multiplication of kinetosomes to be arranged longitudinally on one, more, or all kineties is less than in the logarithmic control. At a first glance the evidence is against this view, because at division the untreated cell is stated (Williams and Scherbaum, 1959) to possess about sixty kinetosomes in kinety n-2, which equals the value for the synchronized cell before division 1. It is only a little less than double the number (2×34) after this division. So, from the data it seems that already after the first synchronous division the kinetosomes must nearly double between divisions, in the synchronous as well as in the untreated control cell; and we are then without explanation of the shortness of the second and third synchronous generations (Table IV). However, the untreated control used by Williams and Scherbaum (1959) may not be a proper control. It represents the early maximum stationary population growth phase, not the logarithmic population in balanced growth, and it is the balanced growth situation towards which the synchronous system regulates.

While a structural system which orients kinetosomes along the longitudinally growing kineties appears capable of continued growth during the temperature shocks, structures which serve to orient kinetosomes at the level of the second mouth appear to be destroyed by heat, cold and by *p*-fluorophenylalanine. It is only when after the heat shocks the orienting system at the level of the second mouth has had a chance to become repaired in a constant and favourable temperature environment that newly formed kinetosomes become oriented and stable. The proteinaceous substance P_1 (or P_2) ('division protein') in the scheme Fig. 26 (see section III-5) could equal the 'morphogenetic material' (Lwoff, 1950) which operates at the level of the second mouth, and perhaps elsewhere in the cell at places where Nature did not supply kinetosomal morphogenetic markers (see also Williams' chapter in this book).

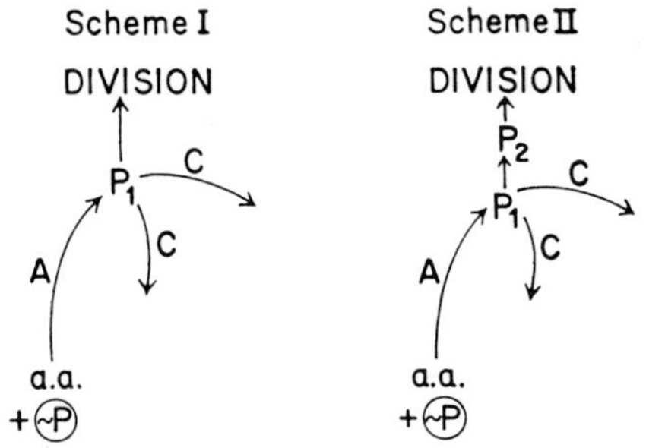

FIG. 26. For explanation, see text. (From Rasmussen and Zeuthen, 1962.)

5. '*Protein Synthesis*', '*Set-back*' and '*Physiological Transition*'

Any final explanation of the induced division synchrony must take into account the many kinds of disturbances (temperature changes, inhibition of protein synthesis, inhibition of aerobic metabolism and of oxidative phosphorylation) which induce full or partial synchrony. It must also take into account that the physical and chemical agents mentioned all produce 'set-backs'. *Without set-backs there can be no temperature-induced synchrony in Tetrahymena!* The set-backs are at the level of synthesis of the postulated division-promoting proteins, or proteinaceous compounds. The schemes of Fig. 26 represent our present interpretations of the events which take place during and ultimately terminate the chemosynthetic phase of the division cycle (Rasmussen and Zeuthen, 1962). Before the 'physiological transition' at 55 min after EH there is net increase in the amount of the proteinaceous compounds P_1 and/or P_2 which is necessary for division to take place. P_1 is continuously produced in reactions A and we have previously thought that it is continuously removed in reactions C. We have suggested (Rasmussen and Zeuthen, 1962) that a normal balance between A and C is disturbed in an unspecified manner by heat and by cold, but in such a way that C is favoured. Dr. John R. Cann now working in our laboratory has made our data the basis of extensive kinetic computations and he suggests that heat and *p*-FPhe (the writer suggests cold, too) slow reactions A and initiate reactions C, which are not normally running. This focuses our attention on scheme II. It is here suggested that all our observations fit the idea that P_2 is a macromolecular structure of which P_1 is only one component. Heat, cold and *p*-FPhe may in various ways (*p*-FPhe presumably by creating short-time starvation for P_1) interfere with the correct assembly of P_2. We need only suggest that P_2, when growing during the period before the physiological transition, collapses fully with the three treatments, and that it cannot be reassembled from its broken parts. The important consequence is that all P_1 made before treatment with the three agents is now spilled into a useless structure. This would be the significance of the side reaction C. There is independent evidence (Nachtwey, personal communication) that in the individual cell the set-back by heat is a sudden event. Regardless of the mechanism by which P_2 is damaged, this structure must later be wholly remade, and this requires new synthesis of P_1, and perhaps of other component parts of P_2. Division seems to be initiated when P_2 attains a critical concentration, according to our scheme II and according to Cann. In physiological terms this happens when the cell becomes insensitive to a number of inhibitors and when the developing oral structures become stable.

Trying to account for the 'set-backs' and thus for the synchrony my former collaborators discuss the matter in thermodynamic terms. Thormar's thesis written in Danish in 1956, quoted in Zeuthen (1958), has now in part been published in the English language (Thormar, 1959; 1962). One conclusion (Thormar, 1959) is: 'The process by which heat effects the mechanism of cell division has a high activation energy. This suggests that increases in temperature above optimum interfere with the development or function of macromolecular structures essential for later cell division'. In the second paper Thormar (1962), stimulated by the kinetic studies of Johnson and Lewin for *Escherichia coli* (1946), analysed the curves shown in Fig. 1 and suggested that the effects of heat and cold on cell multiplication should be explained as reversible heat and cold inactivation of macromolecular structures. Scherbaum (1957a) has further interpreted these data in the restricted sense of enzyme denaturation. Experiments from our laboratory (Zeuthen, unpublished data) supply direct evidence that heat slows protein synthesis (conceivably reactions A, Fig. 26). At 34°C compared to 27°C the rate of amino acid (^{14}C-histidine) incorporation into the PCA precipitable material is reduced to 32–44 per cent (two experiments with washed cells*).

* The comparison is between cells incubated with the label at 27°C for 20 min beginning 50 min after EH, and cells again heated to 34°C at 60 min after EH and incubated for 20 min beginning 50 min after the new rise in temperature.

The rate of increase in dry matter and in total phosphorus has been followed during the heat shocks (30′/30′) which induce the synchrony (Hamburger and Zeuthen, 1960). When shocking begins the net phosphorus uptake is temporarily blocked while the growth in dry weight continues for some time at a decreasing rate. The cells therefore suffer an immediate relative loss in phosphorus. This was observed to reflect also on the nucleic acid (95 per cent RNA) fraction. During continued temperature shocking, there is further relative loss because the rate of phosphorus accumulation continues to be reduced more than the rate of dry weight accumulation (Fig. 27). It was concluded (Hamburger and

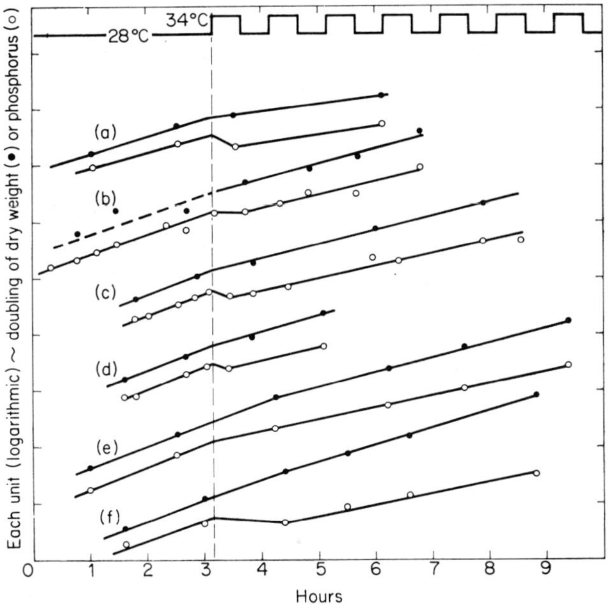

FIG. 27. *Tetrahymena pyriformis* GL. Growth curves for dry matter (●) and for total phosphorus (○) per ml, during constant and changing temperature. The phosphorus determinations are by relative radioactivity counts. Initially there are 2–6 × 10^4 cells/ml. (From Hamburger and Zeuthen, 1960.)

Zeuthen, 1960) that the relative loss in phosphorus suffered during heat shock may be an indicator of more important damage to macromolecules of which phosphorus is part.

Despite the above described results and considerations the detailed chemical nature of the 'set-back', so important for successful synchronization of *Tetrahymena*, remains obscure.

6. *Nuclear Function and Synchronous Division*

It is recalled (section II-7-B) that in *Tetrahymena* synchronized for division there is no synchrony of DNA synthesis. Synchrony of cell division is induced by a 'set-back mechanism' which operates at the level at which proteins or proteinaceous compounds essential for morphogenesis and for division are synthesized. There seems to be no 'set-back' with regard to the synthesis of DNA, and this compound therefore piles during the synchronizing heat treatment. For this reason, we now dare to say, we could not have expected DNA synthesis to be synchronized, and it is shown not to be (Cerroni and Zeuthen, 1962b). RNA also piles during the heat cycles, so we might suspect that the synthesis of this family of compounds is also not synchronized. However, we need to make separation, both at the cytological and at the chemical level, between different RNA-containing structures and chemical entities, and this has not yet been done.

We shall make reference to the results (Cerroni and Zeuthen, 1962a) demonstrated in Table V. The information supplied in this table allows us to argue (1) that the information and mechanisms for post-EH synthesis of essential 'morphogenetic substances' and/or 'division proteins' are contained in the uninhibited fraction of the RNA synthesis, or (2) the information and the mechanisms are already present at the termination of the heat treatment and do not depend on post-EH synthesis of RNA, but only on sufficient time at 28°C. We prefer the last hypothesis, because in the presence of FUDR [and fluorouracil (FU) which in a concentration of 2 m*M* is an inhibitor of ^{3}H-uridine incorporation (Cerroni and Zeuthen, 1962b) but not of division 1 (Zeuthen, 1961a)] any uninhibited RNA synthesis is likely to be fouled up by the inhibitors and thus represent a poor 'messenger' link.

So, at the moment it seems that at the end of the temperature synchronization the 'information' for protein synthesis necessary for *Tetrahymena* to perform morphogenesis and division for one or two generations is available and stored in the cell. One may guess that it is stored in RNA. RNAase (Worthington, protease free) almost erases synchronous divisions (0·1 mg/ml) when incubation is for 10–15 min after EH.* Capacity for continued random division remains (Zeuthen, unpublished data). RNAase is also active when incubation is during the last heat shock. RNAase is thus active also when the population makes no apparent preparation for new

* Inorganic: c medium.

division. Cytoplasmic protein synthesis is sensitive to RNAase, but only moderately sensitive to chloramphenicol (Mager, 1960). 'Mitochondrial' (nuclear plus mitochondrial) syntheses are insensitive to RNAase and highly sensitive to chloramphenicol (Mager, 1960). Inhibition of cell division in synchronous *Tetrahymena* follows the cytoplasmic pattern. We may therefore guess that the 'information' for morphogenesis and cell division is stored in cytoplasmic structures (ribosomes?) of synchronized *Tetrahymena*. (See also Plesner's chapter in this book.)

The stored 'information' seems to be stable to temperature changes but it can express itself only in a constant and favourable temperature environment. Synchronized cells (20′/40′) were transferred to the non-nutrient medium (after shock 6), given a seventh shock, allowed to recover for 35 min at 28°C, and then transferred to 34°C. After 3 hr at this temperature normal synchronous division develops in standard time upon transfer to 28°C, even in the presence of FUDR (10^{-4} *M*) or FU (10^{-3} *M*) (Zeuthen, unpublished data). The 'set-back' by heat is thus not likely to be with respect to, presumably nuclear, products which supply the 'information'.

The chemical nature and the location in the cell of the 'information' deserve as close future interest as do the postulated proteinaceous compounds responsible for morphogenesis and division. More information along these lines may be necessary before we can fully understand the mechanism of the 'set-back', and the mechanism by which synchrony is induced.

IV. Summarizing Discussion

Mechanisms in the synchrony can be separately discussed at the population level, and at the biochemical–biophysical level. The following are the key observations:

I. At the population level this and previous reports have described how the temperature changes induce synchrony by a differential 'set-back-in-time' mechanism. Irrespective of how far progressed towards a defined time point ('physiological transition') in advance of division, in all cells preparation in time for a division is largely undone by the temperature shocks. With respect to preparation of a forthcoming division all cells are 'set back' to become rejuvenated. After the shocks division is prepared and performed in synchrony.

II. As measured by delays of division there is no obvious difference between the effects of heat and of cold shocks. Both treatments can be used for the induction of synchrony.

III. Blockage (full or partial) of protein synthesis has the same delaying effects as has heat and cold.

IV. The three treatments cause disorganization of cortical structures which must be restored and completed before division can take place. This disorganization defines, or it reflects, an equal 'rejuvenation' of all cells in a culture previously random for division. After the treatments the cells mature structurally in synchrony, and they divide in synchrony (see the chapter by Williams, this book).

V. At the time when the last synchronizing heat shock terminates, the contents per cell of both nucleic acids is several times the normal logarithmic value. Preparation, at constant 28°C, for synchronous division occurs in the presence of agents (fluorinated pyrimidines) which reduce the rate of new synthesis of DNA (^{3}H-thymidine incorporation) and of RNA (^{3}H-uridine incorporation) to 15 per cent or less.

VI. DNA synthesis is not synchronized with cell division. At any time after EH only a fraction of the population is engaged in DNA synthesis.

VII. DNA synthesis does not take place at the time of actual cell division.

VIII. The synchronous division breaks down if short treatment is made with RNAase shortly after EH or during the last heat shock.

From these and other observations follow the suggestions:

1. That the preparation of synchronous division, after the temperature treatments, is by synthesis of proteinaceous compounds.

2. During the synchronization procedure the accumulation of specific (thus far unknown) proteinaceous compounds is continuously being reversed ('set back') by the temperature changes with the result that after the temperature shifts (together the 'shocks') the cells are low in, or empty of, specific compounds, which control morphogenesis and division.

3. The nucleic acids (DNA and RNA) are synthesized in the changing temperature environment. In this respect there is no apparent set-back; hence no synchrony is to be expected. For DNA synthesis the absence of synchrony is observed. In the individual cell an alternation between periods of synthesis and non-synthesis of DNA, characteristic of cells from the logarithmic population, may be preserved, perhaps in a slightly twisted form, during the heat shocks and up to a short time before synchronous division. [During the synchronous divisions there is further complication (see 7, below).]

4. While synthesis of DNA and of RNA are conditions for divisions, they are unlikely to be immediate conditions for the synchronous divisions. During the heat shocks the cells have over-matured for division with respect to the amounts they contain of both nucleic acids. After the heat shocks they are to a considerable degree independent, with regard to division, of new nucleic acid synthesis. 'Information' for synchronous division may be stored in RNA.

5. It is only in a constant and favourable temperature environment that the proposed deposited 'information' for the syntheses of specific proteinaceous compounds can express itself and thus produce synchronous morphogenesis and synchronous nuclear and cellular division. Unlike its expression, the 'information' is stable to heat.

6. The temperature shocks act by dissociating the division cycle from the DNA cycle. After the heat shocks the division cycle is synchronized and it is abnormally fast. The DNA cycle is asynchronous and may essentially have preserved the slower logarithmic rate. While all cytoplasmic structural events thus far studied follow the synchronous cycle, it remains to be seen how many nuclear structural and biochemical events follow the cytoplasmic, and how many the DNA cycle. If the 'information' for morphogenesis and for preparation of division is deposited in RNA, then synchronization is only for biochemical and structural events coupled to steps of protein synthesis distal to a late or final step in RNA synthesis.

7. Synchronized cellular events exert control over the asynchronous synthesis of DNA, by interrupting it around division. In the synchronized system the possibility of a reverse immediate control—DNA synthesis triggers division—seems to be eliminated: asynchronous DNA synthesis cannot signal synchronous behaviour of the whole population. DNA synthesis may also in many other cells (see Prescott, this book) normally be limited to periods between divisions, simply because synthesis during division is not possible.

8. Since a nuclear (DNA) and a general (division) cycle, normally coupled, can be separated and made to run at different rates, independent control of the two cycles must be possible. No attempt will be made here to discuss what are the inherent mechanisms controlling the DNA cycle. This problem is presently intensely studied, particularly by the bacteriologists. Here we shall only comment that in a higher cell (*Tetrahymena*) the DNA cycle may be interfered with by signals presumably emitted by the

general cycle. What controls this general division cycle? In mitotic cells division is controlled by extranuclear structures (centrioles, or centrioles plus surrounding island of cytoplasm: centrosome). In the ciliates the structural analogue of the centriole is, and the evolutionary homologue may be, the kinetosome. The kinetosomal apparatus (kineties plus oral structures) is the sum of kinetosomes and underlying orienting structures. Preceding division there is longitudinal growth of the kineties and formation of a new mouth lateral to kinety 1. Both the longitudinal and the lateral growth of the kinetosomal apparatus may be immediate conditions for division, at least in synchronized *Tetrahymena*. The *longitudinal* growth occurs during the period of heat synchronization, so the attached kinetosomes and the underlying *linear* structure must be stable even when growing. The cell therefore in part saves itself the trouble of performing longitudinal growth of the kineties after the synchronization by heat, and this may explain why the synchronous generations are short. Certain developments at the level of the second mouth, occurring between the temperature shocks, are undone by the shocks. We may propose that an underlying orienting structure forms *lateral* to a kinety and that it is unstable while being made. These observations and suggestions may fully or in part account for the induction of the synchrony by 'set-back-in-time' mechanisms. Undoubtedly, both the kinetosomes and the underlying structures contain proteins. So it makes sense to say that division of the synchronized cells is prepared by immediate protein synthesis. The damage done by the temperature shocks may be more general than suggested in this paragraph, but we have no evidence.

V. Conclusions

Studies with the synchronized *Tetrahymena* system have demonstrated dissociability between a cycle in which DNA is alternatively synthesized and not synthesized, and another cycle in which morphogenesis and cell division alternate. In the normal logarithmic cell the two cycles may also be separate, although they seem to be closely coupled, as evidenced by the observation of other workers that DNA synthesis normally occurs between divisions. The coupling of the two cycles (Fig. 28, I) is likely to be mediated by products (DNA and RNA) produced (arrows F) in the first cycle and partitioned (DNA) by, and (RNA) perhaps also used for, the process of cell division which is in the second cycle. Even in the 'normal' cell, moderate dissociability of the two cycles is to be expected and should be looked for. Even under 'normal' environmental conditions populations, more frequently than not, regulate towards a new, usually smaller, average cell size.

The rather dramatic temperature variations used to synchronize *Tetrahymena* populations represent major environmental disturbances. These disturbances stop the division cycle but let the DNA cycle run more or less undisturbed (Fig. 28, II). DNA and RNA continue to pile up. Upon return of the system to a constant and favourable temperature the block of the division cycle is released (Fig. 28, III)

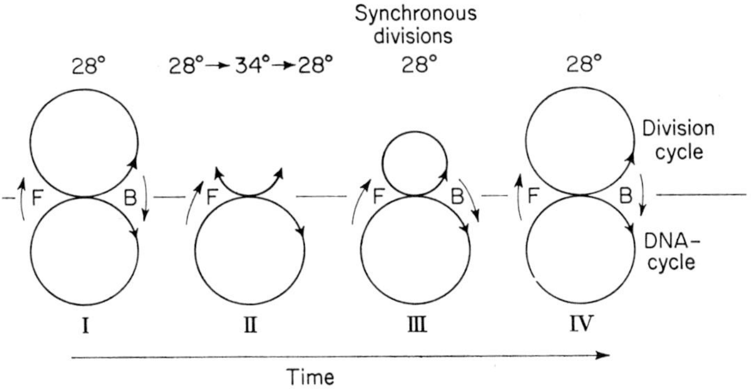

FIG. 28. For explanation, see text.

and products piled in the DNA cycle are partitioned synchronously (DNA) to new cells; perhaps (RNA) they are also consumed in the metabolism which makes the division cycle function. In any event, in this metabolism the accumulated RNA, rather than newly-formed RNA, seems to play a major role in controlling syntheses of proteinaceous compounds necessary for morphogenesis and cell division. The fast synchronous division rate (the circles represent time scales) reduces the amount of DNA and RNA per cell to lower and lower values with the result that for continued division, the cells untimately become dependent on new nucleic acids produced in the DNA cycle. This nuclear 'feeding' control over the cytoplasm brings the two cycles back in phase (Fig. 28, IV), at the same time as the system returns to logarithmic growth and division. The return is then from the fast synchronous to the slower and asynchronous division rate set by the DNA cycle which was never synchronized. The simplest suggestion we can at present make is that the proteinaceous compounds which control the division help make the kinetosomal apparatus, and that this is a condition for the apparatus, upon its completion, to emit the signal, or let a signal pass, to the macronucleus and to the cell body to divide. This signal passes right through to the DNA cycle over which it exerts temporary blocking control (Fig. 28, arrows B).

We have thus suggested that a macronuclear DNA cycle exerts (sometimes rather remote) control over a separate, essentially cytoplasmic (morphogenetic and division) cycle in which protein syntheses are central and also that this second cycle exerts control over the macronuclear cycle which involves alternations between synthesis and non-synthesis of DNA. The first control (F for 'feeding') is positive and stimulatory. The second (B for blocking) is negative and blocking.

Referring to the introductory comments, p. 100, we may finally suggest that the synchronized *Tetrahymena* system resembles, more closely than expected, the fertilized egg which it was meant to supplement as an experimental object. The egg cleaves at a rate which is determined by the cytoplasm, not by the fertilization nucleus. The 'information' for cleavage is thus deposited by the mother, maybe in RNA. Cleavage of the egg (Brachet, 1957) is highly sensitive to the action of RNAase as is synchronous division in *Tetrahymena.*

REFERENCES

BRACHET, J. (1957). *Biochemical Cytology.* Academic Press, New York.

CERRONI, R. E. and E. ZEUTHEN (1962a). Inhibition of macromolecular synthesis and of cell division in synchronized *Tetrahymena. Exptl. Cell Research* **26,** 604.

CERRONI, R. E. and E. ZEUTHEN (1962b). Asynchrony of nuclear incorporation of tritiated thymidine into *Tetrahymena* cells synchronized for division. *Compt. Rend. Trav. Lab. Carlsberg* **32,** 499.

CHRISTENSSON, E. (1959). Changes in free amino acids and proteins during cell growth and synchronous division in mass cultures of *Tetrahymena pyriformis. Acta Physiol. Scand.* **45,** 339.

CHRISTENSSON, E. (1962). Different RNA fractions during cell growth and synchronous division in mass cultures of *Tetrahymena pyriformis. Acta Physiol. Scand.* **54,** 1.

ELLIOTT, A. M. (1959). Biology of *Tetrahymena. Ann. Rev. Microbiol.* **13,** 79.

ELLIOTT, A. M., J. R. KENNEDY and IL JIN BAK (1962). Macronuclear events in synchronously dividing *Tetrahymena pyriformis. J. Cell Biol.* **12,** 515.

FRANKEL, J. (1962). The effects of heat, cold and *p*-fluorophenylalanine in synchronized *Tetrahymena pyriformis* GL. *Compt. Rend. Trav. Lab. Carlsberg* **33,** 1.

HAMBURGER, K. (1962). Division delays induced by metabolic inhibitors in synchronized cells of *Tetrahymena pyriformis. Compt. Rend. Trav. Lab. Carlsberg* **32,** 359.

HAMBURGER, K. and E. ZEUTHEN (1957). Synchronous divisions in *Tetrahymena pyriformis* as studied in an inorganic medium. *Exptl. Cell Research* **13,** 443.

HAMBURGER, K. and E. ZEUTHEN (1960). Some characteristics of growth in normal and synchronized populations of *Tetrahymena pyriformis. Compt. Rend. Trav. Lab. Carlsberg* **32,** 1.

Holz, G. G., Jr. (1960). Structural and functional changes in a generation in *Tetrahymena*. *Biol. Bull.* **118,** 84.

Holz, G. G., Jr. (1961). Actions of inhibitors of lipid metabolism on synchronized *Tetrahymena pyriformis* GL. *Proc. 1st Int. Conf. of Protozool. Prague.*

Holz, G. G., Jr, O. H. Scherbaum and N. E. Williams (1957). The arrest of mitosis and stomatogenesis during temperature induction of synchronous division in *Tetrahymena pyriformis* (Mating Type I Variety 1.) *Exptl. Cell Research* **13,** 618.

Holz, G. G., Jr., L. Rasmussen and E. Zeuthen (1963). Normal versus synchronized division in *Tetrahymena pyriformis*. A study with metabolic analogs. *Compt. Rend. Trav. Lab. Carlsberg* **33,** 289.

Iverson, R. M. and A. C. Giese (1957). Nucleic acid content and ultraviolet susceptibility of *Tetrahymena pyriformis*. *Exptl. Cell Research* **13,** 213

Johnson, F. H. and I. Lewin (1946). The growth rate of *E. coli* in relation to temperature, guanine, and coenzyme. *J. Cellular Comp. Physiol.* **28,** 47.

Kidder, G. W. and V. C. Dewey (1951). The biochemistry of ciliates in pure culture. In *Biochemistry and Physiology of Protozoa* (A. Lwoff, ed.), Vol. I. Academic Press, New York, pp. 323–400.

Lwoff, A. (1923). Sur la nutrition des infusoires. *Compt. Rend. Acad. Sci., Paris* **176,** 928.

Lwoff, A. (1950). *Problems of Morphogenesis in Ciliates*. John Wiley, New York.

Mager, J. (1960). Chloramphenicol and chlortetracycline inhibition of amino acid incorporation into proteins in a cell-free system from *Tetrahymena pyriformis*. *Biochim. et Biophys. Acta* **38,** 150.

Mazia, D. (1961). Mitosis and the physiology of cell division. In *The Cell* (J. Brachet and A. E. Mirsky, eds.), Vol. III. Academic Press, New York.

McDonald, B. B. (1958). Quantitative aspects of deoxyribose nucleic acid (DNA) metabolism in an aminonucleate strain of *Tetrahymena*. *Biol. Bull.* **114,** 71.

Prescott, D. M. (1957). Relation between multiplication rate and temperature in *Tetrahymena pyriformis*, strains HS and GL. *J. Protozool.* **4,** 252.

Prescott, D. M. (1959). Variations in the individual generation times of *Tetrahymena* cells. *Exptl. Cell Research* **16,** 279.

Prescott, D. M. (1960). Relation between cell growth and cell division. IV. The synthesis of DNA, RNA, and protein from division to division in *Tetrahymena*. *Exptl. Cell Research* **19,** 228.

Rasmussen, L. (1963). Delayed divisions in *Tetrahymena* as induced by short-time exposures to anaerobiosis. *Compt. Rend. Trav. Lab. Carlsberg* **33,** 53.

Rasmussen, L. and E. Zeuthen (1962). Cell division and protein synthesis in *Tetrahymena* as studied with *p*-fluorophenylalanine. *Compt. Rend. Trav. Lab. Carlsberg* **32,** 333.

Sato, H. (1960). A morphological study on the mitochondria of *Tetrahymena geleii* W. *Anat. Record* **138,** 381.

Scherbaum, O. (1956). Cell growth in normal and synchronously dividing mass cultures of *Tetrahymena pyriformis*. *Exptl. Cell Research* **11,** 464.

Scherbaum, O. (1957a). Studies on the mechanism of synchronous cell division in *Tetrahymena pyriformis*. *Exptl. Cell Research* **13,** 11.

SCHERBAUM, O. (1957b). The content and composition of nucleic acids in normal and synchronously dividing mass cultures of *Tetrahymena pyriformis*. *Exptl. Cell Research* **13,** 24.

SCHERBAUM, O. H., T. W. JAMES and T. L. JAHN (1959). The amino acid composition in relation to cell growth and cell division in synchronized cultures of *Tetrahymena pyriformis*. *J. Cellular Comp. Physiol.* **53,** 119.

SCHERBAUM, O. H. and M. Levy. (1961). Some aspects of the carbohydrate metabolism in relation to cell growth and cell division. *Pathol. Biol.* **9,** 514.

SCHERBAUM, O. H., A. L. LOUDERBACK and T. L. JAHN (1958). The formation of subnuclear aggregates in normal and synchronized protozoon cells. *Biol. Bull.* **115,** 269.

SCHERBAUM, O. H., A. L. LOUDERBACK and T. L. JAHN (1959). DNA-synthesis, phosphate content and growth in mass and volume in synchronously dividing cells. *Exptl. Cell Research* **18,** 150.

SCHERBAUM, O. and E. ZEUTHEN (1954). Induction of synchronous cell division in mass cultures of *Tetrahymena pyriformis*. *Exptl. Cell Research* **6,** 221.

SCHERBAUM, O. and E. ZEUTHEN (1955). Temperature-induced synchronous divisions in the ciliate protozoon growing in synthetic and proteose–peptone media. *Exptl. Cell Research Suppl.* **3,** 312.

SULLIVAN, W. D. and J. T. SPARKS (1961). The effect of ultraviolet radiation on succinic dehydrogenase activity during different stages of cell division. *Exptl. Cell Research* **23,** 436.

SWANN, M. M. (1954). The control of cell division. In *Recent Developments in Cell Physiology* (J. A. Kitching, ed.). Butterworths, London, p. 185.

TARTAR, V. (1961). *The Biology of* Stentor. Pergamon Press, Oxford–London–New York–Paris.

THORMAR, H. (1956). Temperaturens og temperaturændringers indflydelse på celleformering og cellevækst hos ciliaten *Tetrahymena pyriformis* (Danish); Paper submitted to the Faculty of Science, the University of Copenhagen.

THORMAR, H. (1959). Delayed division in *Tetrahymena* induced by temperature changes. *Compt. Rend. Trav. Lab. Carlsberg* **31,** 207.

THORMAR, H. (1962). Effect of temperature on the reproduction rate of *Tetrahymena pyriformis*. *Exptl. Cell Research* **28,** 269.

WILLIAMS, N. E. and O. H. SCHERBAUM (1959). Morphogenetic events in normal and synchronously dividing *Tetrahymena*. *J. Embryol. Exptl. Morphol.* **7,** 241.

ZEUTHEN, E. (1953). Growth as related to the cell cycle in single-cell cultures of *Tetrahymena pyriformis*. *J. Embryol. Exptl. Morphol.* **1,** 239.

ZEUTHEN, E. (1958). Artificial and induced periodicity in living cells. *Advances in Biol. and Med. Phys.* **6,** 37.

ZEUTHEN, E. (1961a). Synchronized growth in *Tetrahymena* cells. In *Growth in Living Systems* (M. X. Zarrow, ed.). Basic Books, New York, pp. 135–158.

ZEUTHEN, E. (1961b). Cell division and protein synthesis. In *Biological Structure and Function II* (T. W. Goodwin and O. Lindberg, eds.). Academic Press, New York, p. 537.

ZEUTHEN, E. and O. SCHERBAUM (1954). Synchronous divisions in mass cultures of the ciliate protozoon *Tetrahymena pyriformis*, as induced by temperature changes. In *Recent Developments in Cell Physiology* (J. A. Kitching, ed.). Butterworths, London, p. 141.

5

Structural Development in Synchronously Dividing *Tetrahymena pyriformis*

NORMAN E. WILLIAMS, *Department of Zoology, State University of Iowa, Iowa City, Iowa*

CONTENTS

I	Introduction	159
II	Morphology of *Tetrahymena pyriformis*	159
III	Arrest and Reactivation of Morphogenesis	164
IV	Primordium Resorption and Stabilization	168
V	General Problems of Cortical Differentiation	170
VI	Nucleus and Endoplasm	172
VII	Concluding Remarks	173
	References	173

I. INTRODUCTION

Induced division synchrony in *Tetrahymena pyriformis* has renewed interest in problems of structural development in this ciliated protozoan. In *Tetrahymena*, as in other ciliates, there is associated with each division cycle a precise duplication of a highly-ordered, species-specific cytoarchitecture. The involved processes are of interest in themselves, and synchronized cells present certain advantages for this type of study. But of perhaps even greater interest is the indication from studies to date that the division-limiting processes which are affected in synchronized *Tetrahymena* are intimately associated with morphogenetic processes. Before discussing this it is necessary to consider briefly the morphology of the normal cell.

II. MORPHOLOGY OF *Tetrahymena pyriformis*

Tetrahymena is typically pear-shaped, with an oral apparatus located near the narrow anterior end, and a general body ciliation consisting of meridianal rows of cilia (Fig. 1). Other cortical structures are the cytoproct, located near the posterior end on the ventral body surface, and the two contractile vacuole pores lying near the posterior end on the animal's right side. Within the cell there is a roughly spherical macronucleus, and in some strains of *T. pyriformis* there is a small spherical micronucleus as well (Fig. 2). For a complete description of *Tetrahymena*, as well as a review of the older literature, the reader is referred to Furgason (1940).

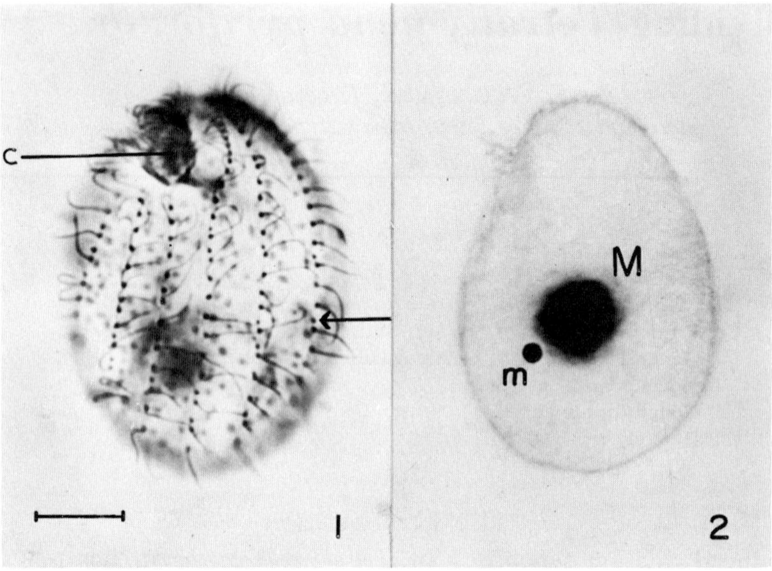

Fig. 1. The major morphological features of *Tetrahymena pyriformis* as revealed by light microscopy. Protargol preparation showing the oral apparatus (C) and the ciliary rows in *T. pyriformis* strain GL. Note the kinetosomes (basal bodies) without cilia found in various regions of the body (arrow). The length of the bar equals 10 μ.

Fig. 2. The major morphological features of *T. pyriformis* as revealed by light microscopy. Feulgen preparation showing the nuclear condition in *T. pyriformis* strain WH-6. Note the large macronucleus (M) and the smaller micronucleus (m). (Photo by courtesy of Dr. George G. Holz, Jr.)

The somatic ciliary rows are integrated morphological units called 'kineties' (Chatton and Brachon, 1935). Each kinety consists of a row of cilium-bearing basal granules, called 'kinetosomes', which are spaced along and to the left side of a common filament, the 'kinetodesma'. This type of structure has been confirmed with the electron microscope (Metz and Westfall, 1954; Pitelka, 1961; Suhama and Yamataka, 1960b). Metz and Westfall (1954) made the observation that the common kinetodesmal fiber is made up of a series of overlapping, tapered kinetodesmal segments, one of which is associated with each kinetosome. Thus the kinety is an assemblage of discrete units, the 'kinetids' (Fauré-Fremiet, 1961), each consisting of a cilium, a kinetosome, and a tapered kinetodesmal segment (Fig. 3). All parts of the kinetid, as well as other features, may be seen in electron micrographs of thin-sectioned *Tetrahymena* (Figs. 4 and 5). In *Tetrahymena* there are eighteen ciliary rows, or kineties, in a typical animal.

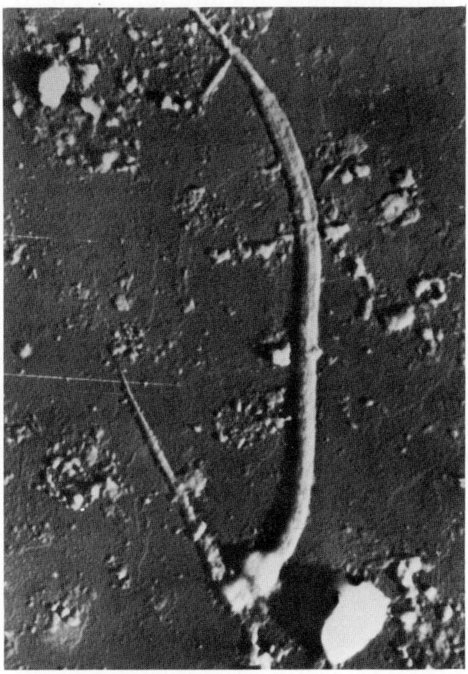

Fig. 3. The major morphological features of *T. pyriformis* as revealed by electron microscopy. The isolated 'kinetid', or unit of kinety structure, prepared by the shadow-cast technique. Note the long cilium, the kinetosome, and the tapered kinetodesmal segment. The swelling above the kinetosome is an artifact. (Photo by courtesy of Dr. Charles B. Metz.)

When *Tetrahymena* is stained according to any of the widely used silver nitrate impregnation techniques, cilia are not stained, but, among other things revealed, kinetosomes and sinuous fiber-like structures called 'silverline fibers' appear. Chatton and Brachon (1935) early recognized that the silverline fibers were independent of the kineties. This has recently been confirmed by electron microscopists, who find that the silverline fibers are pellicular in nature (Pitelka, 1961; Suhama and Yamataka, 1960a). Williams and

Scherbaum (1959) stained *Tetrahymena* with protargol and found that it reveals both cilia and kinetosomes while leaving the kinetodesmata and silverline fibers unstained. Both types of silver staining are useful, but neither reveals all aspects of the kineties.

The oral apparatus consists primarily of four compound ciliary structures, each having a membranous nature, which are associated with a cavity leading into the cell's interior. In Fig. 6 the kinetosomal bases of the four ciliary groupings are indicated. Three of the four groups of cilia constitute a primitive 'adoral zone of membranelles'

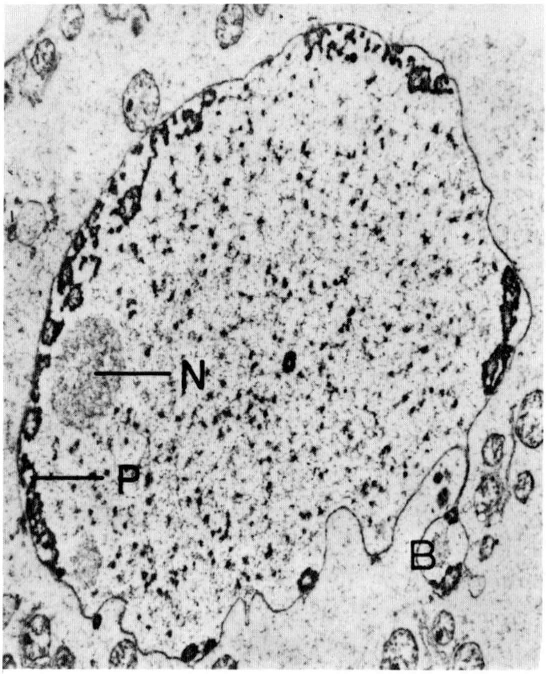

FIG. 4. The major morphological features of *T. pyriformis* as revealed by electron microscopy. Macronucleus, showing peripheral nucleoli (P), central RNA bodies (N), and nuclear bleb (B). (Unpublished photo by courtesy of Dr. Alfred M. Elliott.)

(Furgason, 1940), found along the (animal's) left wall of the buccal cavity. The 'undulating membrane' is a single row of cilia found along the right posterior margins of the buccal overture. A group of 'buccal ridges' which stain with silver nitrate are found adjacent to the undulating membrane and extend down the right wall of the buccal cavity. The 'cytostome' is found at the bottom of the buccal cavity, and below this lies the 'cytopharynx'. Food vacuoles form at the end of the cytopharynx. Recently, fine fibers have been noted in the wall of the buccal cavity and extending past the cytostome and cytopharynx into the endoplasm. These fibers are apparently 'trichites' (Furgason, unpublished data). Although some work with the electron microscope has been reported (Metz and Westfall, 1954; Suhama and Yamataka, 1960b; Williams *et al.*, 1960), the ultrastructure of the *Tetrahymena* oral apparatus is not completely understood.

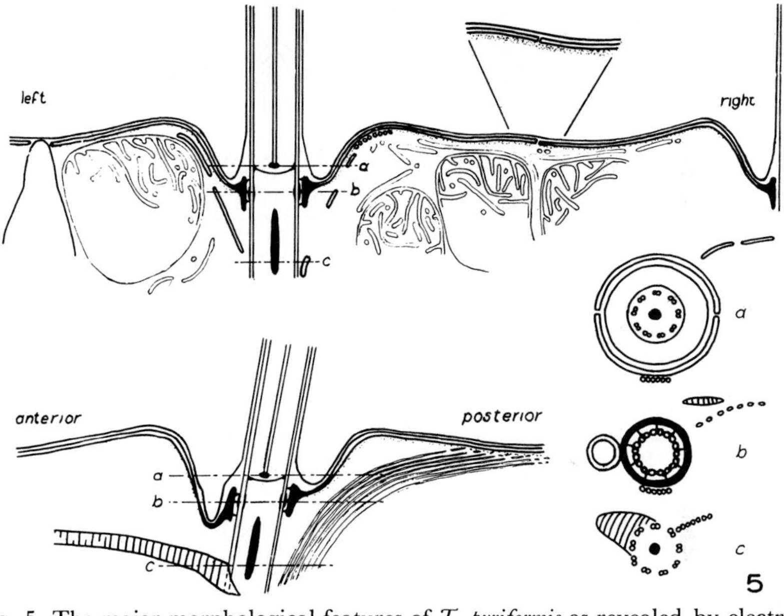

FIG. 5. The major morphological features of *T. pyriformis* as revealed by electron microscopy. Diagrammatic reconstruction of the fine structure of cortical structures in *Tetrahymena* and related ciliates. (By courtesy of Dr. Dorothy R. Pitelka.)

The macronucleus of *Tetrahymena* consists of a nuclear membrane, chromatin granules suspended in nucleoplasm, and numerous Feulgen-negative basophilic bodies located primarily adjacent to the nuclear membrane (Furgason, 1940). The peripheral bodies have been studied recently and they are apparently nucleoli (Elliott *et al.*, 1962; Roth and Minick, 1961). In addition to these, Elliott and coworkers (1962) have found central 'RNA bodies' (Fig. 4) which consist of RNA without protein. Electron micrographs of the micronucleus reveal a membrane, an inner dense mass of chromosomes, and a clear peripheral zone (Elliott *et al.*, 1962; Roth and Minick, 1961).

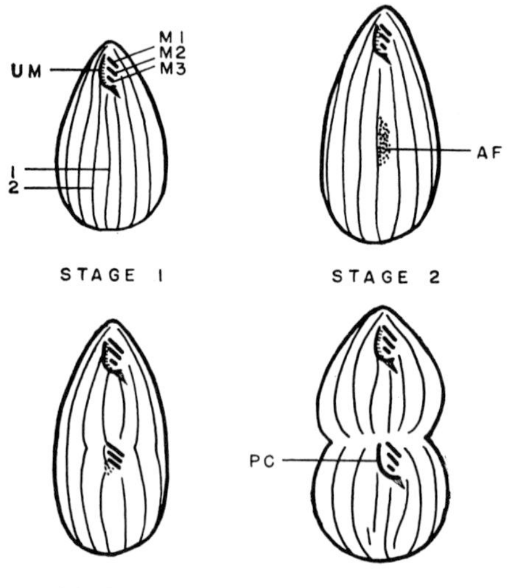

FIG. 6. Division stages of *Tetrahymena*. The four oral 'membranes' are indicated in the drawing of stage 1; M1, M2, and M3 are the threc membranelles, and UM represents the undulating membrane. Stage 2 is marked by the appearance of the oral primordium, the 'anarchic field' of kinetosomes (AF). In stage 3 the primordium develops membranelles, and in stage 4 the oral apparatus completes development (PC) and cleavage occurs.

III. Arrest and Reactivation of Morphogenesis

Structural changes occur in a regular way throughout the entire cell cycle in *Tetrahymena*. During the early part of the cell cycle, the kineties lengthen by means of an increase in the number of kinetids in each. Sometime later, the buccal apparatus of the posterior daughter begins to form (Fig. 6). The initial event in this process is the development of an 'anarchic field' of non-ciliated kinetosomes in the equatorial region of the cell immediately to the left of meridian 1 (the so-called 'stomatogenic meridian'). Sometime late in anarchic field formation, the micronucleus (in micronucleate strains) enters prophase. Next the kinetosomes group into four bands and cilia grow out from them, forming the four oral 'membranes'. Near the end of this period, the macronucleus begins to divide amitotically. Then cytoplasmic constriction and buccal cavity formation take place, after which the daughter cells separate.

The synchronizing treatment* affects this morphogenetic sequence in a specific way. Working with *T. pyriformis* strain WH-6 (mating type I, variety 1), Holz made the observation that, immediately after the synchronizing treatment, all cells had anaphase-like micronuclei (Fig. 7), and the writer discovered that all cells were in the anarchic field stage of stomatogenesis (Fig. 8) (Holz *et al.*, 1957). This condition persisted unchanged for 40–50 min after treatment, at which time micronuclear division and stomatogenesis resumed in all cells. Light microscope studies were subsequently made of the entire sequence of structural changes associated with the cell cycle in synchronously dividing cells. Macronuclear amitosis, anarchic field formation, membranelle and undulating membrane development, contractile vacuole pore formation, and cytoplasmic cleavage were studied in strain GL by Williams and Scherbaum (1959), and in strain WH-6 by Holz (1960); kinety growth and buccal cavity formation were analyzed in strain GL (Williams and Scherbaum, 1959); and micronuclear mitosis was studied in strain WH-6 (Holz, 1960). These studies showed that cells subjected to the synchronizing treatment undergo a general morphogenetic arrest, in which the normal sequence of development is temporarily interrupted at a characteristic point in the cell cycle. Heat synchronization has no differential effect on morphogenesis such that any particular event progresses in time beyond the others, and the structural changes which occur in treated cells prior to the block, and subsequent to continuation of development in the synchronous division, are in all other respects identical with these processes in untreated cells.

Perhaps the simplest hypothesis by which to explain the block to morphogenesis and the subsequent reactivation of developmental processes in heat-synchronized *Tetrahymena* is the one originally suggested by Holz, Scherbaum and Williams (1957). According to this view, the heat treatment would inactivate some substance synthesized in the early part of the cell cycle, which, after accumulating to a threshold level, interacts with the oral and nuclear elements just prior to membranelle formation and chromosomal anaphase. The postulated interaction would then initiate kinetosomal and chromosomal movements necessary to the continued development of the cell, and without which, therefore, morphogenesis stops. Originally the heat-sensitive substance was thought of as enzymatic, but it seems at least as likely that it might be a structural material (cf. Zeuthen, 1958).

* 30-min heat shocks separated by 30 min of recovery at the optimum temperature.

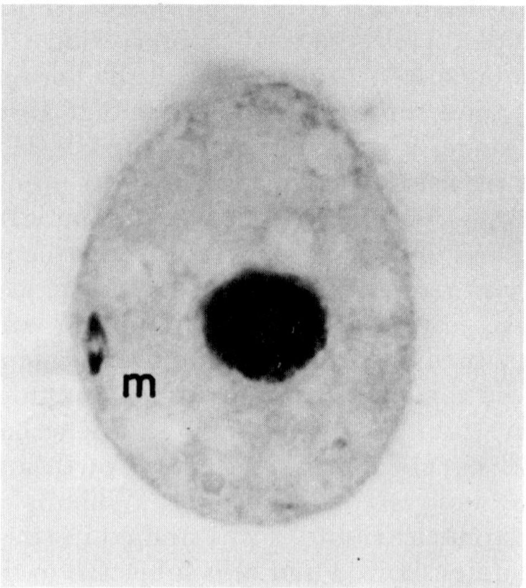

FIG. 7. A Feulgen-stained preparation of a cell immediately after the heat-synchronization procedure. Note the micronucleus (m), arrested in anaphase. (Unpublished photo by courtesy of Dr. George G. Holz, Jr.)

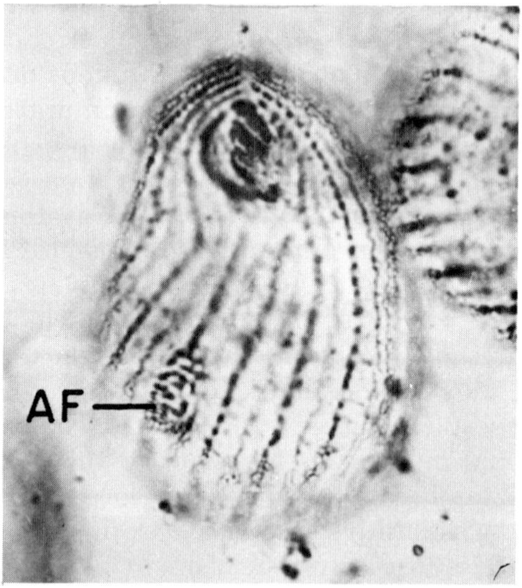

FIG. 8. A silver-impregnated specimen showing morphogenesis arrested in the anarchic field (AF) stage by the heat-synchronization procedure. (Unpublished photo by the author.)

An early suggestion of what physiological processes might be involved in the reactivation of morphogenesis in blocked cells came from Zeuthen's laboratory. As described in chapter 4 of this volume, Rasmussen and Zeuthen (1962) found evidence for a 'division protein', sensitive to certain environmental changes in the early phases of the cell cycle, which becomes insensitive at the 'physiological transition point'. Apparently it is this transition that commits the cell to division. It appeared distinctly possible that events associated with the physiological transition might be involved in reactivation of the interrupted morphogenetic sequence in treated cells. This is *not* so, as is clearly shown by Frankel (1962). He observed the development of the oral primordium in strain GL and found that the cells develop membranelles *prior* to the transition point. Thus, although division protein might be involved, the reactivation of morphogenesis must be triggered by some 'pre-transition' alteration.

Recently Nachtwey and Giese (personal communication) produced evidence for a second critical physiological step in the preparation for division which is independent of the physiological transition, and occurs prior to it. They show that the synchronous division in inorganic medium will be delayed by a fixed amount if the cells are exposed briefly to ultraviolet radiation (115 ergs/mm^2) at any time prior to about 30 min after the synchronizing treatment. About this time the cells become insensitive to ultraviolet, and radiation delivered after about 37 min will not cause division delay. Frankel (1962) has shown that the initiation of membranelle formation in inorganic medium occurs at 40 min after the synchronizing treatment. The close correlation in time between these two events has suggested to Nachtwey (unpublished data) that the physiological changes associated with the transition to ultraviolet insensitivity may be responsible for reactivation of morphogenesis in blocked cells. One difficulty with this interpretation lies in the slight difference in the timing of these two events; the transition is over after 37 min and morphogenesis apparently resumes at 40 min. It might be suggested that morphogenesis actually begins earlier than 40 min but is not detected by light microscope techniques until this time.

Assuming that physiological processes associated with the transition to ultraviolet insensitivity are involved in morphogenesis, the problem is to discover the nature of these processes. A possible mechanism has been suggested by Nachtwey and Giese (personal communication). As a result of studies of the relation of ultraviolet shocks to temperature shocks at various times after synchronization, they suggest that an 'ultraviolet-sensitive substance' interacts with the temperature-sensitive substance, i.e. with division protein, about 30 min after synchronization. The transition to ultraviolet insensitivity and the reactivation of morphogenesis, both occurring at this time, would then presumably result from this interaction. There is some evidence that ultraviolet may directly affect the anarchic field kinetosomes, and the model in its simplest form becomes an interaction of division protein with kinetosomes. This appears to be an hypothesis of some promise. The work was done with amicronucleate *Tetrahymena*, and it would be interesting to see if the behavior of the micronucleus in micronucleate strains could be brought under this concept.

IV. Primordium Resorption and Stabilization

In Zeuthen's laboratory, Frankel (1962) has systematically studied the morphogenetic effects of temperature shocks and exposure to *p*-fluorophenylalanine on synchronized *Tetrahymena*. He has discovered that oral primordia will be either inhibited or resorbed if cells are shocked prior to a 'stabilization point', but will not be affected if shocked after this particular point in the cell cycle. Furthermore, he has established that the stabilization point corresponds exactly with the physiological transition point for the postulated division protein of Rasmussen and Zeuthen.

In the heat-shock experiments it was found that a 5–7 min exposure was required before response was obtained. The responses were of three types, depending upon the extent of primordium development at the time the heat shock took effect. In the first type of response, found in cells which had not yet resumed morphogenesis at the time shock, the onset of this differentiation was delayed but the primordium was otherwise unaffected. In the second type of response, found in cells shocked after morphogenesis had resumed but prior to the time of primordium stabilization, the oral primordium was resorbed and replaced by redifferentiation of a new anarchic field (Fig. 9). According to Frankel, the stabilization point occurs at the time when the three membranelles are fully formed and the undulating membrane is completing its differentiation. The third type of response, which is to ignore the shock, is shown by cells which have passed beyond this point when the heat shock takes effect.

One unusual result was noted. In inorganic media, but not in nutrient media, there was resorption and replacement of oral primordia in some cells which presumably had not yet initiated membranelle formation, i.e. had not resumed morphogenesis. If the resorption response is peculiar to organizing primordia, as indicated above, it may be that late anarchic field stages actually begin to differentiate slightly before this can be detected in the light microscope. Such 'cryptic organization' was postulated above for another reason, and here it would explain the resorption response in a certain number of 'anarchic field' stages. It is not clear why this should occur in inorganic media but not in nutrient media.

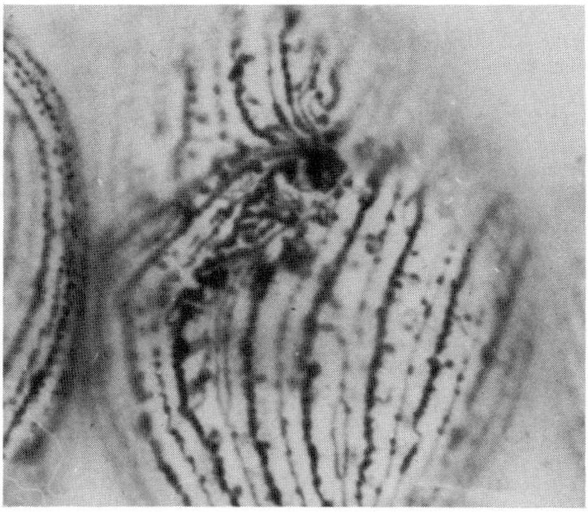

FIG. 9. A silver-impregnated specimen showing the oral apparatus resorbing due to the application of a heat shock. The differentiated membranelles will regress if heat shock is administered just prior to the stabilization point (see text for details). (Unpublished photo by courtesy of Dr. Joseph Frankel.)

The same three basic patterns of response were found when cells were subjected to cold shocks. Some special features should be noted, however. When cold shocks were given, morphogenesis stopped immediately rather than after a delay. Also, in cells which responded by primordium resorption, the cold treatment required 20 min to trigger the process. Thus with cold shocks, but not with heat shocks, halting of morphogenesis and triggering of resorption are separable in time. Frankel points out that this implies an indirect action of cold in bringing about resorption, and Nachtwey's unpublished finding that division delay, which is presumably accompanied by resorption, is developed more quickly at 9°C than at 0·5°C supports this interpretation.

The possibility of a relation between the division protein of Rasmussen and Zeuthen and morphogenetic events prompted Frankel to study the effect of *p*-fluorophenylalanine on oral primordium development. He found that, although low concentrations required longer to act than higher concentrations, the responses to brief exposures to this analog are not different from the responses to heat shock. Cells exposed early delay primordium reactivation, cells exposed somewhat later will resorb and replace the oral primordium, and cells beyond the stabilization point ignore the analog. The fact that diverse agents produce identical effects suggest that in all cases one underlying mechanism is involved. Further, the fact that *p*-fluorophenylalanine is one of these agents suggests that division protein is directly concerned with primordium reactivation, with primordium resorption, and with primordium stabilization. This is consistent with the conclusion of the previous section that interaction of division protein with anarchic field kinetosomes may reactivate blocked primordia in synchronized cells. Both Nachtwey (unpublished data) and Frankel (1962) have suggested what may happen after that. For a short time after reactivation, division protein is thought to enter into mouth-formation processes. During this time effective shocks may act by disturbing reactions supplying the division protein, which in turn would bring the resorptive mechanism into play. If division protein supply is adequate throughout this period, the cells reach stabilization and shocks are without effect. What underlying changes occur at stabilization are obscure, but it is significant that the time when this occurs is precisely the time of physiological transition for the division protein. The two events are likely aspects of the same fundamental process. The morphogenetic work was done on amicronucleate strains of *Tetrahymena*; the question of the behavior of the micronucleus in experiments of the type described remains.

V. General Problems of Cortical Differentiation

Central to the problems of structural development in *Tetrahymena* both untreated and synchronized, as well as in ciliates generally, is the problem of the origin of new kinetosomes. In *Tetrahymena* new kinetosomes are produced in the kineties throughout the early part of the cell cycle prior to oral primordium formation, thus increasing kinety length as the cell grows (Williams and Scherbaum, 1959). Formation of the anarchic field during stomatogenesis also involves kinetosome increase. Several theories for the production of new kinetosomes have been proposed (Ehret and Powers, 1959; Klein, 1927), but the one generally accepted is that kinetosomes are self-reproducing cytoplasmic particles; it is further accepted that they give rise to cilia and to other types of organelles (Chatton, Lwoff and Lwoff, 1929, 1931). Evidence that kinety growth in *Tetrahymena pyriformis* (*Glaucoma piriformis*) involves kinetosome duplication and associated ciliary synthesis was found by Chatton and Brachon (1935). Their cytological preparations suggest that the somatic kinetosomes divide parallel to the long axis of the cell, giving rise to two daughter kinetosomes, one of which retains the old cilium. The daughter kinetosomes then move apart, and the non-ciliated member generates a new cilium. A study of protargol preparations has revealed what appear to be all stages in this pattern of kinety growth (Williams, unpublished data). Some of the stages are illustrated in Fig. 10. Unfortunately, no information is available on the development of the kinetodesmata, but it is probable that new kinetodesmal segments are formed in association with somatic kinetosome duplication and ciliary synthesis. It also seems likely that the kinetodesmal segments, as well as the cilia, are formed under the influence of kinetosomes.

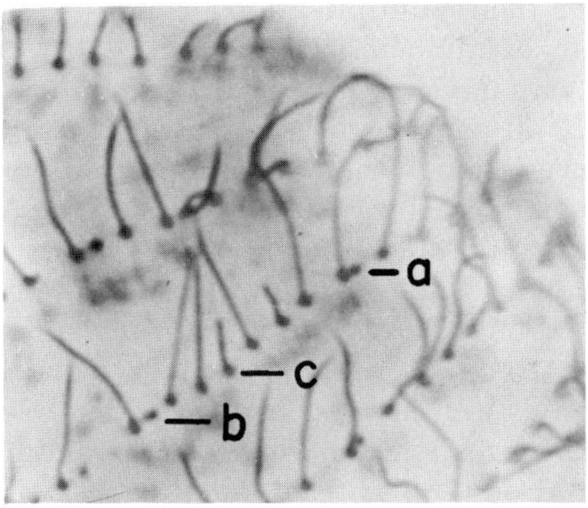

Fig. 10. Protargol preparation showing what appear to be stages in the development of somatic cilia and kinetosomes. New kinetosomes first appear in close proximity to older, ciliated ones (a). Next they move away from the 'parent' (b), and sprout cilia (c). (Unpublished photo by the author.)

The failure of electron microscopists to find dividing kinetosomes may mean that direct duplication does not occur, although ultrastructural evidence for division may yet be found. One alternative possibility is suggested by the work of Mazia, Harris and Bibring (1960). These investigators, working with the mitotic center of the sea urchin egg, have developed a 'generative' model for the self-reproduction of an intracellular body. According to this idea, the intracellular body does not immediately produce a complete copy of itself, but provides a 'germ' or 'seed', which then grows to be equivalent to the parent body. In this connection note the smaller size of the non-ciliated daughter of a recently divided kinetosome in Fig. 1. In any event, the requirements of 'genetic continuity' would be fulfilled by a contribution from the parent to the daughter kinetosome of something less than a fully-formed morphology. It seems clear that, in the kineties of *Tetrahymena* at least, pre-existing kinetosomes play some role in the formation of new kinetosomes. The central problem here, as in the case of ciliary synthesis (and perhaps kinetodesmata formation), is to understand the nature of the kinetosomal contribution.

During anarchic field formation, according to Chatton, Lwoff, Lwoff and Monod (1931), kinetosomes in the mid-region of the right post-oral meridian (the 'strie stomatogene') undergo a number of divisions. It is likely that kinetosomes in the mid-region of the stomatogenic meridian are not different from other somatic kinetosomes, but rather their multiplication in early stomatogenesis is a response to some extrakinetosomal factors. This is suggested by results obtained with *Stentor*, where it was demonstrated that kinetosomes in virtually all parts of the ectoplasm are competent to form anarchic fields, even though the oral primordium normally forms in one particular region (Tartar, 1961).

The sprouting of cilia and fibers by kinetosomes represent other synthetic events about which we know almost nothing. As Grimstone (1961) has pointed out, ciliary fine structure is quite different from kinetosomal fine structure, which leaves us with the problem of pattern formation as well as mass increase. Finally, morphogenetic movements in the cortex, e.g. sinking in of the buccal cavity late in stomatogenesis and position shifts of kinetosomes, will one day have to be explained.

VI. Nucleus and Endoplasm

Structural changes in the nuclei, mitochondria, and cytoplasmic membranes associated with the division cycle have been reported in *Tetrahymena*. Changes in macronuclear granulation during fission were noted in synchronously dividing cells (Holz, 1960), and in untreated cells observed with the electron microscope (Roth and Minick, 1961). Fibers have also been noted within dividing macronuclei (Furgason, 1940; Roth and Minick, 1961). Evidence has been obtained (Elliott *et al.*, 1962) that RNA is transferred from the RNA bodies of the macronucleus to the peripheral nucleoli, which then escape into the cytoplasm, perhaps at some particular stage of the cell cycle. Finally, macronuclear chromatin masses are lost to the cytoplasm during fission in untreated (Furgason, 1940) and in synchronously dividing *Tetrahymena* (Scherbaum *et al.*, 1958). Light microscope observations of dividing micronuclei in synchronized *Tetrahymena* were made by Holz (1960). Elliott (unpublished data) has made electron-microscope observations, and reports the presence of curious centriole-like bodies within the micronucleus during its division.

Structural changes in mitochondria associated with the cell cycle have been noted (Roth and Minick, 1961). The most prominent change is the production of dense intramitochondrial masses at the time of macronuclear division. An increase in the amount, complexity, and localization of endoplasmic reticulum has also been reported in dividing *Tetrahymena* (Roth and Minick, 1961; Williams *et al.*, 1960).

VII. Concluding Remarks

In synchronously dividing *Tetrahymena* we see such morphogenetic problems as kinetosome 'self-reproduction', ciliary and fiber synthesis, oral primordium localization, and intracellular morphogenetic movement. Behind these lie the basic problems of the involvement of the nuclei and unseen organizational factors in the cortex. The fact that morphogenetic processes can be brought into synchrony in populations of *Tetrahymena* should make this excellent material for study. But perhaps the most promising aspects of synchronously dividing *Tetrahymena* from the stand-point of understanding ciliate morphogenesis lie in the phenomena of morphogenetic arrest and reactivation, and in primordium resorption and stabilization. It is clear that in these instances morphogenesis is in intimate relation with the biochemical processes controlling the cell cycle.

Acknowledgments. The author gratefully acknowledges the assistance of Drs Joseph Frankel, E. Zeuthen, D. Stuart Nachtwey, Waldo H. Furgason, Alfred M. Elliott, and George G. Holz, Jr., especially for access to unpublished results and permission to use previously unpublished photographs. The author also wishes to thank Drs Charles B. Metz and Dorothy R. Pitelka for permission to reproduce Figs. 3 and 5 respectively.

References

Chatton, E. and S. Brachon (1935). Discrimination, chez deux Infusoires de genre *Glaucoma*, entre système argentophile et infraciliature. *Compt. Rend. Soc. Biol.* **118,** 399.

Chatton, E., A. Lwoff and M. Lwoff (1929). Les infraciliatures et la continuité génétique des systèmes ciliaires récessifs. *Compt. Rend. Acad. Sci., Paris* **188,** 1190.

Chatton, E., A. Lwoff and M. Lwoff (1931). L'origine infraciliaire et la genèse des trichocystes et des trichites chez les Ciliés *Foettingeriidae. Compt. Rend. Acad. Sci, Paris* **193,** 670.

Chatton, E., A. Lwoff, M. Lwoff and J. L. Monod (1931). La formation de l'ébauche buccale postérieure chez les Ciliés en division et ses relations de continuité topographique et génétique avec la bouche antérieure. *Compt. Rend. Soc. Biol.* **107,** 540.

Ehret, C. F. and E. L. Powers (1959). The cell surface of *Paramecium. Intern. Rev. Cytol.* **8,** 97.

Elliott, A. M., J. R. Kennedy and I. J. Bak (1962). Macronuclear events in synchronously dividing *Tetrahymena pyriformis. J. Cell Biol.* **12,** 515.

Fauré-Fremiet, E. (1961). Cils vibratiles et flagelles. *Biol. Rev. Cambridge Phil. Soc.* **36,** 464.

Frankel, J. (1962). The effects of heat, cold, and *p*-fluorophenylalanine on morphogenesis in synchronized *Tetrahymena pyriformis* GL. *Compt. Rend. Trav. Lab. Carlsberg* **33,** 1.

Furgason, W. H. (1940). The significant cytostomal pattern of the 'Glaucoma–Colpidium' group, with a proposed new genus and species, *Tetrahymena geleii. Arch. Protistenk.* **94,** 224.

Grimstone, A. V. (1961). Fine structure and morphogenesis in protozoa. *Biol. Rev. Cambridge Phil. Soc.* **36,** 97.

Holz, G. G., Jr. (1960). Structural and functional changes in a generation in *Tetrahymena. Biol. Bull.* **118,** 84.

Holz, G. G., O. H. Scherbaum and N. E. Williams (1957). The arrest of mitosis and stomatogenesis during temperature-induction of synchronous division in *Tetrahymena pyriformis* (mating type 1, variety 1). *Exptl. Cell Research* **13,** 618.

Klein, B. M. (1927). Die Silberliniensysteme der Ciliaten. Ihr Verhalten während Teilung und Conjugation, neue Silberbilder, Nachträge. *Arch. Protistenk.* **58,** 55.

Mazia, D., P. J. Harris and T. Bibring (1960). The multiplicity of mitotic centers and the time course of their duplication and separation. *J. Biophys. Biochem. Cytol.* **7,** 1.

Metz, C. B. and J. A. Westfall (1954). The fibrillar systems of ciliates as revealed by the electron microscope. II. *Tetrahymena. Biol. Bull.* **107,** 106.

Nachtwey, D. S. and A. C. Giese. Effects of heat shock and ultraviolet radiation on division of synchronized *Tetrahymena.* Personal communication from Dr. D. S. Nachtwey.

Pitelka, D. R. (1961). Fine structure of the silverline and fibrillar systems of three tetrahymenid ciliates. *J. Protozool.* **8,** 75.

Rasmussen, L. and E. Zeuthen (1962). Cell division and protein synthesis in *Tetrahymena* as studied with *p*-fluorophenylalanine. *Compt. Rend. Trav. Lab. Carlsberg* **32,** 333.

Roth, L. E. and O. T. Minick (1961). Electron microscopy of nuclear and cytoplasmic events during division in *Tetrahymena pyriformis* (Strains W and HAM 3). *J. Protozool.* **8,** 12.

Scherbaum, O. H., A. L. Louderback and T. L. Jahn (1958). The formation of subnuclear aggregates in normal and synchronized protozoan cells. *Biol. Bull.* **115,** 269.

SUHAMA, M. and S. YAMATAKA (1960a). On the cortex of *Tetrahymena*. I. The comparison between the silver-plated structures and the fine structures through the electron microscope in *Tetrahymena vorax*. *Dobutsugaku Zasshi* (*Zool. Mag.*) **69,** 153.

SUHAMA, M. and S. YAMATAKA (1960b). On the cortex of *Tetrahymena*. II. The fine structure of cilia, kinetosomes, and kinetodesmas in *Tetrahymena vorax*. *Dobutsugaku Zasshi* (*Zool. Mag.*) **69,** 158.

TARTAR, V. (1961). *The Biology of* Stentor. Pergamon Press, New York.

WILLIAMS, N. E., E. ANDERSON, R. KESSEL and H. W. BEAMS (1960). Electron microscope observations on synchronously-dividing *Tetrahymena*. *J. Protozool. Suppl.* **7,** 27.

WILLIAMS, N. E. and O. H. SCHERBAUM (1959). Morphogenetic events in normal and synchronously-dividing *Tetrahymena pyriformis* GL. *J. Embryol. Exptl. Morphol.* **7,** 241.

ZEUTHEN, E. (1958). Artificial and induced periodicity in living cells. *Advanc. in. Biol. and Med. Phys.* **6,** 37.

6

Biochemical Studies on Synchronized *Tetrahymena*

Otto H. Scherbaum, *Department of Zoology, University of California, Los Angeles, California*

CONTENTS

I	Introduction	177
II	Amino Acids, Proteins	177
III	Carbohydrates	182
IV	Nucleic Acids	184
V	Phosphates	187
VI	Enzyme Activity and Coenzyme Levels	189
VII	Summary	192
	References	193

I. Introduction

Incubation of *Tetrahymena* mass cultures at certain cycling temperatures induces characteristic changes in size, morphology, and biochemical composition of the cells, which, as a result of the treatment, may undergo one or more synchronized fissions. The methods of treatment and their effect on the cytology of *Tetrahymena*, as well as the action of inhibitors and drugs on these cells, are discussed in other chapters of this book (see Plesner, Williams, Zeuthen). In this chapter, only a few aspects of the present knowledge of the biochemistry of synchronized *Tetrahymena* will be considered. The emphasis will be on evaluation of changes in the chemical constituents of the cells in connection with the temperature treatment and synchronized cell division.

II. Amino Acids, Proteins

The first indication that increase in cell size during the treatment of *Tetrahymena pyriformis* GL is accompanied by synthesis of cellular protein came from studies of the reduced weight of fixed cells (Scherbaum and Zeuthen, 1954). Cells from seven stages of growth (before, during, and after the temperature treatment) were fixed in phosphotungstic acid, twenty to thirty cells were isolated from the sample and pipetted onto a Cartesian diver balance, and their reduced weight was estimated. The results show an approximate 2·8-fold increase in the relative weight of the average cell during eight temperature cycles. A considerable variability of the individual readings can be attributed largely to the small size of the samples ($\leqslant$30 cells) and the fact that they were taken at random from cultures in which cell size is quite variable (Scherbaum, 1956; Scherbaum and Rasch, 1957).

The increase in cellular protein during the temperature treatment has been verified repeatedly (Christensson, 1959; Lee *et al.*, 1959; Scherbaum, James and Jahn, 1959; Scherbaum, Louderback and Jahn, 1959; Singer and Eiler, 1960). Fig. 1 represents data published by Christensson (1959). Total protein and free amino acids increase approximately 2·5 times during the treatment. This result is in good agreement with a later report by Scherbaum and Levy (1961). Christensson separated the proteins into two fractions according to their solubility in a dilute salt solution (0·1 *M* KCl and 0·011 *M* acetic acid). The insoluble proteins increase during the treatment by 2·5 times, while soluble proteins increase 3·2 times. Christensson (1959) found the total protein prior to the treatment to be 17×10^{-12} gram atoms of N/cell. If we assume 16 per cent N in protein, this value is $1{\cdot}49 \times 10^{-3}$ μg protein/cell. Singer and Eiler (1960) reported similar values ($1{\cdot}44$–$1{\cdot}85 \times 10^{-3}$ μg/cell). Higher values were those given by Lee *et al.* (1959), $2{\cdot}6$–$3{\cdot}2 \times 10^{-3}$ μg/cell, and Scherbaum and Levy (1961), $3{\cdot}0 \times 10^{-3}$ μg/cell. These differences may be due, in part, to the length of time the cells were kept in exponential multiplication before the treatment started. For example, the average cell size (and cellular protein is undoubtedly related to size) of *Tetrahymena pyriformis* GL is reduced to approximately one half during 17 hr of exponential multiplication when adequate oxygen supply is maintained (Scherbaum, 1956). The free amino acid content of normal *Tetrahymena* was reported to be 23·2 per cent of the total amino acid content by Wu and Hogg (1956), 12·6–15 per cent by Scherbaum, James and Jahn (1959), and 15 per cent by Christensson (1959). Peptides are 17 per cent of total cellular amino acid content (Wu and Hogg, 1956). This free amino acid pool increases during the treatment in proportion to the total protein of the cells. The treatment is also known to cause changes in the concentration of ethanol-soluble, ninhydrin-positive compounds; these compounds are presumably small peptides (Scherbaum, James and Jahn, 1959).

Several points should be noted with regard to Fig. 1:

(1) At a time when the rate of synthesis of insoluble proteins is already high (after the fourth and fifth heat shocks), the increase in

soluble proteins is still small. Obviously a distinct time lag in the synthesis of soluble proteins occurs.

(2) Protein synthesis is found to stop before division.

(3) After the first synchronized division protein synthesis occurs at lower rates than during the optimum temperature periods in between the heat shocks of the treatment.

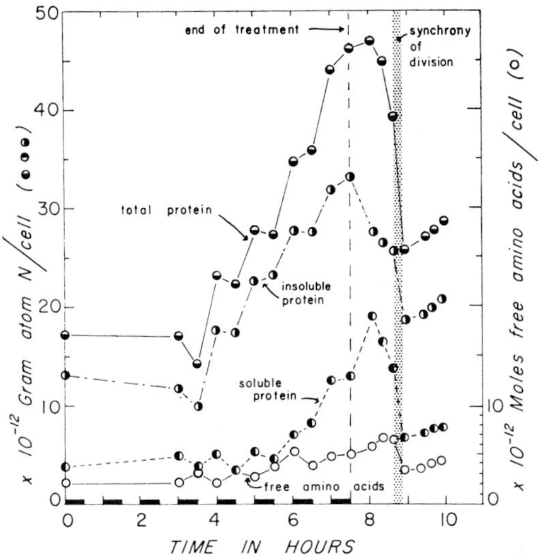

Fig. 1. Free amino acids and proteins in *Tetrahymena pyriformis* GL during and after the synchrony-inducing temperature treatment. The black bars on the abscissa represent the heat shocks (34°C), separated by half-hour intervals of 29°C incubation temperature. (Further explanation in the text.) (Redrawn after Christensson, 1959.)

(4) The rate of protein synthesis is higher at the optimum temperature for growth (29°C) than at the shock temperature (34°C). Christensson concludes that the synthetic activity of the cellular proteins occurs in distinct waves caused by the change in the incubation temperature. It seems that the last heat shocks do not inhibit protein synthesis to the same extent as the fourth and fifth shocks. This progressive loss of the inhibitory effect of the heat shocks on protein synthesis is interpreted by increased protein stabilization or biological adaptation (Christensson, 1959).

(5) Total cell protein does not increase during the first three hours of the treatment in the experiments shown in Fig. 1. It seems very likely that this lack of protein synthesis is not necessarily caused by the heat treatment, but rather the result of the culture conditions:

the heat treatment was started after inoculation at a time when exponential multiplication had not yet begun (Christensson, personal communication to Dr. E. Zeuthen).

(6) During the first half hour after the eighth heat shock, the insoluble proteins decrease by $0{\cdot}5 \times 10^{-3}$ μg/cell and the soluble proteins increase by the same amount. This stoichiometric relationship suggests conversion of insoluble to soluble protein. It is tempting to speculate that breakdown of cellular structures (insoluble proteins) might be a prerequisite for division (Scherbaum, 1960a, b). During the half hour preceding synchrony, the total protein decreases by 5 per cent. Christensson interprets his findings as a block of protein synthesis. This has also been suggested (Zeuthen, 1953; Zeuthen and Scherbaum, 1954) on the basis of respiration studies in single, normal *Tetrahymena* cells.

Mita *et al.* (1959) studied the incorporation of ^{35}S-methionine into proteins of *Tetrahymena pyriformis* W. One representative experiment will be briefly outlined. A 700-ml culture was grown in proteose peptone medium for 11 hr at 29°C. At the end of that time, the population density was 25,000 cells/ml. The culture was aseptically divided into seven cultures of 100 ml each. To each culture, 26·1 μC of ^{35}S methionine was added at the times indicated in Fig. 2. Each culture was incubated for 15 min. The cells were collected by centrifugation at 5°C and washed six times in ice cold 0·4 per cent NaCl.

The protein fraction was obtained, following the procedure of Schneider (1945). Protein was determined by Nesslerization. The results are shown in Fig. 2. The specific activity in the protein fraction decreases during the heat treatment. After the fourth temperature shock, the specific activity is 39 per cent of the activity in the untreated log sample. After the eighth temperature shock, the specific activity is reduced further, to 20 per cent of the log culture's activity.

During the recovery period and the following synchronous division cycles, the activity increases rapidly. 2·5 hr after the end of the treatment it is 45 per cent of the normal control. Mita *et al.* (1959) have reported that the total amount of protein per culture increased 57–64 per cent (in two experiments respectively) during the first 3 hr of the heat treatment. Unfortunately, no direct comparison can be made with Christensson's work, since Mita *et al.* expressed their results per total culture rather than per cell, and it is not known whether some cell multiplication took place during the treatment. Another difficulty in comparing data on the amino acid and protein composition stems from the fact that, even if the cells are washed very carefully before the analysis, foreign peptones and amino acids can be present in the food vacuoles and contribute to the 'protein and amino acid composition' of the cell.

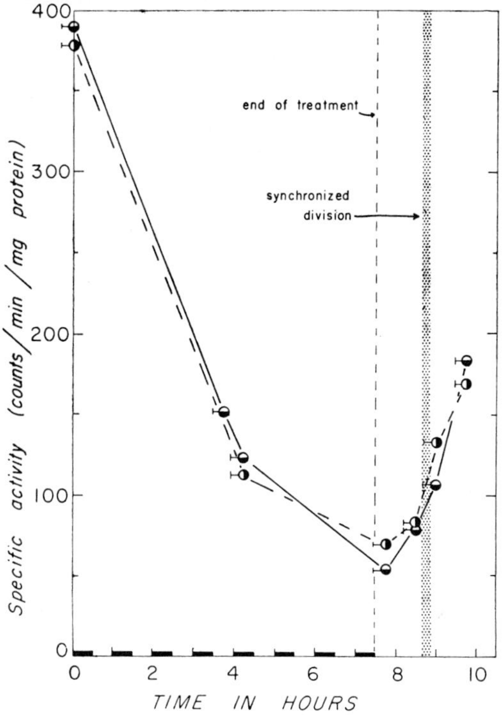

FIG. 2. Incorporation of ^{35}S-methionine into proteins of *Tetrahymena pyriformis* W. The black bars on the abscissa represent the heat shocks (34°C), separated by half-hour intervals of 28°C incubation temperature. Two experiments are shown. The bar with each symbol indicates the time (15 min) of incubation of the culture in ^{35}S-methionine. (Further explanation in the text.) (Data from Mita *et al.*, 1959.)

All analyses published agree that the protein content of synchronized *Tetrahymena* cells increases during the temperature treatment. However, the details of the synthetic pattern in various protein fractions have yet to be verified. It is hoped that use of labeled amino acids will shed some light on the turnover in distinguishable cellular pools in relation to treatment and synchronous multiplication.

III. Carbohydrates

Tetrahymena pyriformis GL, grown on a peptone medium containing less than 0·1 per cent glycogen, synthesizes glycogen in various phases of population growth (Scherbaum and Levy, 1960, 1961). During exponential multiplication, the glycogen content per cell was found to be $1{\cdot}86 \times 10^{-4}$ μg/cell (or 3·6 per cent of dry weight) (Scherbaum and Levy, 1961). In these experiments, the samples were kept at approximately +5°C from their harvesting to their digestion in KOH. If the cells are harvested and dried at 5°C, however, the values can be as low as one-seventh of the figure reported here (Scherbaum and Levy, 1960). After one day of growth, the cellular glycogen was found to be 4·4–9·4 per cent of the dry weight (Scherbaum and Levy, 1961). After twelve days of growth, the glycogen content can be as high as 22 per cent of the dry weight (Manners and Ryley, 1952).

The addition of 0·1 per cent glucose to the culture medium causes a considerable increase in glycogen accumulation. After three days of growth, when the glucose supply is exhausted, the cellular glycogen content can be as high as 50 per cent (Hogg, personal communication).

If cells in the logarithmic or early stationary phase of growth are exposed to the synchrony-inducing temperature treatment for 6·5 hr, glycogen accumulation occurs at a rather constant rate throughout the treatment. A further increase in the glycogen content occurs during the 1·5 hr recovery period—the time interval between the termination of the temperature treatment and the commencement of synchronous multiplication.

During the negative growth acceleration phase, and during the synchrony-inducing temperature treatment, cell multiplication is progressively inhibited; cell growth, however, continues. In both cases the result is the formation of larger cells with excessive amounts of glycogen. Since glycogen is also present in cells of the exponential phase of growth, two interesting questions arise: (1) Is the rate of glycogen synthesis the same in phases of rapid proliferation and no proliferation? and (2) Are different precursors for glycogen synthesis used at these different growth stages? Results pertaining to the first question are shown in Table I. Six 150-ml cultures of *Tetrahymena* were started simultaneously and grown for 23 to 59 hr at 26°C. At each of the times shown in the left column one culture was used in the following way: Two samples were removed, one for cell count and the other for determination of dissolved oxygen by the micromethod of Scholander *et al.* (1955). A few minutes later,

2 μC of 2-^{14}C sodium pyruvate were added to the culture. After 30 minutes' incubation, the culture was harvested and analyzed for glycogen content and incorporation of the label into glycogen. In this particular experiment, the oxygen content dropped rapidly at a population density of approximately 10^5 cells/ml. At the same time, the glycogen content of the cells increased considerably. In stationary phase cultures, i.e. cultures with a population density greater than 10^5 cells/ml, incorporation of pyruvate into glycogen per cell is significantly higher than during exponential multiplication (see Table I). Preliminary observations indicate that acetate might be an even better precursor for glycogen synthesis than pyruvate (Levy and Scherbaum, unpublished data).

TABLE I. Incorporation of ^{14}C pyruvate into glycogen at various population densities of *Tetrahymena*

Time after inoculation (hr)	Cells/ml culture	Dissolved oxygen in mm^3/ml culture	Glycogen (μg/10^6 cells)	Counts/min in glycogen/10^6 cells
23	~30·000	2·20	60	20
26	53·000	1·30	51	15
29	117·000	0·50	131	64
31	121·000	0·27	118	59
35	114·000	0·27	193	93
59	173·000	none detectable	1460	680

Six 500-ml Erlenmeyer flasks, each containing 150 ml medium were inoculated at the same time, and assayed as indicated in the left column. Throughout the experiments the flasks were gently shaken by rotatory movements, with little, if any surface disturbance. Further explanation in the text.

Kamiya (1959) reported changes in the acid-soluble fraction of *Tetrahymena pyriformis* W during the synchrony-inducing temperature treatment. Cells were extracted with cold 10 per cent trichloroacetic acid and the pH was adjusted to 8·2. After addition of barium acetate and ethanol (final concentration 19 per cent), the supernate was analyzed. It contains unknown sugar(s) which decrease markedly during the temperature treatment. It has been suggested that this fraction might be connected with a metabolism which is highly susceptible to elevated temperatures (Kamiya, 1959).

IV. Nucleic Acids

The observation that cell volume, dry weight, and proteins increase during the temperature treatment has been extended to the nucleic acids. DNA synthesis is almost linear during the 6·5 hr of heat treatment. At the end of the treatment, the DNA content is approximately twice the amount found in normal cells (Iverson and Giese, 1957; Scherbaum, 1957). Higher values have also been reported (Zeuthen and Scherbaum, 1954). A further increase (13–25 per cent) occurs between the end of the treatment (6·5 hr) and the first synchronous division (7·8 hr) (Scherbaum, Louderback and Jahn, 1959).

The amount of DNA in single nuclei was estimated microspectrophotometrically (Scherbaum, Louderback and Jahn, 1959). The results are shown in Fig. 3. For each growth stage, fifty randomly-

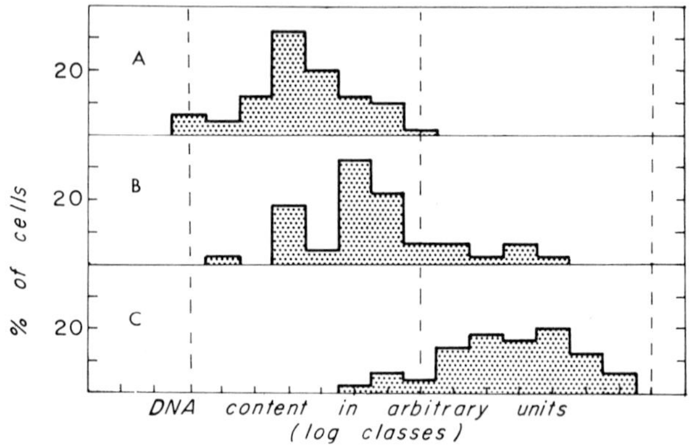

Fig. 3. Histogram of the DNA content in individual cells at the following growth stages: (A) Normal exponential multiplication; (B) after three heat shocks; (C) after seven heat shocks. For each growth stage, the amount of DNA in fifty cells was measured spectrophotometrically. The numerical values for the relative amounts obtained experimentally were converted into log values and grouped into classes of 0·04 log units. The difference between two parallel vertical dotted lines indicates approximate doubling of DNA. (Further explanation in the text.) (Redrawn after Scherbaum, Louderback and Jahn, 1959.)

chosen nuclei were analyzed. The data were pooled in log classes and plotted as frequency histograms. It seems that DNA synthesis occurs in all cells during the treatment. The data do not substantiate the view that DNA synthesis may be arrested by the heat shocks at a specific step in the cell cycle. If this were the case, the percentage of cells in one or two adjacent classes should have increased. The question of whether the DNA synthesis prior to the first synchronous division is necessary for the following synchronous division cycles is discussed in another chapter (see Zeuthen, this book).

The ratio of nuclear volume to DNA content at various growth stages has been determined. For each growth stage, fifty cells were analyzed and the values were averaged (Scherbaum, Louderback and Jahn, 1959). As seen from Table II, the DNA/volume ratio decreases during the treatment and reaches a minimum of 0·56 (as compared to normal log culture, 1·00) half an hour prior to division. The data of Table II indicate that this decrease occurs because the Feulgen-positive material in the nucleus does not increase to the same extent during the treatment as the average nuclear volume. A sharp increase in the DNA/volume ratio to 0·93 was observed during synchronized division. These changes are probably an expression of structural changes which take place in the nucleus during the temperature treatment, synchronized division, and stationary phase of growth.

TABLE II. Nuclear volume and DNA content. For each growth stage the volume and DNA content of fifty cells were measured and averaged

Stage of population growth	Average nuclear volume (arbitr. units)	Average amount of DNA (arbitr. units)	DNA/volume
Normal logarithmic phase	194	13·2	0·068 (1·00)
After three heat shocks	298	16·2	0·054 (0·79)
After seven heat shocks	556	23·7	0·043 (0·63)
1 hr after last shock	663	25·7	0·038 (0·56)
During synchronous division	244	15·5	0·063 (0·93)
2 hr after last heat shock	305	16·3	0·054 (0·79)
Maximum stat. phase (20 hr after inoculation)	118	12·4	0·105 (1·55)

Although the base ratios of RNA and DNA in *Tetrahymena pyriformis* GL appear not to be altered by the treatment which inhibits cell division (Scherbaum, 1957), some macromolecular changes might occur. Table III shows the results of this analysis (Mita and Scherbaum, to be published). DNA was isolated from *Tetrahymena pyriformis* GL by the method of Kay *et al.* (1952), adsorbed on ECTEOLA cellulose columns, and eluted with NaCl of changing concentration as described by Bendich *et al.* (1958). A fraction eluted by 0·5 *M* NaCl appears to increase considerably during the temperature treatment. Further studies are necessary to find out whether this difference in elution pattern is due to the properties of the isolated DNA or to changes caused by the isolation and separation procedures.

TABLE III

Solvent	Growth stage 1 Normal exp. growth	2 End of treatm.	3 1 hr after end of treatm.	4 Max. stat. phase
0·5 *M* NaCl	31·9	77·9	58·8	46·5
2·0 *M* NaCl + 0·1 *M* NH_3	64·5	17·7	41·4	48·5
2·0 *M* NaCl + 1·0 *M* NH_3	2·7	3·1	1·0	2·6
0·5 *M* NaOH	0·9	1·2	3·6	2·4

Chromatography of DNA on columns of ECTEOLA (0·9 × 6 cm; 0·29 m-equiv./g; flow rate about 5 ml/hr of app. 3 mg DNA in each of four growth stages tabulated. Recovery 99·9–104·8 per cent. The figures shown are in per cent of total DNA recovered from the column. Recovery was determined by absorption at λ_{260}.

A peculiarity of macronuclear behavior of *Tetrahymena* is the loss from the nucleus of some of the Feulgen-positive material during amitotic partition of the parent nucleus into two daughter nuclei (Scherbaum *et al.*, 1958). The average volume of the extrusion bodies is relatively constant in the logarithmic and stationary phases of growth ($\sim$ 2·0 μ^3). This value is 0·7 per cent and 1·5 per cent of the macronuclear volume at these two growth phases respectively. However, after synchronous division, the average volume of the extrusion bodies is 12·0 μ^3. This is 2·8 per cent of the macronuclear volume. Attempts to track the fate of these macronuclear extrusions with tritiated thymidine failed (Scherbaum, 1960b). The label is not only incorporated into the cell nucleus, but appears also in the cytoplasm even before cell division has taken place (Scherbaum, 1960b).

The effect of ultraviolet radiation on synchronized cultures of *Tetrahymena pyriformis* W was reported by Iverson and Giese (1957). A dosage of 15 × 10^{13} quanta/mm^2 (=1122 ergs/mm^2) caused the longest division delay. The delay of the first division is the same (25 hr) whether the cells are irradiated before, or at the end of, the synchrony-inducing heat treatment. The delay of the second division is only 30 hr following irradiation at the end of the treatment, but the delay is 100 hr if irradiation occurs before the beginning of the heat treatment. These results show that the cells at the end of the treatment are much more resistant than those irradiated before the treatment. However, when the experiment was repeated using a shorter wavelength of 226 mμ, this marked difference in sensitivity disappeared. The cells exposed to 226 mμ were immobilized quickly. This immobilization was not observed upon irradiation with U.V. of wavelength 265 mμ.

The retardation of division upon irradiation with U.V. of 226 mμ can be reversed by white light. The authors point out that only nuclear effects of U.V. have been found subject to photoreversal and suggest, therefore, that some nuclear activity other than DNA synthesis is being affected by U.V. of 226 mμ (Iverson and Giese, 1957). It is interesting that the second synchronized division is delayed longest upon irradiation (265 mμ) during the terminal stage of the first synchronized division and least upon irradiation (265 mμ) after the end of the heat treatment. This greater resistance at the end of the treatment has been correlated with the greater DNA content at that stage. Iverson and Giese (1957) concluded that the 'the U.V. sensitivity (265 mμ) is inversely proportional to the amount of DNA present per cell and shows a direct relationship with the amount of DNA which a cell must synthesize before division'.

The effect of X-rays upon normal and synchronously dividing *Tetrahymena pyriformis* W has been reported (Ducoff, 1956). The cultures were exposed to 60 kr of X-rays (250 kV, 30 mA, Maxitron unit, hvl 7·0 mm Al). The radiation-induced division delay was approximately one generation time (2·75–3·00 hr). It was the same at the following stages tested: (a) exponential multiplication, (b) before, and (c) after the first synchronous division. The author concluded that the mechanism of this radiation-induced fission block in *Tetrahymena pyriformis* W is *not* inhibition of DNA synthesis (Ducoff, 1956).

V. Phosphates

The uptake of ^{32}P during synchronized division of *Tetrahymena pyriformis* GL and changes of the nucleoside triphosphate content are discussed elsewhere in this book (see Zeuthen, Plesner). Colorimetric determination of orthophosphate (Scherbaum, Louderback and Jahn, 1959) and studies of ^{32}P uptake (Hamburger and Zeuthen, 1960) indicate that, during the heat treatment, cellular dry weight increases more than the total phosphorus content. As a result, the phosphorus/dry weight ratio after 6·5 hr of treatment is 50–85 per cent (mean 77 per cent) of the ratio found in a culture prior to treatment. Such a decrease occurs in all the fractions studied: acid-soluble, lipid, nucleic acid and phosphoprotein (Hamburger and Zeuthen, 1960). The acid-soluble fraction was studied further by incorporation of ^{32}P-orthophosphate and radioautography (Chou and Scherbaum, 1963). In these experiments, 7 l. of sterile proteose peptone medium containing 700 μC of $^{32}PO_4$ (sodium orthophosphate; Abbott Laboratories) were inoculated with 70 ml of a 2-day-old culture. After 16 hours of growth, the first sample (No. 1) was taken at a population density of approximately 50,000 cells/ml. After removal of this sample, the synchrony-inducing heat treatment was started. At the end of the seventh heat shock, the second sample (No. 2) was harvested. Sample No. 3 was taken 1 hr later, 20 min before the division maximum of the first synchronous multiplication. The last sample (No. 4) was collected 24 hr later when the culture had entered the maximum stationary phase of growth. Each sample was concentrated by centrifugation, washed in 0·4 per cent NaCl and distilled water, and extracted with 10 per cent TCA. All these procedures were carried out at +5°C. After removal of glycogen, Whatman No. 1 filter paper was spotted (amounts equivalent to 90 μg phosphorus for each chromatogram) for two dimensional chromatography in the following solvents: first dimension (a) isopropanol-isobutanol, ammonium hydroxide and water; second dimension (b) isopropanol, trichloroacetic acid, ammonium hydroxide and water.

The dried chromatograms were exposed to Kodak no-screen X-ray films for 3 to 5 days. Kodagraph paper was used for making the prints from the developed X-ray films. A chromatogram of the acid-soluble fraction is shown in Fig. 4. The separated compounds have not yet been identified. Attempts are being made to obtain a sharper separation; however, some general suggestions can be made. Complex A is orthophosphate; complex B is the nucleotide fraction which can be adsorbed on and eluted from Norit; complex C shows little mobility in solvent 'b'. If high polymer inorganic polyphosphates are present in the extract they should be found in this complex. By chemical fractionation a Ba precipitate was obtained at pH 4·5 in the acid-soluble fraction. This precipitate shows strong metachromasy and a high content of 7-min phosphates (Chou and Scherbaum, unpublished data); complex X is only present after heat treatment (stage 2) and prior to synchronous fission (stage 3). Recently this complex (X in Fig. 4) was resolved into three separate compounds. Their physico-chemical properties are being studied at present (Chou and Scherbaum, unpublished). Obviously, much more work must be done in order to get a sufficiently sound basis for speculative explanations of these observed changes.

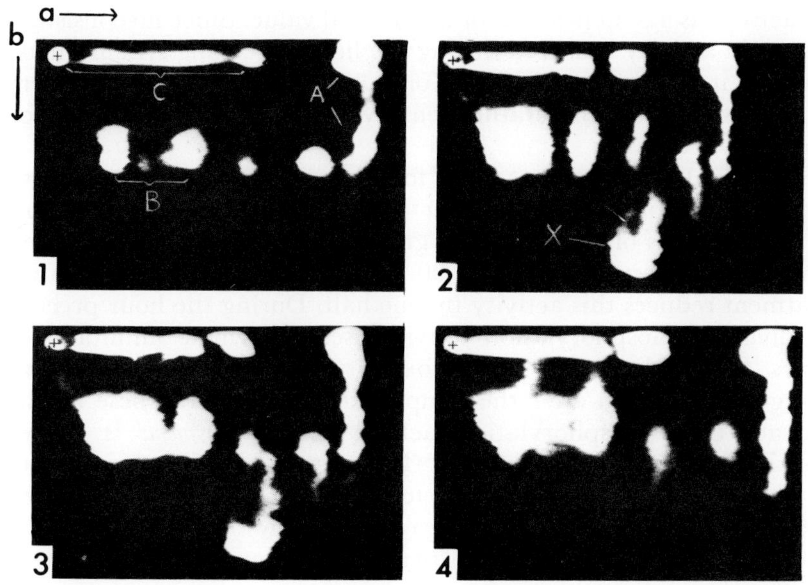

Fig. 4. X-ray film autograph of the acid-soluble fraction of *Tetrahymena pyriformis* GL. The spots show radioactivity of ^{32}P on paper chromatogram developed descending in two dimensions (a, b). The numbers designate the following growth stages: (1) exponential multiplication; (2) end of synchrony-inducing temperature treatment; (3) 1 hr after end of treatment, prior to first synchronized multiplication; and (4) maximum stationary phase. (Further explanation in the text.)

VI. Enzyme Activity and Coenzyme Levels

In homogenates of six strains of *Tetrahymena*, the DPNH oxidase activity was studied as a function of temperature. Mild heating (10 min at 40°C) abolishes the oxidation of DPNH (Eichel and Rem, 1959). Only partial inhibition of the system is observed upon heating at 30°C: 55 per cent after 5 min and 75 per cent after 10 min

(Eichel, 1956). Repeated heating (34°C or 43°C for 30 min) is used for induction of synchronized division in five strains of *T. pyriformis* (GL, GL-R, W, and WH6) (Scherbaum, 1960a). Therefore, homogenates of strain GL were prepared prior to, during, and after the heat treatment and assayed for DPNH oxidase activity to determine the effect of heat upon this system *in vivo* (Nishi and Scherbaum, 1962a). In extracts of cells in the exponential multiplication phase, the DPNH oxidase activity showed values ranging from 3·9 to 6·7 mμmoles oxidized/min/mg protein; at the end of the synchrony-inducing temperature treatment the activity is 42–80 per cent of the normal value. Since no consistent change in activity occurred during the hour preceding synchronized division, it is not clear whether or not the DPNH oxidase system is implicated as a temperature-sensitive-controlling mechanism for synchrony.

Significant changes have been found in phosphorylation activity in heat-synchronized *Tetrahymena* (Nishi and Scherbaum, 1962b). In homogenates of exponentially growing cells, 0·1–0·2 μmoles/mg protein are phosphorylated in 30 min at 23°C. The temperature treatment reduces this activity by one half. During the hour preceding division, phosphorylation activity rises to normal. Simultaneous assays for oxidation of β-hydroxybutyrate show no significant changes during and after the temperature treatment. These results indicate that phosphorylation activity in *Tetrahymena* is rather sensitive to temperature changes. The observed decline in the rate of phosphorylation during the temperature treatment and cell division is in agreement with the report of low ^{32}P orthophosphate incorporation during the same period (Hamburger and Zeuthen, 1960).

The levels of oxidized and reduced pyridine nucleotides were studied in *T. pyriformis* W (Kamiya and Takahashi, 1961) and GL (Nishi and Scherbaum, 1962a). Data for strain GL are shown in Fig. 5. Certain trends in the change of coenzyme levels from one growth stage to another are evident. During the temperature treatment (0–6·5 hr), for example, the levels of DPNH, TPN and TPNH are reduced to 61–75 per cent while the DPN level increases to 106–123 per cent of the initial value. Figure 5 shows a similar pattern of change in the coenzyme levels in duplicate experiments; however, a considerable variation was observed in the degree of change.

The level of oxidized pyridine nucleotides increases 30–40 per cent within 15 min after the end of the heat treatment in *T.*

pyriformis W (Kamiya and Takahashi, 1961). No such change was found in strain GL (Nishi and Scherbaum, 1962a). The DPNH/DPN ratio in exponentially growing cultures of the two strains also differed: it was 1·15 for strain W, but only 0·34 for strain GL.

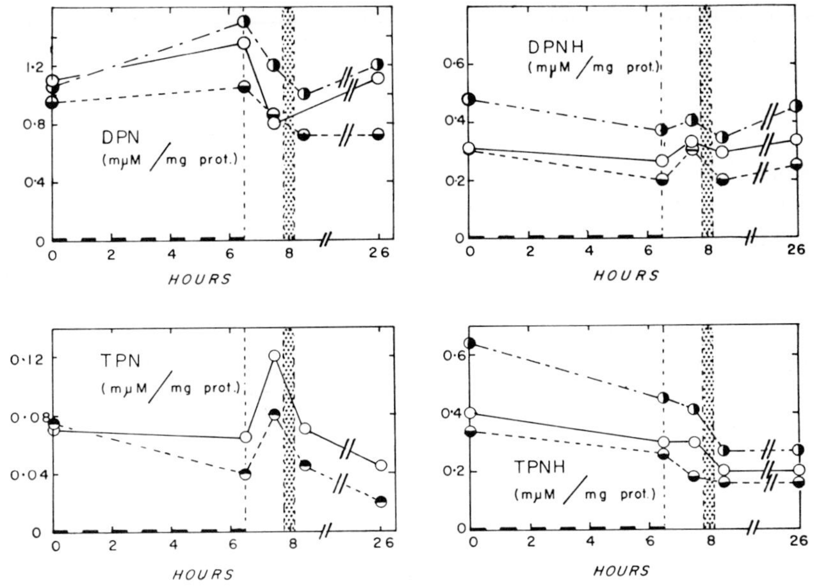

FIG. 5. Coenzyme concentration in *Tetrahymena pyriformis* GL during and after temperature treatment. Two or three parallel experiments are shown for each analysis. The broken vertical lines show the end of the treatment, the dotted area represents the time of synchronized fission.

Of interest is a comparison of the data for *Tetrahymena pyriformis* GL with reports in the literature on the coenzyme ratios found in tissues of higher animals. In *Tetrahymena*, the values for the TPNH/TPN ratio vary from 3·1 (one hour after the end of the treatment) to 7·0 (stationary cells); the DPNH/DPN ratios range from 0·23 (end of treatment) to 0·36 (1 hr after the end of the treatment). These results are comparable to Lowenstein's report that the ratios TPNH/TPN and DPNH/DPN are usually > 5 and < 0·7 respectively (Lowenstein, 1961).

If the ratio (TPN+TPNH)/(DPN+DPNH) is computed for the five growth stages of *Tetrahymena*, the highest ratio is 0·37 (log sample) and the lowest is 0·18 (stationary sample). After the heat treatment, the value is 0·26. This result is interesting in light of the reports that this ratio is 0·37 for rat liver, but only 0·13 in rat kidney (Glock and McLean, 1955). It has been suggested that this high ratio for liver could reflect the highly active synthetic processes in liver, while the low ratio in kidney could be explained on the basis of high demand for energy in kidney function (Klingenberg and Bücher, 1960). *Tetrahymena* obviously represents a system which is capable of both high synthetic capacity and high energy expenditure for ciliary movement and cell division. Thinking in terms of energy expenditure, it may, after all, not be a mere coincidence that cells entering fission show a reduced swimming activity, and the coenzyme ratios (TPN + TPNH)/(DPN + DPNH) could be an expression of the physiological activity of the cells. It is the hope of the author that in the not too distant future we will be able to substantiate these suggestions with more experimental evidence.

VII. Summary

Synchronous cell division can be experimentally induced in mass cultures of *Tetrahymena pyriformis* by certain rhythmic changes in the incubation temperature. Such a treatment (consisting of three to eight temperature shocks) permits cell growth but inhibits cell division almost completely. The biochemical analyses which have been carried out so far on this system are difficult to summarize, but some facets stand out clearly. (1) The increase in cell size during the treatment is paralleled by an increase in total proteins and phosphorus-containing fractions (RNA, DNA, and coenzymes such as DPN and TPN). (2) Cells which have enough DNA to enter division normally continue to synthesize DNA when division is prevented by the heat treatment. In this way, abnormally large amounts of DNA are accumulated in these cells. (3) The base ratios of RNA and DNA seem not to be affected by the treatment. (4) The DNA/nuclear volume ratio decreases during the treatment and prior to synchronous division. (5) The sensitivity to U.V. irradiation (at 265 mμ) is inversely proportional to the amount of DNA present in each cell. This sensitivity was not observed in U.V. light of 226 mμ. (6) X-irradiation causes a division delay which is constant and independent of the amount of DNA present in each cell. (7) Glycogen accumulation occurs during the temperature treatment as well as during the negative growth acceleration phase of a normal culture. (8) Paper chromatographic analyses revealed changes in concentration of phosphorus-containing compounds in cell extracts from heat-treated cells. No estimation of quantity or identification is available. (9) The coenzyme content of *Tetrahymena* (DPN, TPN) is similar to values reported for tissues of higher animals. Relative changes of these compounds during the treatment and prior to synchrony were speculatively related to the physiological activity of the cells.

The work from the author's laboratory was supported in part by Grant RG 6461 from the United States Public Health Service, Grant R-9082 from the National Science Foundation, and Grant 539 from the Cancer Fund of the University of California.

REFERENCES

Bendich, A., H. B. Pahl, G. C. Korngold, H. S. Rosenkranz and J. R. Fresco (1958). Fractionation of deoxyribonucleic acids on columns of anion exchangers; Methodology. *J. Am. Chem. Soc.* **80,** 3949.

Chou, S. C. and O. H. Scherbaum (1963). Temperature-induced changes in phosphorus metabolism in synchronized *Tetrahymena. Biochim. et Biophys. Acta* **71,** 221.

Christensson, E. (1959). Changes in free amino acids and proteins during cell growth and synchronous division in mass cultures of *Tetrahymena pyriformis. Acta Physiol. Scand.* **45,** 339.

Ducoff, H. S. (1956). Radiation-induced fission block in synchronized cultures of *Tetrahymena pyriformis* W. *Exptl. Cell Research* **11,** 218.

Eichel, H. J. (1956). Respiratory enzyme studies in *Tetrahymena pyriformis.* III. Some properties of reduced diphosphopyridine nucleotide oxidase of cell-free homogenates. *J. Biol. Chem.* **222,** 137.

Eichel, H. J. and L. T. Rem (1959). Lactic oxidase of *Tetrahymena pyriformis. Arch. Biochem. Biophys.* **82,** 484.

Glock, G. E. and P. McLean (1955). Levels of oxidized and reduced diphosphopyridine nucleotide and triphosphopyridine nucleotide in animal tissues. *Biochem. J.* **61,** 388.

Hamburger, K. and E. Zeuthen (1960). Some characteristics of growth in normal and synchronized populations of *Tetrahymena pyriformis. Compt. Rend. Trav. Lab. Carlsberg* **32,** 1.

Iverson, R. M. and A. C. Giese (1957). Nucleic acid content and ultraviolet susceptibility of *Tetrahymena pyriformis. Exptl. Cell Research* **13,** 213.

Kamiya, T. (1959). Carbohydrate changes in the acid-soluble fractions of the ciliate protozoon *Tetrahymena geleii* W during the course of synchronous culture. *J. Biochem.* (*Tokyo*) **46,** 501.

Kamiya, T. and T. Takahashi (1961). Mechanism of cell division. I. Changes in pyridine nucleotide and DPNH-cytochrome c reductase in *Tetrahymena geleii* W during the course of synchronous culture. *J. Biochem.* (*Tokyo*) **50,** 277.

Kay, E. R. M., N. S. Simmons and A. L. Dounce (1952). An improved preparation of sodium desoxyribonucleate. *J. Am. Chem. Soc.* **74,** 1724.

Klingenberg, M. and T. Bücher (1960). Biological oxidations. *Ann Rev. Biochem.* **29,** 669.

Lee, K. H., Y. O. Yuzuriha and J. J. Eiler (1959). Studies on cell growth and cell division. II. Selective activity of chloramphenicol and azaserine on cell growth and cell division. *J. Am. Pharm. Assoc., Sci. Ed.* **48,** 470.

LOWENSTEIN, J. M. (1961). Reductions and oxidations in mammalian biosyntheses. *J. Theoret. Biol.* **1,** 98.

MANNERS, D. J. and J. F. RYLEY (1952). Studies on the metabolism of the Protozoa. II. The glycogen of the ciliate *Tetrahymena pyriformis* (*Glaucoma piriformis*). *Biochem. J.* **52,** 480.

MITA, T., T. ONO and T. SUGIMURA (1959). *Zool. Mag., Tokyo* (*Dobutsugaku Zasshi*) **68,** 65 (in Japanese).

NISHI, A. and O. H. SCHERBAUM (1962a). Levels of pyridine nucleotides and DPNH oxidase activity in growing and stationary protozoan cultures. *Biochim. et Biophys. Acta* **65,** 411.

NISHI, A. and O. H. SCHERBAUM (1962b). Oxidative phosphorylation in synchronized cultures of *Tetrahymena pyriformis*. *Biochim. et Biophys. Acta* **65,** 419.

SCHERBAUM, O. (1956). Cell growth in normal and synchronously dividing mass cultures of *Tetrahymena pyriformis*. *Exptl. Cell Research* **11,** 464.

SCHERBAUM, O. (1957). The content and composition of nucleic acids in normal and synchronously dividing mass cultures of *Tetrahymena pyriformis*. *Exptl. Cell Research* **13,** 24.

SCHERBAUM, O. H. (1960a). Synchronous division of microorganisms. *Ann. Rev. Microbiol.* **14,** 283.

SCHERBAUM, O. H. (1960b). Possible sites of metabolic control during the induction of synchronous cell division. *Ann. N. Y. Acad. Sci.* **90,** 565.

SCHERBAUM, O. H., T. W. JAMES and T. L. JAHN (1959). The amino acid composition in relation to cell growth and cell division in synchronised cultures of *Tetrahymena pyriformis*. *J. Cellular Comp. Physiol.* **53,** 119.

SCHERBAUM, O. H. and M. LEVY (1960). Some aspects of the carbohydrate metabolism in relation to cell growth and cell division. *Xe Congrès International de Biologie Cellulaire*. L'Expansion Scientifique Française, Paris (abstract), p. 177.

SCHERBAUM, O. H. and M. LEVY (1961). Some aspects of the carbohydrate metabolism in relation to cell growth and cell division. *Pathol. et Biol. Semaine Hop.* **9,** 514.

SCHERBAUM, O. H., A. L. LOUDERBACK and T. L. JAHN (1958). The formation of subnuclear aggregates in normal and synchronized protozoan cells. *Biol. Bull.* **115,** 269.

SCHERBAUM, O. H., A. L. LOUDERBACK and T. L. JAHN (1959). DNA synthesis, phosphate content and growth in mass and volume in synchronously dividing cells. *Exptl. Cell Research* **18,** 150.

SCHERBAUM, O. and G. RASCH (1957). Cell size distribution and single cell growth in *Tetrahymena pyriformis* GL. *Acta Pathol. Microbiol. Scand.* **41,** 161.

SCHERBAUM, O. and E. ZEUTHEN (1954). Induction of synchronous cell division in mass cultures of *Tetrahymena pyriformis*. *Exptl. Cell Research* **6,** 221.

SCHNEIDER, W. C. (1945). Phosphorus compounds in animal tissues. I. Extraction and estimation of desoxypentose nucleic acid and of pentose nucleic acid. *J. Biol. Chem.* **161,** 293.

SCHOLANDER, P. F., L. VAN DAM, C. L. CLAFF and J. W. KANWISHER (1955). Microgasometric determination of dissolved oxygen and nitrogen. *Biol. Bull.* **109,** 328.

SINGER, W. and J. J. EILER (1960). The biological action of cellular depressants and stimulants. V. The effect of phenylurethane on cellular synthesis by *Tetrahymena pyriformis* GL. *J. Am. Pharm. Assoc., Sci. Ed.* **49,** 669.

WU, C. and J. F. HOGG (1956). Free and nonprotein amino acids of *Tetrahymena pyriformis*. *Arch. Biochem. Biophys.* **62,** 70.

ZEUTHEN, E. (1953). Growth as related to the cell cycle in single-cell cultures of *Tetrahymena pyriformis*. *J. Embryol. Exptl. Morphol.* **1,** 239.

ZEUTHEN, E. and O. SCHERBAUM (1954). Synchronous divisions in mass cultures of the ciliate protozoon *Tetrahymena pyriformis*, as induced by temperature changes. In *Recent Developments in Cell Physiology* (J. A. Kitching, ed). Butterworths, London, pp. 141–157.

7

Nucleotide Metabolism During Synchronized Cell Division in *Tetrahymena pyriformis*

PAUL PLESNER, *Biological Institute of the Carlsberg Foundation, Copenhagen, Denmark*

CONTENTS

I Introduction 197
II Definitions and Terminology 198
III The Timing of Changes in NTP Content 199
1. Material 199
2. The digitonin–luciferase method 199
A. Principle 199
B. Reagents 199
C. Apparatus 199
D. Reaction mixture 201
E. Procedure 201
F. Calibration curves 202
G. Comments 202
3. Results 203
IV Quantitative Estimation of NTP 205
1. Method and materials 205
2. Results 205
V The Components of the NTP Fraction 207
1. Material and methods 207
2. Results 208
VI The Effect of Inhibitors on the NTP Content 209
VII Discussion 209
References 211

I. INTRODUCTION

This chapter deals with the changes in the cellular content of nucleoside triphosphates (NTP) which occur during the synchronized division cycle in *Tetrahymena*.

The unpublished work by Plesner (quoted by Zeuthen, 1958) and by Scherbaum and Seraydarian (quoted by Scherbaum, 1960), which, at best, only bears on the changes in NTP content as affected by the synchronizing heat treatment, will not be included.

In synchronously dividing *Tetrahymena* the nucleotide moiety of the adenosine triphosphate (ATP) molecule turns over at a rate of the order of 200 per minute (Plesner, 1964). This fact calls for the utmost care in preparing the cells for analysis as is discussed in detail in chapter 23 of this book. From the arguments given in that chapter, it also follows that owing to the time error of observation inherent in most analytical methods for NTP, no single method will yield full information on all the following three points:

(1) The correct time relation between changes in morphology and changes in NTP content.

(2) The precise amounts of NTP.

(3) The composition of the NTP fraction.

In the following, which is a condensation of part of the work described by Plesner (1964), each point will, therefore, be discussed separately.

II. Definitions and Terminology

(1) *The division maximum* (defin. chapter 23, I) is the morphological event most easily recognized during an experiment in which available help is engaged in collecting samples as frequently as possible. The time of the division maximum has, therefore, been chosen as a fixed reference time point for defining a relative time scale on which the cellular events can be placed.

(2) *The relative time scale* is denoted T_R. The time from the end of heat treatment to the division maximum is divided into 100 units. (It should be noted that Holz (1960) uses a different reference point (T_R 100), namely the time when sister cells part.)

(3) *The actual time scale* is denoted T_N and is given in minutes.

(4) Both scales begin at the end of heat treatment (EH).

(5) *EH* is the time when the temperature control is switched from the high (34°C) to the low level (28°C). It takes a few minutes before the temperature in the culture reaches the low level.

(6) The admittedly colloquial term 'synchronized *Tetrahymena*' is used to describe a population of *Tetrahymena pyriformis* which has previously been exposed to a heat treatment which induces synchrony of cell division.

(7) *The critical point* is an expression which was introduced by Swann (1955) to indicate the time when the sensitivity of a synchronized cell population to CO and to dinitrophenol decreases sharply as estimated by the delay of the first subsequent division. Hamburger and Zeuthen (1957) found a pattern of changing sensitivity in synchronized *Tetrahymena* similar to that described by Swann for the sea urchin egg. The expression is here used as defined by Swann. The critical point occurs at T_R 67–70.

(8) Macronuclear events (swelling and elongation) starts at T_R 70–75. Stomatogenesis is finished at T_R 90.

(9) The time when fission stages first appear in synchronized *Tetrahymena* (the beginning of division) is less well defined than the division maximum, when fission furrows are fully developed in most cells. In the following, the time of the beginning of division is accepted as occurring five time-units before the division maximum, viz. at T_R 95. The time error ($\pm$ a few units) thus accepted does not influence conclusions based on data described in this article.

(10) *The death time* of the cells is the time when acetone is added to cells collected by centrifugation (see chapter 23, II) or it is the time of lysis in digitonin (see below, this page).

(11) *The time of observation* is identical with the death time.

III. The Timing of Changes in NTP Content

1. *Material*

Tetrahymena cells were synchronized according to the methods described in section I of chapter 23, this book. The cells were taken directly from the experimental culture into the reaction mixture without previous centrifugation.

2. *The Digitonin–Luciferase Method*

This is a modification of the ATP assay with firefly extract described by Strehler and McElroy (1957).

A. Principle

A sample of the experimental culture is pipetted into a mixture of firefly extract and digitonin which in advance has been placed *in situ* in a recording photometer. Upon lysis the ATP liberated reacts with luciferin and the emitted light is recorded. The concentration of digitonin must be so high that all of the cells added lyse simultaneously.

B. Reagents

Crude firefly extract was prepared according to McElroy (1955). The crude extract as used in the assay contains both luciferase and luciferin.

Digitonin was brought into solution by the method described by Seaman (1960). The concentration needed to lyse the cells within 5–10 sec was found to vary. In the experiments a final concentration of 0·1% digitonin procured at Calbiochem was used. The digitonin solutions were never kept for more than one day.

C. Apparatus

A Farrand fluorometer supplied with a photomultiplier tube was used as a photometer. A recorder (Varian G-10) was wired directly to the galvanometer terminals.

A cuvette holder as illustrated in Fig. 1 was inserted in the holder for the secondary filter and ordinary spectrophotometer

FIG. 1. Cuvette holder. Dimensions of the plate: 92 × 55 × 1 mm. The hole in front of the cuvette is 10 × 10 mm.

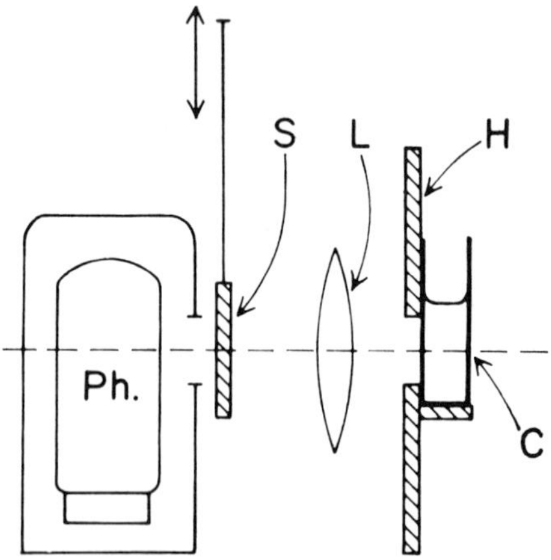

FIG. 2. Position of shutter (S) and cuvette holder (H) in the Farrand apparatus. L: lense; Ph.: photomultiplier tube; C: cuvette.

cuvettes (1 × 1 × 4 cm) were used for the assay. A shutter was placed in front of the photomultiplier tube in order to protect it against light when the lid covering the cuvette holder was removed. The position of the shutter and cuvette holder is shown in Fig. 2.

D. Reaction mixture

Glycyl–glycine buffer, 0·05*M*, pH 7·5	0·300 ml
$MgSO_4$, 1 *M*	0·010 ml
Na_2HASO_4, 1 *M*	0·005 ml
Digitonin solution	0·100 ml
Firefly extract	0·100 ml
Cell suspension	0·500 ml

When cells suspended in inorganic medium were analyzed, proteose–peptone medium (see chapter 23, section I) was substituted for the buffer. This was found empirically to be necessary for good lysis.

E. Procedure

The time schedule of the procedure is shown in Fig. 3a. The first five components were added in the order listed and the cuvette

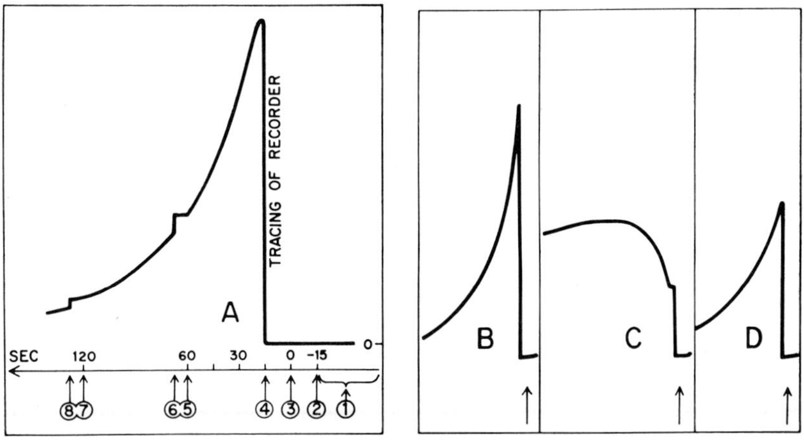

FIG. 3. Digitonin–luciferase method—(A) Diagrammatic representation of time course of assay procedure. (1) Buffer, $MgSO_4$, Na_2HASO_4 and digitonin were added. (2) Firefly extract was added to the reaction mixture and the cuvette was placed *in situ* in the apparatus. (3) The cell suspension was squirted into the reaction mixture and the cover replaced over the cuvette holder. (4) The shutter covering photomultiplier tube was opened. (5) Servo-switch was switched to 'off'. (6) Servo-switch was switched to 'on'. (7, 8) Repeat 5–6. B, C and D: Actual recordings of light emission of (B) 0·7 mμmoles of ATP, (C) 10·5 mμ moles of ADP, (D) 0·92 mμmoles of GTP to a reaction mixture containing lysed *Tetrahymena* cells (as described in the text).

holder carrying the cuvette was slid into position in the apparatus. Fifteen seconds after the digitonin had been added, the cell suspension was squirted into the reaction mixture with a syringe (no needle was used). The cover was then replaced and recording of the light was started 15 sec after the addition of the cells. Forty-five seconds later the servo-switch of the recorder was switched off for 10–15 sec. As the chart continued to move, the reading of the recorder at 60 sec after the addition of the cells was registered as a 'step' on the tracing of the recorder. The 60 sec value thus recorded was used in this work for the estimation of NTP. It was calculated as ATP by reading it against a calibration curve made with ATP (see section of comments).

F. Calibration curves

In each experiment a calibration curve was made by adding ATP to a reaction mixture, which contained digitonin, but no cells. Ideally, ATP should have been added to cells as an internal standard. This was not possible because then ATP should have been added precisely at the time of lysis of the cells. Thus, the procedure only permitted an estimate of the relative content of ATP during an experiment.

G. Comments

The light reaction is specifically dependent on ATP (Strehler and McElroy, 1957). However, in the *Tetrahymena* cells and in the firefly extract enzymes are present which catalyze the transfer of phosphate from other energy-rich compounds to the terminal phosphate group of ADP. This was demonstrated using a reaction mixture in which cells had been added, lysed and, in which the initial light flash had partially subsided. If at this time a substrate for one of the transferring enzymes expected to be present was added, a second light flash was recorded. In this way adenylate kinase and nucleoside diphosphokinase were shown to be present in the mixture of cell lysate and firefly extract. Thus, all the nucleoside triphosphates as well as nucleoside diphosphates contribute to the light reaction in the present system. It can, however, be seen from Figs. 3 B, C and D that the '1-min recording' of the light intensity, measured as described after the addition of ADP or of GTP, is only 6 or 60 per cent of that of ATP, respectively, when calculated on a molar basis.

'The relative content of ATP' in *Tetrahymena* is then a measure of the intensity of the light emitted by the cell lysate at 60 sec after the addition of the cells. The light as recorded, reflects mostly the

ATP present, and only to a lesser extent other compounds which are converted to ATP 1 min after the lysis of the cells. From the chromatographic results, it can be calculated (see p. 209) that those compounds which are converted to ATP during the first minute after lysis of the *Tetrahymena* cells contribute less than 20 per cent to the light intensity measured at 1 min. In the heat synchronized *Tetrahymena* cells the variations in 'the relative content of ATP' are, therefore, mainly the result of variations in the actual content of ATP.

3. *Results*

The timing of the changes in NTP content during the first synchronized division cycle is shown in Fig. 4, which is the original recording obtained from an experiment in which cells were resuspended in inorganic medium between the seventh and the eighth heat shock. Each peak on the curve represents a separate analysis in which the cells were squirted into the reaction mixture at the time given below the analysis. The chart drive motor was stopped between each analysis. It is seen that the NTP content was high immediately after EH. Then a decrease occurred, followed by an increase in NTP content which reached a maximum 10–12 min before the time of the division maximum. On the relative time scale, the maximum NTP content was recorded

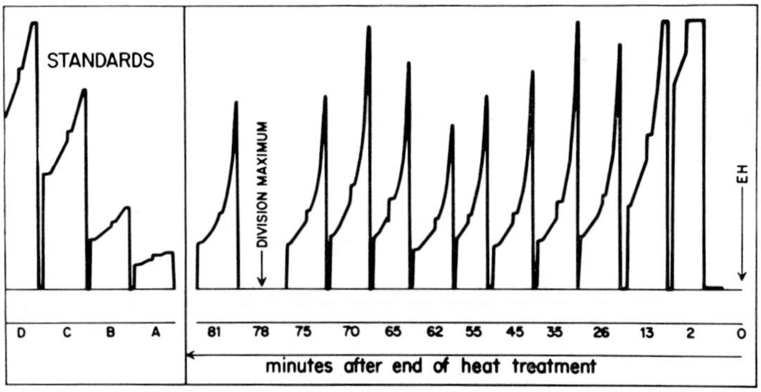

Fig. 4. Actual recording of a series of determinations of the relative content of ATP in *Tetrahymena* cells during the first synchronous division cycle. The time goes from right to left. 0·5 ml of cell suspension was added to a reaction mixture at the time indicated as described in Fig. 3A. The chart drive motor of the recorder was stopped between each analysis. The division occurred in inorganic medium (Type III experiment). The population density was 240·000 cells per ml at EH. A, B, C and D are records of addition of 0·45, 0·88, 1·76 and 2·64 mμmoles of ATP, respectively, to reaction mixtures.

approximately at T_R 85. At T_R 95 the content had decreased again. The NTP content was thus high for a period of 10–20 min preceding the cytoplasmic fission. At the conclusion of the cytoplasmic fission an additional transitory increase was recorded at T_R 110. This increase was defined by only one observation and it is not shown. Fig. 5 shows the recordings from an identical experiment, only the

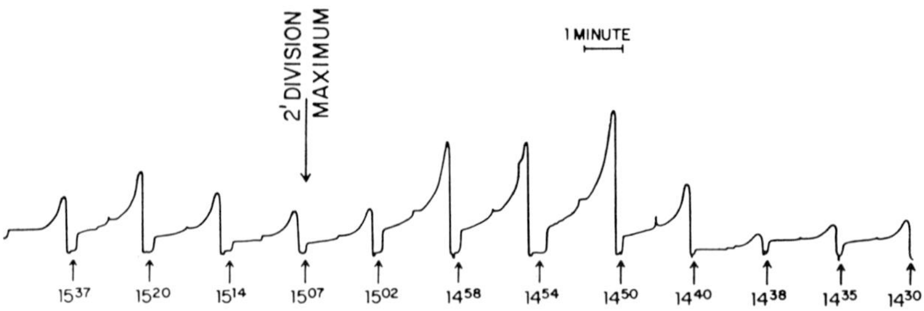

FIG. 5. Actual recording of a series of determinations of the relative content of ATP in washed *Tetrahymena* cells related to the second synchronous division. Experimental details as in Fig. 6. The first synchronous division had occurred at 13^{05}.

NTP content was determined in relation to the second synchronous division. Again, changes in the NTP content preceded the division. The level was high for a period of 10–20 min. The maximum was 17 min prior to the division maximum. At the time of the beginning of the division the content was again low. At the conclusion of cytoplasmic fission an additional peak occurred (13 min after the time of the division maximum). On a relative time scale, fixed by subdividing the time from the first to the second division maximum into 100 units, the pre-division maximum was found at approximately T_R 85, the second maximum was at approximately T_R 110.

In all experiments which showed a degree of synchrony comparable to that described in chapter 4, a pattern of changes in NTP identical to that described above was found *provided* the lysis of the cells was as effective as stated in the description on the digitonin–luciferase method. However, because in each experiment the peak NTP content at T_R 110 was never defined by more than one observational point this finding is not incorporated into the following general conclusions which are based on the results of the presented group of experiments: Preceding the cytoplasmic fission an increase in the content of NTP is consistently observed to start approximately at T_R 70, a maximal content is reached at T_R 85; when cytoplasmic fission starts (at T_R 95) the NTP content has again decreased.

IV. Quantitative Estimation of NTP

1. *Method and Materials*

A quantitative estimate of the NTP throughout the division cycle in synchronized *Tetrahymena* was obtained spectrophotometrically by the phosphoglycerate method in which the participation of ATP in the following sequence of enzymic reactions was used to determine NTP: 3-phosphoglycerate + ATP ⇌ 1,3-diphosphoglycerate + ADP; 1,3-diphosphoglycerate + DPNH + H^+ ⇌ 3-phosphoglyceraldehyde + PO_4^{2-} + DPN^+. The reactions were pulled to the right by trapping the 3-phospho-glyceraldehyde with cysteine. The oxidation of DPNH was followed spectrophotometrically.

The material for the analysis was acetone powder prepared as described in section II of chapter 23. The acetone powder was extracted with cold 5 per cent perchloric acid and the extract was fractionated with barium as described by LePage and Umbreit (1947). The barium-insoluble fraction was dissolved in HCl and used in the assay after neutralization and precipitation of the barium as sulphate. In the hands of the author this procedure gave more reproducible results than when neutralized perchloric acid extracts were used directly. The enzymes and substrates for the assay were obtained from C. F. Boehringer and Soehne, Mannheim, Germany.

2. *Results*

In Fig. 6 results from five experiments are shown using the relative time scale as abscissa. All curves follow a parallel course during the second half of the synchronized division cycle. They all show an increase in NTP preceding the division. Of the five curves shown the two which are on a high general level represent cells kept in proteose peptone. The three which are on a distinctly lower level represent cells which have been resuspended in the inorganic medium as described in section II of chapter 23 this book. The difference between the two sets of curves may reflect the difference in respiratory rate (Hamburger and Zeuthen, 1957) mentioned in chapter 23. It is noteworthy that the absolute increase in NTP, which precedes division, is of the same order both in cells in proteose peptone and in cells in the inorganic medium, i.e. about 0·15 μmoles per 100 mg acetone powder. In the live cells this corresponds to an increase in the average NTP concentration of about 0·25 mM. Thus, for cells in proteose peptone the average NTP concentration fluctuates between 0·30 mM and 0·60 mM. For cells which have been resuspended in inorganic medium, the same fluctuation is between 0·07 mM and 0·30 mM.

On the normal time scale, the occurrence of the division maximum varied from 95 to 160 min after EH in the five experiments. This variation probably mostly reflects differences in the temperature at which the culture was maintained after EH. The purpose of this variation is to demonstrate that on the relative time scale the demonstrated variations are independent of the duration of the first synchronized division cycle.

With the phosphoglycerate method a maximum content of NTP could also be demonstrated prior to the second synchronous division (Plesner, 1964).

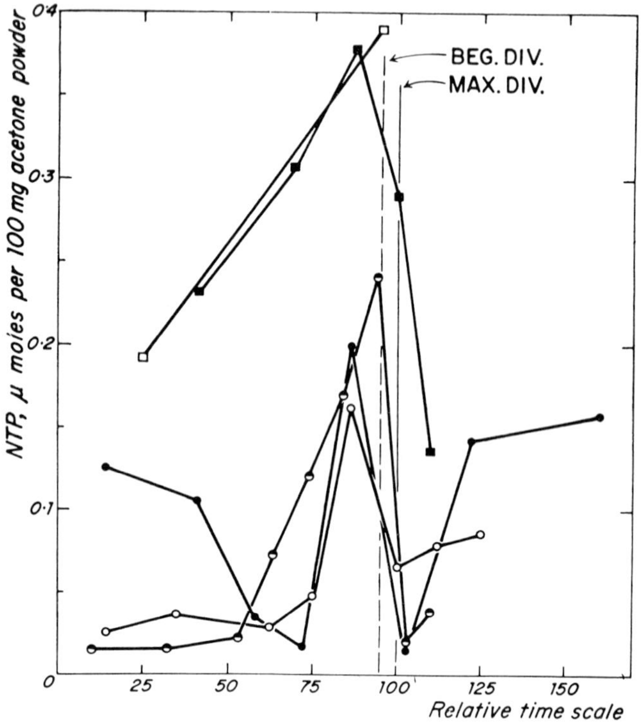

FIG. 6. The content of nucleoside triphosphate per 100 mg acetone powder of *Tetrahymena* cells during the first synchronous division cycle. Phosphoglycerate method. Points from the same experiment are connected with unbroken lines and denoted by identical symbols. Two upper graphs: results from cells dividing in proteose peptone. Three lower graphs: results from cells dividing in the inorganic medium.

The conclusions derived from the experiments described in this section are: The increase in the NTP content which precedes the cytoplasmic division is of the same magnitude in cells which divide in proteose peptone and in cells which divide in the inorganic medium. The increase is approximately 0·15 μmoles per 100 mg acetone powder. Lowest NTP contents are observed just prior to and just after the NTP peak and are about 0·20 μmoles per 100 mg acetone powder in proteose peptone cells and 0·05 μmoles per 100 mg acetone powder in cells which are resuspended in the inorganic medium. The changes in NTP and the occurrence of the division follow the same relative time course within wide limits independent of the duration of the first synchronized division cycle.

V. The Components of the NTP Fraction

1. *Material and Methods*

The components of the NTP fraction were identified by paper chromatography. The nucleotides were extracted from acetone powders with ice-cold 5 per cent perchloric acid and purified by adsorption on Norite and subsequent elution with 50 per cent ammoniacal ethanol. When the total recovery of the nucleotides was better than 50 per cent, then the percentage of recovery of each single nucleotide equalled the total recovery per cent (Plesner, 1964). The nucleotides were first separated by two-dimensional chromatography in (1) 1 *M* ammonium acetate, adjusted to pH 7·6 with ammonium hydroxide (300 ml), 96 per cent ethanol (750 ml) (Paladini and Leloir, 1952), and (2) borate solvent: 1 *M* ammonium acetate, adjusted to pH 9·3 with ammonium hydroxide and saturated with sodium tetraborate (300 ml), 96 per cent enthanol (750 ml) (Plesner, 1955). The spots containing the NTP were then cut out, eluted with water and rechromatographed in 0·1 *M* phosphate buffer, pH 6·8, saturated with ammonium sulphate (1 1.), 20 ml of *n*-propanol (Markham and Smith, 1951). The nucleoside triphosphates were identified by: (1) their R_f values, (2) the identity of the base moiety as revealed after hydrolysis in 1*N* HCl for 1 hr at 100° and chromatography in isopropanol (170 ml), concentrated hydrochloric acid (sp. g. 1·19) (41 ml), water to 250 ml (Wyatt, 1951), (3) the molar ratio between base–ribose–phosphate, (4) the ability to phosphorylate phosphoglycerate in the system described on p. 205 (the enzyme preparations contained nucleoside diphosphokinase).

2. *Results*

The only nucleoside triphosphates which could be identified in the perchloric acid extracts were ATP and GTP. Results from three experiments are shown in Fig. 7.

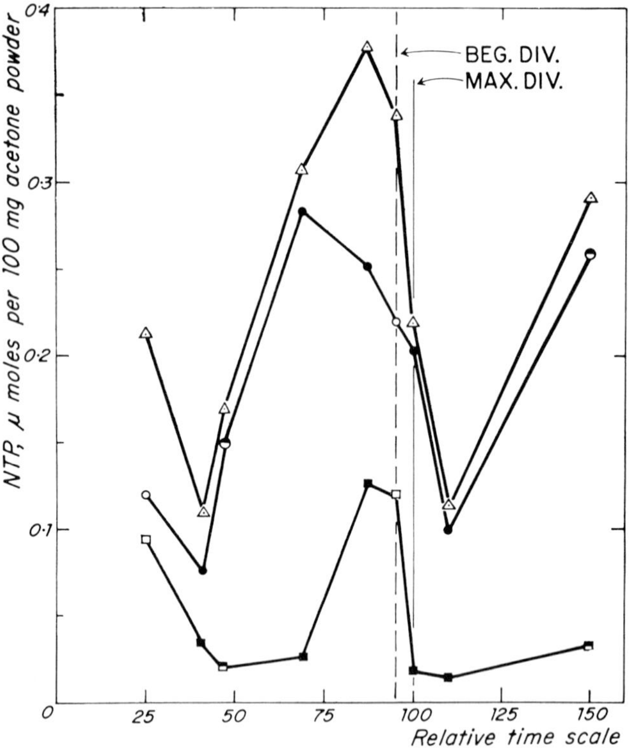

FIG. 7. The content of ATP and GTP determined by chromatography. Combination of data from three experiments in proteose–peptone.

Lower graph obtained by connecting points of observations obtained for GTP.

Middle graph: results for ATP from the same three experiments.

Upper graph: arithmetical sum of lower and middle graphs.

The data in Fig. 7 represent three experiments in proteose peptone. They can be fitted with two curves representing ATP and GTP. It is seen that both ATP and GTP accumulate prior to division. Because of the experimental errors, it is probably not safe to conclude that the maxima of the graphs differ in time.

It is, however, permissible to conclude that the NTP fraction consists of mainly ATP and, to a minor extent, of GTP. The GTP constitutes from one-third to one-fourth of the total amount of NTP at the time when prior to division a maximum content is found. We may also state that if other ribonucleoside triphosphates are present, they can not together represent more than 10 per cent of the total amount of ATP + GTP.

VI. The Effect of Inhibitors on the NTP Content

The effect of chloramphenicol and of *p*-fluorophenylalanine on synchronized *Tetrahymena* has been investigated with the digitonin–luciferase method (Plesner, 1964). The results showed that when a biological effect of the inhibitors was observed as a delayed division maximum, then the *immediate* effect on the ATP level was an increase.

VII. Discussion

The correlation, in time, of changes in NTP with changes in other metabolic parameters of *Tetrahymena* is given in Fig. 8. The curves are meant to illustrate only when events occur, and if a parameter decreases or increases. No quantitative representation is attempted.

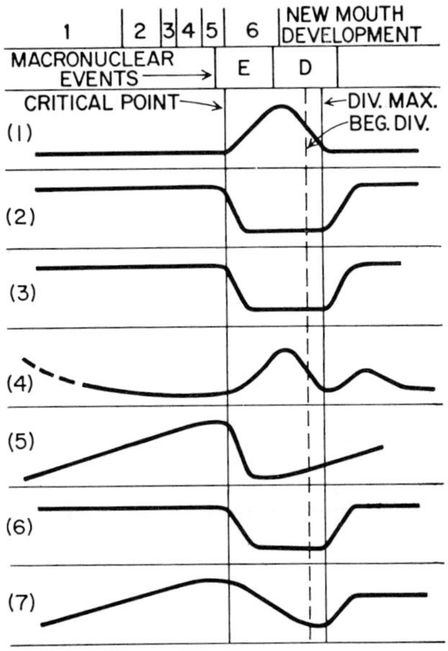

Fig. 8. Qualitative changes in the metabolic patterns during first synchronous division in *Tetrahymena* cells. Interpreted data. Stages of oral development: see text. Macronuclear events: E = elongation, D = division. (1) Rate of RNA synthesis. (2) Rate of DNA synthesis. (3) Rate of protein synthesis. (4) ATP content. (5) Duration of the delay of division caused by DNP. (6) Respiratory rate. (7) Rate of $^{32}PO_4$ uptake. (See text for references.)

The curves are the Author's interpretation of the following information from the literature:

Stages of mouth development (synchronized cells), Frankel (1962).

(1) Loose anarchic field with indefinite margins.
(2) Compact anarchic field with defined outer border.
(3) Formation of anterior part of one or two membranelles.
(4) All the membranelles in adoral zone present; anterior part of undulating membranelle being shaped.
(5) Undulating membrane is completed.
(6) Oral cavity forms.

Macronuclear events (normal cells), Holz (1960).

E: macronuclear elongation.
D: macronuclear division.

Rate of RNA synthesis (normal cells), Prescott (1960); synchronized cells, Plesner (1964).

Rate of DNA synthesis (normal cells), Prescott (1960), McDonald (1958) synchronized cells; Cerroni and Zeuthen (1962), (see, however, chapter 4).

Rate of protein synthesis (synchronized cells), Christensson (1959), Hamburger and Zeuthen (1960).

ATP content (synchronized cells), present work.

Effect of DNP (synchronized cells), Hamburger and Zeuthen (1957).

Respiratory rate (normal and synchronized cells), Zeuthen and Scherbaum (1954).

Rate of $^{32}PO_4$ *uptake* (synchronized cells), Hamburger and Zeuthen (1962).

A detailed discussion of the information contained in Fig. 8 is given by Plesner (1964). Only a summary will be given here.

(1) The decrease in sensitivity to DNP in respiratory rate and in uptake of $^{32}PO_4$ after the critical point is evidence of decreased consumption of NTP during this period. Thus, the increase in ATP prior to the division reflects decreased utilization, not increased rate of synthesis.

(2) The results by Scherbaum, Louderback and Jahn (1959) on the content of phosphates and the author's results (Plesner, 1964) relating to the cellular contents of NMP and of ADP suggest a decrease in pyrophosphorylytic activity after the critical point.

(3) Probably the decrease in the rate of protein synthesis which is shown in Fig. 8 results in the decreased utilization by pyrophosphorolytic cleavage of NTP.

(4) It is proposed that the 'passive' accumulation of NTP exerts a physiological function by creating favourable conditions for the release from the ribosomes of the 'division protein', as defined by Zeuthen (1961).

The theory here proposed has been submitted to various tests by the author and it has thus far been substantiated. The *in vitro* release of proteins was studied using ribosomes from different stages of the synchronized *Tetrahymena* cycle. The stability of the ribosomes to low magnesium ion concentration was evaluated using ultracentrifugal analysis. The *in vivo* release of isotopically-labelled protein from the particles was also studied (Plesner, 1961, 1963).

REFERENCES

Cerroni, R. and E. Zeuthen (1962). Asynchrony of nuclear incorporation of tritiated thymidine into *Tetrahymena* synchronized for division. *Compt. Rend. Trav. Lab. Carlsberg* **32,** 499.

Christensson, E. (1959). Changes in free amino acids and proteins during cell growth and synchronous division in mass cultures of *Tetrahymena pyriformis*. *Acta Physiol. Scand.* **45,** 339.

Frankel, J. (1962). The effects of heat, cold, and *p*-fluorophenylalanine on morphogenesis in synchronized *Tetrahymena pyriformis* GL. *Compt. Rend. Trav. Lab. Carlsberg* **33,** 1.

Hamburger, K. and E. Zeuthen (1957). Synchronous divisions in *Tetrahymena pyriformis* as studied in an inorganic medium. The effect of 2, 4-dinitrophenol. *Exptl. Cell Research* **13,** 443.

Hamburger, K. and E. Zeuthen (1960). Some characteristics of growth in normal and synchronized populations of *Tetrahymena pyriformis*. *Compt. Rend. Trav. Lab. Carlsberg* **32,** 1.

Holz, G. G., Jr. (1960). Structural and functional changes in a generation in *Tetrahymena*. *Biol. Bull.* **118,** 84.

LePage, G. A. and W. W. Umbreit (1947). In *Manometric Techniques and Related Methods for the Study of Tissue Metabolism* (W. W. Umbreit, R. H. Burris and J. F. Stauffer, eds.). Minneapolis, p. 161.

Markham, R. and J. D. Smith (1951). Chromatographic studies of nucleic acids. *Biochem. J.* **49,** 401.

McDonald, B. B. (1958). Quantitative aspects of deoxyribose nucleic acid (DNA) metabolism in an amicronucleate strain of *Tetrahymena*. *Biol. Bull.* **114,** 71.

McElroy, W. D. (1955). Cypridina and firefly luciferase. In *Methods in Enzymology*, Volume II (S. P. Colowick and N. O. Kapland, eds.). Academic Press, New York, pp. 851–856.

Paladini, A. C. and L. F. Leloir (1952). Studies on uridine-diphosphate-glucose. *Biochem. J.* **51,** 426.

Plesner, P. (1955). Methods for studying nucleotide metabolism in synchronized cultures of Protozoa. *Acta Chem. Scand.* **9,** 197.

Plesner, P. (1961). Changes in ribosome structure and function during synchronized cell division. *Cold Spring Harbor Symp. Quant. Biol.* **26,** 159.

PLESNER, P. (1963). Nucleotide metabolism and ribosomal activity during synchronized cell division. In *Cell Growth and Cell Division* (R. J. C. Harris, ed.). Academic Press, New York, pp. 77–91.

PLESNER, P. (1964). Nucleotide metabolism during synchronized cell division in *Tetrahymena pyriformis. Compt. Rend. Trav. Lab. Carlsberg* **34**, 1.

PRESCOTT, D. M. (1960). Relation between cell growth and cell division. IV. The synthesis of DNA, RNA, and protein from division to division in *Tetrahymena. Exptl. Cell Research* **19,** 228.

SCHERBAUM, O. H., A. L. LOUDERBACK and T. L. JAHN (1959). DNA synthesis, phosphate content and growth in mass and volume in synchronously dividing cells. *Exptl. Cell Research* **18,** 150.

SCHERBAUM, O. H. (1960). Possible sites of metabolic control during the induction of synchronous cell division. *Ann. N. Y. Acad. Sci.* **90,** 565.

SEAMAN, G. R. (1960). Large-scale isolation of kinetosomes from the ciliated protozoon *Tetrahymena pyriformis. Exptl. Cell Research* **21,** 292.

STREHLER, B. L. and W. D. MCELROY (1957). Assay of adenosine triphosphate. In *Methods of Enzymology,* Volume III (S. C. Colowick and N. O. Kaplan, eds.). Academic Press, New York, pp. 871–873.

SWANN, M. M. (1955). The mechanism of cell division: The action of 2, 4-dinitrophenol and certain glycolytic inhibitors on the sea urchin egg. *Proc. Roy. Phys. Soc. Edinburgh* **24,** 5.

WYATT, G. R. (1951). The purine and pyrimidine composition of deoxypentose nucleic acids. *Biochem. J.* **48,** 584.

ZEUTHEN, E. (1958). Artificial and induced periodicity in living cells. *Advanc. in Biol. and Med. Phys.* **6,** 37.

ZEUTHEN, E. (1961). Cell division and protein synthesis. In *Biological Structure and Function,* Volume II (T. W. Goodwin and O. Lindberg, eds.). Academic Press, London–New York, pp. 537–548.

ZEUTHEN, E. and O. SCHERBAUM (1954). Synchronous divisions in mass cultures of the ciliate protozoon *Tetrahymena pyriformis* as induced by temperature changes. *Proc. Symp. Colston Res. Soc.* **7,** 141.

8

Induction of Synchronous Division in Amoebae

ROBERT J. NEFF and RUTH H. NEFF,* *Vanderbilt University, Nashville, Tennessee*

CONTENTS

I Introduction 214
II Synchronization of Division in Polyxenic Non-agitated Cultures of *Amoeba proteus* and *Endamoeba terrapinae* 214
III Synchronization of Division in Axenic Non-agitated Cultures of *Acanthamoeba sp.* 216
1. Experimental conditions and procedures 217
2. Temperature cycles 217
A. Temperature parameters of growth 217
B. Cycles between optimal and sublethal temperatures . . 218
C. Cycles between temperatures within the growth range . 220
3. Addition of fresh medium 222
A. Single addition 223
B. Repeated additions 224
IV Synchronization of Division in Aerated Cultures of *Acanthamoeba sp.* 225
1. Experimental conditions 225
2. Temperature cycles 226
A. Temperature shifts between optimal and sublethal temperatures 226
B. Temperature shifts between two temperatures within the growth range 229
3. Effects of aerobic–anaerobic treatment 232
V Evaluation of Results 234
1. Mechanisms of synchrony induction 234
A. Kinetic hypotheses 234
B. Rate-dependent temperature effects 235
C. Structural hypotheses 237
D. The mitotic apparatus 237
2. Nutrition and synchrony induction 239
3. Length of cycle and synchrony induction 239
VI Conclusions and Perspectives 242
References 244

* Unpublished work of the authors, summarized in this article, was supported by grants (E–623, E–1908, RG–6731) from Public Health Service, National Institutes of Health, U.S.A., and by grants from the Natural Sciences Committee, Vanderbilt University. Support is gratefully acknowledged.

I. Introduction

Synchronous division has been experimentally induced in mass cultures of only three amoebae species (Dawson *et al.*, 1937; James, 1959; Neff, 1955; Roslansky, 1958); in none is the degree of induced synchrony spectacular. This fact is surprising since very high degrees of naturally occurring *division synchrony* have been observed in amoebae cultures. In one case—a soil amoeba growing in axenic culture—a spontaneous division synchrony has been observed amounting to as much as 60–70 per cent of the cells in division at one time (Neff, unpublished data). Furthermore, the highest known degrees of naturally occurring *mitotic synchrony* have been observed in amoeboid cell types—the slime molds (Howard, 1932; Lister and Lister, 1925) and the multinucleate amoebae (Kudo, 1947). In these forms, the synchrony of the mitotic stages equals or surpasses that found in the first few divisions of a mass culture of simultaneously fertilized echinoderm eggs. These observations lead the authors to two inescapable conclusions; (1) a near perfect synchrony should be inducible in amoebae cultures, and (2) all of the secrets of synchrony induction in amoebae have not yet been unlocked.

In the amoebae, investigations have largely been channelled into the exploration of treatments that induce synchrony. Because the degree of experimental synchrony attained is low, synchronized amoeba cultures have been used only to a limited extent in the recent studies on the life cycle of the cell (Prescott, 1961). From these explorations and from related experiments, inferences have been drawn as to the controlling mechanisms of cell growth and division.

In this paper, the authors will review the pertinent literature and will also discuss data from their own laboratory. Much of the latter has had a limited circulation; some of it is recent. In the final section, an evaluation of the status of synchrony induction in amoebae will be made, and a structural hypothesis presented.

II. Synchronization of Division in Polyxenic Non-agitated Cultures of *Amoeba proteus* and *Endamoeba terrapinae*

The growth and division of amoebae, like other cells, are functions of temperature. The use of a temperature shift to induce a degree of division synchrony in *Amoeba proteus* was early reported by Dawson *et al.* (1937). Thriving cultures were placed in a refrigerator at 3–4°* overnight, and transferred to 27–29° the next morning; the cultures yielded enriched numbers of dividing amoebae within

*All temperatures given in the text are in centigrade.

2 to 4 hr after returning to the higher temperature. The degree of synchrony was not indicated.

Recently, a similar single cold treatment has been used by Roslansky *et al.* (1958) to induce a high degree of division synchrony in *Endamoeba terrapinae.* Cultures were shocked at 4° for 30 min and rapidly returned to their optimal growth temperature, 25·5°. Following a lag of less than one full generation time (= 7·5 hr), *all* cells divided within a 15-min period. Unfortunately, the cultures enter a 'death phase' following a second post-treatment division.

The first serious attempt to synchronize divisions in *A. proteus* by temperature treatment was made by James (1959). Apparently his working hypothesis was that each specific cellular activity (viz., mitosis, duplication, feeding, movement, etc.) has a characteristic activation energy. Within the temperature range tolerated by the amoeba, any given activation energy may be constant, or temperature dependent. By the judicious experimental manipulation of temperatures and time, the different activities in one cell can be modulated or dissociated to such an extent that the total processes in that cell become 'aligned' with those of all other cells in the culture. In support of this hypothesis, James demonstrated two activation energies for overall growth and division (as defined by generation time); a smaller one above, and a larger one below 20°. He cited other data demonstrating that mitosis has an activation energy which is constant in the same temperature range where those of food capture and movement are temperature dependent. The rates of both the latter activities decrease as the temperature increases.

In his experiments James used polyxenic cultures of *A. proteus* fed on washed *Tetrahymena*. As a synchrony-inducing treatment, he employed temperature cycles which paralleled diurnal fluctuations in temperature, and which permitted one complete division cycle every twenty-four hours. One successful cycle was 12 hr at 18° and 12 hr at 26°. The temperature shifts were slow; about 6 hr was required to go from 26° to 18° and less than 2 hr from 18° to 26°. Although the cell numbers doubled every 24 hr, the degree of synchronization achieved after five cycles was not high. The generation time of the cycled culture was shorter than that predicted from the known generation times of cultures grown at each temperature. The majority of divisions occurred at 26° shortly after the culture was returned to the higher temperature. James reports that cycles using other periods and temperatures also induce synchrony, as long as doubling in numbers is completed in phase with the cycle. He notes also that lower temperatures cause a progressive lag in the period during which divisions occur—while the opposite happens with higher temperatures. The extent to which the synchronization can be attributed to temperature is difficult to assess since the cells are also periodically fed.

James (1959) has reported an interesting discovery in *A. proteus* which may implicate the nucleus directly in the temperature induction of synchrony. Using interference microscopy, he determined the dry mass of the nucleus of cells undergoing temperature shock. If a cell growing at 22° was given a short and rapid cold shock (30 min at 10°) and was then returned to 22°, its nuclear mass decreased by at least 10 per cent. A short heat shock at 35° also resulted in a loss of nuclear mass. The effect resembles that observed by Gruber (1912) and by Prescott (1956b), who found that the volume of the nucleus decreased if part of the cytoplasm were amputated. James concluded that temperature shocks cause a loss of nuclear material which 'may set nuclear synthesis of all cells back to a common starting point'. Changes in cytoplasmic mass were also observed.

Using James' (1954) method, Prescott (1955; 1956a, b; 1959) and Mazia and Prescott (1954) used small cultures of synchronously growing *A. proteus* for studies of physical–chemical events in its life history. Synchronous cultures were achieved in part by selection. Dividing cells were isolated from a partially synchronized culture, and pooled. Such cultures, grown under uniform conditions, show spread in the subsequent time of division even of sister cells; growth and division tend to become random with time.

III. Synchronization of Division in Axenic Non-agitated Cultures of *Acanthamoeba sp.*

Early observations of natural synchrony in axenic cultures of *Acanthamoeba* suggested that a high degree of synchrony might be achieved easily by manipulation of the environmental conditions. This has not proved to be the case. However, many useful data have been accumulated from studies of this organism; they will be summarized in the succeeding sections.

In all experiments on *Acanthamoeba*, growth and division has been followed by counting. Several counting methods have been employed, the details of which are given elsewhere in this volume (chapter 21).

In non-agitated ('stationary') cultures, if the sampling interval was 1 hr or less, the smooth straight growth curve was resolved into a series of abrupt rises and plateaus which frequently lasted about one-fifth of the generation time. If a clearly defined plateau was observed in the growth curve, a plateau was again found at the time when the cell numbers had doubled. These repeated plateaus have been called 'doubling-number plateaus'. Their existence in untreated cultures suggests that a partial growth synchrony exists in many shelf cultures, and that the generation time of individual cells in a culture is constant.

In stationary cultures, the effects of two environmental factors—temperature and food—were studied extensively.

1. *Experimental Conditions and Procedures*

Before attempts at experimental synchronization were made, the optimal growth conditions were defined. They were adopted as standard conditions for the work described in section III.

The growth medium supporting optimal growth was: 1·5 per cent proteose peptone (Difco); 1·5 per cent glucose or maltose; ferrous sulfate, calcium chloride and magnesium sulfate, each 3 mg/l.; potassium phosphate, 0·002M; pH, 5·5.

The most rapid growth occurred at 31–33°; at this optimal temperature, the generation time is 12 hr. Temperature was a variable in many synchronization experiments. In them, the temperature was controlled by suspending cultures in one of several thermostatically regulated water baths. Temperatures of the cultures were changed by shifting them manually from one water bath to another. The temperature equilibration was rapid, since the culture volumes were small.

Agitation of cultures decreased their growth rate. In experiments on stationary cultures, all unnecessary agitation of cultures was avoided and handling minimized by the following procedure. The experimental treatments were imposed on freshly inoculated, relatively large cultures. At the end of the treatment, the large culture was dispensed into small sampling flasks, using sterile techniques. The samples were then placed in a water bath at the chosen temperature, and were not disturbed further. At each sampling interval, duplicates were fixed and counted. The agreement between duplicates was close, even after 90 hr.

2. *Temperature Cycles*

A. Temperature parameters of growth

In preliminary experiments, the growth rates and patterns of amoebae cultures were determined at each of a series of temperatures ranging from 22 to 37°. The generation time was temperature dependent and at all temperatures up to and including 34° it was reproducible within ± 0·5 hr. At 22°, the lowest temperature tested in these experiments, the generation time was 21 hr. The most rapid growth occurred at 31–33°; at this optimal temperature, the generation time was 12 hr. At 35° the generation time was variable ranging from 20 to 28 hr. At 36° and 37°, no divisions occurred. The cell numbers remained constant for at least a day, and then decreased. At these temperatures, the surviving cells became progressively smaller. By the end of four days, at least half of them had disintegrated. The remaining cells were still viable, since, if returned to lower temperatures, they increased in size and divided. They required a considerable period of time to recover. In one case, cells at 37° for 4 days did not resume divisions until 96 hr after they were put at a growth temperature. Thus, 36–37° is the sublethal temperature range for *Acanthamoeba*. The lower temperature limit has not been determined.

B. Cycles between optimal and sublethal temperatures

Shortly after these studies were initiated, Scherbaum and Zeuthen (1954) reported that cycling between optimal and 'sublethal' temperatures in a time cycle slightly shorter than the generation

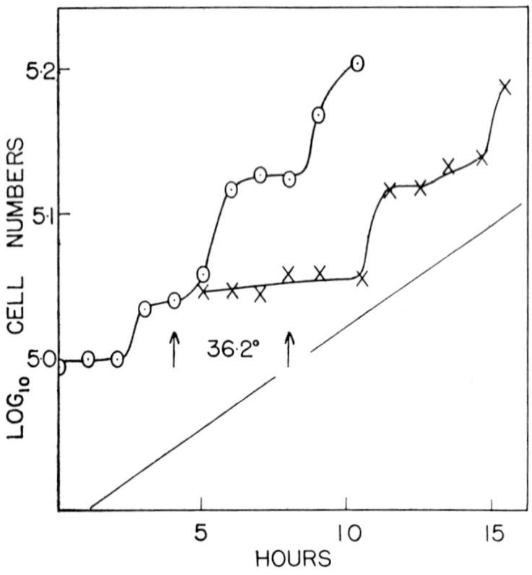

FIG. 1. The effect of a single 4-hr sublethal temperature treatment on growth of a stationary culture of *Acanthamoeba sp.* ⊙ Control; grown at 22° throughout the experiment. × Experimental; grown at 22° for 4 hr, transferred to 36° for 4 hr, and returned to 22° for the next 8 hr. The straight line shows the theoretical growth curve for a stationary culture at 22°; the G.T. is 21 hr.

time resulted in a division synchrony in *Tetrahymena*. Following this lead, the first attempts at synchronization of *Acanthamoeba* tested the effects of temperature cycles between its optimal and sublethal temperatures.

TABLE I. The effects of cycling stationary cultures of *Acanthamoeba* between 32°C and 36°C

Length of cycle (hr)	Time culture incubated at 32·2°	Time culture incubated at 36·2°	Number of cycles	Maximum modification of usual division cycle observed in culture
12	1	11	5	None
12	11	1	4	None
12	6	6	7	None
12	4	8	4	None
12	8	4	5	50% of divisions in 1/3 to 1/2 of generation time (G.T.)[a]
12	8	4	9	50% of divisions in 1/6 of G.T.
12	8	4	11	50% of divisions in 1/4 of G.T.
8	4	4	5	50% of divisions in 1/6 of G.T.
8	4	4	11	50% of divisions in 1/3 of G.T.

[a] The generation time at 32·2°C is 12 hr. All cultures were maintained at 32·2°C during sampling.

(*a*) *Single cycles*. Young cultures were incubated at a temperature, either 22 or 32°, supporting good growth for at least 24 hr, and then transferred to 36° for a time (usually 2–4 hr). After this treatment at 36°, the culture was returned to the original temperature.

Placing actively dividing cells at the sublethal temperature (36°) caused an immediate cessation of divisions. When the cultures were returned to the lower temperature, the division block continued for a period of time roughly equal to the length of the heat treatment. At the end of this period of inhibition, divisions resumed. Thereafter the curves of the treated cultures closely paralleled those of untreated controls. Fig. 1 shows a representative experiment. The experimental and control curves are similar, but the experimental curve is delayed 7 hr. Because of the rises and plateaus observed in both growth curves of stationary cultures, it is difficult to decide whether the *rate* of growth of a culture following a heat treatment is greater than that of the control.

(*b*) *Multiple cycles*. In additional attempts to induce synchrony by sublethal heat treatments, nine multiple-cycle experiments were performed. They are summarized in Table I. Variations of any magnitude were obtained *only* in those experiments where the cultures were treated at 36·2° for 4 hr in either an 8 or 12-hr cycle.

Fig. 2 shows the growth curves of cultures exposed to a cycle of 8 hr at 32·2° followed by 4 hr at 36·2°, i.e., 8 (32°)–4 (36°). It is apparent that repeated sublethal shocks can induce a degree of growth and division synchrony in stationary cultures of *Acanthamoeba*. The initial lag may be attributed in part to the addition of fresh medium, an effect which will be discussed more fully in section III-3. In Fig. 2, the growth of another culture after seven repetitions of a 12-hr cycle—6 (32°)–6 (36°)—is also shown. The latter represents experiments in which little or no synchronization was observed.

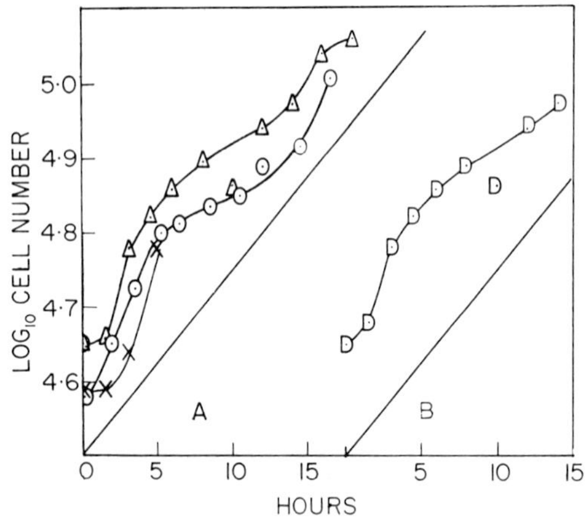

FIG. 2. The effect of multiple temperature cycles on the growth of a stationary culture of *Acanthamoeba sp.*

A. Cycle: 12 hr, 8 (32°)–4 (36°).
⊙ growth at 32° after 5 cycles.
× growth at 32° after 9 cycles.
△ growth at 32° after 11 cycles.

B. Cycle: 12 hr, 6 (32°)–6 (36°).
D growth after 7 cycles.

The straight line shows the theoretical growth rate at 32°; the G.T. is 12 hr.

C. *Cycles between temperatures within the growth range*

A second series of experiments was conducted to determine the effects of rapid temperature shifts between 22° and 32°. Both temperatures support growth adequately.

(*a*) *Single shifts and single cycles.* In the single-shift experiments, cells growing at 22° for 24 hr were transferred to 32° and the growth at the second temperature followed. Control cultures remained at the original temperature. The inverse experiment was also performed. All single-cycle experiments were with cultures growing at

22°; they were transferred to 32° for 4 hr, and returned to 22°. Controls grew at 22° throughout. Counts were initiated at the first temperature shift.

The results of several experiments may be summarized as follows: if the temperature shift coincided with a sharp rise in cell numbers, cell division stopped immediately (within 30 min) and was inhibited for 1 to 3 hr. The inhibition occurred whether the temperature was raised or lowered. The lag plateau was repeated at the doubling number. A 4-hr initial treatment at 32° decreased the generation time from an average of 21 hr to 18 hr. An example of a long initial inhibition following a single shift from 22° to 32° is shown in Fig. 3.

The lags found following all temperature shifts suggested that cells close to division are inhibited from dividing by the change in temperature. The sharp rise following the lag and the appearance of a doubling-number plateau show that the treatment effectively synchronizes part of the population.

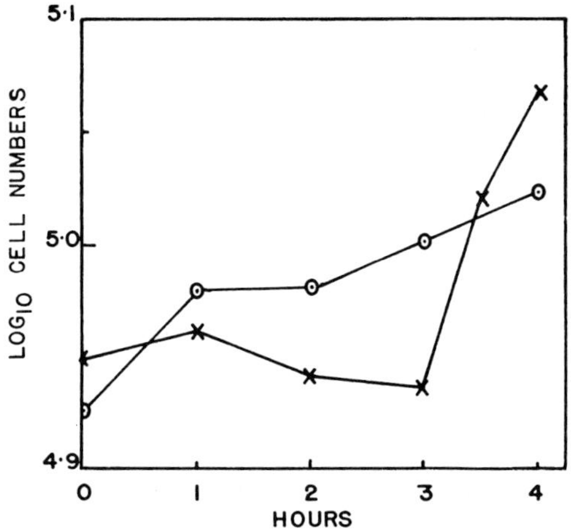

FIG. 3. The effect of shifting a stationary culture of *Acanthamoeba sp.* from 22° to 32°.

⊙ Control; culture at 22° throughout.

× Experimental; culture transferred to 32° at zero time after an initial period of growth at 22°.

(*b*) *Multiple cycles*. In multiple-cycle experiments between 22° and 32°, cultures were grown at 22° and transferred to the optimal growth temperature for 1 hr in each cycle. Two experiments using a longer 32° treatment were performed. The cycled cultures were diluted to counting range just before the final 32° treatment, so that the final treatment occurred with the samples already distributed into the sampling flasks.

The results of ten experiments are summarized in Table II. Of the cycles tested, five repetitions of either a 19-hr cycle—18 (22°)–1 (32°)—or a 20-hr cycle—19 (22°)–1 (32°)—were most effective in inducing synchronous division. The growth curves of cultures which had undergone ten cycles were of the types observed in the controls, indicating that the synchrony imposed by five cycles was lost or cancelled during further repetitions of the cycle. Cycles shorter than 18 hr or longer than 20 hr were ineffective when repeated a moderate number of times (4–6 cycles). All these experiments indicate that the timing of the temperature shifts is very critical.

TABLE II. The effects of cycling stationary cultures of *Acanthamoeba* between 22°C and 32°C

Length of cycle (hr)	Time culture incubated at 22·0°	Time culture incubated at 32·2°	Number of cycles	Maximum modification of usual division cycle observed in culture
16	14	2	6	None
18	12	6	5	50% of divisions in 1/6 of G.T.[a]
18	17	1	5	20% of divisions in 1/10 of G.T.
19	18	1	5	60% of divisions in 1/4–1/5 of G.T.
19	18	1	5	70% of divisions in 1/4 of G.T.
19	18	1	10	None
20	19	1	5	70% of divisions in 1/3 of G.T.
20	19	1	10	None
21	20	1	5	None
22	21	1	4	None

[a] In this experiment, the culture was maintained at 32·2°C during sampling. In all other experiments listed in this table, the temperature during sampling was 22°C.

At 32·2°, the average generation time is 12 hr.

At 22·0°, the average generation time is 21 hr.

3. *Addition of Fresh Medium*

A number of observations on the occurrence of lag plateaus following transfer of cells to fresh medium suggested that the addition of fresh medium produced an effect on a growing culture similar to that produced by a temperature shift. Several experiments were performed to assay the effect of adding fresh food at a constant temperature. Agitation was not adequately controlled in the preliminary experiments (a single addition) but was controlled in all subsequent experiments by subjecting control cultures to the same handling as the experimental ones.

A. Single addition

Portions of a culture growing at 22° were transferred to equal volumes of fresh media. Centrifugation and resuspension in fresh medium were performed at 22° using sterile technique. Results of a representative experiment are shown in Fig. 4. The experimental curves show an initial inhibition lasting 4–6 hr; in each curve, the inhibition plateau occurs at the same cell count level as that of the control culture, demonstrating that no cells were lost in the procedure.

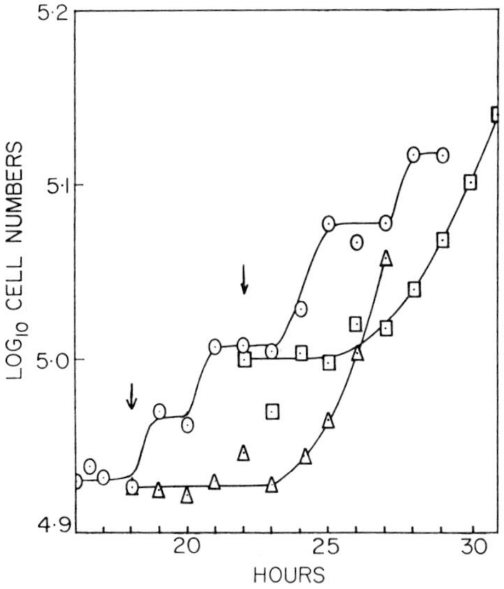

FIG. 4. The effect of replacing medium on the growth of a stationary culture of *Acanthamoeba sp.*

⊙ control; cells transferred to fresh medium at zero time.

△ medium replaced at 18 hr.

⊡ medium replaced at 22 hr.

The most interesting aspect of this experiment is the even and rapid rate of increase of the experimental cultures following the initial lag as compared with the rate of the control. This pattern is interpreted as follows: the addition of fresh medium (or the agitation accompanying the transfer, or both) inhibits those cells on the threshold of division more strongly than other cells in the culture; it forces them to divide at a later time in the doubling cycle with other cells whose divisions would normally occur somewhat later.

Although it is not shown in Fig. 4, in similar experiments where the cultures were followed for at least 24 hr after the addition of new medium, a constant doubling-number plateau was found at 21–23 hr after the addition was made. This fact strongly indicates that the experimental alteration of growth was not transitory, but that the treatment imposed a repeated pattern on the culture for at least one succeeding division. In Fig. 4, a doubling-number plateau is seen in the control curve at 21–23 hr; this results from the inoculation of the culture into new medium at time zero. The permanence of the effect of adding fresh medium to a culture is also supported by the multiple 1 and 2-hr plateaus found rather consistently in stock cultures during the log phase of growth.

B. Repeated additions

The results of single additions of fresh medium to growing cultures suggested that if the additions were repeated at intervals shorter than the initial lag, then divisions might be prevented indefinitely—or at least for as long as the treatment was continued. If the experimental block were removed, an increased degree of synchrony might result.

According to this rationale, experiments were conducted in which medium was replaced at 4, 2 and 1·5-hr intervals. Complete inhibition of the cultures could be maintained no longer than 6–7 hr in any of these treatments. Furthermore, although the agitation resulting from transferring cells never completely inhibited division, the growth rate of a centrifuged control was less than that of an uncentrifuged control. The handling did 'smooth out' the growth curves; except for the initial lag, no plateaus were found.

It is clear that the addition of fresh medium produces an effect on a growing culture similar to that of a rapid temperature shift. However, cells must adapt themselves quickly to the periodic replacement of medium. After a single treatment, a second or repeated addition of medium has little effect on the cells entering a new division cycle. Recent work by Benton in this laboratory indicates that the reaction to adding fresh medium is probably due to the lowering of the internal pH of the cells. With lowered internal pH protoplast contraction occurs as in nitrogen treatment (section IV-3).

IV. Synchronization of Division in Aerated Cultures of *Acanthamoeba sp.*

1. *Experimental Conditions*

Treatments that had successfully synchronized divisions in small stationary cultures were applied to 1-l. aerated cultures, in order to provide large numbers of cells for biochemical studies, and to achieve greater uniformity in the culture. *None* of the treatments were effective. Two differences were noted between the growth curves of the two types of culture. First, the mean G.T. was greatly increased in aerated cultures. At 30° the G.T. was 28 $\pm$ 2 hr for aerated cultures as compared to 12 $\pm$ 0·5 hr for stationary cultures. Second, the sharp rises and short plateaus, so characteristic of stationary cultures, were not observed in aerated cultures.

Accordingly, conditions for maximum growth in mass aerated cultures were investigated. In one series of experiments, the effect of adding new constituents to the growth medium was tested by determining the G.T. at 30° for each constituent added. Eventually a medium was concocted which gave a minimal generation time of 15 $\pm$ 1 hr at 30°. Using this medium, growth curves at several pH's and temperatures were determined at a controlled rate of aeration (2 l./min/l. of medium). A comparison of growth in stationary and aerated cultures is given in Table III. The following differences should be noted; in aerated cultures, the temperature optimum is lower, the pH optimum is broader, and the highest temperature supporting growth is higher. Although it cannot be shown in a table, short plateaus and rises are observed with cultures growing on the improved medium. They are not as sharply defined as those in stationary cultures.

TABLE III. Growth characteristics of amoeba cultures

Culture	Medium	Temperature optimum	Highest growth temp.	pH optimum	Mean generation time at 30°C
Stationary	Original (see text)	31–33°	35°	5 to 6	12 $\pm$ 0·5 hr
Aerated	Improved (see text)	29–30°	36–37°	5 to 7	15 $\pm$ 1 hr

The improved medium for mass aerated cultures is as follows: proteose peptone (Difco), 1·5 per cent; glucose, 1·5 per cent; potassium phosphate, $2 \times 10^{-3}M$; magnesium sulfate, $8 \times 10^{-4}M$;

calcium chloride, $5 \times 10^{-5}M$; ferric citrate, $10^{-6}M$; methionine, $10^{-4}M$; thiamine, 1 mg/l.; vitamin B_{12}, 1 γ/l.; biotin, 200 γ/l. The pH is adjusted to 5·5 and the medium filtered and autoclaved. The role of each constituent cannot yet be defined nor is it known if each is at its optimal concentration. Growth is excellent in this medium. Testing continues while synchrony studies are in progress.

2. *Temperature Cycles*

Induction of synchronous growth and division by temperature changes has been investigated in aerated cultures in the improved medium. Two types of temperature treatment have been investigated. They are shifts between optimal and sublethal temperatures, and shifts in temperature within the growth range. In each category, the effects of single transfers, of single cycles, and of multiple cycles have been tested.

A. Temperature shifts between optimal and sublethal temperatures

(*a*) *Single shifts*. In preliminary experiments, cultures growing at 30° were shifted to a sublethal temperature (36–39°). The results were equivocal. Shifts to 36° sometimes gave a division lag of 12 to 24 hr before the cells started dividing at the 36° rate; in other experiments, no lag occurred. Ultimately, it was found that the result of a shift to a sublethal temperature depends upon the rate of the temperature change. If a shift to 36° required 30 min or less for equilibrium to be established, a lag lasting from 4–12 hr resulted. No lag occurred if equilibration to 36° required 40 min or more.

The results of transferring cultures to 37–39° is also dependent upon the rate of temperature equilibration. Rapid shifts to 37° cause immediate cessation of divisions. If the change is slow, several hours are frequently required before divisions stop.

Cells shifted to any temperature between 36° and 39° show an average volume increase detectable as early as 1 hr after cultures are transferred. The volume of cells transferred to 37–39° continues to increase for 12 hr, occasionally for as long as 24 hr. After this time the volume again decreases and cells begin to break. The initial volume increase may be due to increased pinocytosis. Cells at 39° show many more pinocytosis concentration vacuoles and are more active in movement after 10 hr at the elevated temperature than control cells maintained at 30°. Temperatures above 36° evidently permit food accumulation and stimulate movement while preventing division.

(*b*) *Single cycles*. If cultures growing at 30° are subjected to a single cycle between 30° and 39°, there is always a lag in divisions following the return to the optimal temperature. The results of five experiments using a single cycle are shown in Table IV. The lag is remarkably constant; it is 3–6 hr for periods of growth at the sublethal temperature ranging from 20 min to 10 hr. The degree of synchrony achieved is better for the longer heat treatments (2 to 10 hr). Furthermore, for treatments of 2 hr or more, the generation

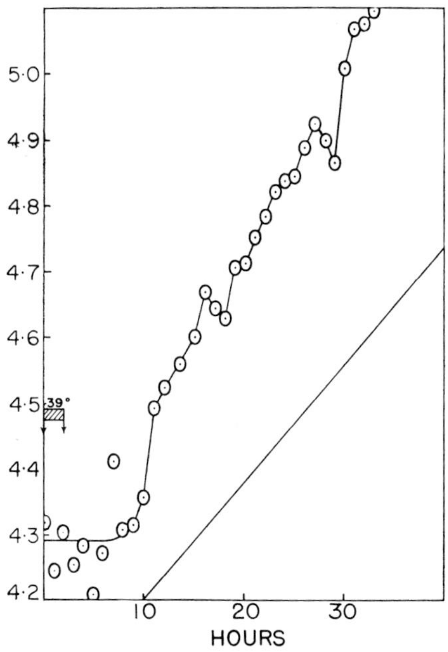

FIG. 5. The effect of a single sublethal temperature treatment on the growth of an aerated culture of *Acanthamoeba sp.*

The culture was inoculated, and allowed to grow for 12 hr at 30°. It was then transferred to 39° for 2 hr, and returned to 20°. It was followed during the sublethal shock and for 30 hr subsequently. The theoretical growth curve for 30° is plotted; the G.T. is 15 hr.

Counts were made with a Coulter Cell Counter, before our procedure minimizing counting errors was perfected. In this experiment, an uninoculated bottle of medium was carried with the culture as a means of controlling particles precipitating from the medium during aeration. Neither cells nor particles were washed before counting. Furthermore, the lower counting threshold was set at 6—a level which we now know excludes some of the smallest cells.

During the experiment, the number of precipitated particles in the uninoculated control increases constantly to a high level. In the culture, the particles are ingested by the cells, and decrease in number as the cells multiply. The failure of the growth curve to show doubling number plateaus at *exactly* 0·3 log unit intervals is attributed to the choice of an unsatisfactory control. The dips at the doubling numbers are due to the high threshold.

TABLE IV. The effect of a single cycle between optimal and sublethal temperatures on growth of aerated soil amoeba cultures

Time (hr) incubated at				Time (min) for culture to reach temperature equilibrium	Division lag (hr) on transfer to		Maximum modification (per cent of cells dividing in a generation time)
30°	36°	39°	30°		High temp.	Low temp.	
13		10	14	30–40	Complete	3–6	86% divide in 0·46 G.T.
13	10		14	30–40	Incomplete	3–6	100% divide in 0·33 G.T.
24		2	29	30–40	Complete	4	35% divide in 0·067 G.T.
24		5	26	30–40	Complete	5	58% divide in 0·13 G.T.
24		0·3	9	less than 10	Complete	3	22% divide in 0·13 G.T.
		0·6					
		1·0					

time after the treatment is shorter than that of the control and plateaus occur in the growth curve at the doubling numbers of the lag. For example, the mean generation time of cells following a 2 or a 5-hr treatment at 39° is 11 hr, as compared to a generation time of 15 hr for control cultures. The 11-hr rate is maintained for at least one additional division.

(*c*) *Multiple cycles*. The effects of twelve different cycles between 30° and a sublethal temperature (36–39°) have been studied. The data are summarized in Table V; a characteristic growth curve is illustrated in Fig. 5. We have found minimal treatments which give complete division inhibition, one such cycle is 1 hr at 30°, 2 hr at 39°, and return to 30°. At the end of eight of these cycles, there is a 4-hr lag in one experiment, a 10-hr lag in a second identical one. However, when the cells resume division, they multiply at approximately the control rate for 30°. A cycle of 3 (30°)–3 (39°) allows slow growth during cycling with little change in rate when tested at the end of eight cycles. Both treatments seem to set back all cells in the culture uniformly. It is surprising that the generation time is not shortened, as it was with a single cycle. Also it is surprising that this cycle failed to inhibit growth while the cycle was operating. Perhaps the cells accommodated to a higher temperature, as observed for stationary cultures, with new food addition. Longer cycles, such as 14 (30°)–10 (39°), which permit doubling in the 14-hr period at 30°, give synchrony, e.g., 90 per cent of the cells dividing in 0·4 of the control's generation time. The synchrony could possibly be improved by shortening the high temperature portion of the cycle. This has not yet been tested.

To summarize—temperature equilibration must be rapid (less than 30 min) if changes in culture temperatures are to affect the growth pattern of cells sufficiently to induce synchronization. For shifts between optimal and sublethal temperatures, the best synchronization is induced by a single cycle with at least 2 hr at the sublethal temperature. No synchronization occurs following multiple cycling unless growth can occur during the lower temperature phase of the cycling treatment. It is concluded that the reduced synchrony observed following multiple cycles represents a cancellation of the results of the early cycles by the later ones.

B. Temperature shifts between two temperatures within the growth range

(*a*) *Single shifts and single cycles*. Fewer experiments have been performed with temperatures below the optimum temperature which still permits growth. The data for single shifts and single cycles are summarized in Table VI. If temperature shifts are rapid, such

TABLE V. The effect of multiple cycles between optimal and sublethal temperatures on growth of aerated soil amoeba cultures

Cycle length (hr)	No. of cycles	Time (hr) incubated at 30°[a]	36°	37°	38°	39°	Time (min) for culture to reach temperature equilibrium	Division lag (hr) on transfer to High temp.	Low temp.	Maximum modification (per cent of cells dividing in a generation time)[b]
24	5	14				10	30–40	Complete	5	90% in 0·4 Generation Time (G.T.).
24	5					8	30–40	Complete	2	55% in 0·27 G.T.
24	2	16	8				30–40	Complete	4	26% in 0·27 G.T.
6	5	3	3				40–50	Incomplete	None	G.T. approximates control; Detectable rises and plateaus during treatment.
6	8	3		3			40–50	Incomplete	2 at end of cycle	78% in 0·73 G.T.; shows a plateau every 0·3 log unit.
6	8	3			3		40–50	Incomplete	2–4 at end of cycle	G.T. approximates control; vague rises and plateaus.
3	8	1			2		40–50	Complete after 2nd cycle	4, 10 at end of cycle	G.T. approximates control beginning 10 hr after last cycle.
12	3	4			8		40–50	Complete	2	G.T. approximates control.
6	4	2		4			40–50	Incomplete	None?	G.T. approximates control.
8	4	2		6			40–50	Complete after 1st 3 hr of 1st cycle	1–3	G.T. approximates control.
12	4	3		9			30–40	,,	4	G.T. approximates control.
12	4	4		8			40–50	?	1–2	G.T. approximates control.

[a] Grown at least 24 hr at 30° before first cycle.
[b] Followed 15–48 hr after last heat treatment.

TABLE VI. The effect of a single cycle between two temperatures within the growth range on the growth of aerated soil amoeba cultures

Time (hr) incubated at					Time (min) for culture to reach temperature equilibrium	Division lag (hr) on transfer to		Maximum modification (per cent of cells dividing in a generation time)
3°	10°	15°	25°	30°		High temp.	Low temp.	
12				Before and after	120 down, 30 up	2	Incomplete slow increase	29% in 0·13 G.T.
	12		Before and after		30–40	?	Incomplete slight increase	52% in 0·37 G.T.
		12	Before and after		30–40	?	Complete by 2 hr	48% in 0·33 G.T. Marked plateaus present.
	Counted for 54		Before		20–30	—	Complete by 2 hr No increase in next 22 hr 22% increase between 22–29 hr	—

as a change from 25° to 15° (within 20 to 30 min), there results a division block which is complete shortly after the first hour, and which lasts for at least 12 hr. If the drop is slow (for example, requiring 3 hr to reach equilibrium), the division rate shifts gradually to the rate for the lower temperature, without detectable lag. The same results are obtained for shifts from 25° to 10°. Rapid rises in temperature, e.g., from 15° to 25°, or from 10° to 25° within 30 min, give complete inhibition of divisions for 1 to 5 hr. The inhibition is followed by growth at an increased rate of division. Plateaus occur close to the doubling numbers. Slow rises of temperature have not yet been studied. Cells in cultures where division is blocked by rapid cooling to 10° and which are kept at that temperature are capable of pinocytosing when examined at 12 and 24 hr after the beginning of the cold treatment. If these cells are examined at 40 hr, they appear moribund, as if the long cold treatment had altered them irreversibly. Cold cells are sampled and examined immediately at 25°.

(*b*) *Multiple cycles.* Seven different cyclic treatments, all 24 hr in duration, have been tested. They are summarized in Table VII. In only one experiment–16 (25°)–8 (10°)—did an effective, but poor, synchronization result after the final cold shock. Further analysis of this limited data suggests that other cycles synchronized small portions of the population, since several rises and plateaus were observed following the last cold shock in four of the experiments.

3. *Effects of Aerobic–Anaerobic Treatment*

Reich (1936) originally reported that growth of a soil amoeba (*Mayorella palestinensis*) could not occur under anaerobic conditions. We have confirmed this fact in *Acanthamoeba sp.* Anaerobiosis produced by bubbling nitrogen through a culture causes an immediate cessation of division. Although the end result of a nitrogen block is the same as that of a sublethal temperature treatment and is reversible, it is accompanied by dramatic structural changes. The cells round up and granules clump around the nucleus. We have called this effect a protoplast contraction (see chapter 21). The mean cell volume does not change under anaerobic conditions. We have made preliminary studies of anaerobiosis as a synchrony inducer.

(*a*) *A single cycle.* The effect of gassing a culture growing at 32° with nitrogen for 6 hr is as follows. The cell numbers remain constant during nitrogen bubbling and for 2–3 hr after the nitrogen is replaced by air. Thereafter, the growth curve of the treated culture mimics the control's plateaus and rises, but is delayed 7–8 hr. Once divisions begin in the treated culture, the increase in cell numbers proceeds

TABLE VII. The effect of multiple cycles between two temperatures within the growth range on the growth of aerated soil amoeba cultures

Cycle length (hr)	No. of cycles	Time (hr) incubated at 30°	25°	15°	10°	Time (min) for culture to reach temperature equilibrium	Division lag (hr) on transfer to High temp.	Low temp.	Maximum modification[a] (per cent of cells dividing in a generation time)
24	4	16		8		30–40	2–4	Complete	40% in 0·3 G.T. after second cold shock; less after last cold shock.
24	12	8		16		30–40	2–4	Complete	20% in 0·13 of G.T. after 9th cold shock.
24	4	12			12	30–40	1–2	Complete	50% in 0·23 G.T. after 2nd cold shock; close to control G.T. after last shock.
24	6	10		14		30–40	2–3	Complete	70% in 0·23 G.T. after 2nd cold shock; synchrony decreases with later cycles.
24	3		12		12	180–200, down 40–50 up	3–4	Incomplete	G.T. approximates control.
24	3		12		12	20–30	1–5	Complete after 1st hour	54% in 0·25 G.T. after 1st cold shock; less after 3rd.
24	3		16		8	20–30	1–2	Complete by 1st hour, or less	26% in 0·15 G.T.

[a] All rates determined at the higher growth temperature.

more rapidly in it than in the controls.

(*b*) *Multiple cycles.* Two multiple-cycle treatments have been performed. In the first, a culture was subjected to twenty cycles of the following 8-hr cycle: 2 hr of nitrogen gassing followed by 6 hr of air bubbling –2 (N)–6 (A). In this experiment counts were made after the end of the twentieth cycle.

The second experiment used a 24-hr cycle; 4 (N)–20 (A). In this case counts were made during the fourth and fifth cycles, and during the tenth and eleventh cycles. In the latter, nitrogen treatments had induced significant increases in synchrony. For example, the maximum degrees of synchrony induced in the two experiments were 30 per cent of the cells dividing in 0·13 per cent of the G.T. and 36 per cent in 0·2 G.T. respectively.

These limited results show clearly that periodic anaerobiosis does induce some division synchrony. Of special interest is the observation that the anaerobic treatment has a direct, easily-recognized and wholly-reversible effect on the cell's structure.

V. Evaluation of Results

In this section, current hypotheses on the mechanism of experimentally induced division synchrony will be examined and evaluated with respect to the amoebae. The relation of nutrition and length of cycle to successful induction of division synchrony will also be discussed.

1. *Mechanisms of Synchrony Induction*

A. Kinetic hypotheses

Both James (1959) and Scherbaum (1960) have proposed hypotheses to explain the synchronizing effects of temperature shifts on cultures in terms of a differential effect of temperature on individual cellular processes. The activation energy hypothesis of James has been discussed in section II. Scherbaum has constructed a complex hypothetical biochemical pathway (the mitogenic channel) which provides energy for the work of cell division. He proposes that the synchronizing effect of a temperature shock is due to its effect on a key temperature-sensitive enzyme, DPNH-oxidase. This enzyme is reversibly denatured at either high or low temperatures.

Neither of these hypotheses can account for the different effects of rapid and slow temperature changes on *Acanthamoeba* cultures. The results presented in section IV-2-B have clearly demonstrated that in *Acanthamoeba*, the *rate* at which a temperature change occurs *is decisive* in determining what effect a given shift will have on a growing culture. *A rapid shift causes an immediate division blockage, and induces a degree of division synchrony; a slow shift results in a gradual transition* to a growth rate characteristic of the new temperature. *No synchronization results.*

If differences in reaction rates or denaturation rates were the only determining factor, one would predict that both rapid and slow shifts would have the same end result—either a complete division blockage, or a gradual shift to a new division rate—once the equilibrium to the new temperature had been established. This prediction is not fulfilled in *Acanthamoeba* cultures. Clearly some additional factor is implicated. The authors propose that a structural factor, similar to that proposed by Zeuthen (1958; see also chapters 4 and 5 in this book), is also involved in division synchrony induction in soil amoebae.

B. Rate-dependent temperature effects

The results of Angerer's (1940) study of the relation of cellular viscosity in *Amoeba proteus* to temperature represents key data in explaining rate-dependent temperature effects on synchronization in *Acanthamoeba*. He determined the effects of various *rates of change* of temperature on the viscosity in *Amoeba proteus*. He reports that a slow shift of temperature (a change of 12° in 200 min) resulted in a slow inverse change of viscosity; the changes in viscosity followed the temperature changes closely, and upon attaining temperature equilibrium, the viscosity remained constant at the new level.

However, a *rapid change* in temperature resulted in a correspondingly rapid change in viscosity. In one example, the temperature was changed from 12° to 26° in 10 min; in this experiment, the viscosity dropped rapidly to a very low level, and within a few minutes shifted to a new and very high level. For the next several hours, the viscosity of the shocked cells fluctuated between these two extreme values, and after 9 to 14 hr reached a constant low value characteristic of cells maintained at 26°. It was concluded that in *A. proteus* the structural bonds of the internal protoplasm were thrown into a slow and uncontrolled rhythmic cycle of association and dissociation by rapid temperature changes; this cycle was not induced by a slow temperature change.

The parallel between Angerer's data and that of the authors is obvious. In *Acanthamoeba*, rapid temperature shifts halt cell division and induce a degree of synchrony upon subsequent growth; slow temperature changes do not. The inference drawn is that rapid temperature shocks cause structural changes in the amoeba which block division and subsequently induce division synchrony.

James' demonstration (described in section II) that there is, in *A. proteus*, a decrease in nuclear mass following a rapid temperature shift supports the possibility of structural change. He also reported an increase in the cytoplasmic volume and apparent mass accompanying the nuclear change, the reason for which is not known. James proposed either a water exchange, a sol–gel transformation, or a molecular–optical phenomenon as possible explanations of the cytoplasmic changes. Unfortunately, the effects of a slow temperature change were not determined.

Although there is no additional evidence on the nature of the changes occurring in amoeba cells as a result of a rapid temperature change there is peripheral evidence that other treatments which produce synchrony in *Acanthamoeba* also produce structural changes. In the aerobic–anaerobic cycling experiment substitution of nitrogen for air in an aerated culture caused a dramatic structural change in the treated cells—a contraction of the cytoplasmic particulates around the nucleus. The structural alteration was completely reversible. There is strong circumstantial evidence that a similar structural change accompanies the addition of new medium to growing cultures, although this change has not been directly demonstrated. It will be recalled that both an anaerobic treatment and the addition of new medium imposed a degree of division synchrony upon *Acanthamoeba* cultures (sections III-3; IV-3).

Other evidence suggesting a structural factor in synchronizing treatments is even more indirect. If we can assume that any treatment which alters the growth rate of a culture may be a potential synchronizing treatment, then we can cite two additional cases where a treatment which increases the generation time also has structural effects. First, Angerer (1936) demonstrated, again in *A. proteus*, that short vigorous mechanical agitation produced severe changes in viscosity similar to those produced by rapid temperature changes. The effects of agitation on *Acanthamoeba* structure has not been determined, but agitation of cultures may increase the generation time by several hours (see Table III). In the second case, an increase in the concentration of calcium in the growth medium of *Acanthamoeba* produces marked structural changes resulting in 'plastic' cells. These cells can be transformed into bizarre spindle shapes by rolling them between slide and coverslip; the imposed shape persists for some time after the stress is removed. The generation time of these cultures may be increased by as much as 60 hr. The importance of calcium in the cell's division machinery has been stressed by Heilbrunn (1954).

C. *Structural hypotheses*

Zeuthen (1958) was the first to recognize clearly the possibility of structural involvement in synchrony induction. Later studies (see chapters 4 and 5) have confirmed the general hypotheses but introduced modifications.

It is obvious that the amoeba data discussed in the preceding section offer strong support for Zeuthen's proposal that structural alterations play a key role in synchrony induction. It—like the *Tetrahymena* evidence—is partially circumstantial.

The authors propose to expand the hypothesis by suggesting that in the amoebae the structural effect which is observed is a change in the mitotic apparatus of the predivision cell.

D. *The mitotic apparatus*

In its preparation for division, a cell completes a number of parallel processes (Swann, 1957), of which energy supply is but one. The physiology of cell division has been elegantly summarized by Mazia (1961). Division may be blocked by inhibition of any one of these parallel processes. As Mazia points out, these are examples of *limiting factors* which may act as control mechanisms in a complex system. However, in preparation for division, there is one process upon which the products of all other preparatory processes depend for division to be complete. This is the assembly and organization of the mitotic apparatus (MA). In general, the formation of the MA is terminal. It is in series with each of the parallel syntheses that occur earlier in the life history of the cell; for this reason, the formation of the MA is the *most vulnerable limiting factor* in cell division. One would predict that the most effective synchronizing treatment would be one that specifically blocked the assembly of, or function of, the MA. If the effect were highly specific for the MA, all preceding parallel processes would go to completion, and the cell would be suspended on the brink of division at the end of a treatment. To what extent are we able at present to relate the effects of successful synchronizing treatments to specific effects on the mitotic apparatus?

In *Acanthamoeba*, three findings strongly suggest that the vulnerable structure is the MA. First, it is evident from a large number of experiments, that the nearer a cell is to division, the more it is inhibited by a temperature shock (sections III and IV; Hamburger and Zeuthen, 1957; Zeuthen, 1958; Thormar, 1959). Second, the fact that results are qualitatively the same whether rapid heat shocks, or cold shocks, or other rapid treatments are used as synchronizing treatments suggests that all affect the same structure.

Third, a sublethal temperature shock, whether it lasts 20 min or 10 hr, produces a constant lag of 3–6 hr. This constant lag suggests that at least 3 hr are needed at optimal temperatures to restore cells to their pre-shock level of preparation and permit divisions to resume. This might represent a measure of the time required for the reorganization and reassembly of a MA destroyed by heat.

Soil amoebae divide by a mitotic process and have a typical MA which can be clearly identified in fixed and stained sections. It is, however, difficult to observe a spindle in living material. It is obscured by the cellular particulates when the cells round up just before dividing.

The proposal that the mitotic apparatus is affected by the synchronizing treatments is given strong support by the mass of early observations on the dissolution or disorientation of the asters and spindles of fertilized eggs or other dividing cells by cold, pressure, ether, colchicine, mercaptoethanol, etc. (Wilson, 1925; Mazia, 1961). Certainly, the MA is extremely labile. The recent work of Holz, Scherbaum and Williams (1957) indicates that repeated sublethal shocks of a micronucleate strain of *Tetrahymena* resulted in a large percentage of the cells being trapped in anaphase at the end of the fifth cycle. It is not clear whether the spindle is 'melted' or 'frozen' by the heat shocks. To cite only one other example the recent work of Inoué (1959) demonstrates that the birefringence of the sea urchin spindle (due to orientation) decreases when the temperature is lowered. The rapidity of the change is unknown.

It is premature to speculate about the nature of the bonds which are broken. As Mazia (1961) points out, many types of bonds are possible in the MA. They are indistinguishable from labile bonds present in interphase cells. Furthermore, although the target structure may be identical for all synchronizing treatments, it is by no means certain that all treatments modify that structure by breaking or altering identical bonds. It can be readily seen that further speculation on this point must await new experimental data.

Finally, it should be pointed out that any hypothesis of synchrony induction based on energy mobilizations *alone*, on enzyme kinetics *alone*, on differences in process rates *alone*, or even on structural disorientation *alone* may be too circumscribed for an adequate exposition of the complex induction process. It is more probable that a good synchronizing treatment will have multiple effects. For example, the Zeuthen group has demonstrated that several well-known metabolic inhibitors of known specificity give setbacks—indicating their potential as synchronizing agents (Zeuthen, 1958;

see also chapter 4 of this book). At least two—nitrogen (chapter 21) and ethyl urethane (Goldacre, 1952)—have dramatic effects on cell structure in addition to their metabolic effects. Perhaps the 'non-specific' effects of a 'specific' inhibitor may be more rewarding to study. Scherbaum *et al.* (1959) report that, in *Tetrahymena* protein, synthesis continues during the heat cycles. Christensson's work (1959) adds further refinement, showing that protein synthesis is intermittent; it occurs during the short time in the multiple cycle when cells are at the optimal (lower) temperature. This result may be explained in terms of 'reversible denaturation' of the protein-synthesizing system; it may also be due to structural isolation of the energy supply from the protein-synthesizing system during treatment, particularly if a structural change limits diffusion of metabolites.

2. *Nutrition and Synchrony Induction*

The experiments described in section IV-1 have demonstrated the fact that a nutritionally-adequate growth medium is a requirement for successful synchrony induction in *Acanthamoeba*. This requirement is similar to that reported by Padilla and James (1960; also chapter 11 of this book) for *Astasia*. Sufficient amounts of both energy-yielding metabolites and precursors for cellular syntheses must be supplied.

We interpret a long generation time in *Acanthamoeba* as diagnostic of a nutritionally incomplete growth medium. In well-nourished cultures, the cells are able to maintain themselves and to continue syntheses of new cell materials and of division machinery units even during a synchronizing treatment, e.g. a heat shock. The net result is an increase in the number of cells possessing a complete complement of *unassembled* MA parts. They are able to commence assembly in phase once the heat treatment is terminated. In incompletely nourished cultures, this is not the case. The temperature shock disrupts assembled MA's. However, it will not permit any accumulation of unassembled MA parts, since deficient cells must use their limited energy supply for maintenance alone at an elevated temperature. The failure of synchronization results from the elimination of any differential effect between synthesis and assembly of MA.

3. *Length of Cycle and Synchrony Induction*

In both *A. proteus* (James, 1959) and *Acanthamoeba*, the most successful synchronizing cycles are long ones, which permit a doubling of cell numbers during the course of a single cycle. This is in striking contrast to *Tetrahymena* where successful synchronizing cycles are short and permit little or no increase in cell numbers during the cycling (Scherbaum, 1960; Scherbaum and Zeuthen, 1954; Zeuthen, 1958). The reasons for the differences between cell types are not completely known.

A comparison of the response of *Tetrahymena* and *Acanthamoeba* to similar synchronizing treatments offers a possible point of attack on the problem. For *Tetrahymena*, the most successful synchronizing cycles are short, i.e., equal to one-third to one-half of the G.T. Usually half of each cycle is at sublethal temperature. Divisions are completely suppressed throughout the cycling; they are characteristically resumed in a large division burst following the completion of the final cycle. In *Acanthamoeba*, short synchronizing cycles—equivalent in terms of G.T. to those used for *Tetrahymena*—are ineffective (see Table V-6 and 3-hr cycles). Only one 3-hr cycle inhibited divisions completely during the cycling. None of the short cycles induced appreciable synchrony.

Why do short synchronizing cycles in *Acanthamoeba* fail to inhibit divisions during temperature cycling? In brief, it is because *Acanthamoeba* cells do not respond to a *repeated* temperature cycle, as they do to a single one. It will be recalled that cells given a single sublethal shock do not divide at the sublethal temperature and show at least a 3-hr delay when returned to the optimal temperature. It was predicted that a repetition of this cycle—3 (37°)–3 (30°)—should prevent cells from dividing for as long as the cycles were continued. At the end of the final cycle, the cells were expected to divide in a synchronized burst as do *Tetrahymena*. When the experiment was performed, this result was not obtained. Instead, cell numbers increased during the synchronizing treatment. The occurrence of division during the cycling indicated that in some way the cells become accustomed to the sublethal temperature, or become less sensitive to the heat shocks, or recover more rapidly during the optimal temperature phase. We have called this phenomenon *accommodation*. The limited data that we have suggests that in multiple 6-hr cycles the major effect is a shortening of the division lag following the second and succeeding heat treatments. This permits some cells to divide during the optimal temperature phase of each cycle—thereby preventing mass synchronization. Accommodation can be eliminated by shortening the period at optimal temperature to 1 hr in a 3-hr multiple cycle. Cycles intermediate to the 3 and 6-hr cycles have not been tested.

We do not know the mechanism of accommodation. We speculate that it is due to an improved rate of recovery during the second and succeeding periods at optimal temperature in multiple cycles, because of the accumulation of precursors and/or energy-rich compounds during each high temperature period. Certainly *Acanthamoeba* enjoys an abundant energy supply at 37 to 39° (at least for the first 10 hr) as shown by greatly increased rates of movement and pinocytosis. In terms of the structural hypothesis, we suggest that accommodation reflects a more rapid restoration or repair of bonds disrupted by the heat shock in the second (and succeeding) cycles than in the initial one. Alternative explanations can be proposed. For example, if the culture contains a small proportion of cells which are inherently more resistant to temperature shock and heat damage, such cells could multiply during the treatment and eventually become dominant in the population. Although such a selective mechanism is possible, it seems less likely, since no growth occurs in cultures placed continuously at 37°, the lowest sublethal temperature for *Acanthamoeba*.

We might expect that a multiple cycle in which accommodation was avoided would produce good synchrony. It did not. In the single experiment where accommodation was circumvented, a long division delay followed the last sublethal treatment. But when divisions resumed, they were at the control rate. We have concluded that the failure of this cycle is due to *cell damage*. Although *Acanthamoeba* cultures have been shown to recover from heat shocks of 2 to 10 hr without apparent residual injury, longer continuous exposure to a sublethal temperature is clearly harmful (sections III-2-A and IV-2-A). We believe that amoebae subjected to repeated short cycles are damaged by the *cumulative* exposure to the elevated temperature (16 hr in the multiple 3-hr cycles tested) when the time at optimal temperature is relatively brief. It is probable that all sublethal treatments impose some irreversible damage on cells. This is in addition to the reversible disruption of bonds by rapid heat shocks discussed in connection with the structural hypothesis. Net synthesis of damaged parts would not be expected at continuous sublethal temperatures or during multiple cycles where a short exposure to the optimal temperature would be insufficient for synthesis to exceed degradation. If this speculation is valid, then one would expect the failure of synchrony in a 3-hr cycle to resemble the failure of synchrony in undernourished cultures.

In the preceding paragraphs we have discussed two factors which we believe explain the failure of short cycles in synchrony induction in *Acanthamoeba*. It is not known whether these factors are peculiar to *Acanthamoeba*, or whether they are of general occurrence. We conclude that both factors—accommodation and cumulative cell damage—must be avoided in the design of successful synchronizing cycles. Indeed, they, together with the generation time, may determine whether a given cell type is more responsive to long or short cycles.

VI. Conclusions and Perspectives

Successful experimental induction of division synchrony requires that a growing culture be subjected to an experimental treatment which will either advance or retard some of the cells in their duplication cycle with respect to the remainder of the cells. It is postulated that the most effective means of achieving this aim is to retard cells by means of a periodically-applied treatment which selectively disrupts or disorients the mitotic apparatus of the cells.

In the amoebae, the successful experimental induction of division synchrony depends upon at least three factors. They are (1) adequate nutrition, (2) length of the synchronizing cycle with respect to generation time, and (3) rate at which the synchronizing treatment is applied. Each factor has been discussed fully in the preceding section.

It is difficult to predict the directions in which studies of synchronous division will move in the next few years. Investigations in several areas may increase our knowledge of division synchrony induction in the amoebae. Several structural aspects of synchrony induction are of special interest to the authors and will be considered briefly below.

The structural hypothesis predicts that the most effective synchronizing treatments will produce detectable changes in the cell's structure—particularly its division apparatus. Since the MA contains no unique bonds, it would be most instructive to determine the effects of agents which change the strength of intracellular bonds (e.g., salt bonds, oxidation–reduction bonds, hydrogen bonds, etc.) on cells preparing to divide. These bonds can be broken or their strength altered by such means as changing internal pH of the cells, by changing the internal oxidation–reduction potential, by temperature shifts, or by certain reagents, e.g., urea. Zeuthen (1958) has proposed an experiment of this genre using colchicine, a spindle dissolver. Indeed, Prescott (1957) has performed some preliminary experiments in this area. He reports that almost any change in the environment (pH, medium changes, etc.) induces some degree of division synchrony in *Tetrahymena*; however, no structural analysis of the effects of these agents was made.

What methods can be utilized to identify structural changes in such experiments? Changes in orientation of submicroscopic structures may be detected and partially identified by polarization microscopy (Inoué and Hyde, 1957; Schmidt, 1937). Intracellular changes which are accompanied by changes in mass may be identified by interference microscopy (James, 1959). Changes in structural rigidity may be inferred from viscosity determinations (Heilbrunn, 1958) or from determinations of pressure sensitivity (Marsland, 1956). Changes in association of subcellular parts may be detected by electron microscopy.

Certain changes considered structural probably would not be useful. One such structural change is size. In *Tetrahymena* (Scherbaum, 1956), and also in *Acanthamoeba* (section IV-2-A), cell size increases during a synchronizing treatment and the larger cells increase proportionally more in volume than do smaller cells. This rather puzzling observation is not difficult to explain from a structural point of view. If one assumes that the treatment induces a gelled structure by formation of additional weak bonds, the flow of diffusable metabolites will be reduced. Even if the probability of breaking bonds per unit volume is the same in large and small cells, the probability of bond breakage per unit time for a large cell is greater than for a small cell. Thus, one would predict greater flow of metabolites in larger cells during treatment.

The effects of *rate of change* in inducing division synchrony and structural change need to be investigated more thoroughly, both for treatments other than sublethal shocks and in organisms other than soil amoebae. Only in *Acanthamoeba* has the effectiveness of a temperature shift been demonstrated to depend upon the rate at which the shift occurred (section IV-2-A). It will be recalled that only rapid shifts were effective in halting divisions and inducing division synchrony. However, it should be noted that in experiments on other organisms reported in the literature, the time required for establishment of temperature equilibrium was short—usually only a few minutes. It is suggested that a comparison of the synchrony resulting from rapid as against gradual transition from one set of conditions to another should be very instructive. One would predict rapid shifts to be most effective, if structural alteration is a factor of universal importance. Second, a comparison of the effects of rapid as against slow shifts in temperature (or other treatment) on the structure of the MA during its formation in late predivision stages should provide direct evidence with respect to its role.

Further, if the structure of the MA is preferentially damaged by synchronizing treatments applied to a randomly dividing culture, one should find increased amounts of MA precursors in cells following a treatment. The pioneering work of Mazia and his colleagues (Mazia, 1955, 1959; Mazia and Dan, 1952; Zimmerman, 1960) on the structure and composition of MA has made an experimental attack on this problem feasible. The major component of the MA (in sea urchin eggs) is a protein of high molecular weight which is difficult to distinguish from other soluble proteins by ordinary chemical means. However, Went (1959a, b) and Went and Mazia (1959) have shown that precursor proteins of MA can be identified even in unfertilized eggs by immunochemical methods. There is no theoretical reason (although there may be technical ones) why the quantity of a MA precursor should not be determined at different times in cells undergoing synchronizing treatments. It is hoped that this approach will be explored in the near future.

Finally, the work on synchrony induction in amoebae has demonstrated that all successful treatments produce an initial division block, before a more highly synchronized division is resumed. Furthermore, the cells grow in volume and mass during the inhibition period. It is of great interest that cultures of amoebae and *Tetrahymena* undergoing synchronization share these characteristics with a highly specialized type of cell—the egg cell. During oögenesis in echinoderms and amphibians, there is a division block, followed by a period of storage and cell growth. When the cells are 'activated' by sperm or a parthenogenetic agent the block is removed, and there follows a near perfect division synchrony. Whether induced synchrony in microorganisms shares common blocking and activating mechanisms with ova remains to be determined. It is suggested that investigations into the division block in oögenesis and into the mechanisms of induced division synchrony may contribute greatly to each other in the near future.

Acknowledgements—We would like to acknowledge the aid of the following people in the new work reported in this paper: S. A. Ray, W. F. Benton, M. Wilborn, H. Z. Park, D. B. Butler, A. T. Morrissey, W. S. Wilborn and J. B. Haynes, Jr.

REFERENCES

Angerer, C. A. (1936). The effects of mechanical agitation on the relative viscosity of amoeba protoplasm. *J. Cellular Comp. Physiol.* **8,** 329.

Angerer, C. A. (1940). The effect of thermal increments on and the subsequent adjustment of the protoplasmic viscosity of *Amoeba proteus*. *Physiol. Zoöl.* **13,** 219.

Christensson, E. (1959). Changes in free amino acids and proteins during cell growth and synchronous division in cultures of *Tetrahymena pyriformis*. *Acta Physiol. Scand.* **45,** 339.

DAWSON, J. A., W. R. KESSLER and J. K. SILBERSTEIN (1937). Mitosis in *Amoeba proteus*. *Biol. Bull.* **72,** 125.

GOLDACRE, R. J. (1952). The action of general anaesthetics on amoebae and the mechanism of the response to touch. *Symposia Soc. Exptl. Biol.* **6,** 128.

GRUBER, K. (1912). Biologische und experimentelle Untersuchungen an *Amoeba proteus*. *Arch. Protistenk.* **25,** 316.

HAMBURGER, K. and E. ZEUTHEN (1957). Synchronous divisions in *Tetrahymena pyriformis* as studied in an inorganic medium. The effect of 2, 4-dinitrophenol. *Exptl. Cell. Research* **13,** 443.

HEILBRUNN, L. V. (1954). The physiology of cell division. In *Trends in Physiology and Biochemistry* (E. S. G. Barron, ed.). Academic Press, New York, pp. 123–134.

HEILBRUNN, L. V. (1958). Protoplasmatologia. In *The Viscosity of Protoplasm.* Monograph. Springer-Verlag, Vienna.

HOLZ, G. G., O. H. SCHERBAUM and N. WILLIAMS (1957). The arrest of mitosis and stomatogenesis during temperature-induction of synchronous division in *Tetrahymena pyriformis*, Mating Type 1, Variety 1. *Exptl. Cell Research* **13,** 618.

HOWARD, F. L. (1932). Nuclear division in plasmodia of *Physarum. Ann. Bot.* **46,** 461.

INOUÉ, S. (1959). Motility of cilia and the mechanism of mitosis. In *Biophysical Science—A Study Program* (J. L. Oncley, ed.). John Wiley, New York, pp. 402–408.

INOUÉ, S. and W. L. HYDE (1957). Studies on depolarization of light at microscope lens surfaces. II. The simultaneous realization of high resolution and high sensitivity with the polarizing microscope. *J. Biophys. Biochem. Cytol.* **3,** 831.

JAMES, T. W. (1954). *Quantitative studies on the ribonucleic acid content in Amoeba proteus in relation to the nucleus and the division cycle.* Ph.D. Thesis, Univ. of Calif., Berkeley.

JAMES, T. W. (1959). Synchronization of cell division in amoebae. *Ann. N. Y. Acad. Sci.* **78,** 501.

KUDO, R. R. (1947). *Pelomyxa carolinensis* Wilson. II. Nuclear division and plasmotomy. *J. Morphol.* **80,** 93.

LISTER, A. and G. LISTER (1925). *Mycetozoa*, 3rd ed. Oliver and Boyd, Edinburgh.

MARSLAND, D. (1956). Protoplasmic contractility in relation to gel structure: Temperature–pressure experiments on cytokinesis and amoeboid movement. *Intern. Rev. Cytol.* **5,** 199.

MAZIA, D. (1955). The organization of the mitotic apparatus. *Symposia Soc. Exptl. Biol.* **9,** 335.

MAZIA, D. (1959). Cell division. *Harvey Lectures* **53,** 130.

MAZIA, D. (1961). Mitosis and the physiology of cell division. In *The Cell*, Vol. III (J. Brachet and A. E. Mirsky, eds.). Academic Press, New York, pp. 77–412.

MAZIA, D. and K. DAN (1952). The isolation and biochemical characterization of the mitotic apparatus of dividing cells. *Proc. Natl. Acad. Sci. U.S.* **38,** 826.

MAZIA, D. and D. M. PRESCOTT (1954). Nuclear function and mitosis. *Science* **120,** 120.

NEFF, R. J. (1955). Physiology of *Amoeba*: With special reference to nuclear physiology and intermediary carbohydrate metabolism in soil amoebae. *Terminal Rept. Public Health Research Grant E*-623.

PADILLA, G. M. and T. W. JAMES (1960). Synchronization of cell division in *Astasia longa* on a chemically defined medium. *Exptl. Cell Research* **20,** 401.

PRESCOTT, D. M. (1955). Relations between cell growth and cell division. I. Reduced weight, cell volume, protein content and nuclear volume of *Amoeba proteus* from division to division. *Exptl. Cell Research* **9,** 328.

PRESCOTT, D. M. (1956a). Relation between cell growth and cell division. II. The effect of cell size on growth rate and generation time in *Amoeba proteus*. *Exptl. Cell Research* **11,** 86.

PRESCOTT, D. M. (1956b). Relation between cell growth and cell division. III. Changes in nuclear volume and growth rate and prevention of cell division in *Amoeba proteus* resulting from cytoplasmic amputations. *Exptl. Cell Research* **11,** 94.

PRESCOTT, D. M. (1957). Change in the physiological state of a cell population as a function of culture growth and age (*Tetrahymena geleii*). *Exptl. Cell Research* **12,** 126.

PRESCOTT, D. M. (1959). Microtechniques in amoebae studies. *Ann. N. Y. Acad. Sci.* **78,** 655.

PRESCOTT, D. M. (1961). The growth duplication cycle of the cell. *Intern. Rev. Cytol.* **11,** 255.

REICH, K. (1936). Studies on the physiology of *Amoeba*. I. The relation between nutrient solution, zone of growth, and density of population. *Physiol. Zoöl.* **9,** 254.

ROSLANSKY, J. D., N. H. BRANCHFLOWER and M. S. HUCKMAN (1958). Temperature-induced synchronization of cell division in cultures of Trophozoites of *Endamoeba terrapinae*. *Anat. Rec.* **132,** 498.

SCHERBAUM, O. H. (1956). Cell growth in normal and synchronously dividing mass cultures of *Tetrahymena pyriformis*. *Exptl. Cell Research* **11,** 464.

SCHERBAUM, O. H. (1960). Synchronous division of microorganisms. *Ann. Rev. Microbiol.* **14,** 283.

SCHERBAUM, O. H., T. W. JAMES and T. L. JAHN (1959). The amino acid composition in relation to cell growth and cell division in synchronized cultures of *Tetrahymena pyriformis*. *J. Cellular Comp. Physiol.* **53,** 119.

SCHERBAUM, O. H. and E. ZEUTHEN (1954). Induction of synchronous cell division in mass cultures of *Tetrahymena pyriformis*. *Exptl. Cell Research* **6,** 221.

SCHMIDT, W. J. (1937). *Die Doppelbrechung von Karyoplasma, Zytoplasma und Metaplasma*, Protoplasma, Monograph, Vol. II. Bornträger, Berlin, pp. 1–388.

SWANN, M. M. (1957). The control of cell division: A Review. I. General mechanisms. *Cancer Research* **17,** 727.

THORMAR, H. (1959). Delayed division in *Tetrahymena pyriformis* induced by temperature changes. *Compt. Rend. Trav. Lab. Carlsberg, Ser. physiol.*, **31,** 207.

WENT, H. A. (1959a). Some immunological studies on the mitotic apparatus of the sea urchin. *J. Biophys. Biochem. Cytol.* **5,** 353.

WENT, H. A. (1959b). Studies on the mitotic apparatus of the sea urchin by means of antigen–antibody reactions in agar. *J. Biophys. Biochem. Cytol.* **6,** 447.

WENT, H. A. and D. MAZIA (1959). Immunochemical study of the origin of the mitotic apparatus. *Exptl. Cell Research, Suppl.* **7,** 200.

WILSON, E. B. (1925). *The Cell in Development and Heredity*, 3rd ed. Macmillan, New York.

ZEUTHEN, E. (1958). Artificial and induced periodicity in living cells. *Advanc. in Biol. and Med. Phys.* **6,** 37.

ZIMMERMAN, A. M. (1960). Physico-chemical analysis of the isolated mitotic apparatus. *Exptl. Cell Research* **20,** 529.

9

Growth and Cell Division of *Chlorella*

HIROSHI TAMIYA, *The Tokugawa Institute for Biological Research, Tokyo, Japan*

CONTENTS

I Introduction 247
Methodological considerations 248
II Various Developmental Stages 257
III Effects of Temperature and Light Intensity 263
IV Photosynthetic and Respiratory Activities 267
V Changes in Content of Various Cell Constituents in Normal Life Cycle. 272
1. Major elements and water 272
2. Protein, lipid and carbohydrate 275
3. Nucleic acids 276
4. Polyphosphates 280
5. Pigments 281
6. Vitamins 282
VI Modifications Caused by Various Environmental Factors . . 283
1. Effect of deprival of each essential mineral nutrient . . . 283
2. Effects of ultraviolet irradiation and 'cold shock' . . . 285
3. Effects of various antimetabolites 286
VII Role of Sulfur in Ripening and Cell Division 291
VIII Competitive Relation Between Growth and Cellular Division . 298
References 302

I. INTRODUCTION*

Since the technique of synchronous culture of *Chlorella* was developed by Tamiya, Iwamura, Shibata, Hase and Nihei in 1953 similar techniques have been widely used in the study of physiological and biochemical events occurring in the life cycle of algae. *Chlorella* is an organism whose growth and cell division can relatively easily be made 'synchronous', in so far as the word synchrony implies, as is usually the case, that during a certain period all the cells in the population grow without performing cell division which later takes place more or less simultaneously in the whole population. The

* Abbreviations used: RNA, ribonucleic acid; DNA, deoxyribonucleic acid; TCA, trichloroacetic acid; PCA, perchloric acid; CoA, coenzyme A; SPN, sulfur-containing peptide-nucleotide compound; SDN, sulfur-containing deoxypentosepolynucleotides.

first observation made by Tamiya, Shibata, Sasa, Iwamura and Morimura (1953) was that when *Chlorella ellipsoidea* was grown under diurnally intermittent illumination (12 hours' light and 12 hours' dark at 26°; light intensity, 16 kilolux), growth in size of cells occurred only in the light, while the cellular division, or separation of daughter cells with the rupture of mother cells, took place predominantly during the dark period.* The average volume of cells was thus large at the end of the light period and small at the end of the dark period. The smaller cells, which were referred to as D cells, had diameters ranging from about 2 to 4·5 μ with a maximum frequency at about 3 μ, while the larger cells, which were called L cells, had diameters of 4·5 to 7·5 μ with a maximum frequency at about 5·5 μ (Tamiya, Shibata, Sasa, Iwamura and Morimura, 1953; Tamiya, Iwamura, Shibata, Hase and Nihei, 1953). Based on these observations, they assumed that the life cycle of *Chlorella* occurs in the following manner:

$$\text{D cell} \xrightarrow{\text{light process}} \text{L cell}$$

$$\text{L cell} \xrightarrow{\text{dark process}} n \times (\text{D cell})$$

where n is the number of D cells released from a single mother cell (division number, see Iwamura *et al.*, 1955). It has been known that the n value varies according to species as well as to culture conditions, being 2, 4, 8, 16, 32 or even more. In many cases of synchronous culture thus far attempted, however, the average division number assumed irregular values between those mentioned above. For example, in the synchronous culture of *Chlorella pyrenoidosa* run by Lorenzen and Ruppel (1960), the average division number fluctuated in a range between 8 and 13, indicating that the population was not homogeneous with respect to the mode of nuclear and cellular division.

Methodological Considerations

In so far as we use the technique of synchronous culture for the purpose of obtaining information about the behavior of a single cell, we have to have a population in which the individual cells behave in a closely similar manner, not only in growth but also in nuclear and cellular division. To find a method by which the

* The first report of this kind was made by Aach in 1952 who on growing *Chlorella pyrenoidosa* under diurnally intermittent illumination (16 hours' light and 8 hours' dark) observed that the cells divided simultaneously soon after the beginning of the light period.

above requirement would be met, Tamiya *et al.* (1961) attempted various culture methods using *Chlorella ellipsoidea* as material. Apparent synchrony of algal growth could be obtained by six different methods, of which one may be referred to as the 'method of programmed light-and-dark regimen', and the rest may be called the 'methods starting from a homogeneous population'. In the former method, which has been widely used by German (Lorenzen, 1959; Lorenzen and Ruppel, 1960; Meffert, 1954; Pirson and Lorenzen, 1958a; Wanka, 1959) and American workers (Schmidt and King, 1961; Sorokin, 1960a, b, c; Sorokin and Krauss, 1959; Sorokin and Myers, 1957; Stange *et al.*, 1960), a random algal culture is subjected to a regular light-and-dark alternation, in which the lengths of light and dark periods are adequately chosen according to the algal species and culture conditions especially temperature and light intensity. The synchrony of growth and cell division is accomplished by repeating the adequate light-and-dark regimen several times, the growth occurring only during the light period and division taking place only, or mostly, in the dark period.

The principle of the other methods is to use, as starting material of a synchronous culture, a cell population made as homogeneous as possible by being separated, by differential centrifugation, from a less homogeneous population. By the method to be described in the method section of this volume, Tamiya *et al.* (1961) and Morimura (1959a) obtained a homogeneous population of small young cells, whose diameters ranged from 2·7 to 3·3 μ with a maximum frequency (more than 80 per cent) at 3·0 μ. Using such cells, which may be referred to as D_s cells (small D cells), synchronous cultures were obtained in various ways which are illustrated diagrammatically in Fig. 1.

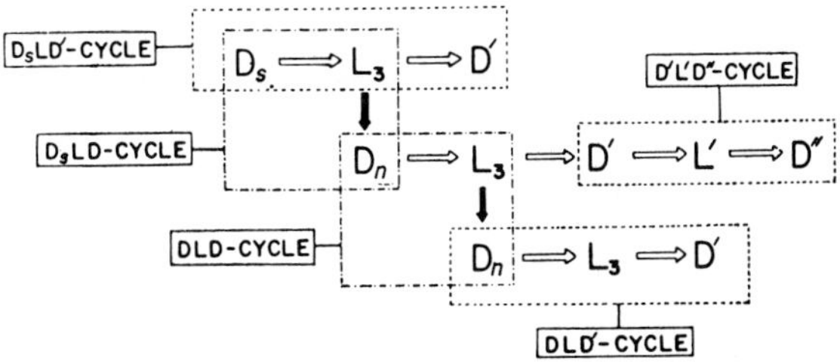

FIG. 1. Diagrammatic representation of various methods of synchronization starting from a homogeneous population (D_s cells) of *Chlorella ellipsoidea* (Tamiya *et al.*, 1961). The white and black arrows indicate the changes occurring in the light and in the dark, respectively. (For further explanation see text.)

(i) D_sLD and D_sLD' cycles. When D_s cells were incubated in the light with supplies of CO_2 and nutrient salts, they grew gradually, and eventually attained the stage which was called L_3, that is the stage of ripening, after which the cells could divide completely when kept in the dark under aerobic conditions. The daughter cells produced from the L_3 cells in the dark have been called D_n cells (nascent D cells; Iwamura, 1955; Iwamura *et al.*, 1955; Tamiya, Iwamura, Shibata, Hase and Nihei, 1953). The cycle starting from D_s and ending, via L_3, with D_n may be referred to as the D_sLD cycle. When, on the other hand, the L_3 cells were left in the light, they also divided synchronously into a number of daughter cells, which were denoted by D′ in distinction from D_n (the D_sLD' cycle).

(ii) DLD and DLD′ cycles. When the D_n cells produced in the D_sLD cycle were incubated in the light they grew and eventually attained the stage corresponding to the L_3 stage of the D_sLD cycle. The cells at this stage divided into D_n cells in the dark and into D′ cells in the light. These cycles were called the DLD and DLD′ cycle, respectively.

(iii) D′L′D″ cycle. The D′ cells produced in the DLD′ cycle appeared to be essentially the same as those produced in the D_sLD' cycle. When incubated in the light, these D′ cells grew into ripened cells and later divided synchronously into a number of daughter cells. The ripened cells and the daughter cells appearing in this case were denoted by L′ and D″, respectively, and the cycle thus obtained was called the D′L′D″ cycle.

As is apparent from the diagram in Fig. 1, only the DLD cycle is a really cyclic process, while the others are pseudo-cyclic in that the daughter cells ultimately produced are different from the starting D cells. The mode of nuclear division occurring in each of these cycles was followed by using the technique of Feulgen staining (Tamiya *et al.*, 1961). It was seen that the D_s and D_n cells were invariably uninucleate, while the L_3 cells were either binucleate or tetranucleate. On being incubated in the dark the L_3 cells ripened further (the process of 'post-ripening'; see below), and after passing through a tetranucleate stage, referred to as the L_4 cells, divided into four uninucleate D_n cells ($n = 4$).

In Fig. 2 the course of the change of relative cell number and average cell volume occurring during the D_sLD and DLD cycles is shown. The cycle starting from D_s cells, for some reason unknown at present, always took about twice as much time, for both its light and dark periods, as did the cycles starting from the D_n cells.

The changes of statistical distribution of cell size and of nuclear patterns occurring in repeated DLD cycles are shown in Fig. 3. As can be seen, the DLD cycle was repetitive, always showing a division number of 4 and with almost no change in the statistical distributions of cell size and nuclear patterns of both D_n and L_3 cells in repeated cycles.

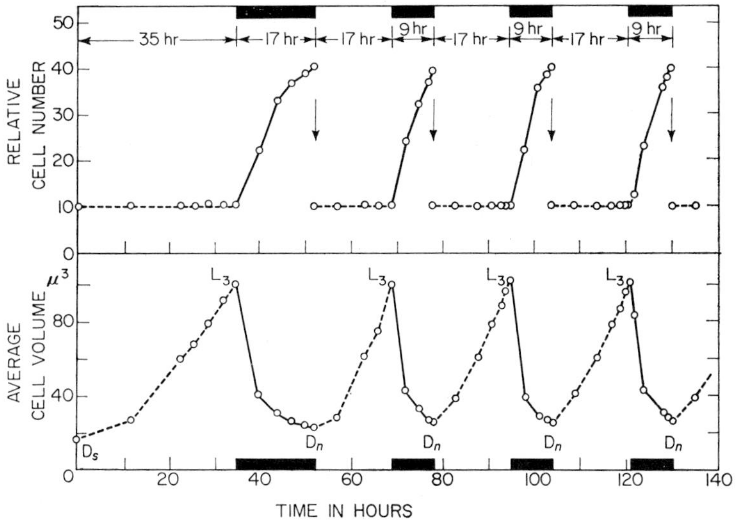

FIG. 2. Changes of relative cell number and average cell volume occurring in D_sLD cycle followed by repeated DLD cycles (*Chlorella ellipsoidea*; temperature: 21°; light intensity: 10 kilolux). Dashed and solid curves show, respectively, the processes occurring in the light and in the dark. Black bars along the abscissa indicate the dark periods. At the end of each dark period the culture was diluted to the original cell concentration (indicated by vertical arrows) (Tamiya *et al.*, 1961.)

The data reproduced in Fig. 4 show the difference between the D_sLD and D_sLD' cycles and that between the DLD and DLD′ cycles. Although in all these cases growth and cell division occurred synchronously with the division number of 4, the division of L_3 cells in the light took place somewhat later than that in the dark, and the D′ cells produced in the light were larger than the D_n cells because of the growth of daughter cells occurring in the light, before the completion of cellular division in the whole population. The D′ cells produced in the light were all initially uninucleate; however, by the time all the mother cells had completed cellular division, some of the daughter cells had undergone nuclear, but not cellular, division. Thus, the D′ cells obtained at the completion of D_sLD' and DLD′ cycles were a mixture of uni-, bi- and tetranucleate cells.

When such D′ cells were incubated in the light, they grew to the stage of L′ cells which were a mixture of bi- and tetranucleate cells; these cells divided synchronously in the light into four D″ cells each. These, like D′ cells, were found to be a mixture of uni-, bi- and tetranucleate cells. It was thus revealed that, in such cases as the D′L′D″ cycle, neither the synchrony in the ordinary sense, nor the uniformity of division can be a criterion of the homogeneity of the mode of nuclear division in algal cells.

With respect to the homogeneity of the modes of nuclear and cellular division of algal cells, the method of programmed light-and-dark regimen gave even worse results (Tamiya *et al.*, 1961).

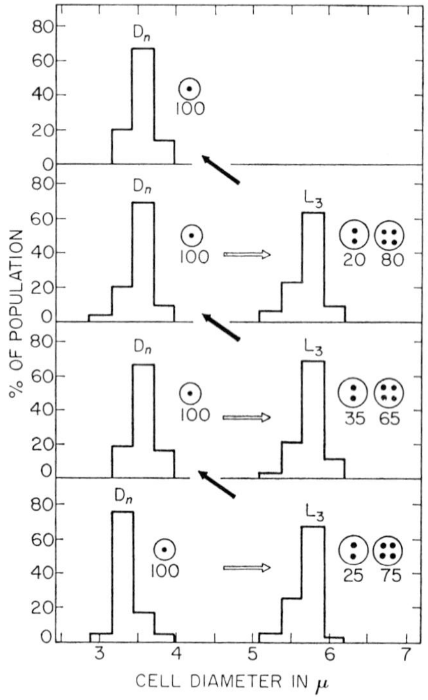

Fig. 3. Changes of statistical distribution of cell size and of nuclear patterns observed in the repeated DLD cycles shown in Fig. 2 (Tamiya *et al.*, 1961). Numerals given under schematic representations of nuclear patterns are the percentages of population showing respective patterns. White and black arrows indicate the changes occurring in the light and in the dark, respectively.

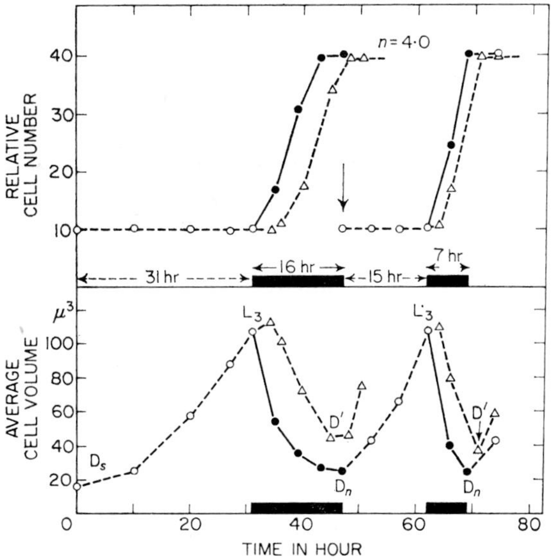

FIG. 4. Changes of relative cell number and average cell volume occurring in the D_sLD, D_sLD', DLD and DLD′ cycles. (*Chlorella ellipsoidea*; temperature: 21°; light intensity: 10 kilolux.) [See the legend of Fig. 2 (Tamiya *et al.*, 1961).]

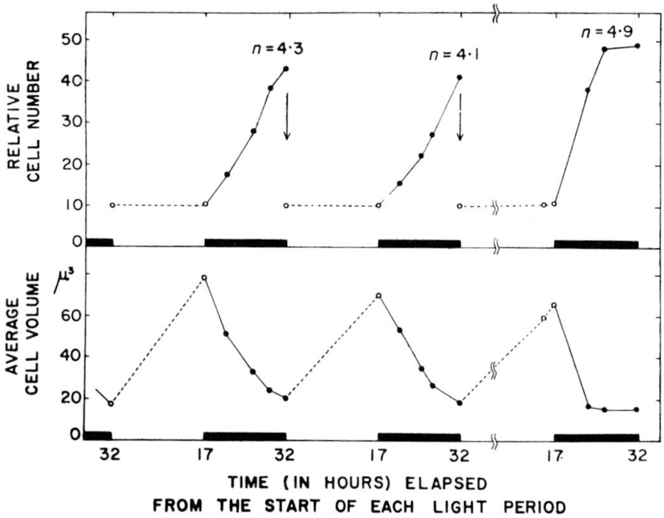

FIG. 5. Changes of relative cell number and average cell volume in the synchronous culture of *Chlorella ellipsoidea* which was obtained by the method of programmed light-and-dark regimen of 17:15 hr (21°, 10 kilolux). [See the legend of Fig. 2 (Tamiya *et al.*, 1961)].

Among the various combinations of light and dark periods applied to the culture of *Chlorella ellipsoidea* at 21°, relatively good results were obtained with the alternation of 17 hours' light and 15 hours' dark (expressed as 17 : 15 hr.). After the light-and-dark cycles had been repeated five times, the synchrony became apparently satisfactory, as shown by the data reproduced in Fig. 5. However, as was also the case in the experiments of Lorenzen and Ruppel (1960), the average division number was not constant, but fluctuated between 4·0 and 4·9 in different cycles. The statistical distribution of cell size was always far broader than was the case in the cycles started from a homogeneous population, and this

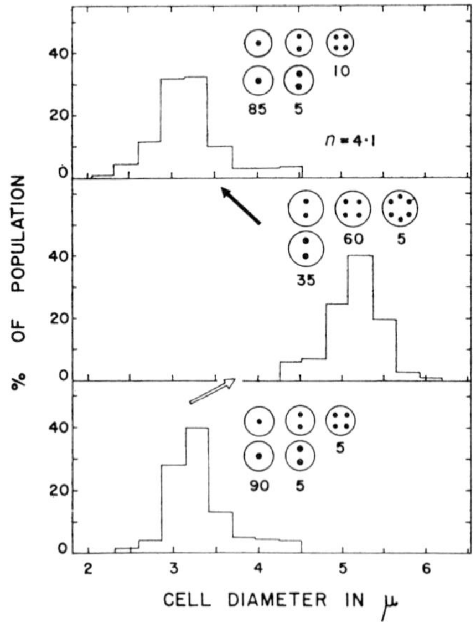

FIG. 6. Changes of statistical distribution of cell size and of nuclear patterns observed in one of the cycles reproduced in Fig. 5. At the end of the light period, about 5 per cent of the population was penta- or hexanucleate. In the figure they were schematically shown as hexanucleate cells. In some other cycles obtained in the same experiment, there appeared a small number of hepta- and octanucleate cells.

situation showed no tendency to improve on further repetition of the cycle. The mode of nuclear division was also very irregular as shown by the data reproduced in Fig. 6. Although the majority of D cells, that is the cells at the end of dark period, were uninucleate and the majority of L cells (the cells at the end of light period) were tetranucleate, some of the former contained two or four nuclei per cell, and some of the latter contained five, six, seven or eight nuclei per cell, a phenomenon which never occurred in the D_sLD or DLD cycles. These results also show that the apparent synchronization of *Chlorella* cells, as can be achieved by the method of programmed light-and-dark regimen, can hardly lay claim to a synchrony on the level of nuclear and cellular division of algal cells.

In order to obtain a synchrony and homogeneity on the level of nuclear division, the choice of adequate algal strains may also be a matter of importance. The strain used by Tamiya and his group has a basic character of dividing into four daughter cells, and that used by Lorenzen and Ruppel (1960) tends to divide into eight or sixteen daughter cells. Sorokin and Krauss (1959) and Sorokin and Myers (1957) reported that a temperature-tolerant strain (7–11–05) of *Chlorella*, which they synchronized by the method of programmed light-and-dark regimen, showed more than forty-fold increase in cell mass (packed cell volume) during the light period. Since it is stated that, at the completion of cell division in the dark, the daughter cells produced from these mother cells comprised up to 99 per cent of the total population, each mother cell must have produced more than forty daughter cells. With a strain showing such an enormous value of division number, it is obviously difficult to obtain a sufficient uniformity in the modes of nuclear and cellular division.

The various methods shown in Fig. 1 have their merits and demerits according to the purpose of the experiment. The events occurring in the DLD cycle are most clear-cut and repetitive, while the D′L′D″ cycle has little to recommend it because of the heterogeneity of the mode of nuclear division occurring in the cycle. The D_sLD cycle is simpler to run than the DLD cycle, and both cycles seem to be identical except that the former requires, for its completion, twice as much time as the latter. In so far as we intend to investigate the mode of growth and cellular division (division number) of algal cells, without going into the problems of nuclear features of the daughter cells produced ultimately, the D_sLD' and DLD′ cycles are more convenient than the D_sLD or DLD cycles. This is especially the case when we study, for instance, the effects of various environmental factors upon the mode of growth and cellular division; for running the D_sLD and DLD cycles we have to know in advance the adequate time of transfer of L_3 cells in the dark, and this is subject to change according to the environmental conditions. In the earlier work of Tamiya, Shibata, Sasa, Iwamura and Morimura (1953), Tamiya, Iwamura, Shibata, Hase and Nihei (1953), Iwamura (1955), Iwamura *et al.* (1955), Nihei (1955, 1957) and Nihei *et al.* (1954), the starting D cells used were not quite homogeneous, being comparable to the D cells which were obtainable by the most adequately programmed method of light-and-dark regimen. Most of their recent investigations (Hase and Mihara, 1960; Hase, Mihara and Otsuka, 1959; Hase, Mihara, Otsuka and Tamiya, 1959b; Hase, Mihara and Tamiya, 1960c; Hase, Morimura, Mihara and Tamiya, 1958; Hase, Morimura and Tamiya, 1957, 1960; Hase, Otsuka, Mihara and Tamiya, 1959; Mihara, 1961; Miyachi and Tamiya, 1961; Morimura, 1959a, b; Otsuka, 1961a) have been performed by the D_sLD or D_sLD' method.

In connection with the term 'average division number' (or simply, 'division number'), used by Tamiya's group, a remark seems to be in order concerning a similar term used by some other workers (Pirson and Lorenzen, 1958a). The value n used by Tamiya's group means the average number of daughter cells liberated from single mother cells. When the division in the whole population is complete (that is when all the cells in the population perform cellular division, which can be clearly recognized when the changes of statistical distribution of cell size are such as those illustrated in Fig. 3), the division number can be calculated by simply dividing the cell number counted after the completion of division by that counted before the division. When the division is incomplete, and this can only be recognized by following the change of statistical distribution of cell size, the division number is calculated by: $n = (d' - d)/(l - l')$, where d, l, and d', l' are, respectively, the number of 'D' and 'L' cells before and after the cell division. At any rate, in order to determine the division number, both the cell number and the statistical distribution of cell size before and after the cell division have to be known.

In the experiments of German workers (Lorenzen and Ruppel, 1960; Pirson and Lorenzen, 1958a), the process of cell division is, in most cases, followed by comparing the cell numbers before and after the cell division, irrespective of whether the division took place completely or only incompletely in the whole population. In their papers, increases in cell number are expressed as a percentage of the cells found before cell division. According to the term 'Teilungsschub' used by the German authors, we shall call the ratio of cell numbers before and after (complete or incomplete) cell division the 'Schub-number' and denote it by s, which is equal to n only when the division is complete. It becomes increasingly smaller than *n* in proportion to the degree of incompleteness of the cell division occurring in the population.

II. Various Developmental Stages

That the *Chlorella* cells, when well ripened in the light, can perform cell division in the dark under aerobic conditions has been known since the early work of Tamiya, Shibata, Sasa, Iwamura and Morimura (1953). The capacity of cells for dividing in the dark was found to develop gradually with the advancement of ripening of cells in the light (Iwamura, 1955; Iwamura, Hase, Morimura and Tamiya, 1955; Tamiya, Iwamura, Shibata, Hase and Nihei, 1953). The state of affairs in the DLD cycle of *Chlorella ellipsoidea* is shown in Fig. 7. Based on these and many other observations to be described later, Hase, Morimura and Tamiya (1957) and Morimura (1959a) have distinguished seven different stages—designated as D_n, D_a, $D \sim L$, L_1, L_2, L_3 and L_4—in the life cycle of *Chlorella ellipsoidea*. The characteristics of each of these stages are as follows:

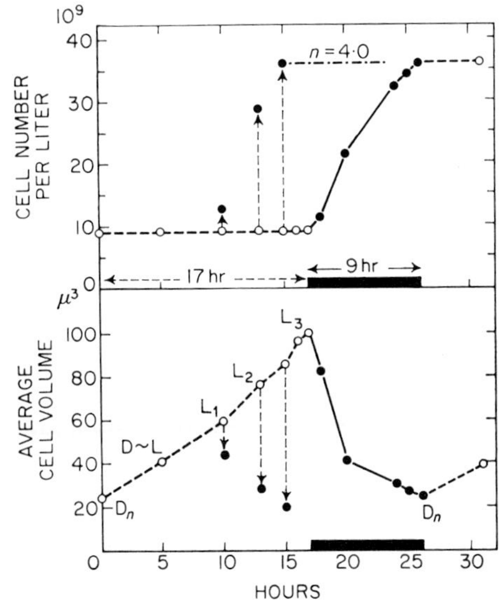

Fig. 7. Progress of ripening during the light period in the DLD cycle of *Chlorella ellipsoidea* (the third DLD cycle shown in Fig. 2). At various stages during the light period, aliquots of cell suspension were taken out and incubated in the dark for 40 hr (at 21°). Solid circles indicated by vertical (dotted) arrows show the values obtained after the dark incubation (Tamiya *et al.*, 1961).

D_n cells (nascent D cells), which are the young cells newly produced (in the dark) from mother L cells. They are incapable of cell division in the dark.

D_a cells (active D cells): the most photosynthetically active cells which are derived from D_n when illuminated; incapable of cell division in the dark.

$D \sim L$ cells: cells of transient stage between D and L; also incapable of cell division in the dark.

L_1 cells (unripened L cells), which are large in size, but not yet ripe enough to perform cell division when incubated in the dark.

L_2 cells (half-ripened L cells), which can only incompletely divide when kept in the dark.

L_3 cells (ripened L cells), which can divide completely when kept in the dark.

L_4 cells (fully-matured L cells), which are at the stage immediately prior to the cell division.

Changes in the nuclear patterns of the cells during the transformation of D_s cells to L_3 cells, and of L_3 cells to D_n cells are shown in Figs. 8 and 9 (Tamiya *et al.*, 1961) Until the stage of $D \sim L$, the cells remained uninucleate, but at the L_1 stage some of the cells became bi- or tetranucleate. The uni- and binucleate L_1 cells had either larger or smaller nuclei. At the L_2 stage, the proportion of bi- and tetranucleate cells further increased, while all uninucleate cells contained larger nuclei. On incubating the L_2 cells in the dark, the daughter cells partially produced were all uninucleate, while the cells which remained incapable of cell division were all binucleate, having either larger or smaller nuclei. From the change of cell number and cell size of L_2 cells, measured before and after the dark incubation, it was observed that some of the L_2 cells which could divide in the dark had each given rise to four daughter cells.*

The L_3 cells were 65–95 per cent tetranucleate, and only 5–35 per cent binucleate. In the binucleate L_3 cells, each nucleus was larger than that in the D_s or D_n cell, or those in the tetranucleate L cell. When incubated in the dark (Fig. 9), all L_3 cells were transformed into tetranucleate L_4 cells, which in turn divided

* In their earlier papers, Iwamura (1955) and Iwamura, Hase, Morimura and Tamiya (1955) reported that as ripening of L-cells progressed their division number also increased to some extent—from 4·1 to 5·3. This was found to be incorrect; the errors in calculating *n* values were significant at that time, when the synchrony of the cell population was not as good as today.

completely into uninucleate D_n cells with the division number of 4. The course of development and division of nuclei was entirely the same in the D_sLD and DLD cycles, although the proportion of uni-, bi- and tetranucleate cells at the transient stages, L_1 and L_2, where the synthesis of DNA and nuclear division occurred most actively, were subject to change in repeated experiments.

These observations show that in both D_sLD and DLD cycles of *Chlorella ellipsoidea*, the single nucleus contained in a D cell first grows larger, then divides into two smaller nuclei which, after becoming larger, each divide into two, thus giving rise to the tetranucleate L_4 cell. It is interesting to note that, in these cycles, the cellular division never occurs before the cell enters the stage

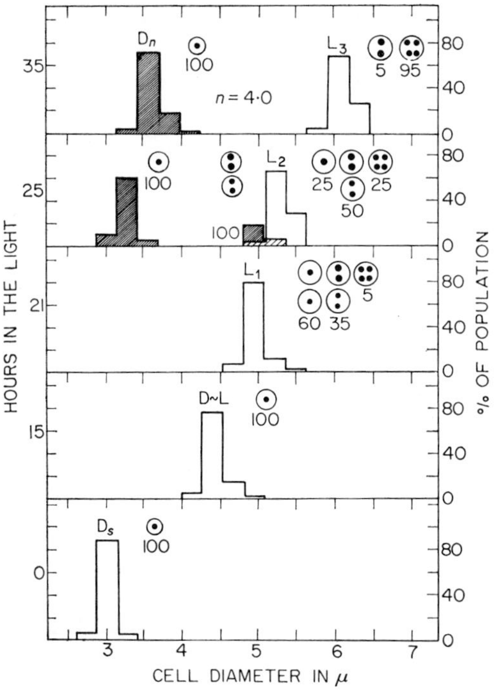

FIG. 8. Changes of statistical distribution of cell size and of nuclear patterns which occurred during the transformation of D_s cells into L_3 cells (*Chlorella ellipsoidea*; 21°, 10 kilolux, Tamiya *et al.*, 1961). Essentially the same result was obtained during the transformation of D_n cells into L_3 cells. Shaded polygons show the distribution of cell size obtained after the dark incubation of L_2 and L_3 cells. In the case of L_2 cells, the dark incubation caused a splitting of distribution polygon; the smaller cells newly produced in the dark were all uninucleate, while those remaining undivided were all binucleate (Tamiya *et al.*, 1961).

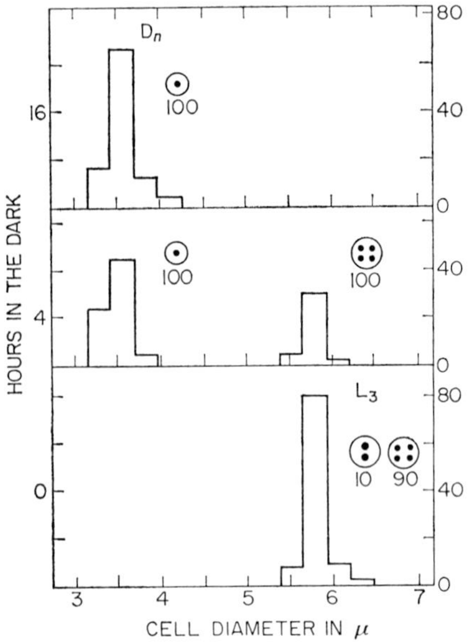

FIG. 9. Changes of statistical distribution of cell size and of nuclear patterns observed during the transformation of L_3 cells into D_n cells in the dark. (*Chlorella ellipsoidea*; 21°. Tamiya *et al.*, 1961.)

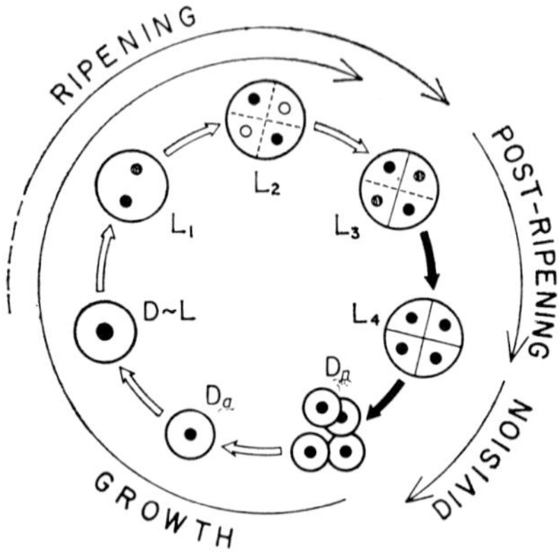

FIG. 10. Schematic representation of the life cycle of *Chlorella ellipsoidea* (Tamiya *et al.*, 1961). The white arrows indicate the light-dependent processes, while the black arrows show the transformations occurring independently of light. (For further explanations see text.)

of the tetranucleate L_4 cell, and the latter always divides into four uninucleate D_n cells and never into two binucleate cells.

The general course of events occurring in the DLD cycle of *Chlorella ellipsoidea* is schematically illustrated in Fig. 10 (Tamiya *et al.*, 1961; also Hase, Mihara and Tamiya, 1961a; Morimura, 1959a). The process of growth, or increase in total cell mass, occurs from the D_n stage until around the L_3 stage. The process of ripening, in which various synthetic reactions occur in preparation for the formation of daughter cells, seems to occur somewhat later than the growth process and it involves light-dependent and light-independent phases. These are designated as ripening and post-ripening, respectively, in the figure.* The phase of post-ripening is followed by the cellular division which is also a light-independent process.

In their study of *Chlorella pyrenoidosa*, synchronized by the method of programmed light-and-dark regimen (16 : $\overline{1}$2-hr, 30°, 9 kilolux), Lorenzen and Ruppel (1960) distinguished three developmental phases, I, II and III. According to them, Phase I starts from the beginning of the light period and ends at the eighth hour of the same period. This phase is defined by the fact that the cells are not capable of cell division when incubated in the dark. Apparently, this phase corresponds to the stages from D_n to L_1 cells in the scheme presented above. Phase II extends from the eighth to the fourteenth hour in the light period, during which time the cells, when transferred to the darkness, divide either incompletely or completely but with n values smaller than normal (normally: n = 8–13). This phase is, therefore, partly similar to the L_2 stage of Tamiya *et al.* (1961). Phase III lasts 5 hr, extending from the fourteenth hour in the light period to the third hour of the dark period. During this period, the cells, in a manner similar to those of L_3 and L_4 cells, divide completely, with the normal division number when incubated in the dark. In these experiments using *Chlorella pyrenoidosa*, Lorenzen (1958) observed that toward the end of Phase I, the nuclei of cells became temporarily unstainable with the Feulgen reagent. This had never happened in *Chlorella ellipsoidea* used by Tamiya *et al.* (1961), who could count the number of nuclei quite clearly at all stages of the normal algal life cycle.

The question as to the manner in which the cell elements other than the nucleus, such as chloroplasts and mitochondria, would divide during the algal life cycle, has not yet been studied in detail

* The phase of post-ripening has been called 'maturing' in the papers of Hase *et al.* (1960a) and Morimura (1959a).

using the technique of synchronous culture. With synchronized cultures of *Chlorella ellipsoidea*, Hase and Kushida (unpublished experiments) took electron-microscopic photographs of ultra-thin slices of cells at different developmental stages, and found that in the normal algal life cycle the division of chloroplasts occurs almost keeping pace with the nuclear division. Of interest in this connection are the cytomorphological and cytochemical observations made recently by Řetovský and Klášterská (1961b) with *Chlorella pyrenoidosa* and *Chlorella vulgaris*, var. *viridis*, although the materials they used were utterly asynchronous cultures. They state that in young algal cells the chondriome forms a coherent zone at the periphery of cells, and that with the ripening of cells the peripheral zone disappears and new filamentous mitochondria are formed, and find their way by active movement into the inner space of cells. According to these authors (Řetovský and Klášterská, 1961a), the divisions of nuclei, chloroplasts and mitochondria are immediately followed by a formation of wall-less, mobile 'proautospores' which, after formation of walls, become immobile autospores within the mother cell, and in the mature autospores the

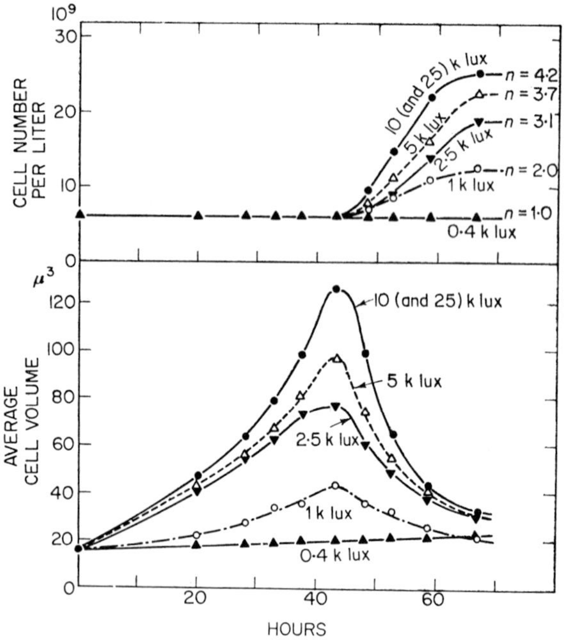

FIG. 11. D_5LD' cycles of *Chlorella ellipsoidea* run at 16° under illumination with lights of varying intensities (Morimura, 1959a).

chondriome again spreads over the periphery of the cell. Confirmation of these statements with well synchronized cultures of *Chlorella* is awaited with interest.

III. Effects of Temperature and Light Intensity

Using the D_5LD' method mentioned in a previous section, Morimura (1959a) studied the effects of temperature and light intensity upon the life cycle of *Chlorella ellipsoidea*. Fig. 11 shows the results of experiment in which lights of varying intensities (0·4, 1·0, 2·5, 5·0, 10 and 25 kilolux) were applied keeping the temperature constant at 16°. As was expected, the rate of growth was higher at higher light intensities, but irrespective of the difference in the growth rate, the incipient cell division at L_3 occurred simultaneously at all light intensities except at 0·4 kilolux, in which case the cells continued to grow very slowly without dividing during the whole course of the experiment. The volume of L_3 cells at the stage of incipient cell division was greater at higher light intensities. The time elapsed from the stage of incipient cell division to the stage of completion of cell division was also found to be independent of the light intensity. The fact that the division number was largely dependent on the light intensity was interesting; namely, it was 2·0 at 1 kilolux and about 4 at saturating intensities (10 and 25 kilolux), while intermediate values (3·1–3·7) were obtained under the light of intermediate intensities.

Entirely different phenomena were observed when the temperature was varied (9°, 16° and 25°), keeping the light intensity constant at a saturating level of 10 kilolux. As may be seen from the results shown in Fig. 12, the rate of growth and the rate of post-ripening and division were temperature-dependent, both being higher at higher temperatures. Despite such a profound difference in rates, the division number remained the same (4·0–4·1) at all temperatures tested.

Assuming the four developmental phases presented in Fig. 10, namely, growth, ripening, post-ripening and division, Morimura (1959a) analyzed kinetically the data presented in Figs. 11 and 12. By comparing the dependencies on the light intensity of the rates (at 16°) of 'growth' and 'ripening' with that of photosynthesis, measured in short-term manometric experiments, it was deduced that the three processes probably involve the same photic reaction, as far as the dependency of rate upon the light intensity is concerned, although their dark reactions are profoundly different from each other in nature. Analysis of the temperature-dependencies of the processes in question has revealed that the dark reactions involved in growth and ripening, as well as the processes of post-ripening and division, are—especially at lower temperatures—more temperature-dependent than the dark reaction of photosynthesis. The temperature-dependency of the post-ripening and division was most pronounced; it showed an apparent activation energy of as high as 43 kcal at lower temperatures. That the light-independent process involved in the growth of *Chlorella* has an apparent activation energy as high as 49 kcal has already been reported by Tamiya, Iwamura *et al.*, in 1953. Recently it was stated by Sorokin (1960d) that in a high-temperature strain of *Chlorella* the process of cell division has a greater activation energy than the process of accumulation of cell material.

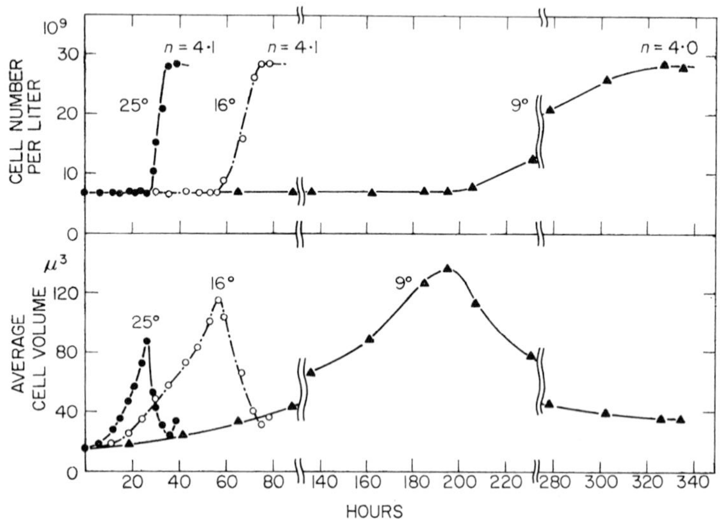

FIG. 12. D_sLD' cycles of *Chlorella ellipsoidea* run at different temperatures (9°, 16 and 25°) under illumination with light of saturating intensity (10 kilolux) (Morimura, 1959a).

A phenomenon which deserves special attention from the physiological point of view is that, as shown in Fig. 11, the L cells grown under different light intensities, at a fixed temperature, began to divide simultaneously giving rise to different numbers of daughter cells according to the light intensity. Similar 'isochronism' of cellular division with different division numbers has repeatedly been observed when the algal cells were subjected, at a fixed temperature, to various environmental conditions, such as deprivation of phosphate, nitrate, or potassium from the medium (Hase, Morimura and Tamiya, 1957), or the addition of some anti-metabolites to the medium (Tamiya, Morimura and Yokota, 1962). The number of daughter cells is no doubt determined by the quantity of nuclear substances which have been formed during the ripening process preceding the cellular division. The phenomenon observed implies that the process of cellular division itself occurs largely independently of the quantity of nuclear substances as well as of the total cell mass formed previously. It may be natural to assume that the occurrence of cellular division is incited by some factor or factors which is formed during the process of ripening. Experimental results show that the process of formation of such an inducer substance or substances is a light-independent, but temperature-dependent, process.

Of interest in this connection are the observations made by Pirson and Lorenzen (1958a) using *Chlorella pyrenoidosa* synchronized by the method of programmed light-and-dark regimen. By varying the lengths of the light and dark periods, temperature and light intensity, they obtained synchronous cultures of multifarious features. They noticed, however, that the occurrence of incipient cell division, irrespective of whether it proceeded to completion or not, was determined by the time of commencement of the light period. For instance, in an experiment at 23°, 5 kilolux, in which the length of the dark period was fixed at 12 hr and the length of light period was varied by 8, 12, 16, 20 and 24 hr, the cell division, mostly incomplete, always started 20 hr after the beginning of each light period. The Schub-number (s) was, however, largely dependent upon the length of the light period, being only 2·5 at $8:\overline{12}$, 4·8 at $12:\overline{12}$, 7·5 at $16:\overline{12}$—only in this case the division was almost complete (Lorenzen, 1959)—6·2 at $20:\overline{12}$ and 5·8 at $24:\overline{12}$. With a fixed light-and-dark regimen of $12:\overline{12}$, they changed the intensity of the light applied during the light periods. The time which elapsed from the start of each light period until the onset of cell division (mostly incomplete) was 24 hr in the experiment performed at 20°, and averaged 20 hr in the experiments performed at 25° and 30°. The light intensity applied during the light period of $12:\overline{12}$-regimen also affected the Schub-number profoundly; for instance at 30°, it was only 1·9 at 2 kilolux, 2·9 at 4 kilolux, 6·4 at 9 kilolux and 7·6 at 15 kilolux.

Based on these observations, Pirson and Lorenzen (1958a) concluded that for the onset of cell division, regardless of its completeness or otherwise, the lengths of alternating light and dark periods have no essential significance, but that the time of beginning of the light period determines the time of incipient division. According to these authors, the onset of light period acts as a pace-maker (Zeitgeber) giving an impulse to some endogenous machinery provoking a definite rhythm of the process of cell division.

The endogenous component predicted by Pirson and Lorenzen (1958a) and Lorenzen (1957) seems to have some similarity to the inducer substance or substances predicted by Morimura (1959a), except that the action, or formation, of the former is stated to be a temperature-independent process, while that of the latter has been proved to be a temperature-dependent process. Clarification of the chemical nature of these factors and of their mode of action is a matter of utmost interest. As will be described later, studies on the role of sulfur in the processes of ripening and division of algal cells seem to offer some important clue in solving this highly interesting problem.

Concerning the effect of light intensity upon the cell division of *Chlorella*, Wanka (1959) has also made some observations which seem to be partly in accordance and partly in conflict with the data obtained by Morimura (1959a). *Chlorella pyrenoidosa* was synchronized by the method of programmed light-and-dark regimen (16:8, 20°, 2·5–3·0 kilolux), and at a certain stage the cells were placed under continuous illumination of varying intensities. Unfortunately, the algal population used by Wanka was so nonhomogeneous that the cultures showed, in the diurnal light-and-dark alternation, the so-called group synchrony (Lorenzen, 1957, 1958), that is, different groups of cells divided alternately every other day, each group requiring two light-and-dark periods for completing one life cycle. Moreover, the cultures were run by supplies of ordinary air without enrichment of CO_2. Consequently, the cells synchronized by the light-and-dark alternation showed, when placed under continuous illumination, two steps of cellular division with different features. Inspection of his data reveals that, in the first step, the division occurred, contrary to the statement made by Wanka himself, almost simultaneously at different light intensities, and that the Schub-number was larger at higher light intensities. Strangely enough, an entirely different phenomenon was observed in the second step of cellular division; this time, the higher the light intensities the earlier was the onset of cellular division and the smaller was the Schub-number. The observations made by Wanka are unintelligible also in other respects; for instance, he reported that under continuous illumination the average cell volume became larger at lower light intensities, a situation which is quite the reverse of what has been observed by other authors (Myers, 1946; Tamiya, Shibata *et al.*, 1953).

IV. Photosynthetic and Respiratory Activities

Already in their earliest study with the synchronous culture of *Chlorella*, Tamiya, Iwamura *et al.* (1953) observed that the D cells showed a stronger photosynthetic activity—both under strong and weak light—and weaker, endogenous, respiratory activity than the L cells. Nihei *et al.* (1954) found that the photosynthetic activity was maximum at the D_a stage and minimum at the stage around L_2 and L_3, while the respiratory activity was maximum at the stage around L_2 and L_3, and minimum at the D_a stage (see also Sasa, 1961). Using $^{14}CO_2$ as a tracer according to the method of Calvin and Benson (1948), they also followed the changes of the light-induced CO_2-fixing capacity (fixation occurring immediately after the pre-illumination) and of the background CO_2-fixation, that is, fixation in the absence of pre-illumination. It was found that the background CO_2-fixation was stronger in L cells than in D cells, whereas the light-induced ^{14}C-fixing capacity was much stronger in D cells than in L cells. It was noteworthy that the latter capacity, measured in the presence of phosphate, temporarily dropped to almost the zero level at the stage around L_2 and L_3.

A somewhat different observation was reported recently by Stange *et al.* (1960) who measured the short-term $^{14}CO_2$-fixation, after 2 and 5 min in the light, by partially-synchronized *Chlorella pyrenoidosa*. They observed that the photosynthetic ^{14}C-fixation per unit volume of packed cells increased with the progress of cell development. The fate of the ^{14}C assimilated in algal cells appeared to be more or less different according to the developmental stage. In later stages of cell development, prior to the cell division, the proportion of radioactivity in the 80 per cent and 20 per cent ethanol-insoluble material was higher than it was during the earlier developmental stages. The progress of cell development was also accompanied by an increase of ^{14}C incorporated in sucrose and a decrease of ^{14}C incorporated in alanine, threonine, aspartic acid and citrulline.

In the case of *Chlorella ellipsoidea* studied by Nihei *et al.* (1954), the weakening of the photosynthetic activity in L cells was more pronounced in CO_2-uptake than in O_2-output; consequently, the ratio of O_2/CO_2 attained as high a value as 3·3 at the stage around L_2 and L_3, whereas the ratio was unity in D cells as well as in L_4

cells (see also Sasa, 1961). Nihei (1955) discovered that the high O_2/CO_2 ratio obtained at around the $L_2 \sim L_3$ stage was due to the photochemical evolution of O_2 associated with a process of phosphate assimilation. As is clearly demonstrated by the experiment reproduced in Fig. 13, the cells at the $L_2 \sim L_3$ stage, in the absence of CO_2,

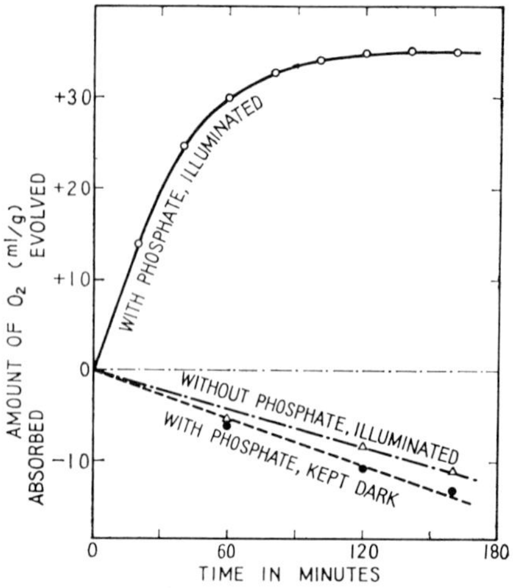

FIG. 13. Effect of phosphate on the process of light-induced O_2 evolution shown by the cells (*Chlorella ellipsoidea*) at a stage around L_2 and L_3 (Nihei, 1955). All experiments were performed under deprival of CO_2 (at 25°).

showed a positive photochemical O_2 evolution in the presence of phosphate and not in its absence (tartrate buffer being used instead of phosphate buffer), a phenomenon which was never observed with D cells. Moreover, the fact that the photochemical process in question occurred competitively with the photochemical CO_2 assimilation is of interest. The data presented in Fig. 14 show that, in the presence of CO_2, the process of O_2 evolution itself proceeded at the same rate in phosphate and tartrate buffers. In the absence of phosphate, CO_2 absorption occurred at the same rate as O_2 evolution, showing a photosynthetic quotient of unity. When phosphate was present, the CO_2-absorption was distinctly suppressed so that the apparent photosynthetic quotient became well above unity. The role of phosphate in this case is, at least outwardly, similar to that of various Hill reagents, causing, in competition with CO_2, a photochemical reduction of some cellular material. The behavior of phosphate in this interesting process was investigated by Nihei (1955) in some detail, and it was found that the enhanced photochemical O_2

liberation in the presence of phosphate was accompanied by an increased assimilation of phosphate which was accumulated in algal cells mainly in the form of polyphosphate (see below).

Comparison of D and L cells for their chlorophyll content and rates of photosynthesis at sufficiently low light-intensities is of special interest in relation to the problem of quantum efficiency of photosynthesis. With D and L cells Nihei *et al.* (1954) measured, on the one hand, the chlorophyll content (denoted by C) in per cent of dry weight of cells, and, on the other, the light-limited rates of photosynthetic O_2 output and CO_2 uptake (denoted by α_O and α_C, respectively) in ml per kilolux of light and per hr and g (dry weight) of cells. In so far as chlorophyll represents the principal factor in causing the light absorption by algal cells, the ratio α/C may be regarded as representing the relative quantum efficiency of photosynthesis. The experiment showed that the values of α_O/C and α_C/C for D cells were 1·9 and 2·7 times, respectively, those for the L cells, indicating that the quantum requirement of photosynthesis of D cells is considerably smaller than that of L cells.

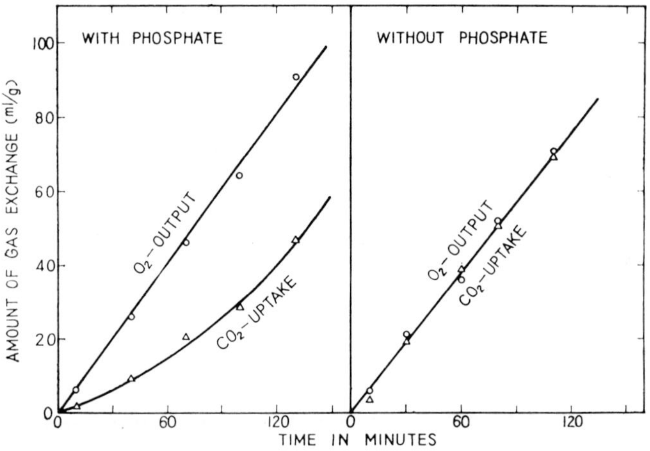

FIG. 14. Effect of phosphate on the photochemical gas exchange (O_2 output and CO_2 uptake), in the presence of CO_2, shown by the cells (*Chlorella ellipsoidea*) at a stage around L_2 and L_3 (Nihei, 1955). While O_2 output was not affected by phosphate, CO_2 uptake was distinctly lowered in the presence of phosphate (temperature: 25°).

A number of reports have been made by German and American workers concerning the change of photosynthetic activity observed during the life cycle of their *Chlorella* strains. Direct comparison of their data with those of Tamiya's group appears, however, to be somewhat difficult because of the differences in algal strains used and in the methods of synchronization adopted. In their experiment with *Chlorella pyrenoidosa* synchronized by the method of programmed light-and-dark regimen ($16{:}\overline{12}$, 30°, a condition which was stated to give a complete synchrony of division), Pirson and Lorenzen (1958b), and Lorenzen (1959) observed a change of photosynthetic activity similar to that reported by Nihei *et al.* (1954). However, they maintained that the phenomenon might have been due more to the external conditions of light-and-dark alternation—temperature, the length of the dark period and the intensity of light applied during the light period—than to the change of endogenous conditions inherent to the algal life cycle. Lorenzen (1959) observed, for instance, that at the temperature of 20° the rate of photosynthetic O_2 evolution, on a dry weight basis, decreased during the light period and increased during the dark period, while at 30° the increase and decrease occurred during the first and latter halves, respectively, of the light period. In another paper by Pirson and Lorenzen (1958a) it was reported that under different conditions of light-and-dark alternation considerable changes occurred in the Schub-number as well as in the degree of incompleteness of cell division, indicating that the populations synchronized under different light-and-dark conditions varied widely in their degree of heterogeneity. Obviously, the data obtained with heterogeneous populations can hardly give correct information about the life cycle of algal cells.

Using the method of programmed light-and-dark regimen, Sorokin (1958, 1960a–d, 1961) followed the change of photosynthetic activity of his high temperature *Chlorella* strain. In the $9{:}\overline{15}$-hour regimen at 39° (see Sorokin and Myers, 1957), the peak of photosynthetic activity of algal cells was observed at around the third hour of the light period, after which it gradually decreased and reached the lowest level immediately before the cell division (Sorokin, 1957, 1960c). On measuring the photosynthetic activity of the most active cells, under saturating light at 39°, 3 hr after the start of light period, Sorokin (1960a) observed that the activity decreased abruptly after about 1 hr, the decrease being more pronounced in CO_2 uptake than in O_2 evolution, with a result that the O_2/CO_2 ratio increased to the large value of 10. On further observation, both the photosynthetic activity and the O_2/CO_2 ratio recovered to their original levels. The temporary decrease of photosynthetic rate such as occurred in the 3-hr old cells was not observed with the cells of later developmental stages, which showed constantly decreasing rate during the photosynthesis experiment (Sorokin, 1960a, b). The phenomena observed with the 3-hr old cells, i.e. the temporary decrease of photosynthetic rate and a striking increase of O_2/CO_2 ratio, seem to be similar to those observed by Nihei (1955), Nihei *et al.* (1954) with the cells at the stage around L_2 and L_3 of *Chlorella ellipsoidea.* It should be noted that the high-temperature *Chlorella* used in Sorokin's experiments has a very high rate of growth and ripening, and that his measurement of photosynthetic activity was performed by suspending the algal cells in 0·02 *M* KH_2PO_4. Probably, on shaking the 3-hr old cells in this medium for 1 hr, the cells had entered a stage corresponding to $L_2 \sim L_3$ of *Chlorella ellipsoidea* and performed the photochemically induced phosphate assimilation with a suppression of CO_2-uptake. Repetition of Sorokin's experiments with due consideration of Nihei's observations seems to be relevant.

The *Chlorella* strain used in Sorokin's experiments appears to be different from *Chlorella ellipsoidea* also in the mode of change of respiratory activity during the life cycle. Sorokin and Myers (1957) observed with their strain that both endogenous and glucose respiration rose immediately after the start of the development of young cells in the light, soon reached a maximum, then underwent a slow decline, and eventually, by the time when the matured cells ruptured into daughter cells, reached the lowest level. The course of endogenous respiration, measured in the darkness, was also different according to the developmental stage. Young cells taken at the end of the dark period (0-hr cells) had a low respiration rate which decreased only slowly with time. Cells of intermediate developmental stages (1 and 3-hr cells) had the highest starting respiration rate, which underwent a rapid decline as soon as the cells were darkened. In the cells well advanced in their development (6 and 8-hr cells), the starting respiration rate was lower, but this rate remained stable for almost 3 hr of keeping the cells in darkness.

With *Chlorella ellipsoidea*, whose endogenous respiration is strongest in L_2 or L_3 cells and weakest in D_a cells, Nihei *et al.* (1954) observed that the respiratory quotient CO_2/O_2 always remained around unity during the whole course of cell transformation. With the same alga, Sasa (1961) found that the rate of exogenous glucose respiration (the excess over endogenous respiration of the respiration observed in the presence of exogenous glucose) was, unlike the rate of endogenous respiration, highest at the D stage and lowest, practically zero, at the stage around L_2.

V. Changes in Content of Various Cell Constituents in Normal Life Cycle

1. *Major Elements and Water*

As stated in one of the earlier sections, the growth of *Chlorella ellipsoidea* occurs from the stage of D_n cells until the stage of L_3 cells, after which the cells do not require light in performing further ripening (post-ripening) followed by cellular division. During the course of such a life cycle various nutrient elements are assimilated, each element absorbed not necessarily keeping pace with the photosynthetic assimilation of carbon. The time relation of assimilation of each element to that of photosynthetic carbon assimilation may be studied by determining the percentage of each element in the cell dry matter at various stages of cell development. In Fig. 15 are shown the contents of N, P and S in per cent of dry weight of cells at different developmental stages of *Chlorella ellipsoidea* (Hase,

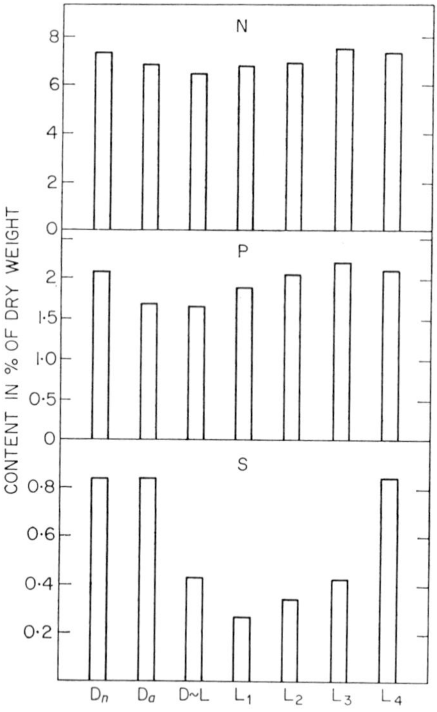

Fig. 15. Percentages (on a dry weight basis) of nitrogen, phosphorus, and sulfur as found in cells of *Chlorella ellipsoidea* at different developmental stages (Hase *et al.*, 1957).

Morimura and Tamiya, 1957). As may be seen, the percentage of N is relatively constant but shows a slight decrease during the growing phase. The decrease in the percentage of P occurring in the growing phase is more distinct. Quite different is the change of sulfur content which decreases considerably as the cells transform themselves from the D to L stage, and increases markedly as L_1 changes, via L_2 and L_3, into L_4. These results show that the assimilation of N and P lags more or less behind the assimilation of C starting from D_n cells, while the assimilation of S is almost nil during the growth phase and occurs most actively during the phases of ripening and post-ripening (see also Hase, Otsuka, Mihara and Tamiya, 1959; Iwamura, Hase, Morimura and Tamiya, 1955; Iwamura, 1955; Tamiya, Iwamura *et al.*, 1953). By similar experiments it was revealed that the contents (in per cent of dry weight) of K, Mg and Fe in algal cells also decrease more or less during the stages from D_n to L_1, and increase sooner or later during the later stages (Kanazawa, 1958). The courses of assimilation of various elements, as inferred from the results of these experiments, are schematically illustrated in Fig. 16.*

Nihei *et al.* (1954) compared the dry matters of D cells and L cells (around the L_2-stage) of *Chlorella ellipsoidea* for their contents in C, H, O and N according to the principle proposed earlier by Tamiya (1935, 1942) in his studies on the metabolism of fungi. In Fig. 17 the results obtained are reproduced, together with the analytical data for P, K, S, Mg and Fe obtained by Hase *et al.* (1957) and Kanazawa (1958). It is noteworthy that the relative content of oxygen was far lower in D cells than in L cells. On an ash-free basis, the elementary composition of 100 g (dry weight) of algal cells was found to be $C_{4 \cdot 36}\ H_{7 \cdot 57}\ O_{1 \cdot 93}\ N_{0 \cdot 60}$ for D cells, and

* Concerning the changes of N and P contents (in per cent of dry weight) occurring in the algal life cycle, somewhat different results were obtained by Müller (1960) for *Scenedesmus obliquus* (a green alga closely related to *Chlorella*), which was synchronized by the method of programmed light-and-dark regimen (14:10-hr, 25°, 10 kilolux). In this case, the N content was lowest at the end of the light period, increased remarkably during the dark period and attained a maximum level soon after the beginning of the light period. The P content also increased during the dark period and decreased or remained constant during the light period. With a synchronous culture of *Chlorella pyrenoidosa* Schmidt (1961) observed that the time and amount of phosphate absorbed by the cells during growth differed according to the sort of nitrogen source applied. When ammonium was the sole nitrogen source, the cells not only absorbed a larger quantity of phosphate than cells cultured on nitrate nitrogen, but they absorbed it earlier in cellular development. The rate of phosphate absorption decreased during nuclear and cellular division on both nitrogen sources.

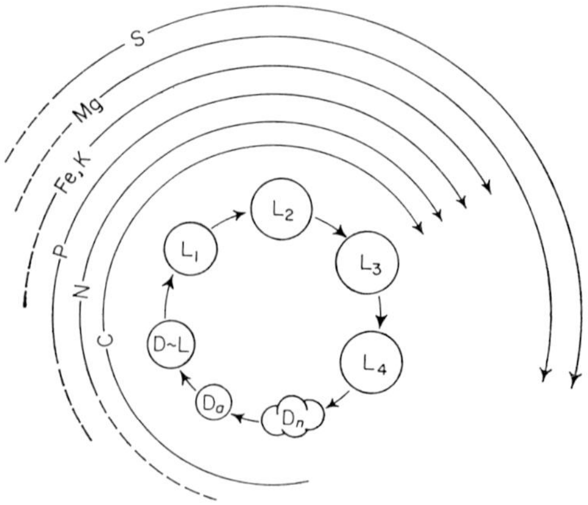

FIG. 16. Diagrammatic representation of the modes of assimilation of various nutrient elements during the life cycle of *Chlorella ellipsoidea* (according to the data of Hase *et al.*, 1957, and Kanazawa, 1958).

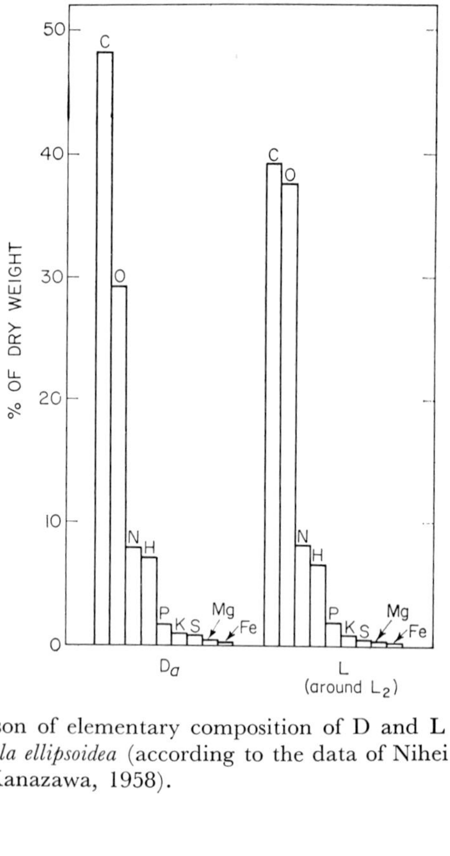

FIG. 17. Comparison of elementary composition of D and L cells (around the L_2 stage) of *Chlorella ellipsoidea* (according to the data of Nihei *et al.*, 1954, Hase *et al.*, 1957, and Kanazawa, 1958).

$C_{3\cdot53}\ H_{7\cdot23}\ O_{2\cdot56}\ N_{0\cdot63}$ for L cells. The 'combustion quotient' (CQ) according to Tamiya (1935, 1942) may be calculated by the formula: $CQ = 4n_C/(4n_C + n_H - 2n_O - 3n_N)$, where n_C, n_H, n_O and n_N are relative numbers of carbon, hydrogen, oxygen and nitrogen atoms, respectively. As was discussed in detail by Tamiya, the magnitude of CQ is a measure of the oxido-reductive state of organic substance as a whole, the value $CQ = 1$ indicating the oxido-reductive state equivalent to that of carbohydrate, and the values $CQ > 1$ and $CQ < 1$ signifying the states that are more oxidized and more reduced, respectively, than the carbohydrate level. Analytical data show that CQ was 0·906 for D cells and 0·985 for L cells, indicating that the former are, as a whole, in a much more reduced state than the latter. This result is in harmony with the observation described in an earlier section that the photosynthetic quotient O_2/CO_2 of algal cells, which is in general around unity, showed an abrupt increase to a value of more than 3 at the stage around L_2 and L_3. When the D cells grow into L cells by performing photosynthesis with an O_2/CO_2 quotient of unity, the cell mass increases about four-fold, the consequence of which will naturally be that the L cells show a combustion quotient near unity. Toward the end of the process of ripening, such L cells undergo reductive processes, accompanied by assimilation of phosphate, which is manifest in the O_2 production in excess of the CO_2 uptake. After this reductive process, L cells perform further maturing and eventually give rise to D cells, which show a combustion quotient much lower than unity.

With a synchronous culture of *Chlorella ellipsoidea*, Yuhara and Hase (1961) followed the change of the ratio of cell dry weight to average cell volume, a ratio which may be regarded as showing the relative content of water in fresh algal cells. They found that the ratio decreased during the earlier stages of growth until about L_1, then increased considerably during the process of ripening, and decreased again as the cells proceeded to the phase of post-ripening. Thus, the process of ripening is accompanied not only by the active intake of various nutrient minerals but also by an active absorption of water in algal cells.

2. *Protein, Lipid and Carbohydrate*

Hase, Morimura and Tamiya (1957) investigated the changes in contents of the bulk of protein, lipid and carbohydrate in synchronized cells of *Chlorella ellipsoidea*. In terms of percentage of dry weight, the content of protein, like that of total nitrogen, remained relatively constant (47–54 per cent) showing only a slight decrease at earlier stages of the growing phase. A somewhat different result was obtained with the same alga by Miyachi and Tamiya (1961), according to whom the percentage of protein decreased progressively until the stage of L_2 cells, after which it increased until the stage of incipient cell division.

The percentage of carbohydrate (2–23 per cent) and lipid (19–29 per cent) showed irregular changes, decreasing temporarily at stages L_1 or L_2 and again decreasing markedly at L_4 and D_n. The decrease of these substances, and especially of carbohydrates at the L_4 stage (2 per cent) seems to be due partly to their conversion to other cell constituents and partly to their consumption as substrates of respiration in the dark (see also Lorenzen and Ruppel, 1960). Among the carbohydrates, starch forms the major part throughout the whole course of the life cycle of *Chlorella ellipsoidea*; sucrose and unidentified polysaccharides, which are contained in considerably smaller amounts, show a temporary fall in quantity at earlier stages of L cells. By the technique of paper chromatography, the existence of various monoses (glucose, fructose, xylose, galactose, ribose, triose, etc.) and their phosphate esters were demonstrated, but their changes in quantity occurring during the life cycle have not yet been investigated. Concerning the change of the lipid fraction, it was observed that a temporary fall of phosphorus content and a temporary increase of iodine number occurred at earlier stages of L cells.

Recently, Northcote *et al.* (1959) demonstrated the presence of glucosamine in the hydrolyzate of cell walls of *Chlorella*. According to Mihara (1961), the substance is derived from chitosan which forms one of the constituents of the wall of algal cell. With synchronized cultures of *Chlorella ellipsoidea*, Mihara found that the increase in content of glucosamine occurred in three steps; first, in proportion to the growth of D cells into larger cells; second, corresponding to the formation of autospores within ripened cells; and third, parallel to the growth of newly-born daughter cells. At the stage around L_2, an abrupt and temporary stop in the increase of glucosamine occurred, indicating the occurrence of some change in the property—probably, in permeability—of the cell wall at that stage. It has already been mentioned earlier that a most active uptake of water as well as of some nutrient ions occurs around the same stage.

3. *Nucleic Acids*

Changes in the RNA and DNA content of synchronized *Chorella ellipsoidea* were investigated by Iwamura (1955) and Iwamura *et al.*

(1955) who obtained the results illustrated in Fig. 18. The white bars in the figure represent the contents per cell of the substances which were analyzed immediately after the harvest of cells from the illuminated culture, and the hatched bars represent the data which were obtained for the respective cells after they had been kept in the dark for a sufficient length of time. Comparison of the white and hatched bars shows clearly that, by incubation of cells in the dark, RNA decreased, except in the case of D_n, while DNA increased, especially in a large measure in L cells. Throughout the course of the algal life cycle, the synthesis of RNA proceeded fairly in parallel with the synthesis of protein, except that during the transformation of L_3 cells into L_4 cells RNA decreased, while protein increased slightly (see also Lorenzen and Ruppel, 1960). On the other hand, DNA synthesis took place only insignificantly during the growing phase and occurred rapidly during the phases of ripening and post-ripening. These findings are quite in line with the data presented in Figs. 3, 8 and 9 showing the change of nuclear patterns during the course of algal life cycle.

The black bars in the upper part of Fig. 18 show the values of n, from which it may be seen that, by and large, there exists a

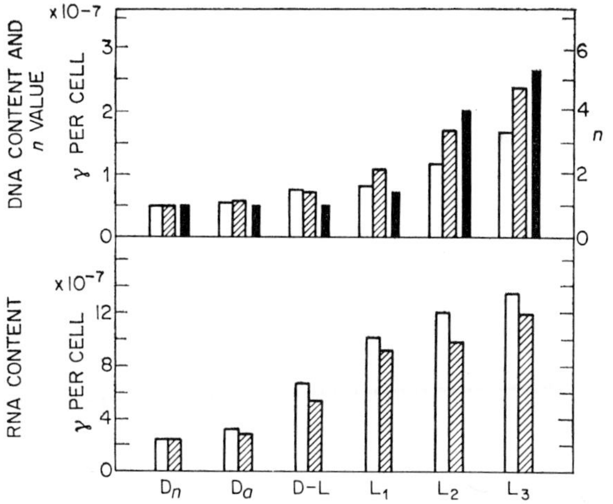

FIG. 18. Contents (per cell) of DNA and RNA, and the n values found at different developmental stages of *Chlorella ellipsoidea* (Iwamura, 1955; Iwamura *et al.*, 1955). White and hatched bars represent, respectively, the values found before and after incubation of cells in the dark. Black bars show the division number obtained after the dark incubation. $n = 1$ means that the cells did not divide on incubation in the dark. The somewhat large n value ($= 5{\cdot}3$) obtained for L_3 cells resulted from insufficient homogeneity of the synchronized population.

proportionality between the *n*-value and the DNA content obtained after the dark incubation.

Essentially the same results as above were later reported by Lorenzen (1958), Lorenzen and Ruppel (1960) and Pirson, Lorenzen and Koepper (1959) who synchronized *Chlorella pyrenoidosa* by the method of programmed light-and-dark regimen. The report of Lorenzen (1958), however, that at a certain stage, probably corresponding to the L_1 stage according to Tamiya *et al.*, the nuclei of algal cells became unstainable with the Feulgen reagent was not corroborated by Tamiya *et al.* (1961) in their experiment with *Chlorella ellipsoidea.*

By fractionating the algal cells into acid-soluble, RNA and DNA fractions, Iwamura and Myers (1959) investigated the base composition and its quantitative changes occurring during the algal life cycle. For the DNA fraction they observed marked changes in content (percentage on molar basis) of thymine and guanine, while those of adenine and cytosine remained constant. The increase of guanine occurred during the growth phase, while that of thymine took place during the ripening phase. During the growth phase, the RNA fraction showed a relatively higher content of guanine and a lower content of cytosine and adenine, while the relation was reversed during the ripening phase. The content of uracil in the RNA fraction appeared to be practically constant throughout the course of the life cycle. In the acid-soluble fraction, adenine and uracil were dominant; the percentages of guanine and cytosine remained constant during the life cycle, that of adenine decreased during the ripening phase, while that of uracil showed a reverse trend.

The fact that the relative contents of purine and pyrimidine bases change more or less during the algal life cycle was attributed by Iwamura (1960) to the existence of different forms of RNA and DNA in algal cells. By subjecting the algal cells (*Chlorella ellipsoidea*) to disruption by a French pressure cell followed by differential centrifugation, he separated three fractions: (i) precipitate obtained at 20,000*g*, 10 minutes; (ii) precipitate obtained at 105,000*g*, 90 minutes; and (iii) supernatant. These fractions were found to contain different forms of RNA and DNA. The analytical data of these three fractions are presented in Tables I and II. Table I shows that practically all the chloroplasts went into Fraction (i), and that most of the DNA and proteins were distributed in Fraction (i) and supernatant. As may be seen from Table II, the DNA in these two fractions as well as the RNA in the three fractions showed more or less different base compositions. Using ^{32}P as a tracer, Iwamura

TABLE I. Chemical analyses on the three cell fractions of *Chlorella ellipsoidea* (Iwamura, 1960). Cell material used: 3·6 g in dry weight

	Amount of material (mg)			
	In fraction			In whole homogenate
	(i)	(ii)	(iii)	
Chlorophyll	50	0·03	0·00	55
RNA	22	27	8·4	61
DNA	3·1	0·36	3·5	7·5
Protein	510	85	200	820

(1960) has demonstrated that, under growing conditions in the light, the DNA in Fraction (i) performs far more active turnover of P than does the DNA in the supernatant. The former DNA is probably that contained in the chloroplasts and the latter that of nucleus. Iwamura (unpublished experiments) also observed that in the process of algal life cycle the two DNAs are synthesized at somewhat different paces.

The behavior of DNA-P and RNA-P during the life cycle of *Chlorella ellipsoidea* was also studied by Miyachi and Tamiya (1961), who grew ^{32}P-labeled D cells synchronously in a cold normal medium, and followed the change of proportion of ^{32}P and ^{31}P in DNA and RNA in algal cells. They found that during the earlier stages of synchronous culture the P in DNA was mostly ^{32}P, and ^{31}P was vigorously incorporated only after the stage around L_2. On the other hand, RNA actively incorporated exogenous ^{31}P from the

TABLE II. Base compositions of the nucleic acids isolated from the three cell fractions of *Chlorella ellipsoidea* (Iwamura, 1960)

Nucleic acid	Base composition	Range of mole per cent in fraction (i)	(ii)	(iii)
RNA	Uracil	23–24	23–24	21–24
	Adenine	24–27	23–27	18–24
	Cytosine	23–25	22–24	26–30
	Guanine	24–30	27–30	25–32
DNA	Thymine	22–28		19–21
	Adenine	24–28		21–23
	Cytosine	24–28		27–30
	Guanine	20–28		28–30

start of the synchronous culture. They also observed that the ^{32}P in labile forms in nucleotidic compounds, as well as in the acid-soluble fraction, decreased in earlier stages of growth, attained a maximum at around the L_2 stage and then decreased gradually.

By investigating the role of sulfur in the formative metabolism of *Chlorella*, Hase, Mihara, Otsuka and Tamiya (1959b), Hase, Mihara and Tamiya (1960a, b, 1961a), Hase, Otsuka, Mihara and Tamiya (1959) discovered some new forms of sulfur-containing polynucleotides which seem to play important roles in the processes of ripening and cellular division of algal cells. These compounds will be dealt with later in Section VII.

4. *Polyphosphates*

As has been mentioned earlier, Nihei (1955) observed that, at a certain stage of L cells, phosphate was assimilated concomitantly with a photochemical evolution of oxygen. The assimilated phosphate was found to be mainly in the hot TCA-soluble fraction which was obtained after the cells had been extracted with cold TCA (5 per cent) and alcohol–ether (3:1) mixture. In the hot TCA-soluble fraction he found a certain form of polyphosphate, whose existence in *Chlorella* cells was also reported by some other authors (Holzer, 1951; Wintermans, 1954, 1955; Schmidt and King, 1961). Recent experiments performed by Miyachi and Tamiya (1961) have revealed that *Chlorella* cells contain at least three forms of polyphosphate, which they named polyphosphates A, B and C. The former two (A being acid-soluble, and B acid-insoluble and soluble at pH 8–10) seem to correspond, respectively, to the acid-soluble and -insoluble metaphosphates obtained by Wiame (1949) from yeast, and polyphosphate C (insoluble either in acid or at pH 8–10, but extractable by treatment with 2 *N* KOH at 37° for 16–20 hr) seems to correspond to Langen and Liss' 'polyphosphate 4' found in brewer's yeast (1958). The polyphosphates thus far obtained by other workers (Holzer, 1951; Nihei, 1955; Schmidt and King, 1961; Wintermans, 1954, 1955) from *Chlorella* cells are thought to be mixtures of polyphosphates B and C.

Miyachi and Tamiya (1961) found that the contents (in per cent of dry weight) of both polyphosphates B and C increased during the growing phase, attained maxima at around the L_2- (in the case of polyphosphate B) or L_3-stage (in the case of polyphosphate C), after which they decreased rapidly during the process of cellular division. Using the temperature-tolerant strain (7–11–05) of *Chlorella*, Schmidt and King (1961) observed that the hot TCA-soluble polyphosphate accumulated at a certain stage of growth and ripening of cells. According to Schmidt (1961), the synchronized cells of *Chlorella pyrenoidosa* contain granules which are stainable with toluidine blue, a stain specific for polyphosphate. Young cells contained one or two staining granules, while their number increased to 20 or more per cell at a later stage of cell development.

In Fig. 19 are compared the quantities of polyphosphates A, B and C (in terms of their labile P) with those of various other phosphate compounds existing in cells of *Chlorella ellipsoidea;* analyses were made with ^{32}P-labeled cells consisting mainly of D cells. As may be seen, the quantities of labile P in the polyphosphates far exceed those of nucleotidic labile P, suggesting the importance of the polyphosphates as reservoirs of energy-rich phosphate in algal cells.

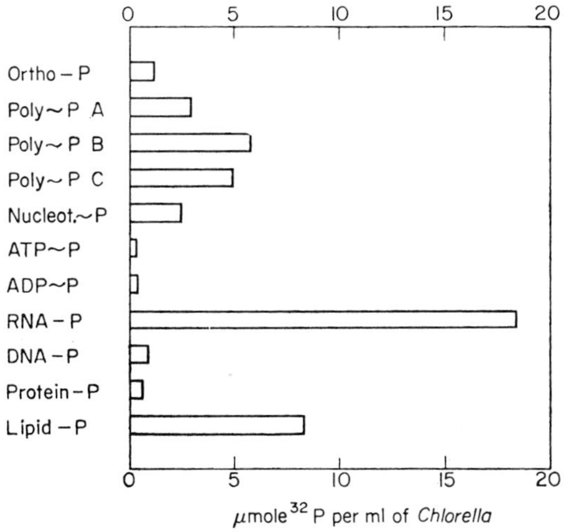

FIG. 19. Distribution of phosphorus in various compounds in cells (mostly D cells) of *Chlorella ellipsoidea* (Miyachi and Tamiya, 1961). Compounds were separated from uniformly ^{32}P-labeled algal cells and their P contents were determined by measuring the radioactivity.

5. *Pigments*

With a synchronous culture of *Chlorella ellipsoidea*, Otsuka (unpublished experiments) followed the contents in cells of chlorophyll and β-carotene as they changed during the course of the algal life cycle. In terms of per cent of dry weight, the chlorophyll content was highest at the D_a stage and lowest at the L_1 stage (less than a half of that shown by D_a cells). Equally low chlorophyll content was found for $D \sim L$, L_2 and L_3 cells, but at the L_4 and

D_n stages the content increased considerably, approaching almost the level in D_a cells, indicating the occurrence of chlorophyll formation in the dark (see also Tamiya, Iwamura, Shibata, Hase and Nihei, 1953). The fact that the assimilation of Mg continued until the stage of transformation of L_3 cells into L_4-cells in the dark (see Fig. 16) is undoubtedly related to the process of chlorophyll formation at this stage. During the whole course of the life cycle, the ratio of chlorophylls *a* to *b* appeared to remain constant at about 3:1. The fact that the L cells show lower chlorophyll content than the D cells seems to be in accord with the earlier observation made by Myers (1946) that the chlorophyll content (in per cent of dry weight) of *Chlorella* was lower in larger cells grown under strong light, than in smaller cells grown under dim light. At variance with these observations is the report made by Lorenzen (1959) that the chlorophyll content in 'synchronized' *Chlorella pyrenoidosa* remained almost constant throughout the whole course of the algal life cycle.

The percentage content of β-carotene was highest at the D_a stage, and remained almost constant at all other stages, showing no increase during the transformation of L_3 into L_4 in the dark. The ratio chlorophyll: β-carotene (in weight) was highest in L_4 and D_n cells (around 60:1), while it was around 30:1 in other stages of cell development.

6. *Vitamins*

With *Chlorella ellipsoidea* grown synchronously, Morimura (1959b) investigated the changes in content of various vitamins during the algal life cycle. In terms of p.p.m. of cell dry weight, the contents of niacin, biotin, inositol and choline remained almost constant throughout the life cycle, while those of pantothenic acid, folic acid, thiamine, riboflavin and vitamin B_6 complex were found to decrease more or less markedly during the growing phase and increase at the stages of ripening. The content of *p*-aminobenzoic acid increased only at an early stage of ripening, and that of ascorbic acid increased only at the stages where photosynthesis occurred most actively. The ranges of concentrations of various vitamins as they changed during the algal life cycle are shown in Table III.

It is interesting to note that *p*-aminobenzoic acid and folic acid, which are known to be involved in the synthesis of purines and pyrimidines, are formed mainly during the process of ripening, and that the synthesis of vitamin B_6 complex which is known to be an important factor in the amino acid metabolism (decarboxylation, transamination, etc.) occurred actively at later phases of ripening. The increase of thiamine and riboflavin in the L stages may also be related to the enhanced formative metabolism occurring in those stages. Recently, Otsuka (1961b) found an interesting phenomenon concerning the fate of free and bound pantothenic acid in *Chlorella* cells whose cell division was controlled by limited supply of S and N sources. Her observations will be dealt with later in connection with the problem of the role of sulfur in the process of algal life cycle.

TABLE III. Contents of various vitamins in the cells of *Chlorella ellipsoidea* (Morimura, 1959b)

	Content (μg/g)
Thiamine	10–23
Riboflavin	23–37
Niacin	112–125
Ascorbic acid	1000–3200
Biotin	0·19–0·23
Inositol	1600–2100
Choline	2200–2500
p-Aminobenzoic acid	12–24
Folic acid conjugate	22–47
Vitamin B_6 complex	0·3–2·5
Pantothenic acid	3·5–8·6

VI. Modifications Caused by Various Environmental Factors

1. *Effect of Deprival of Each Essential Mineral Nutrient*

To know how the course of the algal life cycle is affected by deprival of each essential mineral nutrient may be the first step in studying the role played by each element in the life of algal cells. With synchronous cultures of *Chlorella ellipsoidea*, Tamiya, Iwamura, Shibata, Hase and Nihei (1953) observed that the earlier phase of growth of D cells was not affected markedly by the deprival of whole nutrient salts from the culture medium, while the subsequent course of events was more or less markedly affected by the nutrient deficiency. They also observed that the transformation of L_3 cells into L_4 cells in the dark was not markedly affected by deprival of phosphate from the medium, but was somewhat retarded on deprival of nitrate, and was most strongly suppressed in distilled water.

The effect of deprival of each major element (N, P, Mg, K and S) from the normal nutrient medium was studied in detail by Hase *et al.* (1957) with *Chlorella ellipsoidea*. They observed that, on deprival of N (nitrate) or P (phosphate), the increase of average cell volume and packed cell volume proceeded apparently normally until about the stage of D $\sim$ L cells, but that thereafter the growth was more or less suppressed, the effect being more pronounced in the absence of N than in the absence of P. Although the cells grown in the absence of N or P were smaller than the normal L cells, they divided almost simultaneously with normal L cells ('isochronism' of cell division; see an earlier section). The division number was, however, smaller than normal, being 2·4 in the case of N deficiency and 3·5 in the case of P deficiency. The daughter cells produced in the absence of N were considerably smaller than normal daughter cells and were strongly etiolated, indicating the suppression of chlorophyll formation. The daughter cells produced in the P-free medium looked almost normal as far as the size and chlorophyll content were concerned, but were almost incapable of performing subsequent growth, to say nothing of the subsequent cell division. The same was true with the etiolated daughter cells produced in the absence of N.

In the absence of Mg, the life cycle, in its first round, proceeded apparently normally with the same division number as that of control culture; however, there was some time lag in the occurrence of cell division, and the daughter cells produced were strongly etiolated and sluggish in growth, being incapable of performing further cell division. When potassium was absent in the medium, D cells grew almost normally until about the stage of D $\sim$ L or L_1 cells, after which the growth became sluggish and the cells began to divide isochronously with the control culture showing an n-value somewhat smaller than normal. The resulting daughter cells could grow only poorly and were unable to perform further division, although they looked apparently normal both in size and in chlorophyll content.

In all the cases described above, the D cells could grow to some extent and then divided, although the resulting daughter cells were, in different ways, unhealthy. Quite different was the situation when S (sulfate) was omitted from the culture medium. In this case, the D cells grew apparently normally up to the L_1 cells, but at this stage the cells were completely stalemated and there occurred no division at all. This fact together with the observation, described in an earlier section, that the assimilation of sulfur occurs most actively during the ripening phase points strongly to an important role played by sulfur in the processes of ripening and cell division. Being instigated by these observations Hase, Mihara, Otsuka and Tamiya (1959b); Hase, Mihara and Tamiya (1960a, b; 1961a); and Hase, Morimura, Mihara and Tamiya (1958); Hase, Otsuka, Mihara and Tamiya (1959) conducted a series of investigations on the role of sulfur in the process of algal life cycle, and they discovered a number of new facts which we shall discuss in detail later.

2. *Effects of Ultraviolet Irradiation and 'Cold Shock'*

Sasa (1961) investigated, using synchronous cultures of *Chlorella ellipsoidea*, the effect of ultraviolet irradiation upon various activities of cells at different developmental stages in their life cycle. Cell activities investigated were photosynthesis, respiration (with or without addition of exogenous glucose), overall growth (survival ratio), mode of growth and cell division as well as the formation of nucleic acids. The u.v. light applied was 2537 Å in wavelength and 30 $\mu W/cm^2$ in intensity, and its dose was varied by changing the duration of exposure.

The most u.v. sensitive activity was the overall growth as measured by the colony count, and in this respect the irradiation applied at the L_2 stage was more inhibitive than that given at the D stage. The next most vulnerable was the photosynthetic activity (measured both under saturating and rate-limiting light intensities), the sensitivity being the same in the D and L cells. The most resistant toward u.v. irradiation was the endogenous respiration of D cells followed by their respiration using exogenous glucose as substrate. As mentioned already, the L_2 cells were found to be unable to use exogenous glucose as substrate of respiration, but their endogenous respiration was considerably stronger than that of D cells. The u.v. sensitivity of the endogenous respiration of L cells was the same as that of the glucose respiration of D cells.

When D cells were u.v. irradiated immediately before the start of synchronous culture, growth and cell division as well as the formation of DNA and RNA were retarded in proportion to the u.v. dose applied. The division number was normal (around 4) at lower u.v. doses of 1–2 min irradiation, but was reduced to a half (about 2) at a higher dose of 4-min irradiation. When, during the synchronous culture, 1-min u.v. irradiation was applied at various stages of the ripening phase, L_1, L_2 and L_3, the division was retarded, but the cells, after attaining an abnormally large size, divided into about eight daughter cells each. If the irradiation was given at the L_4 stage, the division number was practically unmodified ($n = 4{\cdot}5$), although the division was somewhat retarded compared with that of the control culture. When a 1-min irradiation was given at the L_2 stage, there occurred an apparent stimulation of DNA and RNA formation, a phenomenon which corresponds to the production of a larger number of daughter cells than was the case in control cultures. Thus, the cells which were moderately u.v. irradiated at different stages of synchronous culture were able to complete their life cycle maintaining a certain degree of synchrony, but later marked irregularities occurred in the growth of cells, indicating that a certain portion of irradiated cells were unable to grow normally after one or more rounds of the life cycle. The enhancement of DNA and RNA formation, as well as the increase in the division number caused by the 1-min irradiation given at around the L_2 stage, might be a phenomenon of anomalous stimulation leading to the death of some of the cells involved.

Pirson *et al.* (1959) reported that the cells of *Chlorella pyrenoidosa*, which were synchronized by the method of programmed light-and-dark regimen ($16:\overline{12}$ hr, 30°, 9 kilolux), showed, at a certain stage of cell development, a marked sensitivity toward a cold shock, that is, exposure of cells to a temperature of 4° for 2 hr in a light of 9 kilolux. When the shock was given at the beginning or at the end of the light period the Schub-number was somewhat lower than that in the untreated control. When it was given 7 to 8 hr after the beginning of illumination, the subsequent cell division was halted completely, and a considerable loss of chlorophyll occurred.

Presumably, this sensitive stage corresponds to the L_1 stage according to Tamiya *et al.* The effect observed by Pirson *et al.* may rather be interpreted as the breaching effect of light occurring at lower temperatures, which has been known to be more harmful to L cells than to D cells (Tamiya, Iwamura, Shibata, Hase and Nihei, 1953). In fact, Pirson *et al.* observed that when the cells at the sensitive stage were given the low-temperature treatment in the dark, they could divide normally, suffering only some decrease in the Schub-number. The German authors (Pirson *et al.*, 1959) remarked that the cells at the sensitive stage showed the maximum photosynthetic activity and at the same time their nuclei became unstainable with the Feulgen reagent (Lorenzen, 1958). As has been mentioned already, both these statements do not apply to the *Chlorella ellipsoidea* studied by Tamiya *et al.*

3. *Effects of Various Antimetabolites*

Extensive studies have been performed by Tamiya *et al.* (1962) on the effects of various metabolic inhibitors upon the life cycle of *Chlorella ellipsoidea*. Synchronization was effected by the D_sLD' method (at 21°; light intensity, 10 kilolux), and the substances tested were more than fifty in number including: (i) those known to affect the growth of higher plants, such as indoleacetic acid and gibberellin, (ii) various antibiotics, such as penicillin and chloramphenicol, (iii) some anti-amino acids, such as DL-α-methyl-glutamic acid and β-2-thienyl-serine, (iv) some anti-vitamins, such as desoxypyridoxine and sulfanilamide, and (v) various substances known to affect nucleotide metabolism, such as 8-azaguanine and 5-bromouracil.

As far as the tests were made for one D_sLD' cycle of the algal cells, the following substances were found to give no, or almost no, effect even at the highest concentrations tested (in most cases 10^{-2} *M*); kinetin, colchicine, gibberellin A_3, DL-α-methyl-glutamic acid, DL-methionine-DL-sulfoxide, desoxypyridoxine, DL-desthiobiotin, pantoyl taurine, pyridine-3-sulfonic acid, 8-azaadenine, 8-aza-2,6-diamino purine sulfate, 5-nitrouracil.

The possibility remains that some of these substances did not permeate through the membrane of algal cells. With the rest of the substances tested, the following five types of modification were observed. Tests were made, in most cases, at four concentrations of ten-fold dilutions: 10^{-2}, 10^{-3}, 10^{-4} and $10^{-5}M$.

(1) By a ten-fold difference in concentration of substance, the life cycle was affected in an 'all-or-nothing' manner; in other words, there was no or almost no effect at a certain concentration, while at the ten-fold higher concentration the growth of D cells was completely suppressed. The substances exerting this kind of effect were indoleacetic acid, 2,4-dichlorophenoxyacetic acid, aureomycin, sarkomycin, penicillin G, terramycin, nitromin, acriflavin and benzimidazole.

(2) At a certain applied concentration, the growth was not affected and the cell division also took place apparently normally and completely, but with division numbers larger than normal ($n = 6$–8). The substances showing such an effect were: maleic hydrazide, DL-β-phenyl-lactic acid, β-2-thienylserine, D-leucine, allyl glycine, 6-azathymine (at lower concentrations), 8-aza-hypoxanthine, 2-thiocytosine, 6-mercaptopurine and 5-bromouracil. As an example, the results obtained with maleic hydrazide are presented in Fig. 20. By Feulgen staining it was revealed that the mother cells produced in the presence of $10^{-3}M$ maleic hydrazide contained six to eight nuclei per cell.

(3) Growth of cells was not, or only slightly, affected, while the cell division was markedly retarded, or completely or partially halted according to the concentration applied. (When the cell division was partially arrested, only the Schub-number—and not the division number—was determined.) When the complete cell

division took place with retardation, the n-value was, in most cases, the same as that of the control. Examples of substances showing such an effect were dihydrostreptomycin, β-2-thienylalanine, indole, sulfanilamide, sulfathiazole, hexachlorocyclohexane, 6-azathymine (at $10^{-2}M$), 8-azaguanine, 8-azaxanthine and 2-thiobarbituric acid.

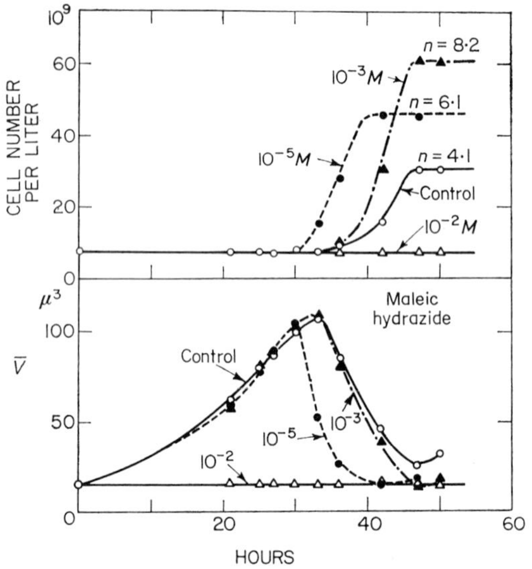

Fig. 20. Effect of maleic hydrazide of different concentrations upon the life cycle of *Chlorella ellipsoidea* (D_sLD' cycle; 21°, 10 kilolux). $\bar{V}$: average cell volume. (Tamiya *et al.*, 1962.)

The results obtained with 8-azaguanine and dihydrostreptomycin are reproduced in Figs. 21 and 22. In the presence of $10^{-4}M$ azaguanine, partial cell division occurred in two steps, of which the earlier one was isochronous with the division of control cells. At $10^{-3}M$, the initial rate of growth was almost the same as that in normal culture, but the cells continued to grow to an enormous size without being able to perform cell division. By the Feulgen staining it was found that the abnormally matured cells contained, at first, a multiple number of irregularly shaped nuclei, which, however, disappeared later and the whole cell seemingly became uniformly stained. A similar phenomenon was observed with the cells whose division was inhibited by $10^{-2 \cdot 5}M$ dihydrostreptomycin. In the presence of $10^{-4}M$ dihydrostreptomycin, the cells grew considerably in size and the daughter cells produced (with the division number of 4·2) were also much larger compared with those produced in the control culture.

(4) Growth was more or less suppressed and the cell division was either retarded, or completely or partially halted according to the

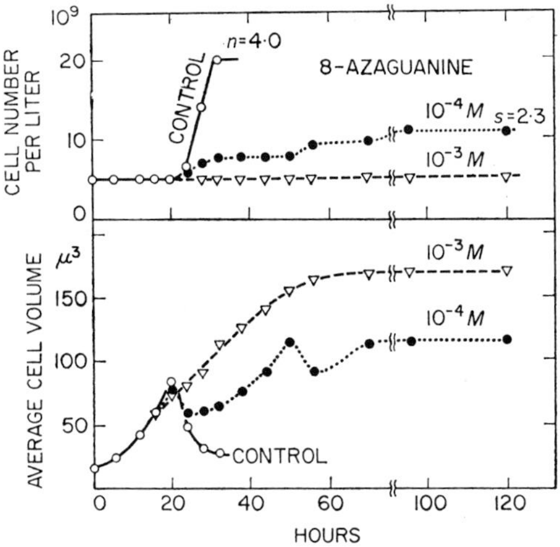

FIG. 21. Effect of 8-azaguanine of different concentrations upon the life cycle of *Chlorella ellipsoidea* (D_sLD' cycle; 21°, 10 kilolux).

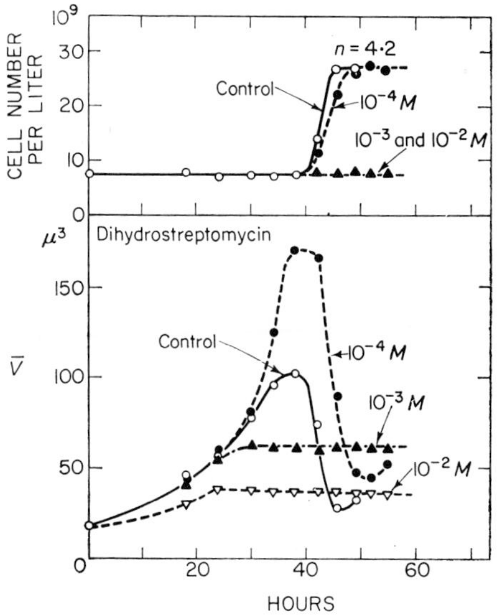

FIG. 22. Effect of dihydrostreptomycin of different concentrations upon the life cycle of *Chlorella ellipsoidea* (D_sLD' cycle; 21°, 10 kilolux).

concentrations applied. When complete cell division took place with retardation, the division number was either different from, or the same as, that of the control.

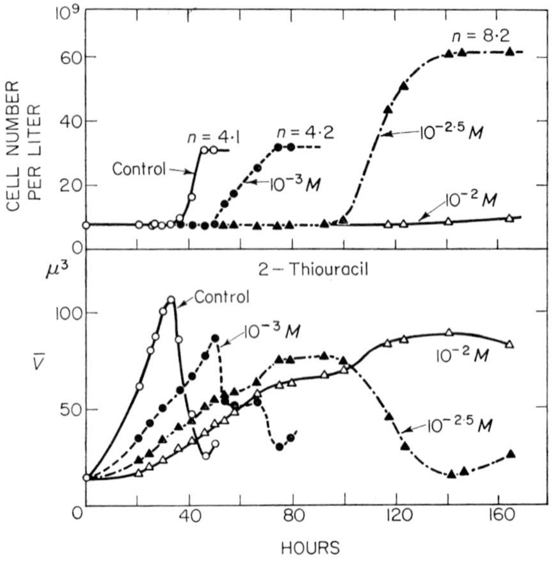

FIG. 23. Effect of 2-thiouracil of different concentrations upon the life cycle of *Chlorella ellipsoidea* (D_5LD' cycle; 21°, 10 kilolux).

Division numbers greater than 4 were obtained at certain concentrations of allyl glycine and 2-thiouracil. Division numbers around four were obtained at certain concentrations of chloramphenicol, oxythiamine, 2-thiouracil, nitrogen mustard, 6-azauracil, 5-methyl-4-oxy-2-thiopyrimidine and caffeine. Division numbers smaller than 4 were obtained at certain concentrations of aminopterin and oxythiamine.

The results obtained with 2-thiouracil are shown in Fig. 23.

(5) Growth was completely suppressed, while the cell division took place. This type of effect was observed only with chloramphenicol at the concentration of 10^{-2}–$10^{-2.5}M$. At lower concentrations, this antibiotic exerted an effect described under (4), as shown by the data presented in Fig. 24. Chloramphenicol is known to be an effective inhibitor of protein synthesis, and the results obtained with 10^{-2}–$10^{-2.5}M$ of the substance indicate that the substance, while inhibiting the protein synthesis, did not affect the formation of nuclear substances. The cells treated with $10^{-2}M$ of the

antibiotic were stained with the Feulgen reagent, and it was found that the nucleus, which had been clearly visible in the starting D_s cells, became hardly recognizable before the onset of cell division. The extremely small daughter cells produced in this case were unable to grow when transferred, after washing, to the normal culture medium.

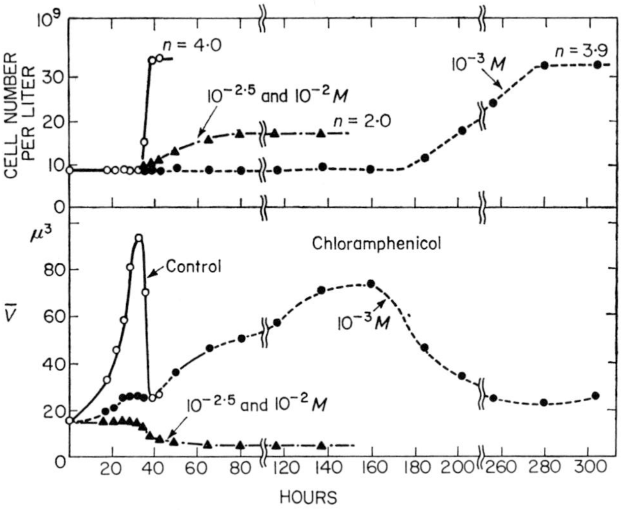

Fig. 24. Effect of chloramphenicol of different concentrations upon the life cycle of *Chlorella ellipsoidea* (D_sLD′ cycle; 21°, 10 kilolux).

The modes of action of the substances studied by Tamiya *et al.* (1962) were indeed quite multifarious; substances showing manifestly similar effects may not necessarily act with the same biochemical mechanism. Certainly much remains to be investigated as to the biochemical mechanism of the abnormal processes caused by these inhibitor substances.

VII. Role of Sulfur in Ripening and Cell Division

As has been described already, *Chlorella* cells assimilate sulfate most strongly during the process of ripening, and on deprival of sulfate from the medium the D cells grow only partially and remain incapable of performing cell division. These facts indicate that sulfur plays some important role in the processes of ripening and division of algal cells. When the cells whose formative metabolism has been halted by the sulfur deficiency are supplied with sulfate, cells resume their process of ripening, but the subsequent course of events is markedly different according to whether the cells are placed under photosynthesizing or non-photosynthesizing conditions (Hase, Morimura *et al.*, 1958; Hase, Otsuka *et al.*, 1959). The state of affairs is illustrated schematically in Fig. 25. When the D cells were

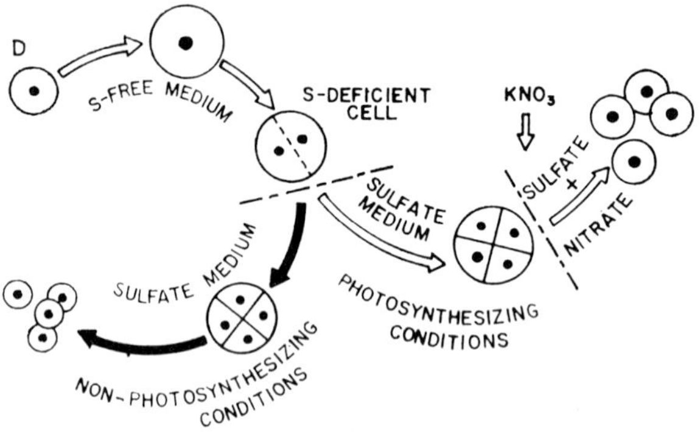

FIG. 25. Schematic representation of the events occurring in the S-N-controlled synchronous culture of *Chlorella ellipsoidea* (Hase *et al.*, 1958). The white and black arrows indicate, respectively, the processes occurring under photosynthesizing and non-photosynthesizing conditions. The dots in each cell represent nuclei which were detected by the Feulgen staining. Broken lines drawn in the figure indicate that the cell development did not proceed further under the respective conditions.

incubated in an S-free medium, under photosynthesizing conditions, they grew in mass to some extent with their nuclei dividing into two, but the cells were stalemated at this stage performing no further development. After washing with distilled water, the cells at this stage were suspended in a solution containing K_2SO_4 only ($10^{-2}M$), and then placed under photosynthesizing conditions, on the one hand, and under non-photosynthesizing conditions (in the light without provision of CO_2), on the other. Under non-photosynthesizing conditions, the cells performed, without increase in total cell mass, DNA synthesis and nuclear division, and eventually divided into four small daughter cells each. Under photosynthesizing conditions, on the contrary, the S-starved cells suspended in the K_2SO_4 solution grew further, synthesizing DNA and undergoing nuclear division, but the cells were stalemated at this stage without being able to divide. Such cells could be induced to cell division only when a nitrogen source (nitrate) was supplied to them simultaneously with sulfate. The last step, cellular division, was found not to occur when nitrate alone was supplied to the cells.

It thus became clear that sulfur is indispensable, not only for the DNA synthesis, but also for nuclear and cellular division, and that for the cellular division under photosynthesizing conditions a supply of both sulfur and nitrogen is required. The characteristic symptoms caused by S starvation were found to be relieved not only by provision of sulfate but also by supply of some organic sulfur compounds such as cysteine, cystine, methionine, reduced glutathione, meso-lanthionine or inactive cystathionine (Hase, Mihara and Tamiya, 1961b). The oxidized form of glutathione was found to be ineffective in this respect, which is thought to be due to its impermeability through the membrane of algal cells.

The arrest of cellular development caused by the effect of S deficiency was found to be accompanied by an appreciable decrease of carbohydrate and a marked increase of lipid in algal cells (Otsuka, 1961b), while the content (in per cent of dry weight) of bulk protein plus soluble amino acids remained about the same as that in normal cells (Hase and Mihara, unpublished experiments). When the S-starved cells were transferred to a medium containing sulfate alone under photosynthesizing conditions, the depleted level of carbohydrate returned to normal, the lipid content showed a tendency to increase further, and the percentage content of protein plus soluble amino acids decreased. Under such conditions the cells did not incorporate the sulfur into their lipid fraction to any significant extent. When, under the same condition, the S-starved cells were incubated in a medium containing both sulfate and nitrate, there occurred an appreciable S incorporation in the lipid fraction, the lipid content returned to normal, and the cells became able to perform cellular division.*

The fact that, as shown in Fig. 25, different phenomena occurred under photosynthesizing and non-photosynthesizing conditions indicates that there exists a competitive relationship between the photosynthetic process and cell division, which we shall discuss later.

* Benson *et al.* (1959) discovered that a certain form of sulfolipid is contained in *Chlorella* cells as well as in higher plants, and a similar substance was found to be present in normal cells of the *Chlorella* strain used by the Tokyo group (Hase, Otsuka *et al.*, 1959). The findings of Hase *et al.* mentioned above indicate that the lipid that had accumulated in the S-starved cells contained almost no trace of the sulfolipid in question. It is evident that the lipids accumulated in the S-starved cells were somewhat different in their chemical composition from those contained in normally grown cells, and conceivably, the appearance of the sulfolipid might be an indication of, or an initial step toward, the normalization of the quantity and quality of the lipid in algal cells.

Using ^{35}S as a tracer, the sulfur metabolism involved in the processes of ripening and cellular division of normal and S-N-controlled synchronous culture of *Chlorella ellipsoidea* has been studied in detail by Hase, Otsuka *et al.* (1959) and Hase, Mihara *et al.* (1959b; 1960a, b; 1961a). By separating the cell material into three fractions—extracts with 70 per cent boiling ethanol, with cold TCA (10 per cent) and the residue—they found that, in the normal synchronous culture, the sulfur actively assimilated during the post-ripening stage was accumulated in the ethanol and TCA-soluble fractions, while the sulfur content in the residual, proteinous, fraction, which was about 60–70 per cent of the total and contained sulfur mostly in the forms of cyst(e)ine and methionine, remained relatively constant during the life cycle (Hase, Otsuka *et al.*, 1959). The most remarkable change was in the S content in the TCA extract, although it occupied only a few per cent of the total S. It increased rapidly in the L_2–L_4 stages where the processes of DNA formation and nuclear division occurred most actively. In the ethanol extract they found a certain peptide-like substance, which appeared, both in normal and S-N-controlled cultures, only before the onset of cellular division. This substance did not contain sulfur, although its formation required the simultaneous presence of sulfur and nitrogen sources.

In further experiments using more elaborate methods of fractionation of cell material, Hase *et al.* found two S-containing nucleotidic compounds which seem to play important roles in the processes of ripening and division of algal cells. The S-containing substance appearing in extracts with cold acids (TCA or PCA) was revealed to be a peptide nucleotide complex which contains S in the peptide moiety in the form of cyst(e)ine and two unidentified compounds (Hase, Mihara and Otsuka, 1959; Hase, Mihara, Otsuka and Tamiya, 1959a, b; Hase, Morimura *et al.*, 1958; Hase, Otsuka *et al.*, 1959). The peptide moiety contains, besides cyst(e)ine, glutamic acid, glycine, aspartic acid, serine, arginine, alanine, etc., while the nucleotide moiety contains a ribose-type sugar and adenine, uracil and guanine as the major bases. In the following this complex compound may be referred to as the SPN or sulfur-containing nucleotide peptide compound. Another S-containing compound was found in the DNA fraction which was obtained by extracting the algal cells with hot PCA according to the methods either of Ogur and Rosen (1950) or of Schmidt–Thannhauser–Schneider (Volkin and Cohn 1954). This compound (Hase, Mihara and Tamiya, 1960a, b) contains no peptide moiety, and when subjected to the zone electrophoresis, at pH 5 in potato starch medium, it is separated into three nucleotidic components, of which one contains some purine derivatives but no sulfur, while the other two contain some pyrimidine derivatives and are associated with sulfur. The sulfur in these components appears to be in some form other than cyst(e)ine, methionine and sulfate, and is probably bound to deoxyribose of DNA or to some other deoxypentose polynucleotides (Hase, Mihara and Tamiya, 1960b; 1961a). For convenience of reference this compound may be denoted by SDN or S-containing deoxypentose polynucleotides in the future.

It was found that, when the S-starved cells were fed with ^{35}S-labeled sulfate and kept under non-photosynthesizing conditions to render them capable of cell division (see Fig. 25), SDN increased considerably during the process of cellular division.* When, on the other hand, ^{35}S-labeled sulfate was supplied to the S-starved cells under the photosynthesizing condition—a condition in which the cells grow to some extent, synthesizing DNA and performing nuclear division, but without being able to perform cellular division —^{35}S went in a large amount into the protein fraction and SDN was formed only in a negligibly small amount. A marked formation of SDN together with that of protein occurred only when nitrate, together with ^{35}S-sulfate, was supplied to the S-starved cells, which were then capable of cellular division (Hase, Mihara and Tamiya, 1960b; 1961a). These observations point strongly to the importance of SDN in the occurrence of cellular division in *Chlorella*. It has been mentioned earlier (Fig. 11) that to explain the isochronism of cellular division occurring under different light intensities, at a given temperature, one has to assume the existence of some factor, or factors, which, largely independent of the processes of growth and ripening, induces the processes of post-ripening and cellular division. It is tempting to assume that SDN might have some causal relation to this factor or factors (Hase, Morimura and Tamiya, 1960).

While SDN seems to have a significance in relation to the process of cellular division, SPN appears to bear an important part in the synthetic processes of nucleic acids and proteins. During the course of the normal life cycle of *Chlorella*, the quantity of SPN increases before and during the process of nuclear division (Hase, Otsuka *et al.*, 1959). When S-starved cells were supplied with sulfate under non-photosynthesizing conditions, S was incorporated into SPN, at

* It was noticed that the relative quantities of the three nucleotidic components in SDN, which are separable by the zone electrophoresis, changed to some extent during the course of cell development (Hase, Mihara and Tamiya, 1961a).

first gradually and later rapidly. The most active incorporation was observed at the onset of cellular division, after which the S content in SPN decreased gradually (Hase, Mihara and Tamiya, 1960a).

In Fig. 26 the zone electrophoretic patterns of SPN are shown, in terms of absorption at 260 mμ and ^{35}S-radioactivity, obtained from the cold TCA-soluble fraction of ^{35}S-labeled D and L_3 cells as well as of the S-starved cells (Hase, Mihara and Tamiya, 1960c).

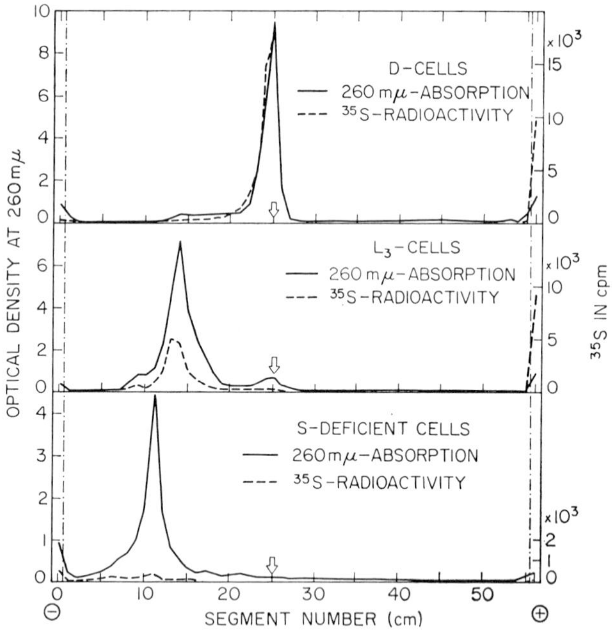

Fig. 26. Electrophoretic patterns of the cold TCA extracts obtained from ^{35}S-labeled D, L_3 and S-starved cells of *Chlorella ellipsoidea* (Hase, Mihara and Tamiya, 1960c). Medium: starch in acetate buffer of pH 3·6 with an ionic strength of 0·1. The arrows show the place of application of the original sample. Experiment lasted 38 hr at 3°.

The samples obtained from D and L_3 cells showed some difference in rates of migration, indicating a slight difference in their properties. In both cases, however, the peaks for ^{35}S and 260 mμ absorption, as well as the peak of ninhydrin reaction obtained after hydrolysis, were closely associated, whereas the sample obtained from the S-starved cells showed a distinct peak of the 260 mμ absorption but with only a slight ^{35}S-activity and with no detectable reaction of peptide (Hase, Mihara and Tamiya, 1960a, c). It is interesting to note that the S- and peptide-free sample obtained from the S-starved cells showed a u.v. absorption spectrum almost identical with those of SPNs obtained from D and L_3 cells (Hase, Mihara and Tamiya, 1960c). Conceivably, the S-starved cells had been able to synthesize, to some extent, the nucleotidic moiety of SPN complex, and on being supplied with sulfate, the cells started to synthesize the S-containing peptide, which then combined with the pre-existing nucleotides to form the SPN complex.

The supply of sulfate to S-starved cells, under non-photosynthesizing conditions, caused not only the increase of S-containing peptide moiety of SPN, but also further increase of nucleotidic moiety. There was, however, a certain difference in the time course of the increase of the two components. The increase of the nucleotidic moiety, as followed by the u.v. absorption of the cold PCA-soluble fraction, occurred earlier than the ^{35}S-incorporation; the quantity of the former attained a maximum level before the onset of cell division, while the ^{35}S in SPN attained its maximum during earlier stages of cell division. At later stages of cell division, however, both ^{35}S and nucleotide moiety decreased gradually.

The changes of the relative quantities of ^{35}S, that is of the S-containing peptide moiety, and of nucleotidic moiety in SPN in the S-controlled culture of *Chlorella* seem to reveal the behavior of this substance in the process of cell development of the alga. The gradual increase of S in SPN occurring soon after the provision of sulfate to S-starved cells was found to proceed approximately in parallel with the increase of RNA in algal cells. The active formation of both peptide and nucleotidic moieties of SPN occurring during the period before and at the initial stage of cell division was found to correspond to the stage in which DNA was synthesized at a considerable rate. Gradual decrease of both u.v.-absorbing and S-containing components at later stages of cell division seems to indicate that both nucleotides and peptide moieties of the SPN complex are transformed to some other substances with the progress of formative metabolism of algal cells. In fact, the gradual decrease of S in SPN at later stages of cell division proceeded approximately keeping pace with the gradual increase of S-incorporation into SDN as well as into the residual, proteinous, fraction (Hase, Mihara and Tamiya, 1960a). These observations may suggest that SPN is, most probably, acting as a pool or intermediate in the synthetic processes of nucleic acids, proteins and perhaps also of some other constituent cell substances (Hase, Mihara and Tamiya, 1961a). It should be added that SPN seems to be of widespread occurrence in microorganisms. A compound almost exactly identical with SPN of

Chlorella has been obtained from the cold acid-soluble fraction of two species of yeast, *Saccharomyces formosensis* and *S. cerevisiae* (Hase, Mihara, Otsuka and Tamiya, 1959a; Hase and Mihara, 1960).

Concerning the role of sulfur in the algal life cycle, another interesting observation was recently made by Otsuka (1961b) who followed the fate of pantothenic acid in algal cells during the course of S-N-controlled synchronous culture. Quantitative analyses were made of both free and combined forms of pantothenic acid, which is known to be, in combination with β-mercaptoethylamine and an adenosine derivative, a constituent of CoA. It was found that when the S-starved cells were provided with sulfate alone under photosynthesizing conditions, so that the cells grew to some extent without being able to divide, the free form of pantothenic acid increased as was the case in normal cells at their ripening stage, but the formation of the combined form was considerably suppressed. During the same period the quantity of lipid which had been accumulated by the effect of S-starvation (Otsuka 1961a; see above) did not return to the normal level. When the S-starved cells were able to undergo cell division by being supplied with sulfate under non-photosynthesizing conditions, or with sulfate plus nitrate under photosynthesizing conditions, the combined pantothenic acid (CoA) increased remarkably and the surplus lipid disappeared. In view of these results and considering the fact that CoA is a sulfur-containing substance, it may be inferred that in the process of cell division of *Chlorella* sulfur plays an important role, also by virtue of the biocatalytic action of CoA.

VIII. Competitive Relation Between Growth and Cellular Division

As has been observed in an earlier experiment by Iwamura (1955) and Iwamura *et al.* (1955), the cells of *Chlorella ellipsoidea* at their stage of L_2 can divide, although only incompletely, when incubated in the dark, while under photosynthesizing conditions they grow further and start to divide completely after attaining the stage of L_4 cells. Essentially the same phenomenon was observed later by Lorenzen and Ruppel (1960) using *Chlorella pyrenoidosa* as material. These results indicate that at the stage of L_2 the cells find themselves, so to speak, in a competitive relation between the process of photosynthetic growth and the light-independent process of cell division. Tamiya *et al.* (1961) demonstrated that the competitive relation between the two processes lasts until an earlier stage of L_3 cells but ceases at a later stage of L_3 cells. The state of affairs is illustrated by the data presented in Fig. 27. When the L_3 cells at an earlier stage, denoted by I, were transferred to darkness, cells started to divide almost immediately, whereas, when left in the light, they grew further and began to divide after a time lag of about 3 hr.*

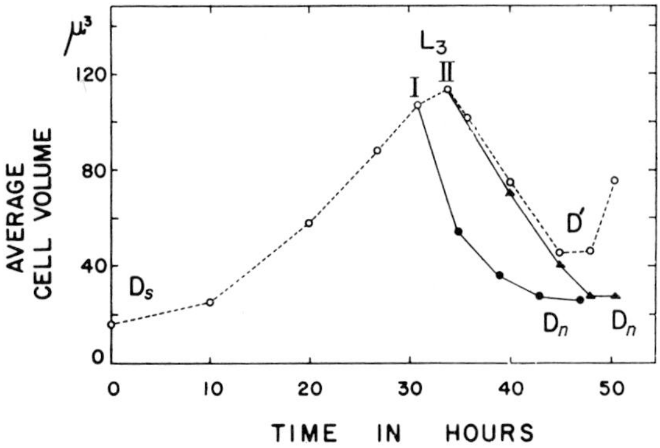

FIG. 27. Division (in dark and light) of L_3 cells of different degrees of ripening (*Chlorella ellipsoidea*; temperature: 21°; light intensity: 10 kilolux). Dashed and solid curves show, respectively, the processes occurring in light and in dark. In all cycles the division number was 4·0.

When the cells at this later stage (denoted by II) were transferred to darkness, they divided simultaneously with those kept in the light, showing no effect of illumination except that the daughter cells produced in the light were, because of the growth occurring during the dividing phase, somewhat larger than those produced in the dark. It is apparent that until Stage I there had been a competition between the processes of growth and cellular division, and that the competition came to an end at the later stage of L_3 cells. When competition ceases, the process of cellular division *per se* could occur without being affected by illumination.

Studies of Hase, Mihara and Tamiya (1960a; 1961a) on the role of sulfur compounds and of Otsuka (1961b) on the fate of combined pantothenic acid in the process of ripening and division of *Chlorella* cells have thrown some light on the biochemical cause underlying the competitive phenomenon in question. As has been described

* Using synchronous cultures of a high temperature strain of *Chlorella*, Sorokin (1960d) observed that the division of ripened cells was retarded by the effect of illumination. When such cells were transferred to darkness at 39°, cell division occurred in 3–4 hr, whereas if they were left in the light the time necessary for the cell division was prolonged to 8 to 9 hr.

already, the S-starved *Chlorella* cells, which have been stalemated at a certain stage of formative metabolism, behave differently under photosynthesizing and non-photosynthesizing conditions in their nutritional requirements for the resumption of nuclear and cellular division. Under non-photosynthesizing conditions the supply of sulfate alone was sufficient to induce the S-starved cells to ripen further (post-ripening) and to further cellular division, while under photosynthesizing conditions simultaneous supply of sulfate and nitrate was necessary for bringing about similar results. The most plausible assumption to account for this fact is that the S-starved cells require, for further ripening and cellular division, both S and N sources, and that under non-photosynthesizing conditions the requirement for the N-source is covered with some endogenous nitrogenous substance(s), while under photosynthesizing conditions this endogenous N substance would be consumed for the synthesis of some cell material not directly related to the process of ripening and cellular division, so that an exogenous supply of the N source, besides that of the S source, becomes necessary for the occurrence of further process of ripening and division.

It has been revealed (Hase, Mihara and Tamiya, 1960a; 1961a) that the mode of S metabolism changes appreciably, both qualitatively and quantitatively, depending upon whether the S-starved cells are supplied with sulfate under photosynthesizing or non-photosynthesizing conditions. A remarkable event occurring under photosynthesizing conditions was that a large amount of sulfur was incorporated into protein, and only a little went into the sulfur-containing deoxypentosepolynucleotides (SDN), which have been shown to play some important part in the process of cellular division. No doubt, both protein and SDN require N and S sources for their formation. Under non-photosynthesizing conditions, the nitrogen required for the synthesis of SDN would be taken from some endogenous N source(s), while under photosynthesizing conditions this source would be preferentially used for the protein synthesis, making it unavailable for the formation of SDN. The state of affairs is schematically illustrated in Fig. 28, in which the letters S and N outside the circles (representing algal cells) indicate, respectively, sulfate and nitrate supplied to the surrounding medium, and the N inside the circles means the endogenous nitrogenous substance(s) present in the S-starved cells. The white arrows represent the process of protein formation and the black ones the process of formation of SDN. Under photosynthesizing conditions and in the absence of exogenous N (Cell I), most of the endogenous N source is used, together with the S source supplied to the medium, preferentially for the protein synthesis, which results in an increase of total cell mass but not in the occurrence of cellular division because of the lack of SDN. On further provision of the N source (nitrate) (Cell II), the exogenous N and S sources are used not only for the synthesis of proteins but also for the formation

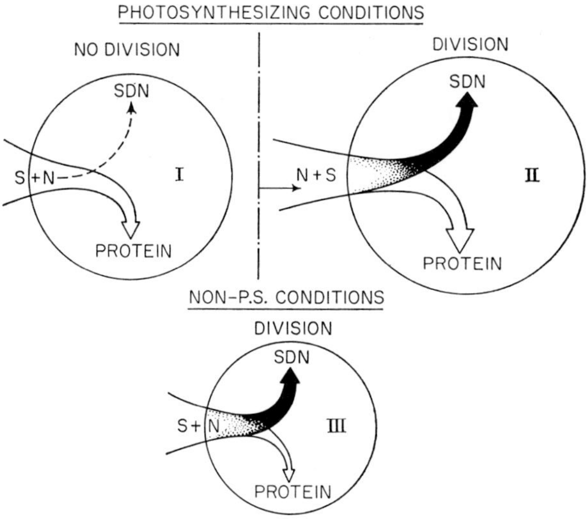

Fig. 28. Schematic representation of the competitive relations between protein synthesis and the formation of sulfur-containing deoxypentosepolynucleotides (SDN). (For explanations see text. Hase, Mihara and Tamiya, 1961a.)

of SDN, which enables the cells to perform cellular division. Under non-photosynthesizing conditions and in the presence of exogenous S (Cell III), the protein synthesis can occur only slightly, so that a large portion of the endogenous N source can be used for the formation of SDN, thus enabling the cells to divide. The observations made by Otsuka (1961b) on the fate of combined pantothenic acid (CoA) in the S-N-controlled culture of *Chlorella* indicate that a similar competitive relation also exists between the synthesis of protein and of CoA, which also seems to be an important S-containing substance functioning in the process of cell division of algae.

It has repeatedly been reported that the cell division of green cells is more or less suppressed by the effect of illumination, a phenomenon which is often construed in terms of 'photo-inhibition' of cell division (cf. Sorokin and Krauss, 1959). At least in the case of *Chlorella*, this interpretation is misleading; it is a manifestation of competition for certain cellular materials between growth and cellular division, the former being a light-requiring process and the latter a light-independent process.

REFERENCES

AACH, H. G. (1952). Über Wachstum und Zusammensetzung von *Chlorella pyrenoidosa* bei unterschiedlichen Lichtstärken und Nitratmengen. *Arch. Mikrobiol.* **17,** 213.

BENSON, A. A., H. DANIEL and R. WISER (1959). A sulfolipid in plants. *Proc. Natl. Acad. Sci., U.S.* **45,** 1582.

CALVIN, M. and A. A. BENSON (1948). The path of carbon in photosynthesis. *Science* **107,** 476.

HASE, E. and H. KUSHIDA; unpublished experiments.

HASE, E., and S. MIHARA; unpublished experiments.

HASE, E. and S. MIHARA (1960). Further notes on the nature of sulfur-containing peptide–nucleotide complex obtained from yeast. *J. Gen. Appl. Microbiol.* (*Japan*) **5,** 221.

HASE, E., S. MIHARA and H. OTSUKA (1959). An additional note on the nature of sulfur-containing peptide–nucleotide complex obtained from *Chlorella* and yeast. *J. Gen. Appl. Microbiol.* (*Japan*) **5,** 43.

HASE, E., S. MIHARA, H. OTSUKA and H. TAMIYA (1959a). New peptide–nucleotide compounds obtained from *Chlorella* and yeasts. *Biochim. et Biophys. Acta* **32,** 298.

HASE, E., S. MIHARA, H. OTSUKA and H. TAMIYA (1959b). Sulfur-containing peptide–nucleotide complex isolated from *Chlorella* and yeast cells. *Arch. Biochem. Biophys.* **83,** 170.

HASE, E., S. MIHARA and H. TAMIYA (1960a). Role of sulfur in the cell division of *Chlorella,* with special reference to the sulfur compounds appearing during the process of cell division. I. *Plant and Cell Physiol.* **1,** 131.

HASE, E., S. MIHARA and H. TAMIYA (1960b). Sulfur-containing deoxypentose polynucleotides obtained from *Chlorella* cells. *Biochim. et Biophys. Acta* **39,** 381.

HASE, E., S. MIHARA and H. TAMIYA (1960c). Some data on the nature of sulfur-containing peptide–nucleotide compounds obtained from *Chlorella* cells at different developmental stages in their life cycle. *J. Gen. Appl. Microbiol.* (*Japan*) **6,** 61.

HASE, E., S. MIHARA and H. TAMIYA (1961a). Role of sulfur in the cell division of *Chlorella,* with special reference to the sulfur compounds appearing during the process of cell division. II. *Plant and Cell Physiol.* **2,** 9.

HASE, E., S. MIHARA and H. TAMIYA (1961b). Capabilities of various sulfur substances in inducing the cell division of *Chlorella* arrested by sulfur deficiency. *J. Gen. Appl. Microbiol.* (*Japan*) **7,** 173.

HASE, E., Y. MORIMURA, S. MIHARA and H. TAMIYA (1958). The role of sulfur in the cell division of *Chlorella. Arch. Mikrobiol.* **32,** 87.

HASE, E., Y. MORIMURA and H. TAMIYA (1957). Some data on the growth physiology of *Chlorella* studied by the technique of synchronous culture. *Arch. Biochem. Biophys.* **69,** 149.

HASE, E., Y. MORIMURA and H. TAMIYA (1960). A short remark on the role of sulfur in the cell division of *Chlorella. J. Gen. Appl. Microbiol.* (*Japan*) **6,** 68.

HASE, E., H. OTSUKA, S. MIHARA and H. TAMIYA (1959). Role of sulfur in the cell division of *Chlorella*, studied by the technique of synchronous culture. *Biochim. et Biophys. Acta* **35,** 180.

HOLZER, H. (1951). Photosynthese und Atmungskettenphosphorylierung. *Z. Naturforsch.* **6** (b), 424.

IWAMURA, T. (1955). Change of nucleic acid content in *Chlorella* cells during the course of their life cycle. *J. Biochem.* (*Tokyo*) **42,** 575.

IWAMURA, T. (1960). Distribution of nucleic acids among subcellular fractions of *Chlorella. Biochim. et Biophys. Acta* **42,** 161.

IWAMURA, T.; unpublished experiments.

IWAMURA, T., E. HASE, Y. MORIMURA and H. TAMIYA (1955). Life cycle of the green alga *Chlorella* with special reference to the protein and nucleic acids contents of cells in successive formative stages. *Ann. Acad. Sci. Fennicae, Ser. A*, II. In Honour of Artturi I. Virtanen; *Biochem. Nitrogen* **60,** 89.

IWAMURA, T. and J. MYERS (1959). Changes in the content and distribution of the nucleic acid bases in *Chlorella* during the life cycle. *Arch. Biochem. Biophys.* **84,** 267.

KANAZAWA, T. (1958). Synchronous culture of *Chlorella* with special reference to the processes of assimilation of potassium, magnesium and iron. *J. Gen. Appl. Microbiol.* (*Japan*) **4,** 102.

LANGEN, P. and E. LISS (1958). Über Bildung und Umsatz der Polyphosphate der Hefe. *Biochem. Z.* **330,** 455.

LORENZEN, H. (1957). Synchrone Zellteilungen von *Chlorella* bei verschiedenen Licht-Dunkel-Wechseln. *Flora* (*Jena*) **144,** 473.

LORENZEN, H. (1958). Periodizität von Nuklealreaktion und Kernteilung in *Chlorella. Ber. Deut. Botan. Ges.* **71,** 89.

LORENZEN, H. (1959). Die photosynthetische Sauerstoffproduktion wachsender *Chlorella* bei langfristig intermittierender Belichtung. *Flora* (*Jena*) **147,** 382.

LORENZEN, H. and H.-G. RUPPEL (1960). Versuche zur Gliederung des Entwicklungsverlaufs der *Chlorella*-Zelle. *Planta* **54,** 394.

MEFFERT, M.-E. (1954). Ein Beitrag zur Wirkung diskontinuierlicher Belichtung auf die Massen- und Individuenvermehrung von *Chlorella vulgaris. Arch. Mikrobiol.* **20,** 410.

MIHARA, S. (1961). Change in glucosamine content of *Chlorella* cells during the course of their life cycle. *Plant and Cell Physiol.* **2,** 25.

MIYACHI, S. and H. TAMIYA (1961). Distribution and turnover of phosphate compounds in growing *Chlorella* cells. *Plant and Cell Physiol.* **2,** 405.

MORIMURA, Y. (1959a). Synchronous culture of *Chlorella.* I. Kinetic analysis of the life cycle of *Chlorella ellipsoidea* as affected by changes of temperature and light intensity. *Plant and Cell Physiol.* **1,** 49.

MORIMURA, Y. (1959b). Synchronous culture of *Chlorella.* II. Changes in content of various vitamins during the course of the algal life cycle. *Plant and Cell Physiol.* **1,** 63.

MÜLLER, H.-M. (1960). Über Veränderungen der stofflichen Zusammensetzung von *Scenedesmus obliquus* (Turp.) Krüger im Licht-Dunkel-Wechsel. *Naturwissenschaften*, **47,** 453.

MYERS, J. (1946). Culture conditions and the development of the photosynthetic mechanism. III. Influence of light intensity on cellular characteristics of *Chlorella. J. Gen. Physiol.* **29,** 419.

Nihei, T. (1955). A phosphorylative process, accompanied by photochemical liberation of oxygen, occurring at the stage of nuclear division in *Chlorella* cells. I. *J. Biochem.* (*Tokyo*) **42,** 245.

Nihei, T. (1957). A phosphorylative process, accompanied by photochemical liberation of oxygen, occurring at the stage of nuclear division in *Chlorella* cells. II. *J. Biochem.* (*Tokyo*) **44,** 389.

Nihei, T., S. Sasa, S. Miyachi, K. Suzuki and H. Tamiya (1954). Change of photosynthetic activity of *Chlorella* cells during the course of their normal life cycle. *Arch. Mikrobiol.* **21,** 155.

Northcote, D. H., K. J. Goulding and R. W. Horne (1959). The chemical composition and structure of the cell wall of *Chlorella pyrenoidosa, Biochem. J.* **70,** 391.

Ogur, M. and G. Rosen (1950). The nucleic acids of plant tissues. I. The extraction and estimation of desoxypentose nucleic acid and pentose nucleic acid. *Arch. Biochem.* **25,** 262.

Otsuka, H. (1961a). Changes of lipid and carbohydrate contents in *Chlorella* cells during the sulfur starvation, as studied by the technique of synchronous culture. *J. Gen. Appl. Microbiol.* (*Japan*) **7,** 72.

Otsuka, H. (1961b). Change of pantothenic acid content in *Chlorella* cells during the sulfur starvation and subsequent restoration, as studied by the technique of synchronous culture. *J. Gen. Appl. Microbiol.* (*Japan*) **7,** Suppl. 1, 353.

Otsuka, H.; unpublished experiments.

Pirson, A. and H. Lorenzen (1958a). Ein endogener Zeitfaktor bei der Teilung von *Chlorella. Z. Botan.* **46,** 53.

Pirson, A. and H. Lorenzen (1958b). Photosynthetische Sauerstoffentwicklung von *Chlorella* nach Synchronization durch Licht-Dunkel-Wechsel. *Naturwissenschaften* **45,** 497.

Pirson, A., H. Lorenzen and A. Koepper (1959). A sensitive stage in synchronized cultures of *Chlorella. Plant Physiol.* **34,** 353.

Řetovský, R. and I. Klášterská (1961a). Study of the growth and development of *Chlorella* populations in the culture as a whole. V. The influence of magnesium sulphate on autospore formation. *Folia Microbiol.* **6,** 115.

Řetovský, R. and I. Klášterská (1961b). Study of the growth and development of *Chlorella* populations in the culture as a whole, VI. Basophilia and oxido-reduction relationships in *Chlorella* cells. *Folia Microbiol.* **6,** 127.

Sasa, T. (1961). Effect of ultraviolet light upon various physiological activities of *Chlorella* cells at different stages in their life cycle. *Plant and Cell Physiol.* **2,** 253.

Schmidt, R. R. (1961). Nitrogen and phosphorus metabolism during synchronous growth of *Chlorella pyrenoidosa. Exptl. Cell Research* **23,** 209.

Schmidt, R. R. and K. W. King (1961). Metabolic shifts during synchronous growth of *Chlorella pyrenoidosa. Biochem. et Biophys. Acta* **47,** 391.

Sorokin, C. (1957). Changes in photosynthetic activity in the course of cell development in *Chlorella. Physiol. Plantarum* **10,** 659.

Sorokin, C. (1958). The effect of the past history of cells of *Chlorella* on their photosynthetic capacity. *Physiol. Plantarum* **11,** 275.

Sorokin, C. (1960a). A reversible inhibition of photosynthesis in synchronized cultures of algae. *Nature* **185,** 933.

Sorokin, C. (1960b). Injury and recovery of photosynthesis in cells of successive developmental stages. The effects of light intensity. *Physiol. Plantarum* **13,** 687.

SOROKIN, C. (1960c). Injury and recovery of photosynthesis. The capacity of cells of different developmental stages to regenerate their photosynthetic activity. *Physiol. Plantarum* **13,** 20.

SOROKIN, C. (1960d). Kinetic studies of temperature effects on the cellular level. *Biochim. et Biophys. Acta* **38,** 197.

SOROKIN, C. (1961). Time course of oxygen evolution during photosynthesis in synchronized cultures of algae. *Plant Physiol.* **36,** 232.

SOROKIN, C. and R. W. KRAUSS (1959). Maximum growth rates of *Chlorella* in steady-state and in synchronized cultures. *Proc. Natl. Acad. Sci., U.S.* **45,** 1740.

SOROKIN, C. and J. MYERS (1957). The course of respiration during the life cycle of *Chlorella* cells. *J. Gen. Physiol.* **40,** 579.

STANGE, L., E. L. BENNETT and M. CALVIN (1960). Short-time $^{14}CO_2$ incorporation experiments with synchronously growing *Chlorella* cells. *Biochem. et Biophys. Acta* **37,** 92.

TAMIYA, H. (1935). Le bilan materiel et l'énergétique des synthèses biologiques. *Actualités sci. et ind.* **214.**

TAMIYA, H. (1942). Atmung, Gärung und die sich daran beteiligenden Enzyme von *Aspergillus. Advances in Enzymol.* **2,** 183.

TAMIYA, H., T. IWAMURA, K. SHIBATA, E. HASE and T. NIHEI (1953). Correlation between photosynthesis and light-independent metabolism in the growth of *Chlorella. Biochim. et Biophys. Acta* **12,** 23.

TAMIYA, H., Y. MORIMURA and M. YOKOTA (1962). Effects of various antimetabolites upon the life cycle of *Chlorella. Arch. Mikrobiol.* **42,** 4.

TAMIYA, H., Y. MORIMURA, M. YOKOTA and R. KUNIEDA (1961). Mode of nuclear division in synchronous cultures of *Chlorella*: Comparison of various methods of synchronization. *Plant and Cell Physiol.* **2,** 383.

TAMIYA, H., K. SHIBATA, T. SASA, T. IWAMURA and Y. MORIMURA (1953). Effect of diurnally intermittent illumination on the growth and some cellular characteristics of *Chlorella. Carnegie Inst. Wash. Publ. No.* 600, 76.

VOLKIN, E. and W. E. COHN (1954). Estimation of nucleic acids. In *Methods of Biochemical Analysis* (D. Glick, ed.), Vol. 1, p. 287, Interscience, New York.

WANKA, F. (1959). Untersuchungen über die Wirkung des Lichts auf die Zellteilung von *Chlorella pyrenoidosa. Arch. Mikrobiol.* **34,** 161.

WIAME, J. M. (1949). The occurrence and physiological behaviour of two metaphosphate fractions in yeast. *J. Biol. Chem.* **178,** 919.

WINTERMANS, J. F. G. M. (1954). On the formation of polyphosphates in *Chlorella* in relation to conditions for photosynthesis. *Proc. Koninkl. Ned. Akad. Wetenschap.* Ser. C. **57,** 574.

WINTERMANS, J. F. G. M. (1955). Polyphosphate formation in *Chlorella* in relation to photosynthesis. *Mededeel. Landbouwhoogeschool, Wageningen* **55,** 69.

YUHARA, T. and E. HASE (1961). A brief note on the absorption of water by *Chlorella* cells during the process of their life cycle. *J. Gen. Appl. Microbiol.* (*Japan*) **7,** 70.

10

Phased Cell Division in the Marine Dinoflagellates*

J. Woodland Hastings and Beatrice M. Sweeney,
Division of Biochemistry, University of Illinois, Urbana, Illinois; and Biology Department, Yale University, New Haven, Connecticut

CONTENTS

I	Introduction	307
II	The Endogenous Nature of the Rhythm of Cell Division	309
III	Properties of the Endogenous Rhythm	313
IV	Mitotic Cycle in *Gonyaulax*	317
	References	320

I. Introduction

It has been known for a long time that in nature there occurs what might be called a partial synchrony of cell division. In certain organisms cell divisions take place only at a particular time of day (Gough, 1905; Jörgensen, 1911). These cases do not constitute true synchrony because all cells in the population do not divide each day. Since divisions are confined to certain times of day, it is more accurate to describe the phenomenon as phased cell division. Recent studies of the marine dinoflagellates (Sweeney and Hastings, 1958) indicate that cell division in these organisms is controlled by an endogenous 'biological clock' or diurnal rhythmic mechanism. A consideration of these rhythms is relevant to the general problem of the induction and control of synchrony of cell division (Zeuthen, 1958). Knowledge of how control is mediated in diurnal rhythms is of special interest in reference to the biochemical steps involved in cell division. In addition, it is possible that with experimental manipulation the rhythmic organisms could constitute good synchronous experimental material for the study of the mitotic cycle.

Some of the earliest observations of phased cell division in nature were reported in the dinoflagellates by Gough (1905) and by Jörgensen (1911), who found that members of the genus *Ceratium* divide only at night. This genus is very convenient for the detection

* The authors wish to acknowledge the support of the National Science Foundation and the Office of Naval Research.

of cell division because the theca, which is extended anteriorly and posteriorly into long horns, is split obliquely at division. Recently divided cells may therefore be easily distinguished because they possess horns of unequal length. Using this method of detection, Gough found that as many as 70 per cent of the individuals in plankton hauls at 3.30 a.m. had recently divided, whereas such cells were rare at other times of day.

The same phenomenon was later described and studied in cultures of various other dinoflagellates. Braarud and Pappas (1951), using *Peridinium triquetrum* growing under natural illumination, showed that divisions occurred only between 5 a.m. and 1 p.m., with a

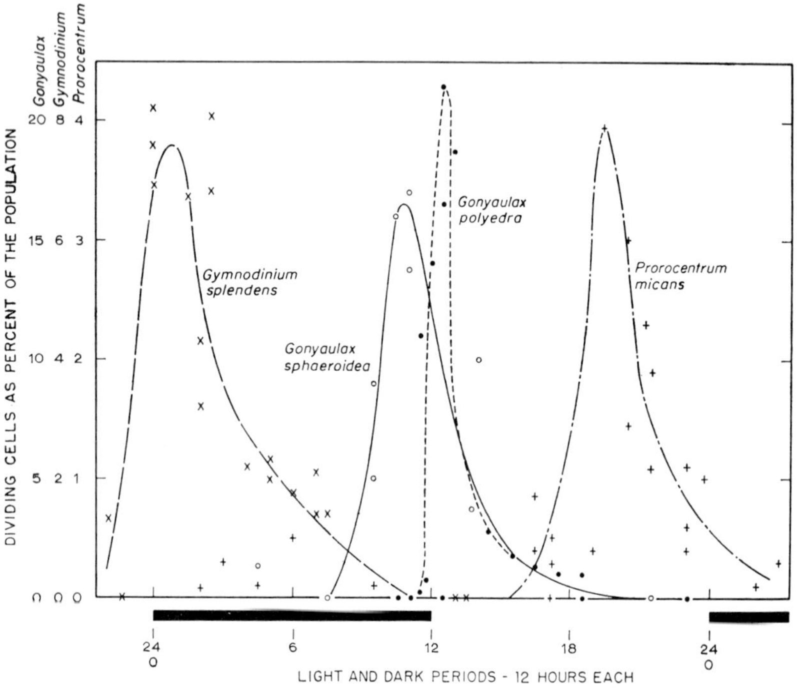

FIG. 1. The time at which cell divisions occur in various dinoflagellates growing under the 12-hr light–12-hr dark cycle indicated on the abcissa: (X) *Gymnodinium splendens;* (O) *Gonyaulax sphaeroidea;* (●) *Gonyaulax polyedra;* (+) *Prorocentrum micans.* Since recently divided cells of these species remain attached, forming a pair for a short time after division, the occurrence of cell division was determined by counting pairs. The number of pairs present at any time in a population was determined by counts with living material, since pairs are more easily detected when the cells are motile than when they are fixed. (From Sweeney, 1959.)

maximum at about 9 a.m. A similar phasing of cell division has been encountered in a variety of photosynthetic dinoflagellates investigated by the authors, including *Gonyaulax polyedra* (Sweeney and Hastings, 1958), *Gymnodinium splendens*, *Prorocentrum micans* and *Gonyaulax sphaeroidea* (Sweeney, 1959). It is significant to point out that the actual clock time at which the maxima occur differs in different species, as illustrated in Fig. 1.

This property of 'natural' phasing in cell division is not confined to the dinoflagellates; it is also known to occur in some other algae (Sweeney and Hastings, 1962) including *Chlamydomonas* (Bernstein, 1960), as well as in higher plants (Stålfelt, 1921) and animals (Carleton, 1934; Halberg, Bittner and Smith, 1958). Our discussion here will be confined to the dinoflagellates.

If cell division occurs at only one time of day, then the growth curve should resemble a staircase, i.e., the cell number should rise rather abruptly at the time of division and be relatively constant at other times. In fact this was found to be the case with cultures maintained under alternating light and dark periods of 12 hr each (Fig. 2). Note that the slope of the curve (dotted line) is a measure of the *average* generation time, i.e., the average growth rate.

In this same figure it may be noted that the rhythm of cell division is inhibited if the cells are grown and maintained under conditions of constant bright light (900 ft-candles). This is a typical property of rhythms (see below).

II. The Endogenous Nature of the Rhythm of Cell Division

The observation that cell divisions occur at a particular time of day does not necessarily constitute evidence for a persistent endogenous rhythm. Cell division might be triggered by some environmental change, in light for example. Such a rhythm is indicated, however, when the periodicity is found to continue under conditions of constant light and constant temperature. In *Gonyaulax polyedra*, cell divisions continue to be phased even after the cells have been kept in constant conditions for as long as 15 days, as illustrated in Fig. 3. Such persistence occurs under a relatively low level of illumination (approximately 100 ft-candles). At higher light intensities the rhythm is inhibited (Fig. 2).

It is important to distinguish clearly between a cell division rhythm and phased cell division which results from a true synchrony of mitosis. In the latter case the period is, of necessity, equal to the generation time. In the cell division rhythm in *Gonyaulax*, where the period is about 24 hr, the generation time of a given cell may have any value ranging from 24 hr up to 6 or 8 days—although it is always a value which is close to a multiple of 24. In consequence, the *average* generation time in cultures may vary widely, and does not necessarily equal a multiple of 24 hr (Fig. 3).

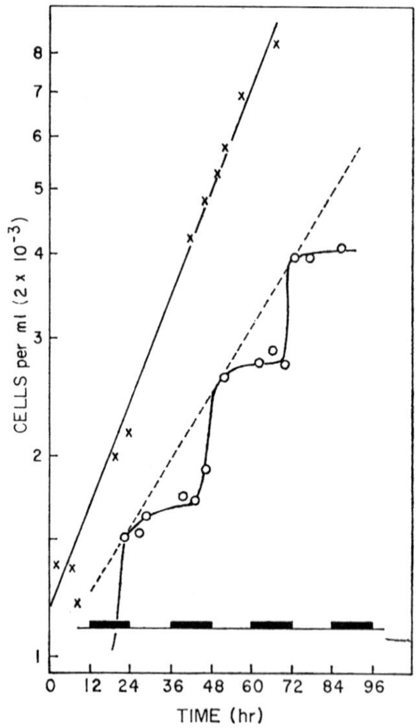

FIG. 2. Increase in the cell population of *Gonyaulax polyedra* in continuous light (×), and in a 24-hr light–dark cycle (○), the dark periods being indicated by black bars along the abcissa. Light intensity, 900 ft-candles: temperature, 21·5° ± 1°C. Average generation time in continuous light, 1 day; in light–dark cycle, 1·5 days.

This particular point was investigated and verified by determining the time between divisions of single cell isolates. Daughter cells from a division were isolated in microdrops in capillary tubes and were observed regularly thereafter to determine the time interval until the subsequent division. The cells were maintained on a 24-hr cycle: 12 hours' light, 12 hours' darkness. Although the manipulations and observations were technically rather difficult, it is clear that the values for the generation time of individual cells clustered around $n \times 24$ hr. Figs. 4 and 5 illustrate these experiments. Under these conditions, the most frequently encountered generation time was 48 hr. The measured generation times fall into normal distributions having a relatively small spread. Generation times of 24 ($n = 1$) and 72 ($n = 3$) were less common than 48 ($n = 2$). Generation times differing greatly from $n \times 24$ hr were not found. The progeny from a rapid division (24 hr) did not show any marked tendency to divide with a shorter average generation time than other cells; that is, it was not possible to select and/or establish rapidly—or slowly—dividing clones.

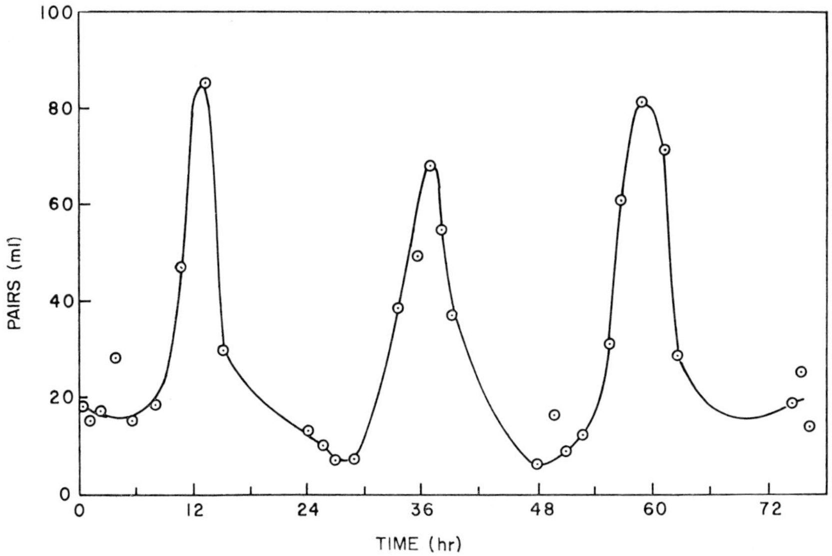

Fig. 3. The number of paired cells present in a cell suspension of *Gonyaulax polyedra* on the twelfth, thirteenth and fourteenth days of continuous dim light (100 ft-candles) at 19°C. *G. polyedra* was grown in a light–dark cycle with light from 1100 to 2300, and was transferred to continuous dim light at 1100 (at the end of a dark period). The measurements shown were started twelve days later. The average generation time between the tenth and fourteenth days was 3·8 days.

When isolated cells were maintained under continuous bright light (900 ft-candles) rather than under a 24-hr light–dark cycle, the generation times were then found to vary widely and showed no tendency to be $n \times 24$ hr (Figs. 4 and 5, dotted lines). As mentioned previously, constant bright light inhibits rhythmicity in cell populations. It is possible that this could be due to asynchrony between the various members of a population, each member of which might be remaining rhythmic. The absence of any evidence of a 24-hr rhythm in isolated cells in continuous bright light provides evidence for a true loss of rhythmicity. Asynchrony is not a possible explanation for the observed inhibition of rhythmicity under these conditions. This same conclusion was also reached from experiments on the rhythm of photosynthesis studied in single isolated cells of *Gonyaulax* (Sweeney, 1960).

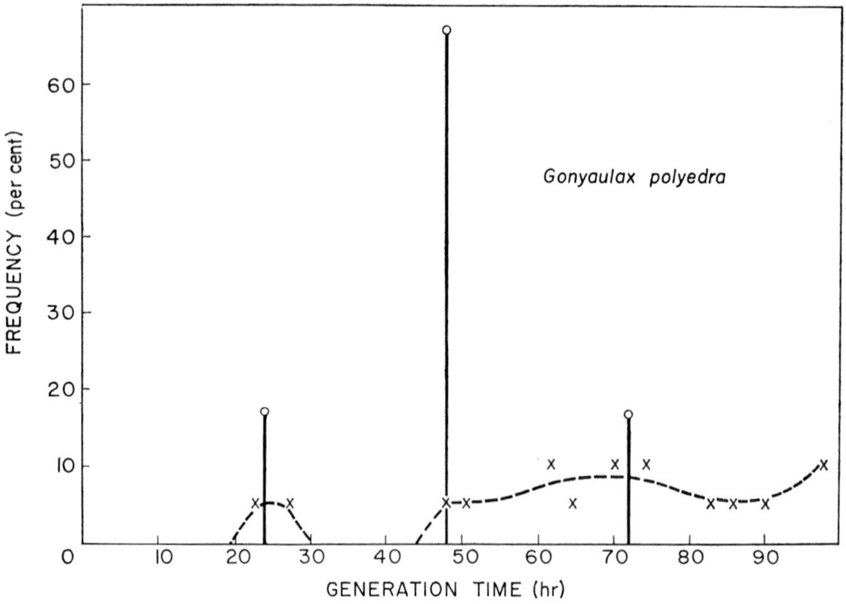

FIG. 4. The frequency of different generation times measured for a number of isolated cells of *Gonyaulax polyedra* under a light–dark cycle (——O——), or in continuous bright light (- - - X - - -). This figure illustrates that cells on a 24-hr cycle have generation times equal to a multiple of 24, whereas cells grown under continuous bright light do not.

Under conditions of continuous bright light, very short generation times (6–8 hr) are occasionally encountered in *Gymnodinium* (Fig. 5). The daughter cells from such divisions are extremely small and often do not divide again for several days. The significance of this extreme irregularity is not understood.

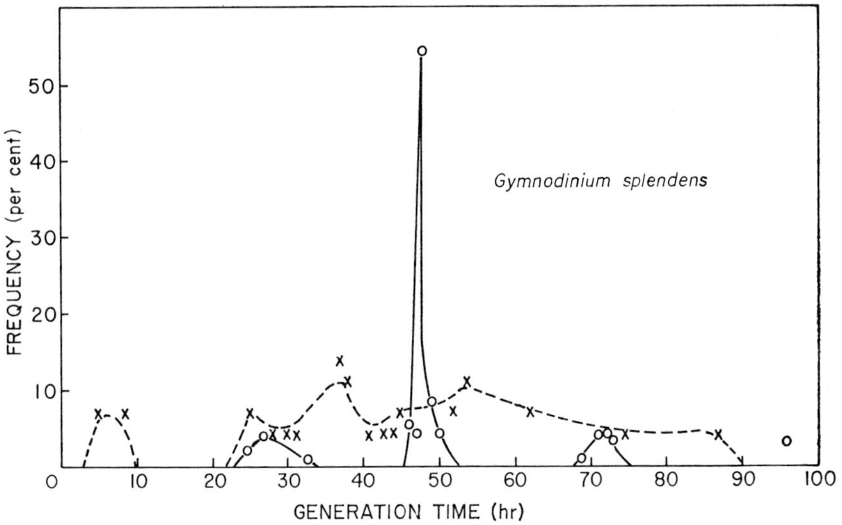

FIG. 5. The frequency of different generation times measured for a number of isolated cells of *Gymnodinium splendens* under a light–dark cycle (——O——), or in continuous bright light (- - X - -). As in *Gonyaulax polyedra* (Fig. 4), cells on a 24-hr cycle show generation times which are approximate multiples of 24, while cells in continuous light do not. Note the very short generation times encountered in continuous bright light.

III. Properties of the Endogenous Rhythm

The cell-division rhythm has properties which are very similar to the other persistent diurnal rhythms (bioluminescence, photosynthetic capacity) which have been studied in *Gonyaulax* (Hastings and Sweeney, 1958; Hastings, Astrachan and Sweeney, 1961). There are also striking similarities to rhythms in a variety of other organisms (Pittendrigh and Bruce, 1957; Hastings, 1959; Sweeney and Hastings, 1960).

In nature the period of the cell division rhythm is exactly 24 hr, i.e., directly related to the light–dark changes of the solar day. Under constant conditions however, the period of the rhythm differs slightly from 24 hr, suggesting that the rhythm is no longer coupled to a periodicity in the environment. Under constant conditions the observed period has been referred to as the 'natural' or 'free-running' period (Pittendrigh, 1960).

By subjecting *Gonyaulax* cultures to light–dark cycles with periods which are less than 24 hr, it has been possible to 'drive' or 'entrain' the rhythm so that a maximum occurs once every cycle. Thus with a

16-hr cycle (8 hr light, 8 hr darkness; Fig. 6) a maximum in the number of recently divided cells occurs every cycle at about the fifth hour of the light period.

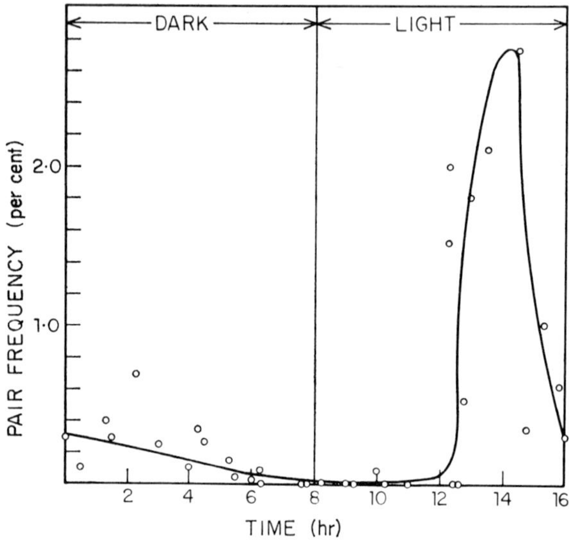

FIG. 6. This figure illustrates that cell division in *Gonyaulax polyedra* may be phased to periods shorter than 24 hr by growing the cells on the appropriate light cycle. In this experiment, cultures were maintained on a 16-hr cycle and inspected at frequent intervals for pairs. The pair frequency found (ordinate) is plotted against the time in the cycle when the observation was made.

Under these conditions the *average* growth rate is about the same as with a 24-hr cycle. First, this indicates that there appears to be no deleterious effects upon growth of cycles of 'abnormal' length. This has been found to be true in a larger number of experiments with cycles of other than 24-hr duration. Secondly, it necessarily follows that generation times in such a culture are equal to $n \times 16$ hours, and that the culture as a whole is actually somewhat more heterogenous with respect to synchrony of division than where the cycle length equals 24.

Various attempts were made to achieve a greater synchrony in cultures. One method which we tried was to *increase* the duration of a complete light–dark cycle to values greater than 24 hr. This was not successful; the times at which cell divisions occurred as determined by the occurrence of pairs showed a greater spread and the phasing was much less clear. Cycles having a duration appreciably greater than 24 hr (28 hr or more) fail to 'entrain' the rhythm in the fashion discussed above (Fig. 6) in connection with cycles with periods less than 24 hr.

Another way in which we attempted to achieve a greater degree of synchrony was by changing the duration of the dark period within a 24-hr cycle. With longer light periods and shorter dark periods (for example 16 hr light and 8 hr dark), the average generation time of the cultures was indeed slightly shortened, but it was still far from 24 hr. Furthermore, with dark periods shorter than 6 hr, the rhythmicity tended to be lost altogether, as in continuous bright light. Consequently this approach does not appear to have promise.

However, future studies may reveal ways whereby synchronization may be improved. In order to maintain a near-perfect synchronization, it would clearly be necessary to adjust conditions of light and temperature so as to achieve an *average* generation time precisely equal to some specific multiple of the period of the diurnal rhythm. Since we had not been able to increase the growth rate sufficiently to allow n to equal one, we attempted to adjust n to equal two, and to start the division cycle in all cells from the same point at the same time, by prior starvation of the cells in darkness. This was not successful. Attempts have not been made to use inhibitors or temperature shocks for this purpose.

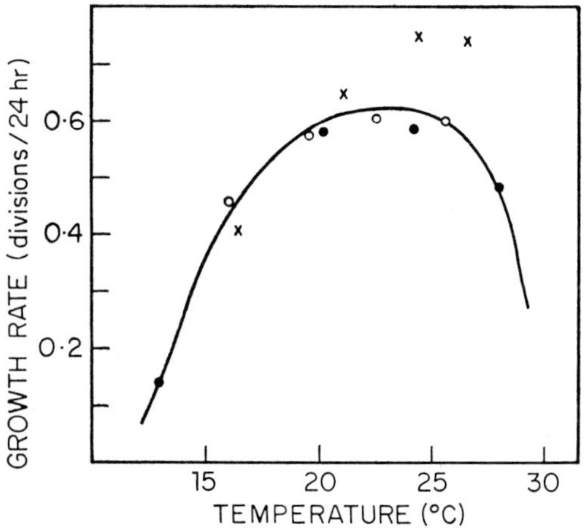

Fig. 7. The growth rate of *Gonyaulax polyedra*, determined at several temperatures, shows an optimum around 25°C. Note that the rate of growth is expressed as divisions per day.

The evidence is clear that the 'natural' period is not a characteristic which may be 'learned' or 'impressed' on the organism by external light cycles. Cultures may be maintained for several years on cycles shorter than 24 hr. Whenever they are returned to constant dim light, the rhythm reverts to the natural period of approximately 24 hr (Hastings, 1959).

One of the most striking and interesting features of biological clocks is the feature of temperature-independence (Hastings and Sweeney, 1957; Sweeney and Hastings, 1960). Accuracy is clearly an important aspect of a clock which must operate satisfactorily under a variety of diverse experimental conditions. The temperature-independence of biological clocks refers to the lack of large differences

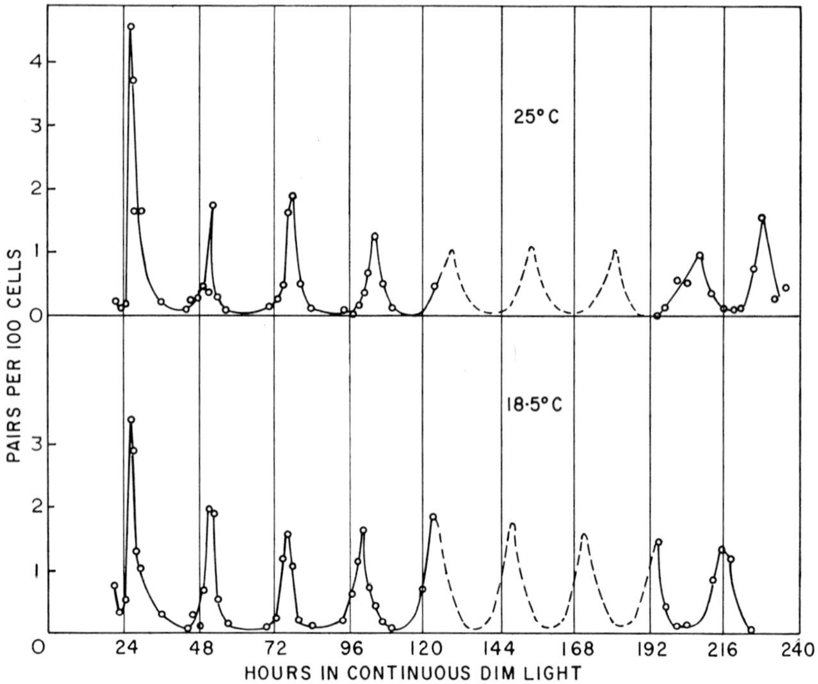

Fig. 8. This figure illustrates the persistence of the endogenous cell-division rhythm in *Gonyaulax polyedra* under conditions of continuous dim light. On the ordinate is plotted the percentage of paired cells present in a cell suspension transferred from a 24-hr light cycle to continuous dim light (200 ft-candles) at the beginning of a light period (zero time). The two temperatures were used to illustrate the near temperature independence of the period, which was 23·9 hr at 18·5°C and 25·4 hr at 25°C, while the average generation time was 4·2 days and 3·8 days respectively (from Sweeney and Hastings, 1958).

in the period of the rhythm when the organism is maintained under constant conditions but at different temperatures. This does not mean, for example, that the *average* rate of cell division is unaffected by temperature. In fact temperature does affect growth rate (measured in terms of divisions per unit time; Fig. 7). Since the *period* of the cell division rhythm is relatively unchanged at different temperatures, the interpretation is that the statistical distribution of cells falling into various cycle lengths (Fig. 4) would be different at each temperature.

The precise measurement of the period of a rhythm requires a relatively good assay method and a reasonably long-term experiment. Measurements of this kind have been made with the rhythm of cell division in *Gonyaulax* (Fig. 8). The notable feature is the slight lengthening of the period at higher temperatures, suggesting that a slight overcompensation was occurring. Since any temperature-compensating mechanism carries the possibility of imprecision, over-compensation is not an unexpected phenomenon. Other rhythms in *Gonyaulax* were found to respond to temperature differences in an essentially identical way (Hastings and Sweeney, 1959).

In general, the phase of endogenous rhythms may be reset by exposure to light. This has been demonstrated for the rhythm in luminescence in *Gonyaulax* (Hastings and Sweeney, 1959) where irradiation for three hours at the time of maximum light emission serves to alter the phase of the rhythm by a full 180° (12 hr). The amount of phase shift which is obtained in such experiments is strongly dependent upon the time of day when the treatment is given.

When cells are given such an exposure to light and the cell division rhythm is studied, it is also found to be reset in a manner essentially identical with that of luminescence.

IV. Mitotic Cycle in *Gonyaulax*

In view of the naturally-occurring phasing of cell division in dinoflagellates, a fuller understanding of the cytological and, if possible, cytochemical events of the mitotic cycle in *Gonyaulax* are of importance. Mitosis in dinoflagellates shows certain somewhat unusual features (Borgert, 1910; Jollos, 1910; Grell and Wohlfarth-Bottermann, 1957; Skoczylas, 1958). The nuclei in these cells are large and show a fibrillar structure even at interphase. Mitotic spindle fibers are indistinct. In *Gonyaulax* the changes in the nucleus associated with cell division were studied in Feulgen-stained

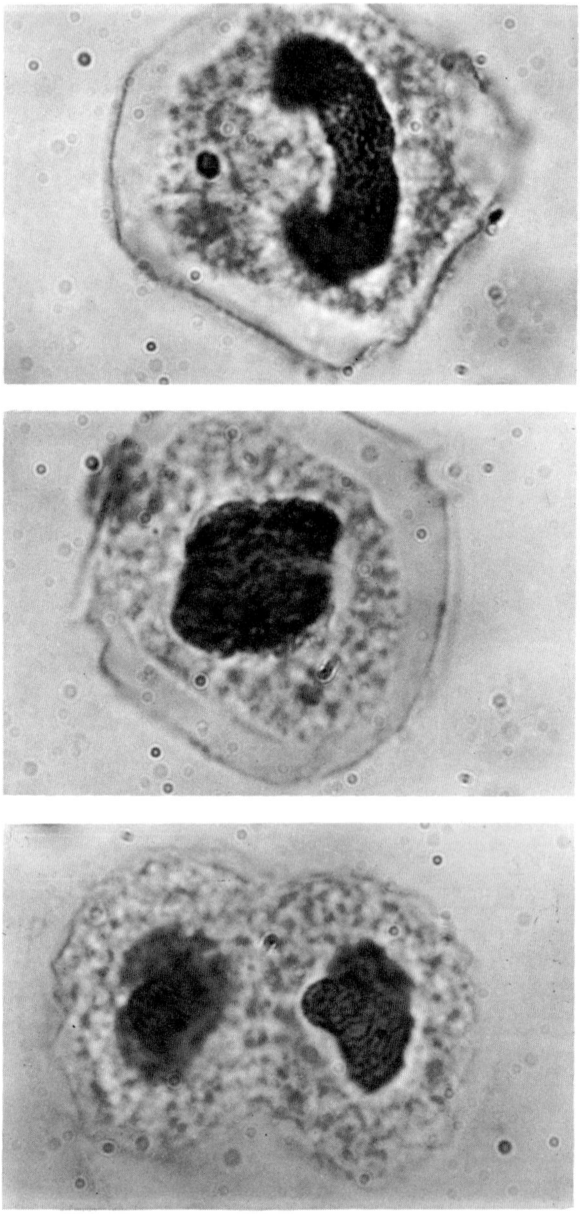

(a)

(b)

(c)

FIG. 9. Nuclear division in *Gonyaulax polyedra* stained by the Feulgen reaction. (a) crescent-shaped resting nucleus showing typical granular structure; (b) nucleus just prior to separation showing fibrillar structure; and (c) nuclei just after separation into two. (Photographs by Dr. Priscilla Roslansky.)

preparations (Fig. 9)*. The interphase nucleus is sickle-shaped. When cell division begins, the crescent thickens and its arms come together to form an ellipse. Just prior to division the normal nuclear fibrils are much thicker and arranged in parallel rows. The nucleus pulls apart, like a skein of yarn, with no visible spindle, and the two daughter nuclei round up as the cell divides and reform into crescent-shaped resting nuclei. In cultures in alternating light and darkness early division figures are seen about 1 hr before the beginning of the light period. The maximum number of dividing nuclei occurs about 1 or 2 hr later, while the peak in paired nuclei follows about $\frac{1}{2}$ to 1 hr thereafter. The entire division process requires 5–6 hr.

The use of cells in which the division time is related to a light–dark cycle might very well allow us to obtain a 'natural' synchrony close to 100 per cent. As we have seen, with *Gonyaulax* the outlook is not very promising at the present time. The best we have achieved is the division of about 65 per cent of the cells in each successive 24-hr period. For a sophisticated biochemical study of the mitotic cycle, this is clearly not sufficient.

Other cells, such as *Chlamydomonas* (Bernstein, 1960), may show rhythmic cell division. Thus far, however, knowledge of these systems is insufficient to allow their use for the purpose of studying the chemical events of the mitotic cycle.

An unusual case of natural synchrony coupled with phased cell division occurs in certain dinoflagellate species which form chains, such as *Gonyaulax monilata*. The culture conditions and the amount of shaking can effect the length to which chains ultimately grow; we have noted that sixteen and thirty-two-celled chains are generally the longest observed without special precautions. In this organism, the divisions in a particular chain are always in relatively close synchrony. Although in the longest chains some difference in the time of division in the two halves is sometimes evident, the synchrony exhibited in chains is considerably greater than that observed in the progeny of a single isolated cell in a non-chain-forming species such as *Gonyaulax polyedra*.

In conclusion then, natural phasing of cell divison via an endogenous diurnal rhythmicity is not uncommon, and is well documented in the dinoflagellates. At present, however, it is not possible to achieve complete synchrony of cell division in the dinoflagellates by the exploitation of this rhythmicity.

* Findings with Feulgen-stained *Gonyaulax* cells were confirmed and extended by Dr. Priscilla Roslansky.

REFERENCES

BERNSTEIN, E. (1960). Synchronous division in *Chlamydomonas moewusii*. *Science* **131,** 1528–1529.

BORGERT, A. (1910). Kern- und Zellteilung bei marinen Ceratium Arten. *Arch. Protistenk.* **20,** 1–46.

BRAARUD, T. and I. PAPPAS (1951). Experimental studies on the dinoflagellate *Peridinium triquetrum* (Ehrb.) Lebour. *Avhandl. Norske Videnskaps-Akad. Oslo. I. Mat.-Naturv. Kl.* No. 2, 1–23.

CARLETON, A. (1934). A rhythmic periodicity in the mitotic division of animal cells. *J. Anat.* **68,** 251–263.

GOUGH, L. H. (1905). Report on the plankton of the English Channel in 1903. Mar. Biol. Assoc. U.K. International Fishery Investigation. *First Report on Fishery and Hygrographical Investigations in the North Sea and Adjacent Waters* (*Southern Area*). 1902–1903. London, 325–377.

GRELL, K. G. and K. E. WOHLFARTH-BOTTERMANN (1957). Licht- und elektronenmikroskopische Untersuchungen an dem Dinoflagellaten *Amphidinium elegans* n. sp. *Z. Zellforsch.* **47,** 7–17.

HALBERG, F., J. J. BITTNER and D. SMITH (1958). Mitotic rhythm in mice, mammary tumor milk agent and breast cancer. *Proc. Am. Assoc. Cancer Research* **2,** 305.

HASTINGS, J. W. (1959). Unicellular clocks. *Ann. Rev. Microbiol.* **13,** 297–312.

HASTINGS, J. W. and B. M. SWEENEY (1957). On the mechanism of temperature independence in a biological clock. *Proc. Natl. Acad. Sci., U.S.* **43,** 804–811.

HASTINGS, J. W. and B. M. SWEENEY (1958). A persistent diurnal rhythm of luminescence in *Gonyaulax polyedra*. *Biol. Bull.* **115,** 440–458.

HASTINGS, J. W. and B. M. SWEENEY (1959). The *Gonyaulax* clock, pp. 567–584. *Photoperiodism and Related Phenomena in Plants and Animals* (R. B. Withrow, ed.), Am. Assoc. Adv. Sci., Washington, D.C.

HASTINGS, J. W., L. ASTRACHAN and B. M. SWEENEY (1961). A persistent daily rhythm in photosynthesis. *J. Gen. Physiol.* **45,** 69–76.

JOLLOS, V. (1910). Dinoflagellatenstudien. *Arch. Protistenk.* **19,** 178–206.

JÖRGENSEN, E. (1911). *Die Ceratien.* Eine Kurze Monographie der Gattung *Ceratium* Shrank. *Internat. Rev. gesamt. Hydrobiol. Hygrogr.* Biol. Suppl. Ser. II, 1–124.

PITTENDRIGH, C. S. (1960). Circadian rhythms and the circadian organizations of living systems. *Cold Spring Harbor Symp. Quant. Biol.* **25,** 159–184.

PITTENDRIGH, C. S. and V. G. BRUCE (1957). An oscillator model for biological clocks. In *Rhythmic and Synthetic Processes in Growth* (D. Rudnick, ed.), University Press, Princeton, N.J., pp. 75–109.

SKOCZYLAS, O. (1958). Über die Mitose von *Ceratium cornutum* und einigen anderen Peridineen. *Arch. Protistenk.* **103,** 193–228.

STÅLFELT, M. G. (1921). Studien über die Periodität der Zellteilung und sich daran anschliessende Erscheinungen. *Kunst. Svenska Vetersk. Handb.* **62,** 1–114.

SWEENEY, B. M. (1959). Endogenous diurnal rhythms in marine dinoflagellates. *Intern. Ocean Congress* (M. Sears, ed.), Am. Assoc. Adv. Sci., Washington, D.C., pp. 204–207.

SWEENEY, B. M. (1960). The photosynthetic rhythm in single cells of *Gonyaulax polyedra. Cold Spring Harbor Symp. Quant. Biol.* **25,** 145–148.

SWEENEY, B. M. and J. W. HASTINGS (1958). Rhythmic cell division in populations of *Gonyaulax polyedra. J. Protozool.* **5,** 217–224.

SWEENEY, B. M. and J. W. HASTINGS (1960). Effects of temperature upon diurnal rhythms. *Cold Spring Harbor Symp. Quant. Biol.* **25,** 87–104.

SWEENEY, B. M. and J. M. HASTINGS (1962). Diurnal, tidal and lunar rhythms in the algae. In *Physiology and Biochemistry of the Algae* (R. Lewin, ed.), Academic Press, New York, pp. 687–700.

ZEUTHEN, E. (1958). Artificial and induced periodicity in living cells. *Advances in Biol. and Med. Phys.* **6,** 37–73.

11

Induced Division Synchrony in the Flagellates*

THOMAS W. JAMES, *Department of Zoology, University of California at Los Angeles, Los Angeles, California*

CONTENTS

I	Introduction	323
II	Natural Diurnal Mitotic Activity: a Scheme for Laboratory Studies	324
III	Choice of Organisms for Synchronization Studies	325
IV	The Design of Cycles to Synchronize Cell Division	326
V	Effects on the Fission Time	330
VI	Criteria for Division Synchrony in Flagellates	332
VII	The Synchronous Cycle	333
VIII	The Uptake Rate and Localization of Thioglycolate	337
IX	Biochemical Investigations over the Synchronized Cycle	341
X	Respiration and Photosynthesis of Synchronized Flagellates	344
XI	Inhibitor Effects on the Synchronized Cell Cycle of Flagellates	345
	References	347

I. INTRODUCTION

Even if one examines but a few of the many protozoan organisms that exist on this planet, it is impossible not to be impressed with the consistent cellular design which is their basic theme. The flagellates as a group attract one's attention for a variety of reasons. The most obvious fact is that they contain representatives which are chemorganotrophs as well as photosynthetic autotrophs. In this sense, they illustrate the unity between the animal and plant world. But even more compelling is the fact that these cells can assimilate a few essential inorganic ions and, either through photosynthesis or the utilization of a simple organic compound like acetate or ethanol, can construct facsimiles of themselves which are yielded at the time of cell division. This entire process is not only remarkable but challenging since it taxes the imagination to a degree that makes our investment of thought in giant computers appear puny by comparison. One might even fancy that the passage from acetate and ions to a completed and functioning cell represents in a sense a

* The work described in this paper was supported in part by Grant 5666 from the National Science Foundation, Washington, D.C., and by Grant 536 from the California Cancer Research Committee.

recapitulation of the genesis of life. These cells contain machinery that permit them to carry on all functions with only the most rudimentary energetic precursors. At the same time, unlike bacteria, their size allows access to their cytology over their life cycle. The study of their division thus becomes highly attractive since their biology permits their growth on a well-defined medium in relatively dense cultures. Furthermore, the synchronization of their division and growth can be approached from many points of view. The two methods which are reported in this chapter, i.e., synchronization by means of an applied temperature cycle and synchronization by the use of a light cycle, are a result of recognizing some of the details of their biology.

II. Natural Diurnal Mitotic Activity: A Scheme for Laboratory Studies

The most persistent cycle in nature is the diurnal light cycle which is accompanied by a slightly out-of-phase temperature cycle. It is hard to imagine how cells in the course of their evolution could have ignored this physical relationship and the limitations it imposed. One indication that they did not is that many different cells show a distinct tendency to divide at night. Some well-known cases are cited by Leedale (1959) in his study of the periodicity of mitosis and cell division in the *Euglenineae*. He examined fifteen green and five colorless species in his own work and was able to show that the green species confine their mitosis to the dark period. The colorless species did not show this periodicity. His bibliography cites some thirty papers on the tendency of cells to show diurnal mitotic periodicity. These citations include work on both photosynthetic and nonphotosynthetic unicells, plant tissue cells, and animal tissue cells. Displaying diurnal mitotic activity does not, however, require that the cells divide once each day.

The first supposition that one is tempted to make is that this diurnal variation in mitotic activity is consistent with the basic ecology of day and night. Given a free living cell with a generation time of the order of 24 hr, one assumes that diurnal variations in parameters which profoundly affect the rates of activity in biological systems will make an impression on the growth and division processes of the cell and lead to ordering them in time. This is one way of interpreting this fact of nature. It is not, however, the only way since an alternate view would arise from asking the now famous circular question 'Which came first, the chicken or the egg?' In this instance it would be, 'Which came first, the cell or the diurnal light and temperature cycle?' Barring interplanetary biotic contamination, the answer is, of course, the cycle. It thus appears that cells have had to live with this daily repetition of events through their entire evolutionary history. With this in mind, one is driven to ask, 'How important were such cycles to the early formative periods of cellular origin and evolution?' Only an armchair biologist could answer this question, but it behooves us to recognize this focus of cosmic events on the biology of early cells (Kavanau, 1947).

III. Choice of Organisms for Synchronization Studies

Our decision to use the flagellates, *Astasia longa*, *Euglena gracilis* var. *bacillaris*, and *Polytomella agilis* rested to a considerable extent on the possibility of finding some relationship between natural synchrony and laboratory-induced synchrony using cycles of temperature and light that were consistent with natural conditions. The production and study of division synchrony in cell populations hinge on the use of a dependable cell system. The dependability is a function of the cell species and the environment which it demands for consistent and uniform growth. The specific characteristics of this group are highly favorable.

We also asked, 'What would be the most favorable type of nutrition for a cell if it were possible to bring it into a synchronized state.' It was obvious that a cell which could be grown on a defined medium was desirable, but what should one look for—a cell that is chemorganotrophic with a minimum of organic requirements or a cell that requires a spectrum of organic nutrients? If a cell with a simple nutrition were selected, there would be the advantage of observing synthetic changes over the synchronized cell cycle against a background free of interfering organic precursors. If the other were chosen, namely a cell that required a complex of organic components, specific control would be gained over certain organic pathways by the mere act of deleting one such requirement from the medium. Furthermore, using Woolley's postulate (1950) that an antimetabolite is most effective in the case where the related metabolite must be supplied exogenously, the effective use of inhibiting antimetabolites could be expected. The choice cannot be made without the sacrifice of specific desirable properties that are present in both types. The best system would be some cell type in which biochemical mutants could be obtained. Little has been done to obtain mutants of these flagellates but attempts are now under way using replicate techniques. A third unscientific basis for a choice rests on economics since large batch or pilot plant production of synchronized cells can run to considerable sustained expense where amino acids, nitrogenous bases, vitamins, and other precursors are needed in large quantities. The defined medium which we used for growing flagellates is that devised by Myers and Cramer (1952). Its content, with modifications, is given in Table I. Our decision to use flagellates was tempered by these considerations.

TABLE I. Cramer–Myers' medium (modified) (1952)

Constituents	Concentrations (mg/l.)	Constituents	Concentrations (mg/l.)
$(NH_4)_2HPO_4$	1000	$ZnSO_4 \cdot 7H_2O$	0·4
KH_2PO_4	1000	$Na_3MoO_4 \cdot 2H_2O$	0·2
$MgSO_4 \cdot XH_2O$	200	$CuSO_4 \cdot 5H_2O$	0·02
$Na_3C_6H_5O_7 \cdot 5H_2O$	800	Sodium acetate (anhydrous)	5000
$CaCl_2$	20		
$Fe_2(SO_4)_3 \cdot XH_2O$	3	Vitamin B_1	0·01
$MnCl_2 \cdot 4H_2O$	1·8	Vitamin B_{12}	0·0005
$CoSO_4 \cdot 7H_2O$[a]	1·5	pH 6·8	

[a] Substituted for $CoCl_2 \cdot 6H_2O$ used at 1·3 mg/l.

Astasia longa (*Jahn*) was the subject of our first efforts to synchronize flagellates (James and Padilla 1959; James, 1960; Padilla 1960; Padilla and James, 1960), and it followed a scheme similar to our previous work with *Amoeba proteus* (James, 1954; 1959). *Astasia* is sufficiently large to study its cytology by microscopic methods. It has a mitosis in which chromosomes are seen as distinct entities (Leedale, 1958a). The chromosome number is perhaps similar to the other strains of this species. If Pringsheim (1952) is right in his proposal that *Astasia longa* is derived from *Euglena gracilis* through some natural process of bleaching or mutation, it should have a chromosome number of about 45 (Leedale, 1958a). The fact that various other species of *Astasia* contain eyespots but no chlorophyll suggests that *Astasia longa*, which contains neither of these characters, may have been the result of several aberrations in its progenitors. These relationships provided us with the interesting possibility of examining the synchronization of both *Astasia longa* and *Euglena gracilis*.

IV. The Design of Cycles to Synchronize Cell Division

The methods which have been used to synchronize the division of flagellates will be covered in detail in chapter 22 of this book. It might be worth while briefly to recount the method which was used initially on other cells (James, 1954, 1959) and has now been used to bring both *Astasia longa* and *Polytomella agilis* into division synchrony. It should be recognized that the method may have general applicability. Another method was used initially by Zeuthen and Scherbaum (1954) and by Zeuthen (1958).

A temperature cycle that will bring a cell population into synchronous division is of itself an indication of certain limitations and characteristics of a given cell species. The role of the cell mass, oxygen consumption, and other characteristics of cell growth at various temperatures in the design of a temperature cycle will be covered in the chapter mentioned above. In this consideration we will emphasize the use of the incubation temperature-generation time curve in cycle design. Fig. 1 is a curve of this relationship obtained with *Astasia longa*. Data of this kind can readily be obtained on almost any system.

The first question that one encounters in designing a temperature cycle is, 'What temperatures should be used and how long should each be allowed to act?' Prior to answering this, data must be obtained that will permit one to draw a relatively smooth curve such

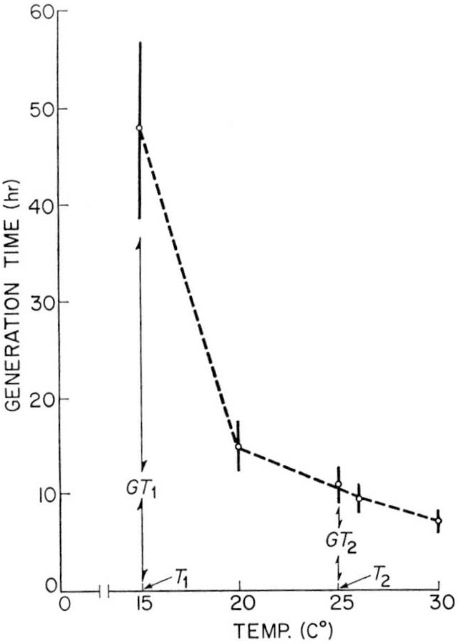

FIG. 1. The relation between generation time and incubation temperature of *Astasia longa* grown on Cramer–Myers' medium. GT_1, GT_2 and T_1, T_2 refer to generation times and temperatures used in the design of a temperature cycle.

as shown in Fig. 1 which relates the average generation time of the cell to the temperature at which it is incubated. If a smooth curve of this function can be obtained, it is possible to make a reasonable guess as to what temperatures will work. There are several reasons why one cannot expect a perfect answer. They will be dealt with later. The first decision that must be made is how long should the complete cycle be. This must be done with a knowledge of the range of generation times over a range of temperatures. The cycle length, CL, will be the sum of the two time periods γ_1 and γ_2 at temperatures of T_1 and T_2 respectively.

$$CL = \gamma_1 + \gamma_2 \tag{1}$$

The duration of γ_1 and γ_2 cannot be greater than the generation times GT_1 and GT_2 at temperatures T_1 and T_2 respectively.

As a first approximation, the fraction of the generation time completed in each time period will be γ_1/GT_1 and γ_2/GT_2. The sum of these two fractions can be set equal to one, indicating that one generation will be completed during the total cycle.

$$\gamma_1/GT_1 + \gamma_2/GT_2 = 1 \tag{2}$$

Using these formulas and Fig. 1, a cycle is constructed by first selecting γ_1 and T_1. Then from the graph, GT_1 is determined and the fraction γ_1/GT_1 is formed. Since the value of γ_2 has been fixed by (1), it can be substituted in (2), which can be solved for GT_2. The temperature T_2 to be used for γ_2 units of time can then be obtained from the graph. Of course, if either γ_1 or γ_2 is made too small or approaches zero, synchrony will not occur.

This operation yields a cycle which is only an approximation since it is based on the assumption that various subperiods of the cell cycle have the same Q_{10}, a supposition that is known not to be the case (James, 1959). The use of this approach on the cells mentioned above has proved effective. In these instances a slight decrease of the high temperature or a decrease of the lower temperature has been sufficient to bring the system into synchrony. Adjustments of times can also be utilized to accomplish the same end. Fig. 2 illustrates the construction of a cycle for *Astasia longa*. In exploring several different cycles on *Polytomella agilis*, Cantor, in our laboratory, has been able to show that, if the low temperature is one at which the cell has near infinite generation time, the cells will still double in the warm period in less time than would be expected at that temperature in logarithmic growth. He has used a cycle of 18 hr at 10°C and 6 hr at 25°C. The generation time at 10°C is so long that it is not measurable; while at 25°C it is 7 hr.

The minimum doubling time on this cycle is 4 hr in the warm period. This experiment gives clear indication that events in the cold period are essential to the functioning of the cycle.

The general impression that one gains from working with this method is that the burst of cell divisions will tend to occur prematurely when the times and temperatures calculated by the formula are used. We have taken this to mean that the cell cycle is not inhibited relative to what would be predicted from our knowledge of the effect of temperature on the generation time. If this effect can be shown to be real in other cases, it should be explored more extensively since it would provide some evidence that the temperature cycle must be adjusted or tuned for maximum effectiveness when used with a specific cell type. The use of (2) also provides a measure of this tendency once a synchronous cycle has been established. Substituting the actual time periods and the generation times found at the temperatures used, i.e., GT_1 and GT_2, into (2) will yield an answer of less than, equal to, or greater than 1. If it is less than 1, the cycle may be assumed to be acceleratory: while if it is greater than one, it would be inhibitory. If the result is equal to one, it can be assumed that no effect occurs.

At first glance this method of appraising the cycle may appear quite attractive, but a short excursion into some of the associated problems will indicate why caution must be exercised in drawing conclusions from any such findings. First, generation times as seen in Fig. 1 can show a range of variability at any given temperature, and even in the best defined systems one may find an improvement in the growth rates as experience is gained in handling a specific organism. This may be the result of adaptive changes in the organism if not in the experimenter himself. For this reason, Padilla and Cook in chapter 22 show how an alternate method based on dry weight production can be used to formulate a cycle. Second, the question arises as to what degree one can make a comparison between temperature-adapted cells and cells that have been subjected to acute changes in temperature even when the latter are maintained in balanced growth (Barner and Cohen, 1955) from one generation to the next. Studies on the effectiveness of a cell to produce material of a given state of free energy from specific substrates and at a given temperature would be necessary before any real comparisons of this kind could be made. The extent of oxidation of acetate and ethanol have been measured in a close relative of *Astasia longa*, i.e., a bleached strain of *Euglena gracilis* var. *bacillaris* (Wilson and Danforth, 1958). Pathways for these processes (Danforth, 1961) have also been examined but this is only a first step in this direction. Heat production and substrate utilization studies such as those carried out by Blum, Podolaski, and Hutchens (1951) on the flagellate *Chilomonas paramecium* would also be essential for drawing comparisons between cells maintained under different growth conditions. Consequently, rigorous comparisons will have to be based on the thermodynamics of cell growth and division using 'open system methods'. This approach, however, is still in its infancy.

V. Effects on the Fission Time

Cycle design cannot be separated from properties of a cell that are related to the division process. To illustrate this, it is necessary to recount the steps that led to the discovery of the need for organo-sulfur compounds to synchronize *Astasia*. Padilla, while working in our laboratory, discovered that *Astasia* grown on a proteose peptone broth showed a fission index (ratio of cells in cleavage to total

Table II. Comparison of generation time, fission time, and fission index in *Astasia longa* grown in various media at 15° and 25°C

Medium	Incubation temperature (°C)	Generation time (hr)	Fission index	Average fission time[a] (min)
2% Proteose peptone	15	52·4 ± 7·3	0·75 ± 0·16	31
	25	11·5 ± 2·7	0·94 ± 0·10	9
Cramer–Myers only	15	42·7 ± 4·5	2·59 ± 1·17	84
	25	12·8 ± 2·3	3·79 ± 1·52	46
Cramer–Myers, cysteine and methionine	15	61·9 ± 7·1	1·58 ± 0·20	81
	25	13·0 ± 2·6	1·62 ± 0·49	11

[a] Calculated by fission time equation $FT = GT \dfrac{\log (FI + 1)}{\log 2}$

The generation time (GT), fission index (FI), and fission time (FT) were measured for cells in the logarithmic phase of growth.

number of cells) which was considerably less than found on Cramer–Myers' medium despite the fact that the generation times on the two media were not markedly different. This led eventually to the derivation of the formula given in the chapter on age distribution which permitted us to calculate the time spent in fission from knowledge of the generation time and the fission index. Table II shows the results of the initial experiments which indicate that the fission time is considerably reduced on the broth medium. The study of Voegtlin and Chalkley (1930), which shows that sulfhydryl compounds markedly reduce the division time in *Amoeba proteus*, suggests a possible source of this difference. The extent to which cysteine and methionine shorten the fission time in *Astasia longa* is shown in Table II. The lowest value at 25°C is on proteose peptone (Difco) and averages 9·0 min, the highest is on Cramer–Myers' medium at 46 min, and Cramer–Myers' plus cysteine and methionine reduces the time to 11 min.

TABLE III. Effects of compounds on fission time (James, 1960)

Compound ($10^{-3}M$)	Fission time: Experiment (Cramer–Myers' + Compound) (hr)	Fission time: Control (Cramer–Myers') (hr)
Methionine	0·18	0·34
Cysteine	0·14	0·34
Glutathione	0·24	0·44
Ascorbic acid	0·19	0·44
Thioglycolate ($10^{-6}M$)	0·22	0·38
Thioglycolate ($10^{-3}M$)	0·19	0·38
Serine	0·34	0·28

Following this discovery, Cramer–Myers' media supplemented with 10^{-3} molar sulfhydryls were tested to see if decreased fission times could be obtained. (See Table III.) The experiments proved that any one of several of these compounds could be used (James, 1960). Cysteine, methionine, glutathione, and thioglycolate all gave this effect. Thioglycolate at a concentration of 10^{-6} molar was effective but slightly better results were obtained at 10^{-3} molar. Ascorbic acid also decreased the fission time, but serine did not. Since thioglycolate is similar in structure to acetate the possibility that the compound was indeed entering the cells was strengthened on the grounds that some common pathway may be involved in its uptake.

Before we report on the synchrony that was obtained, it will be informative to discuss the tentative goals which we sought to achieve. They should be kept in mind while reviewing the results.

VI. Criteria for Division Synchrony in Flagellates

The work on flagellates can best be evaluated on the basis of a number of criteria which were set down as ideals to be hopefully achieved.

1. To obtain a cell population in which the doubling of cell number was confined to the shortest possible period relative to the generation time of the cells.
2. To maintain the cell cycle as nearly intact as possible by keeping it in a one to one relationship with the environmental cycle.

   This would require that the cells divide once in each environmental cycle retaining their capacity to divide in the following cycle at the same relative cycle time.
3. To secure a doubling of the cell number once and only once in each cycle.
4. To keep cells on a repeating cycle in a continuous culture maintained by constant withdrawal and replacement of one-half the culture with new medium at a specific time in the cycle.

The reasons for holding to these criteria are both biological and technical. The technical difficulties encountered in the development of cultures which fulfilled these conditions are presented in chapter 22 (Padilla and Cook) on methods, together with refinements that have considerable importance in biochemical studies. The biological reasons for holding to this pattern of culturing are linked to the nature of the cell cycle. As a guide to the use of the first criterion, it was felt that the ratio between the time of doubling and the length of the synchronized cycle should be as small as possible and tend to approach in size the ratio of cell fission time to the generation time. Relating these ratios offers considerable difficulty when a temperature cycle is used. For example, if cell fission lasts 20 min and the generation time is 10 hr, the best burst time that could be expected would be 1/30 of the 24-hr temperature cycle or about three-quarters of an hour, but one can hardly compare the cycle time to the generation time when the temperature is cycled. The degree to which the second criterion can be successfully met rests on the renewal of medium and the continuation of the temperature cycle on a repetitive basis and is met if the cells retain their capacity to divide each cycle without a diminution in the number that go through division. Results of this kind are given in the chapter on methods. The requirement that there should be a doubling of cells each cycle is reasonable on the grounds that these cells in nature do not divide twice in succession without an intervening interphase of considerable duration relative to the mitotic period. In the case of *Chlamydomonas* which Bernstein (1960) has synchronized, the cells undergo three divisions in succession, and consequently this behavior would have to be taken into account.

VII. The Synchronous Cycle

An example of the synchrony that was first obtained with *Astasia longa* on proteose peptone medium using a cycle of 15°C for 15 hr and 25°C for 8 hr with half-hour transition periods is shown in Fig. 2.

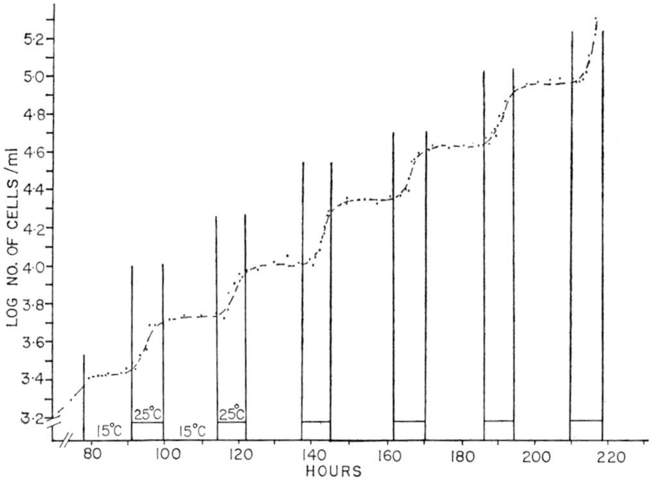

Fig. 2. Growth pattern in synchronous culture of *Astasia longa* maintained on a cycle of 16 hr at 15°C and 8 hr at 25°C. Half-hour transition periods are not shown. Cells were grown on 1 per cent proteose peptone; similar patterns of synchrony are obtained on supplemented Cramer–Myers' medium. (From James, 1960).

Comparable results were consistently obtained in sulfhydryl supplemented Cramer–Myers' medium. In the absence of sulfhydryls, synchrony could not be obtained. Shortening of the transition periods has resulted in a decrease in the burst time or the time of doubling. The shape of the steps in Fig. 2 at the time of cell division has been used to analyze the results by a simple statistical technique (James, 1960). The steps are sigmoid, and therefore the point of inflexion on the step is the time at which the division rate is maximum. The first derivative of this portion of the curve gives an estimate of this time. The burst is probably a normal distribution.

Although this assumption of normality cannot be defended on rigorous grounds, it provides a means of estimating the time of maximum division activity, and furthermore it allows an estimate of the first standard deviation, i.e., a measure of the range of time over which 68 per cent of the divisions occur, or whatever other confidence limit one cares to set. The easiest way of making such an estimate is to reduce the change in cell number to a percentage of the total change, and, by plotting it on the ordinate of probability or probit paper, the time corresponding to 50 per cent completion of the division can be determined on the abscissa. One standard deviation can also be read off the abscissa over the range covered by one probit unit from either side of the 50 per cent point, i.e., probit five. This range can be used as a measure of the degree of synchrony provided a complete doubling of the cell number has occurred in the burst of divisions. To illustrate these results, the data of Fig. 2 have been analyzed in this manner and are presented

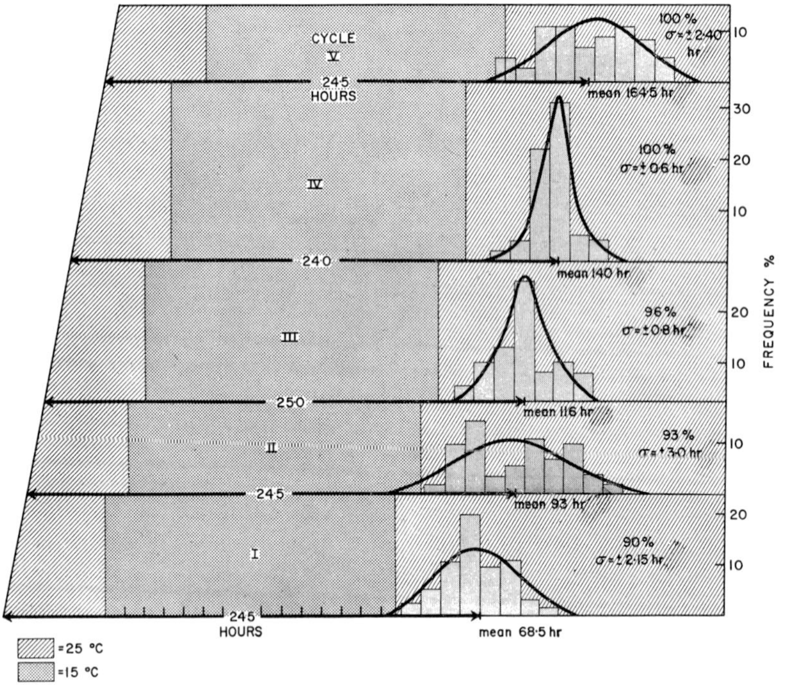

FIG. 3. The analysis of successive division bursts in a culture of *Astasia longa* grown on 1 per cent proteose peptone (Difco). (From James, 1960).

as histograms and fitted distribution curves in Fig. 3. In it, successive cycles are shown and various parameters of the cycles are indicated. The time between successive mean bursts, the standard deviation (σ) of each in hours, and the percentage of cells that completed division in each burst are given. The variation in the spread of each successive burst gives information on how the dispersion of division synchrony may vary from cycle to cycle. It is obvious that this approach can be used not only to monitor the division burst but also could be applied to the change in any constituent over the synchronized cycle to assign a time of maximum synthetic rate. If, however, the range of time over which it increases overlaps the range of the division bursts, adjustments must be made to prevent the burst from being distorted. If the change is expressed per ml of culture, the mean times can be used to estimate the time between synthesis and cell division.

TABLE IV. Burst times and burst duration of synchronized *Astasia longa* (James, 1960)

Cycle	Mean of burst time at culture ages (hr)	σ (hr)	Time in warm period to the mean (hr)	Time between bursts (hr)
I	71·4	± 3·4	4·4	
				24·6
II	96·0	± 2·3	5·0	
				24·0
III	120·0	± 2·5	4·0	
				23·0
IV	143·0	± 1·2	5·0	
				23·2
V	166·2	± 1·2	4·2	
				24·3
VI	190·5	± 1·2	4·5	
				24·5
VII	215·0	± 1·0	5·0	

Grown on Cramer–Myers' medium plus cysteine ($0{\cdot}58 \times 10^{-3}M$) and methionine($1 \times 10^{-5}M$).

The use of sulfhydryl compounds as supplements to Cramer–Myers' medium and their effect on the duration of the synchronous burst of divisions has been appraised in this manner. Some of the data from the work on temperature-synchronized *Astasia longa* (Padilla, 1960; Padilla and James, 1960) and light-synchronized *Euglena gracilis* (Cook, 1960; Cook and James, 1960) have been analyzed by the probit method and the results are given in Tables

IV and V. The method of synchronizing *Euglena gracilis* will be covered in the chapter on methods for flagellates. The data in Table IV obtained on *Astasia* show that the supplemented medium gives consistent division synchrony in which the duration of the burst decreases from cycle to cycle becoming progressively shorter down to σ of $\pm$ 1 hr.

TABLE V. Light-synchronized *Euglena gracilis* on 24-hr cycle (James, 1960)

	Cramer–Myers' medium		Cysteine and methionine supplemented	
Cycle	Mean age at burst (hr) $\pm \sigma$	Inter-burst time (hr)	Mean age at burst (hr) $\pm \sigma$	Inter-burst time (hr)
I	35·0 $\pm$ 2·8		35·0 $\pm$ 1·5	
		25·2		24·6
II	60·2 $\pm$ 2·6		59·6 $\pm$ 1·5	
		24·3		23·7
III	84·5 $\pm$ 3·0		83·3 $\pm$ 1·3	
		24·0		24·3
IV	108·5 $\pm$ 2·7		107·6 $\pm$ 1·6	

Cycle of 16 hr light (130 ft-candles) and 8 hr dark.

Table V shows a comparison of light-synchronized *Euglena gracilis* in the presence and absence of cysteine and methionine supplemented Cramer–Myers' medium. The light cycle used in this case consisted of 16 hr of light at 130 ft-candles and 8 hr of darkness. The cells divided in the dark. Temperature was held constant at 20°C. In these studies the carbon source was CO_2. It can be seen that these compounds at a concentration of $0{\cdot}58 \times 10^{-3}M$ for cysteine and $1 \times 10^{-5}M$ for methionine decrease the duration of the division burst by about 50 per cent. Synchronized cultures of both of these organisms have been maintained for extended periods by daily dilution with fresh nutrients.

Light cycles have been used to synchronize other photosynthetic organisms that will be considered elsewhere in this book. Another green flagellate in which synchrony has been induced by a light cycle is *Chlamydomonas moewusii*. This work has been carried on by Bernstein (1960), and in it a 12-hr light, 12-hr dark regime was used. Since this organism undergoes two or three divisions in rapid succession the product of each burst leads to more than simple doubling in number. In fact, the data show that an eight-fold increase occurs in each cycle provided the population density is not too high. Smaller increases of less than doubling occur on successive cycles as the population exhausts the medium.

VIII. The Uptake Rate and Localization of Thioglycolate

Thus far we have discussed two experimental findings showing that sulfur compounds, particularly sulfhydryls, have an effect that is associated with the fission process. First, these compounds decrease the fission time, and, second, their addition to cultures of *Astasia* permit their synchronization or, as in the case of *Euglena*, they shorten the synchronized burst of divisions. These effects would indicate that the compounds are operating on the same process and in some way accelerating them. The effect does not appear to change the generation times of the organisms appreciably but this is undoubtedly because the generation times are estimated in hours while the fission times are estimated in minutes. A 50 per cent decrease in fission time will be such that it would change the generation time by less than 3 per cent. To gain further information about the effect of sulfur, experiments were carried out on the rate of uptake of sulfur-labeled thioglycolate over the synchronized cell cycle (James, 1961). The results of such experiments are given in Figs. 4 and 5. The rates of uptake were measured on aliquots of cells removed from the synchronizing apparatus and maintained in ^{35}S thioglycolate for 1 hr at a temperature of 15°C in one series and at 26·5°C in another series. In other words, the uptake was measured at both temperatures regardless of the temperature at which the cells were being maintained at the time of removal. Rate measurements were made by placing a washed and counted number of cells on Millipore SS filter discs, and after drying they were counted with a Geiger counter. Incubation of 1-hr duration at a series of points in the cycle was established as a period in which linear uptake was maintained. Figs. 4 and 5 illustrate the way in which the rate of uptake changes as a function of the synchronized cell cycle. The range of one standard deviation for each time is designated by the length of the bar. At both incubation temperatures, the rate of uptake increases at a time prior to the burst of cell divisions. The increase in rate in both cases is greater than a doubling. Thus, it must be assumed that this change in the rate of uptake over the synchronized cycle is a cell property that persists irrespective of the temperature at which the rates are measured.

These findings are supported by additional work on the localization of sulfur label from thioglycolate using radioautography. Although autographs over the synchronized cycle give only qualitative hints of a difference in rate of uptake, they do show a spatial localization of the sulfur label in the anterior portion of the organism. The degree of localization can be appraised from the photograph seen as Plate 1 of Fig. 6. Furthermore, cells in the act of fission show that the anterior portions retain the label (Plate 2 of Fig. 6).

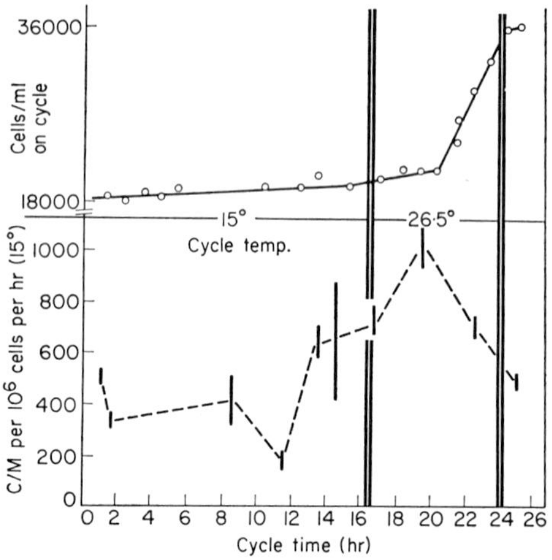

FIG. 4. Upper curve shows the increase in cell number in the parent synchronized culture using the temperature cycle indicated. Lower curve shows the rate of uptake of radioactive thioglycolate by cells removed from the cycle and incubated at 15°C for 1 hr at various times in the cycle. The length of the bars give ± one standard deviation at each test point. (From James, 1960.)

The cytology of *Astasia* is similar to other flagellates. The organism has a single flagellum. It is attached at the anterior end through an opening which communicates with the reservoir. The flagellum bifurcates and attaches to basal granules on the posterior wall of the reservoir. Division of this organism consists of a series of events which begin with a migration of the cell nucleus from a mid-central region to an anterior position at the back wall of the reservoir. Nuclear division occurs in this position and is marked by an elongation of the endosome or nucleolus. The chromosomes, according to Leedale (1958b), are in contact with the nuclear membrane. The nuclear membrane stays intact throughout the division process. The elongation of the endosome and the surrounding nuclear membrane leads to the separation of the two sets of chromosomes. Fission begins at the anterior tip of the cell and can be identified by a slight indentation which continues longitudinally. In the final stage of fission the two new organisms are attached only by their posterior tips. The breakage of this attachment constitutes the completion of cleavage.

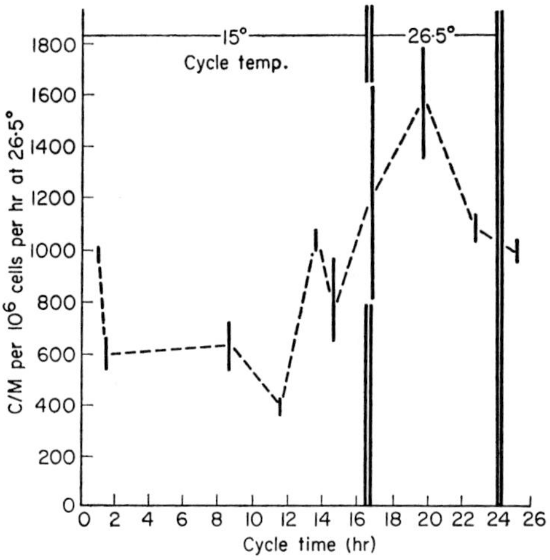

FIG. 5. The rate of uptake of ^{35}S thioglycolate by cells removed from the cycle and incubated at 26·5°C. (From James, 1960.)

Since the radioactive label appears to be superimposed on the anterior region, it may be associated with the reservoir or the basal granules which are the point of attachment of the bifurcated flagellum. The sulfur label is not removed by washing with ethanol and therefore seems to be well fixed to the structure. The possibility that it is in a nuclear structure is excluded to some extent by the fact that it shows a localization in the anterior tip prior to the time the nucleus moves forward.

The fate of the sulfur label is of primary importance. It is ultimately incorporated into proteins. The recent studies on synchronized *Astasia* by Padilla and Blum (1961) have shown that 80 per cent of the label is found in the protein fraction. The relationship between structure and the position of the label cannot be assigned unequivocally. If the basal granules are homologous to the centrioles of other cells, the accumulation of sulfur would be expected in these structures in the light of the work by Kawamura (1960) in which it was shown that the centrospheres of marine eggs are rich in SH groups. Leedale

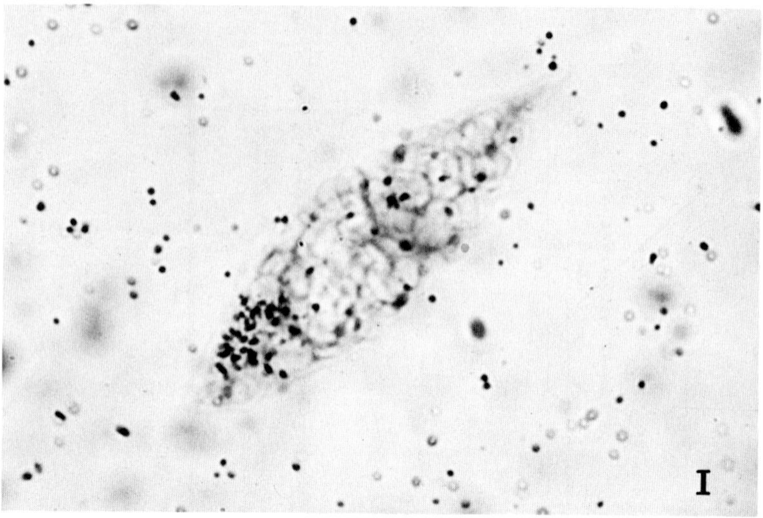

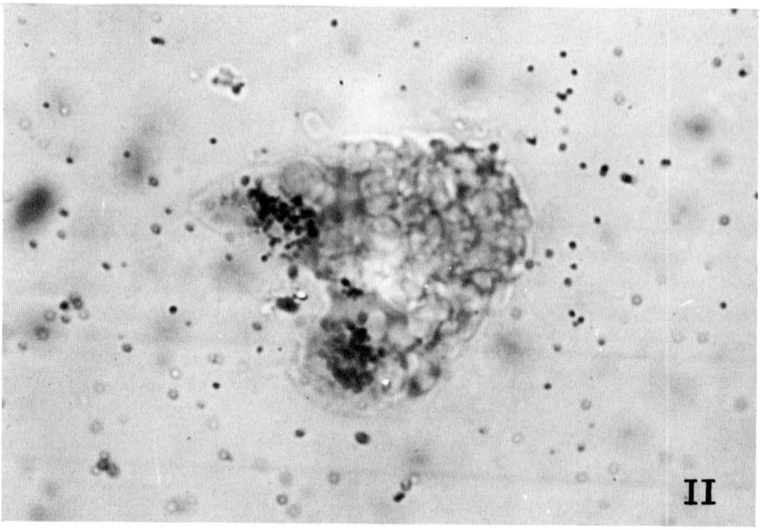

FIG. 6. Plate I—Radioautograph of a predivision (19 hr) *Astasia longa* showing ^{35}S label at anterior end.

Plate II—Radioautograph of a division figure of *Astasia longa* showing label in region of nuclear division. (From James, 1960.)

(1958b), however, maintains that nuclear division in the *Euglenineae* is not related to the basal granules. This is in contrast to the illustration of this relationship given by Lackey (1934) in a colorless flagellate *Distigma*. The possibility that the sulfur may be associated with a peptide–nucleotide fraction such as found by Hase *et al.* (1960) on light-synchronized *Chlorella* is also not excluded.

The change in the uptake rate with the cell cycle could be linked to the increase in the transport system such as postulated by Lark and Lark (1960) to explain their findings on the changes of methionine transport rate in synchronized cultures of the bacterium *Alcaligenes fecalis*. They consider this phenomenon to be related to a doubling of transport mechanisms in the cells. In *Astasia longa* the doubling of the reservoir occurs prior to cytokinesis and thus might be related to transport of materials via this route into the cytoplasm.

IX. Biočhemical Investigations over the Synchronized Cycle

When it was realized that cell division could be synchronized in a population of cells (Hotchkiss, 1954; James, 1954; Scherbaum and Zeuthen, 1954), it was at first considered to be a discovery which would permit one to lay hold of the problem of cell division and with a few well-planned experiments sketch out the basic plan that a cell follows in growth and division. The life history of a cell as charted through microscopic observations gives one the impression that the neat sequence of events must be accompanied by an equally neat series of biochemical transformations, and therefore a one to one correlation with structural changes should be evident. This attitude, together with a faith in the fundamental similarity of cells from one biological kingdom to the next, has met with sobering contradictions of facts, such as the differences in the course of deoxyribonucleic acid synthesis over the cell cycle (McDonald, 1958; Prescott, 1960; Scherbaum *et al.*, 1959; Walker and Mitchison, 1957). Findings of this kind have prompted some to relegate the techniques of synchronization to the role of a clever, but not too useful, stunt.

It was thought that all that need be done was to carry out analysis for the change in cellular constituents over the synchronized cycle, and each in sequence would display its importance to cell division as the cell proceeded to grow and divide. Despite the fact that this is still one of the most logical approaches to the problem, it is plagued with serious difficulties. The first objection that can be levied against synchronizing procedures is that the operation which is effective in synchronizing divisions is itself the source of non-steady state conditions that do not permit an easy access to rate phenomena, and thus the true relationship between components is lost. But what is often missed in this argument is that any relationship which can be distorted and still leaves the cell capable of division is a relationship that is not essential to cell division. The argument against non-steady state systems is often brought against the methods which use temperature changes to bring about division synchrony. There is little doubt that the temperature-synchronized cell cycle is not identical to what might be called the normal cell cycle obtained under steady state growth conditions. Awareness of this fact is a poor reason to abandon such methods or to insist that other methods are superior. The use of filtration methods (Abbo and Pardee, 1960; Anderson and Pettijohn, 1960; Maruyama and Yanagita, 1956), where the selection of size classes becomes the basis for obtaining a synchronized population, appears ideal until it is recognized that cell size is poorly correlated to the generation time of a cell. The variation of generation times in a population of cells is a statistical fact (Powell, 1956), as is the variation of size of cells at the time of division (James and Read, 1957). Consequently, one is faced with an uncertainty irrespective of the method employed, and it becomes apparent that patterns of synthesis in the growth and division of a cell type must be explored by alternate methods. In any event, more attention should be paid to the criteria used to evaluate the degree of synchrony whatever method is used. Too often a few wiggles in a growth curve which do not represent an acute doubling in cell number are the subject of extensive biochemical studies.

It would be a pleasure to report that a single species of flagellate had been examined by two alternate methods, but this is not the case. However, the two flagellates that have been investigated in parallel are very close relatives despite the fact that one, *Euglena gracilis*, is a photosynthetic autotroph, while the other, *Astasia longa* is a chemorganotroph. As shown previously, both of these organisms have been synchronized to the same degree of refinement, one by light and the other by temperature on 24 hr cycles. Analyses for the change in similar biochemical constituents have been carried out on both cells over the synchronized cell cycle. The cycle of temperature used on *Astasia* was 7·5 hr at 26·5°C and 16·5 hr at 15°C. Each time contains a transition period of less than 30 min. The light cycle used on *Euglena gracilis* was given in an earlier section of this chapter.

The patterns of synthesis of a number of cellular constituents are all placed on a common scale of amounts and the corresponding cycle times are indicated in Fig. 7. Despite the fact that these curves are idealized from the real data, they can be accepted with a high degree of confidence since the same general picture has been obtained repeatedly on both cells. The time course of change for ribonucleic acid (RNA), deoxyribonucleic acid (DNA), and cell number in *Astasia* (upper section of Fig. 7) has not been resolved beyond the range delineated on the graph. Work by Blum and Padilla (1962) gives a better indication of the separation between the nucleic acid increase and the time of cell number increase. (See chapter 22.) The earlier data presented in Fig. 7 suggests that the nucleic acid synthesis precedes cell division by approximately an hour when the above cycle is used. Padilla (1960) also used longer warm periods which caused the onset of RNA synthesis to occur approximately 3 hr before cell divisions begin. Protein synthesis takes place in a stepwise manner, reaching a first plateau at about the twelfth hour of the cold period. It is completed several hours later during the warm period. This same pattern is also followed for the rate of sulfhydryl uptake as

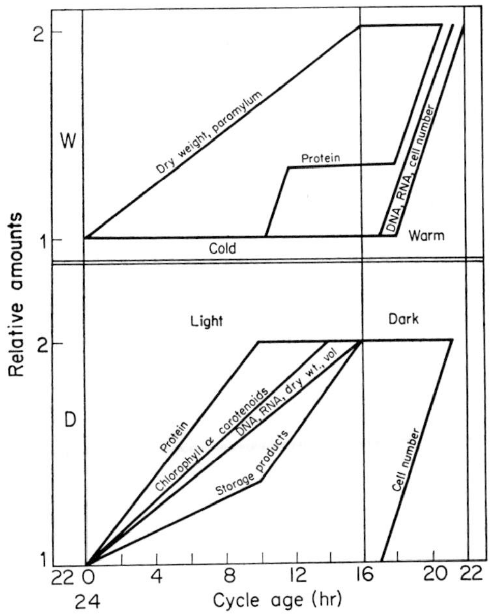

FIG. 7. Upper section—The sequence of synthesis of cellular constituents over the temperature-synchronized cell cycle of *Astasia longa*. The relative amounts over the cycle and the warm and cold periods are indicated. Lower section—The sequence of synthesis of cellular constituents in the light-synchronized cell cycle of *Euglena gracilis*. The light and dark periods are indicated.

shown previously in Figs. 5 and 6. Dry weight and paramylum synthesis show a linear increase in the cold period.

In light-synchronized *Euglena gracilis* (Cook, 1960; 1961a, b; Cook and James, 1960) (lower section of Fig. 7) a completely different pattern of synthesis is followed. Because nucleic acid increase is linear from the onset to completion of the light period, protein synthesis occurs in the first half of the light period, while chlorophyll *a* and the carotenoids take an intermediate linear rate of synthesis. Storage products, the main fraction of which is paramylum, increase at a slow rate in the early portion of that period. Cell number increase occurs only in the dark period. This is consistent with other work by Cook (1960) in which he has demonstrated conclusively that white light of low intensity inhibits growth rates of a variety of organisms both green and colorless. Control aliquots of *Astasia longa* removed from the synchronizing apparatus will go through a complete doubling at the temperature of the warm period provided they are kept in total darkness. Aliquots kept in low intensity light do not double in number. The nucleic acids and other constituents have also been investigated by Bernstein in *Chlamydomonas* (personal communication). He notes that during the first light period RNA, DNA, protein, and dry weight all increase exponentially with RNA synthesis beginning somewhat earlier. RNA synthesis, however, is then interrupted for about 2 to 4 hr at the beginning of the dark period while the other components continue to increase exponentially. After this, RNA resumes its exponential increase.

X. Respiration and Photosynthesis of Synchronized Flagellates

In the first work on the synchronization of *A. longa* (Padilla and Wilson, unpublished data) measurements of the respiration rates in the cold and warm period led us to propose that during the cold period there was a gradual increase in the respiratory machinery per cell. Subsequent work (Wilson and James, unpublished data) has shown that better controlled conditions are necessary because this transient change apparently depends on oxygen tension in the cell culture vessel.

The exact reason for this type of behavior is now being explored by Wilson using an oxygen electrode. This instrument has several advantages over the Warburg respirometer in work of this kind. First, the time of equilibration can be greatly reduced; second, the cells will go through a complete division burst in the respirometer vessel, a condition which we have not been able to duplicate in the Warburg; third, the respiration rates can be easily measured as a function of oxygen tension. This work has shown that respiration rate changes can take a variety of routes over the synchronized cycle depending on the substrate concentration and gas conditions in the parent culture (Wilson, unpublished data).

Preliminary studies of photosynthesis over the synchronized cycle have been carried out by Cook (1960) on *Euglena gracilis* and the rate appears to be constant in the light period as indicated by the uptake of ^{14}C bicarbonate. Incorporation into specific entities is now being explored. Bernstein (personal communication) is also carrying on studies of photosynthesis in *Chlamydomonas*. Results on this organism will be interesting since it is an obligate photosynthetic autotroph.

XI. Inhibitor Effects on the Synchronized Cell Cycle of Flagellates

Little has been done with inhibitors on synchronized flagellates. There are, however, several interesting short studies that would indicate more should be done. For example, the effect of thioglycolate on the shortening of the fission time as mentioned previously indicates that the similarity in structure of acetate and this compound could be one reason for its ease of entry and its effect. Therefore the effect was explored on ethanol grown cells. As a control, acetate and ethanol grown cells were subjected to mercaptoethanol because of its similarity to ethanol. Ethanol grown cells show the same response to the thioglycolate that acetate grown cells do, and mecaptoethanol is inhibitory in both media. Another series of experiments were carried out using the cysteine analogue allylglycine. No definite effect could be shown with this compound in terms of a decrease in the number of cells that go through the division burst when it was added in the cold period. Comparable experiments were made with chloramphenicol at saturating concentrations. Even when added at the beginning of a 24-hr cycle, the population went through a complete division burst paralleling the control culture.

Padilla and Blum (1961) have shown an interesting effect with 8-azaguanine. This compound completely interrupts the division burst when added at any time in the first 10 hr of the cold period, and becomes less effective in stopping division when it is added at later times in the cold period. If it is added in the warm period, the cells divide and there is no effect on the RNA or protein synthesis per cell. Logarithmic growth occurs normally, however, when measured at either of the two temperatures that were used to synchronize the cells. These authors interpret the results to mean that there is a guanine nucleotide dependent step in the cold period. This work is an excellent demonstration of the existence of cold period processes that are essential to division.

Astasia longa and *Euglena gracilis*, being close relatives with a common ancestor, would be expected to display similar patterns of synthesis over their growth and division cycle, but to establish this fact one would have to determine these patterns under comparable steady state conditions. This is difficult, if not impossible, to carry out by present techniques. We can state, however, that when different methods are used to synchronize their cell growth and division, different patterns arise. Two points of view can be taken: (1) these differences are a result of inherent fundamental differences in these organisms, or (2) they are the result of the synchronizing procedure. If the second view is taken, it will tend to support the existence of data in the literature in which different patterns of synthesis are reported for the same cell (McDonald, 1958; Prescott, 1960; Scherbaum and Rasch, 1957; Walker and Mitchison, 1957). It is not the purpose of this chapter to rationalize such differences but a few concluding remarks should be made to specify the postulates that have grown out of our work to date.

One cannot regard the activities and patterns of synthesis of cells like the flagellates without recognizing the applicability of the concept of equifinality (von Bertalanffy, 1950) to the cycle of cell growth and division. To paraphrase this idea, one might suppose that the cell on the verge of division is in a state which contains the end products from a host of synthetic pathways each essential to the act of division. It is not difficult to suppose that the routes by which this end has been accomplished are of little or no importance to an action that may occur once the final state has been achieved. Indeed, the only stratagem that the cell need follow would be to decide whether or not the essential units have all been completed. The routes by which these have been completed could take a variety of forms, each of which may have a priority on energy and precursors depending on the environment under which it has carried out the synthetic steps. The statistical variation of cell properties, such as their generation times and their range of volumes, is an indication of this flexibility of the cell. The action of any given synchronizing procedure would be to change the relative priorities of pathways and regulate their sequence of completion. Another sequence obtained by a different treatment would still permit the cells to divide. The value of synchronizing procedures may turn out to be that through their application we will come to recognize not the fixity of one biochemical sequence to another in growth and division, but their capacity to employ alternate pathways. A fitting analogy to this capacity of a cell would be a chess game where checkmate is the final state. The cell has many gambits by which it can finish out the game. Who can say which move won the game?

REFERENCES

ABBO, F. E. and A. B. PARDEE (1960). Synthesis of macromolecules in synchronously dividing bacteria. *Biochim. et Biophys. Acta* **39,** 478.

ANDERSON, P. A. and D. E. PETTIJOHN (1960). Synchronization of division in *Escherichia coli. Science* **131,** 1098.

BARNER, H. D. and S. S. COHEN (1955). Synchronization of division of a thymineless mutant of *Escherichia coli. Federation Proc.* **14,** 177.

BERNSTEIN, E. (1960). Synchronous division in *Chlamydomonas moewusii. Science* **131,** 1528.

BLUM, J. J. and G. M. PADILLA (1962). Studies on synchronized cells: The time course of DNA, RNA, and protein synthesis in *Astasia longa. Exptl. Cell Research* **28,** 512.

BLUM, J. J., B. PODOLASKI and J. O. HUTCHENS (1951). Heat production in *Chilomonas. J. Cellular Comp. Physiol.* **37,** 403.

COOK, J. R. (1960). Light-induced division synchrony of *Euglena gracilis*, and the effect of visible light on related cells. Ph. D. Thesis, Univ. Calif., Los Angeles.

COOK, J. R. (1961a). *Euglena gracilis* in synchronous division. I. Dry mass and volume characteristics. *Plant and Cell Physiol.* **2,** 199.

COOK, J. R. (1961b). *Euglena gracilis* in synchronous division. II. Biosynthetic rates over the life cycle. *Biol. Bull.* **121,** 277.

COOK, J. R. and T. W. JAMES (1960). Light-induced division synchrony in *Euglena gracilis* var. *bacillaris. Exptl. Cell Research* **21,** 583.

DANFORTH, W. F. (1961). Oxidative assimilation of acetate by *Euglena.* Carbon balance and effects of ethanol. *J. Protozool.* **8,** 152.

HASE, E., S. MIHARA and H. TAMIYA (1960). Role of sulfur in the cell division of *Chlorella*, with special reference to the sulfur compounds appearing during the process of cell division. I. *Plant and Cell. Physiol* **1,** 131.

HOTCHKISS, R. D. (1954). Cyclical behaviour in pneumococcal growth and transformability occasioned by environmental changes. *Proc. Natl. Acad. Sci., U.S.* **40,** 49.

JAMES, T. W. (1954). Quantitative studies on the ribonucleic acid content in *Amoeba proteus* in relation to the nucleus and the division cycle. Ph.D. Thesis, Univ. Calif., Berkeley, Calif.

JAMES, T. W. (1959). Synchronization of cell division in Amoebae. *Ann. N. Y. Acad. Sci.* **78,** 501.

JAMES, T. W. (1960). Controlled division synchrony and growth in protozoan microorganisms. *Ann. N. Y. Acad. Sci.* **90,** 550.

JAMES, T. W. (1961). Uptake and localization of thioglycolate in the synchronized flagellate *Astasia longa*. *Pathologie-Biologie* **9,** 510.

JAMES, T. W. and G. M. PADILLA (1959). Physiological and size changes in protozoan cells in response to incubation temperatures. In *Proc. of the 1st Nat. Biophys. Con.* (H. Quastler and H. J. Morowitz, eds.), Yale University Press, New Haven, p. 694.

JAMES, T. W. and C. P. READ (1957). The effect of incubation temperature on the cell size of *Tetrahymena pyriformis*. *Exptl. Cell Research* **13,** 510.

KAVANAU, J. L. (1947). Some physico-chemical aspects of life and evolution in relation to the living state. *Am. Naturalist* **81,** 161.

KAWAMURA, N. (1960). Cytochemical and quantitative study of protein-bound sulfhydryl and disulphide groups in eggs of *Arbacia* during the first cleavage. *Exptl. Cell Research* **20,** 127.

LACKEY, J. B. (1934). Studies in the life histories of *Euglenida*. IV. A comparison of the structure and division of *Distigma Proteus Ehrenberg* and *Astasia Dangeardi Lemm*. *Biol. Bull.* **67,** 145.

LARK, K. G. and C. LARK (1960). Changes during the division cycle in bacterial cell wall synthesis, volume, and ability to concentrate free amino acids. *Biochem. et Biophys. Acta* **43,** 520.

LEEDALE, G. F. (1958a). Mitosis and chromosome numbers in the *Euglenineae* (*Flagellata*). *Nature* **181,** 502.

LEEDALE, G. F. (1958b). Nuclear structure and mitosis in the *Euglenineae*. *Arch. Mikrobiol.* **32,** 32.

LEEDALE, G. F. (1959). Periodicity of mitosis and cell division in the *Euglenineae*. *Biol. Bull.* **116,** 162.

MARUYAMA, Y. and T. YANAGITA (1956). Physical methods for obtaining synchronous cultures of *Escherichia coli*. *J. Bacteriol.* **71,** 542.

McDONALD, B. B. (1958). Quantitative aspects of deoxyribose nucleic acid (DNA) metabolism in an amicronucleate strain of *Tetrahymena*. *Biol. Bull.* **114,** 71.

MYERS, J. and M. CRAMER (1952). Grown and photosynthetic characteristics of *Euglena gracilis*. *Arch. Mikrobiol.* **17,** 384.

PADILLA, G. M. (1960). Studies on the synchronization of cell division in the colorless flagellate *Astasia longa*. Ph.D. Thesis, Univ. Calif.

PADILLA, G. M. and J. J. BLUM (1961). Inhibition of synchronized cell division by 8-azaguanine. *Abstracts 1st Ann. Meeting Am. Soc. Cell Biology*, p. 160.

PADILLA, G. M. and T. W. JAMES. (1960). Synchronization of cell division in *Astasia longa* on a chemically defined medium. *Exptl. Cell Research* **20,** 401.

POWELL, E. O. (1956). Growth rate and generation time of bacteria, with special reference to continuous culture. *J. Gen. Microbiol.* **15,** 492.

PRESCOTT, D. M. (1960). Relation between cell growth and cell division. IV. The syntheses of DNA, RNA, and protein from division to division in *Tetrahymena*. *Exptl. Cell Research* **19,** 228.

PRINGSHEIM, E. G. (1952). The loss of chromatophores in *Euglena gracilis*. *New Physiologist* **45,** 52.

SCHERBAUM, O. H., A. L. LOUDERBACK and T. L. JAHN (1959). DNA synthesis, phosphate content and growth in mass and volume in synchronously dividing cells. *Exptl. Cell Research* **18,** 150.

SCHERBAUM, O. H. and G. RASCH (1957). Cell size distribution and single cell growth in *Tetrahymena pyriformis* GL. *Acta Pathol. Microbiol. Scand.* **41,** 161.

SCHERBAUM, O. H. and E. ZEUTHEN (1954). Induction of synchronous cell division in mass cultures of *Tetrahymena pyriformis*. *Exptl. Cell Research* **6,** 221.

VON BERTALANFFY, L. (1950). The theory of open systems in physics and biology. *Science* **111,** 23.

VOEGTLIN, C. and H. W. CHALKLEY (1930). The chemistry of cell division. I. The effect of glutathione on cell division in *Amoeba proteus*. *Public Health Repts. U.S.* **45,** 3041.

WALKER, P. M. B., and J. M. MITCHISON (1957). DNA synthesis in two ciliates. *Exptl. Cell Research* **13,** 167.

WILSON, B. W. and W. F. DANFORTH (1958). The extent of acetate and ethanol oxidation by *Euglena gracilis*. *J. Gen. Microbiol.* **18,** 535.

WOOLLEY, D. W. (1950). *A Study of Antimetabolites.* Wiley, New York, pp. 82–118.

ZEUTHEN, E. (1958). Artificial and induced periodicity in living cells. *Advances in Biol. and Med. Phys.* **6,** 37.

ZEUTHEN, E. and O. H. SCHERBAUM (1954). Synchronous divisions, in mass cultures of the ciliate protozoon *Tetrahymena pyriformis*, as induced by temperature changes. In *Recent Developments in Cell Physiology* (J. A. Kitching ed.), *Proc. 7th Symp. Colston Research Soc.*, Butterworths, London, p. 141.

12

Division Synchrony in Yeasts

D. H. Williamson, *The Cell Biology Department, John Innes Institute, Bayfordbury, Hertford, Hertfordshire, England*

CONTENTS

I	Introduction	351
II	Special Problems in Synchronizing Yeast Cultures	352
III	'Natural' Synchrony	357
IV	Artificially Induced Synchrony	358
	1. Methods based on starvation	358
	2. Synchronous division induced by X-rays	364
V	The Mechanisms Underlying Synchronizing Processes	366
VI	Physiological Behaviour of the Synchronized Yeast Cell	368
	1. The lag phase	368
	2. The cell cycle	369
	A. Growth in volume and mass	369
	B. Nucleic acids	371
	C. Protein synthesis and total nitrogen content	372
	D. Oxygen consumption	373
	E. Cell size and generation time	375
VII	Conclusions	376
	References	378

I. Introduction

The yeasts possess certain attributes that have long made them a rewarding subject of study. Most of them grow abundantly on comparatively simple media, some requiring no more than a suitable mixture of salts, ammonium ions and sugar. Their thick cell wall endows them with considerable mechanical robustness, while their moderate size facilitates both manipulation and microscopic examination. Consequently yeasts are found to have an increasing use as experimental material for the study of cell biology, and it is not surprising that current interest in the concept and techniques of synchronous division should have found expression among this group of organisms. Indeed, it would be surprising if this were not so, for the readiness with which certain strains become slightly synchronous simply on ageing in culture, or on starvation, has been known for many years.

Nevertheless, collectively, the yeasts are by no means ideal subjects for this field of enquiry, for many species exhibit properties which seem wholly incompatible with easy synchronization. This is particularly true of the majority which reproduce by budding, for this mode of division is often associated with an undesirable heterogeneity in cell size and other characters. Equally formidable problems are raised by the common tendency to form chains or loose aggregates of cells, which hinders the making of cell counts. However, these difficulties have been overcome by several authors who, with but one exception, have used strains of the budding yeast *Saccharomyces cerevisiae* as the subject of their studies.

Vegetative reproduction by budding seems to be restricted to yeasts and certain other fungi, and at first sight this mode of division may seem to have little in common with the more familiar cellular act of dividing by binary fission. In fact however, in their overall physiology, these two reproductive processes are fundamentally similar. Nevertheless, the budding habit, *per se*, confers upon the yeasts certain general properties which are not found, or at least are not so marked in other cell types, and which have a direct bearing on the problems of inducing and maintaining synchrony. These aspects of the budding process are considered in more detail in the following paragraphs.

II. Special Problems in Synchronizing Yeast Cultures

Stationary phase cultures of most strains of *Saccharomyces cerevisiae* invariably contain a number of single 'resting' organisms free of attached buds or daughter cells. The proportion of such individuals varies widely among different strains as well as with the type of medium employed, but under appropriate conditions, mature cultures of certain strains may exist almost entirely in this form.

When a resting cell of this type is inoculated into a growth medium, it undergoes the characteristic sequence of morphological changes outlined in Fig. 1. The first bud emerges only after an interval which, though variable, is generally of the order of 1–2 hr. Outwardly there is little change in the cell during this period, for its mass and volume normally remain largely unchanged, and little or no synthesis of macromolecules takes place. Presumably, however, some preparation for active growth is carried out, and by loose analogy with the behaviour of conventional random cultures, this preparatory period will be referred to as the 'lag phase'.

Once the synthesis of macromolecules starts, a need presumably arises for some increase in the volume of the cell, and this is met by the formation at one point on the cell wall of a tiny outgrowth or bud, which progressively expands, being filled all the time with cytoplasmic materials. Meanwhile the parental nucleus is duplicated, and at a late stage in the cell cycle, one of the resulting nuclei moves into the bud. Soon afterwards, by the deposition of a double layer of cell wall across the narrow 'neck' where the bud first appeared, the cell cycle is completed and the parent and fully grown bud become physiologically independent of one another.

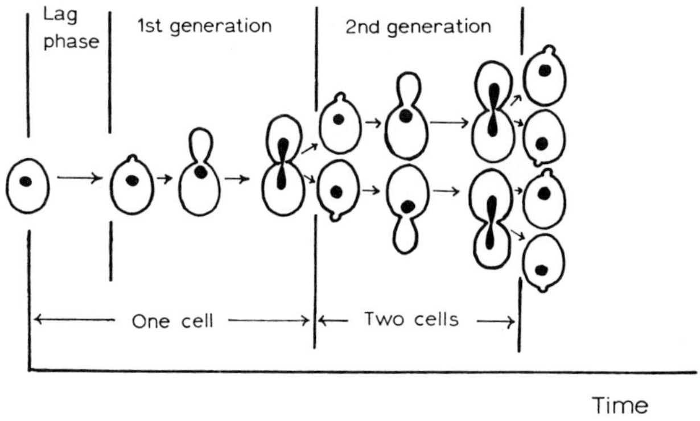

FIG. 1. Diagrammatic representation of the growth and division of an initially resting cell on inoculation into a growth medium. The central body in each cell represents the nucleus.

In practice, precise determination of the moment of division is often difficult. The cell wall laid down between parent and bud is not always easily discernable, even with phase contrast optics, and the neck between them changes little during the growth period of the cell (Bradley, 1955). Thus the onset of division is not marked, as in many other cell types, by some distinctive change such as the appearance of an equatorial furrow, and from an early stage in the development of the bud, the growing organism superficially resembles two adherent individuals. This would be of little consequence if the daughter and parent cells separated physically as soon as division occurred. Unfortunately this is rarely the case, for in most strains the two units tend to remain in firm physical contact for an appreciable period after division, and in extreme, though by no means rare instances, this leads to the formation of chains. Obviously such behaviour poses a serious problem for the observer wishing to distinguish and count whole cells, and various devices have been employed to overcome it. For instance, the bud is commonly counted as a whole cell as soon as it reaches a certain arbitrary size relative to its parent. Provided they are clearly defined, conventions of this type at least have the merit of allowing the demonstration of periodic growth in a population. However, they may be misleading if the arbitrarily chosen point of 'division' does not coincide with the actual moment at which parent and daughter cells become physiologically independent.

Fortunately, recent developments have provided what appears to be a satisfactory solution of this problem, at least in the case of actively growing cells. This rests on the fact that a close time relationship exists between the completion of division and the appearance of the next generation of buds. This was first noted by Burns (1956), who examined several strains of *Saccharomyces cerevisiae*. He found that, provided at least one new (second generation) bud was present, the components of a parent–daughter pair could be readily separated in an undamaged state by micromanipulation. In the absence of such a bud, separation in this way proved impossible. Williamson and Scopes (1960) confirmed this observation, using a gentle rolling action between slide and coverslip to part the cells and, more recently, brief exposure to a weak source of ultrasonic vibrations (100 W, 650 kc/sec) has been found to achieve the same effect (Scopes, personal communication). On applying this ultrasonic treatment to cell pairs in samples taken from synchronously dividing cultures, it has been found that prior to the emergence of a second generation bud, the only effect, even of prolonged exposure, is to cause the contents of the cell to coagulate; the parent and daughter components still remain firmly attached. However, once either component has formed a new bud (and it need be only just emerging) a few seconds treatment is sufficient to separate the parent and daughter units in a superficially undamaged condition. Thus there are good grounds for believing that the appearance of a second generation bud signals the completion of division and, as indicated in Fig. 1, this convention will be used throughout this chapter. Apart from the ease with which an emerging bud can be detected under the phase contrast microscope, this convention has the merit of being comparatively free from subjective errors, and is based on a change which seems closely related to the actual process of division. It must however be stressed that its use is restricted to cells growing actively in an adequate medium, for it is only in these circumstances that division and bud-formation are so closely linked. In a deficient or nearly exhausted medium, a considerable interval may occur between these two events, and a different criterion must then be used to assess the completion of division.

This brings us to another important aspect of the growth of yeast populations. It appears that in the extreme conditions of a culture approaching exhaustion, the interval between division and the emergence of the next generation of buds becomes so extended that many of the individuals resulting from divisions in the final stages of the culture's growth entirely fail to form new buds. Accordingly, the stationary phase population may contain a large proportion of mature single cells, all of which are effectively at the same stage of the cell cycle, but are prevented from further development by the absence of nutrients. On transfer to a fresh medium, however, these cells are fully capable of forming buds, and do so with a degree of synchrony which has been used to good effect by several authors.

However, the synchrony of these stationary phase populations leaves much to be desired, for although the emergence of the first generation of buds may be appreciably in phase, the first division is often largely asynchronous (Ogur *et al.*, 1953). This may be partly attributable to the presence of a proportion of cells having markedly longer lag times than the majority (Williamson and Scopes, 1960). Why this should be so is uncertain but it may be significant that the lag period displayed by a mature resting cell seems to be related to its size, and variation in cell size is a striking feature of stationary phase cultures of most strains of *Saccharomyces cerevisiae* (Lindegren and Haddad, 1954; Williamson and Scopes, 1961c) (Fig. 2). Since the smaller cells, which have a longer lag period than their larger counterparts, are relatively deficient in ribonucleic acid and protein (Williamson and Scopes, 1961c), it is conceivable that they may have to make good this deficiency before they can produce their first bud, and this may account for the long lag time they display. In any event, this diversity in lag times certainly limits the synchrony attained on transferring these resting populations to fresh medium and, as described later, the synchrony of the system can be appreciably enhanced by removing the smaller cells.

The size variation in the fully grown cultures is, in itself, of considerable interest, for it stems from an aspect of cellular growth that is peculiar to budding yeasts. Broadly speaking, the volume of an individual yeast cell is governed by the size of its comparatively tough cell wall. It is true that the volume of the cell may vary slightly in response to changes in environmental osmotic pressure (Drabble *et al.*, 1907). It may also increase slightly with successive buddings (Mortimer and Johnston, 1959) or as a result of growth in unusual circumstances (Williamson and Scopes, 1961b). In general, however, the only way in which a marked increase in volume can occur is by the formation of a bud, and when this happens, the size of the parental part of the cell remains substantially unchanged (Bayne-Jones and Adolph, 1932; Lindegren and Haddad, 1954). Ideally, division only occurs when the mass of the entire cell (parental unit plus bud) has reached exactly double that of the parent and, in these

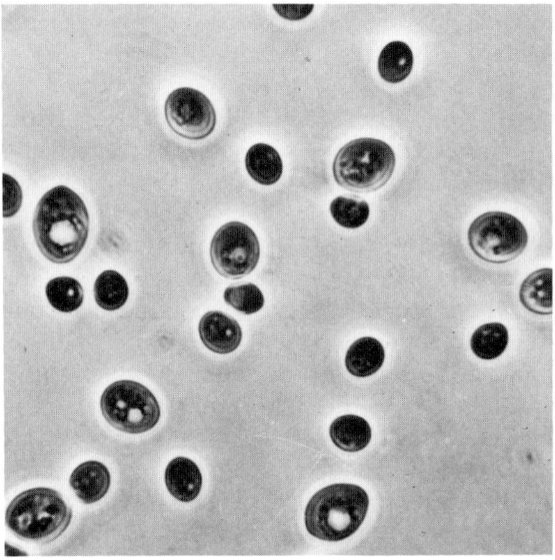

FIG. 2. Fully grown culture of *Saccharomyces cerevisiae* (National Collection of Yeast Cultures, strain 239) showing typical variation in cell size. (Phase-contrast.)

circumstances, the two cells resulting from division are of equal size (Fig. 3). However, in practice, such balanced growth is something of a rarity, and is probably achieved for extended periods only in very dilute cultures or in the steady state environment of a chemostat (Novick and Szilard, 1950). More often, defects in the environment prevent the cell from doubling its mass prior to division, and since the size of the parental part of the cell is largely unaffected by environmental conditions, division in these circumstances inevitably leads to the production of a daughter cell which is smaller than its parent. Such unbalanced growth may be seen whenever actively growing cells begin to run short of nutrients (Lindegren and Haddad, 1954) and its occurrence as a growing population enters the stationary phase accounts for the diversity in the size of resting cells referred to above.

Of more immediate interest is the fact that a similar degree of heterogeneity in size may result from the occurrence of unbalanced growth in synchronously dividing cultures. The possibility has to be considered that this heterogeneity is accompanied by a proportionate variability in the physiological activities of different cells in the population. This point will be further discussed below, but it is at any rate clear that the uniformity of a synchronously dividing yeast culture will be best maintained by using conditions which promote balanced growth, and the desirability of seeking such conditions needs no further comment.

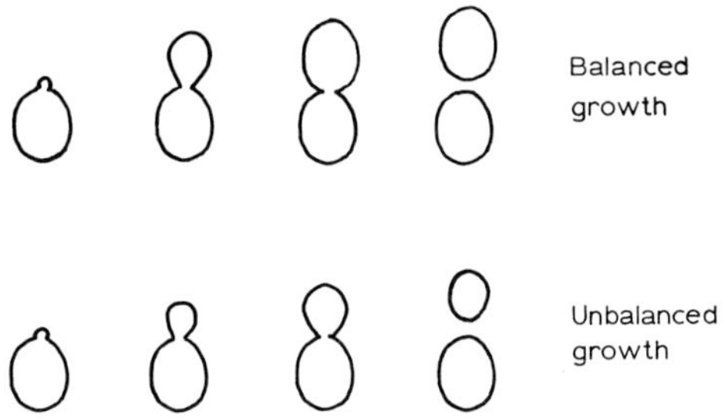

FIG. 3. Diagrammatic illustration of the difference between balanced and unbalanced growth. Balanced growth entails a doubling in the size of the cell prior to division, and the resulting daughter cell is the same size as its parent. When unbalanced growth occurs, however, the cell fails to double its size, and since the parental part of the cell neither shrinks nor expands, the resulting daughter individual is smaller than its parent.

III. 'NATURAL' SYNCHRONY

The tendency of many strains of *Saccharomyces cerevisiae* to enter a potentially synchronous condition simply on ageing in culture has already been mentioned. In the sense that no special treatment is required to bring about this condition in suitable strains, it may be regarded as an example of 'natural' synchrony. The condition of these cultures is of course only apparent on inoculation into fresh medium, and for this reason they are referred to as being potentially, rather than actually, synchronous.

These cultures are of interest from two points of view. In the first place, most of the procedures devised for artificially inducing synchrony in yeasts make use of similar resting populations, and their action possibly depends therefore on amplifying the inherently synchronous tendency of this type of material. Secondly, although the synchrony of these cultures is relatively poor, their early use revealed for the first time the broad outlines of the biochemical changes that accompany the yeast cell's transition from a resting to an actively growing state.

Thus Lindegren and Haddad (1953) reported that the cells underwent only slight changes in density and volume during the lag phase, and their conclusion that little or no synthesis of macromolecules occurred during this period has since been substantially confirmed.

Pursuing the theme of macromolecular synthesis, important observations were made relating to the formation of deoxyribonucleic acid (DNA). It was found that the cell's resistance to the lethal effects of X-rays (Beam *et al.*, 1954) increased abruptly as soon as its first bud appeared, and it is inferred that this might have resulted from simultaneous nuclear changes possibly involving a sudden synthesis of DNA. Substantial evidence for the actual occurrence of such 'stepwise' synthesis of this material was provided by the direct analyses of similar cultures performed by Ogur *et al.* (1953). Despite the relatively poor synchrony of their populations, their results strongly suggested that the DNA content of the cell abruptly doubled at about the time it formed a bud, and this has since been confirmed (Williamson and Scopes, 1960).

The mechanism underlying the attainment of the naturally synchronous state in these mature yeast cultures is by no means understood, though the solution may be found along the lines indicated earlier. In any event it is fairly certain that lack of one or more nutrients plays an important role. It is probably no coincidence that starvation was among the first avenues to be explored as a means of artificially synchronizing yeast populations and, despite the difficulties encountered by some of the earlier authors (Beam *et al.*, 1954; Burns, 1954; Campbell, 1957), starvation, in one guise or another, forms the basis of all the more important procedures in use today.

IV. Artificially Induced Synchrony

1. *Methods Based on Starvation*

Four procedures of varying efficacy have been devised for synchronizing yeast populations by controlled starvation. Although the end product, a potentially synchronous population of resting cells, is the same in each case, the procedures themselves differ in certain important details.

The first technique to be described is that of Hayashi and Shichiji (1959). The entire cell content of a specially prepared 48-hr culture of *Saccharomyces cerevisiae* was transferred into fresh medium at a density of about 10^8 cells per ml. In the course of a few hours' incubation, the number of cells increased about three-fold, and as the glucose in the medium became depleted, the culture approached a resting state. A further period of incubation ensued, during which the culture was continuously supplied with glucose, and the proportion of cells bearing young buds gradually declined. When only about 2 per cent of the cells were in a budding condition, the supply of glucose was stopped, and after a further 8-hr incubation, the population was found to have attained a satisfactory degree of potential synchrony. The data presented by the authors indicated that the disappearance of buds during the period of incubation with added glucose was accompanied by a slight increase in cell numbers. This situation would be explained if the environmental conditions permitted existing buds to grow to maturity, while inhibiting the formation of new ones. If this were so, the success of this treatment presumably rests on the operation of the same factors that bring about the natural synchrony of mature cultures described above.

Unfortunately, judged by the period occupied by the emergence of the first generation of buds, the degree of synchrony attained in these cultures was not very impressive. Moreover, only a small part of the growth curve was presented, so that it is impossible to judge how well synchrony was maintained. Nevertheless, the authors successfully used their system to study various aspects of carbon metabolism during the lag phase and first generation period, and their results will be referred to below.

A more effective procedure, substantiated by extensive cell counts, was that devised by Sylvén *et al.* (1959). The details of this technique are presented elsewhere in this volume, but it is of some interest that, unlike the other starvation methods, this process made use of cells from a comparatively young culture (24 hr). These were washed and incubated for about 3 hr in a salts solution containing succinate but no sugar. During this period their content of ultraviolet-absorbing substances and protein diminished, and they passed into a resting, potentially synchronous, state. The mode of action of this process is obscure, though the authors implied that succinate was essential. In any event, on inoculation in a suitable medium, the treated cells passed through a lag phase and one or two divisions with a degree of synchrony that proved adequate for resolving the course of changes in certain enzymic activities of the dividing individuals (see chapter 13).

Both the remaining procedures in the starvation category made use of resting cells of a strain of *Saccharomyces cerevisiae* in which parent and daughter individuals tend to separate fairly readily at, or soon after, the completion of division (Williamson and Scopes, 1960, 1962). Stationary phase cultures of this strain in suitable media comprise individuals which are homogeneous in the sense that they are nearly all free of attached buds, but which nevertheless vary markedly in size (Fig. 2). The factors leading to this situation in mature yeast cultures have already been discussed. It might be added, however, that the smaller cells in these particular populations form a distinctly homogeneous group which can be separated from the remaining organisms by differential centrifugation and other means (see chapter 27). That the two groups isolated in this way are not merely adjacent fractions of a single normally distributed population is apparent from the size distribution of this population (Fig. 4) as well as by the existence of striking differences between

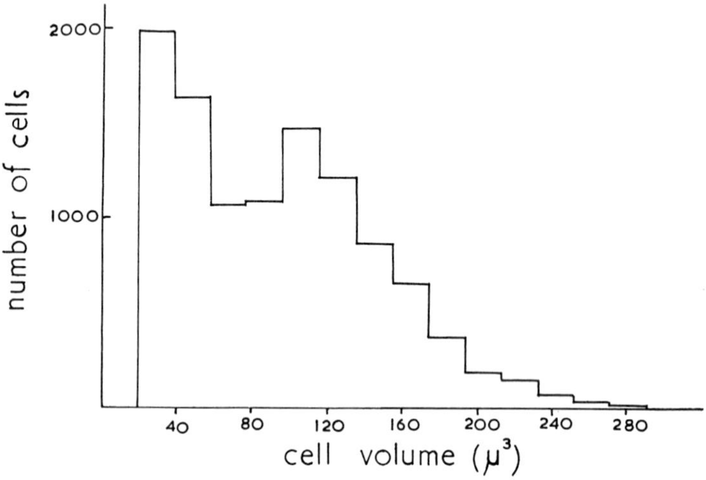

FIG. 4. Size distribution of cells from a fully grown culture of *Saccharomyces cerevisiae* (N.C.Y.C., strain 239). The measurements were made with an electronic particle counter (Coulter Electronics Ltd., St. Albans, Hertfordshire). (Previously unpublished observations of Williamson and Scopes.)

the two groups with respect to macromolecular composition and biological behaviour (Williamson and Scopes, 1960, 1961c). In the present context, the most important of these differences was the finding that, on transfer to fresh medium, the small cells behaved as a homogeneous group and displayed a mean lag period which was about twice as long as that shown by the larger cells (Williamson and Scopes, 1960). This observation prompted the thought that the potential synchrony of the stationary phase population might be improved by selectively removing the smaller cells. This has since been found to be the case, though it will be clear from Fig. 5 that, in itself, this selection is comparatively ineffective. However, initial attempts to achieve it led to the discovery that the potential synchrony of the population could be further enhanced by subjecting the cells to alternate periods of feeding and starvation, conditions

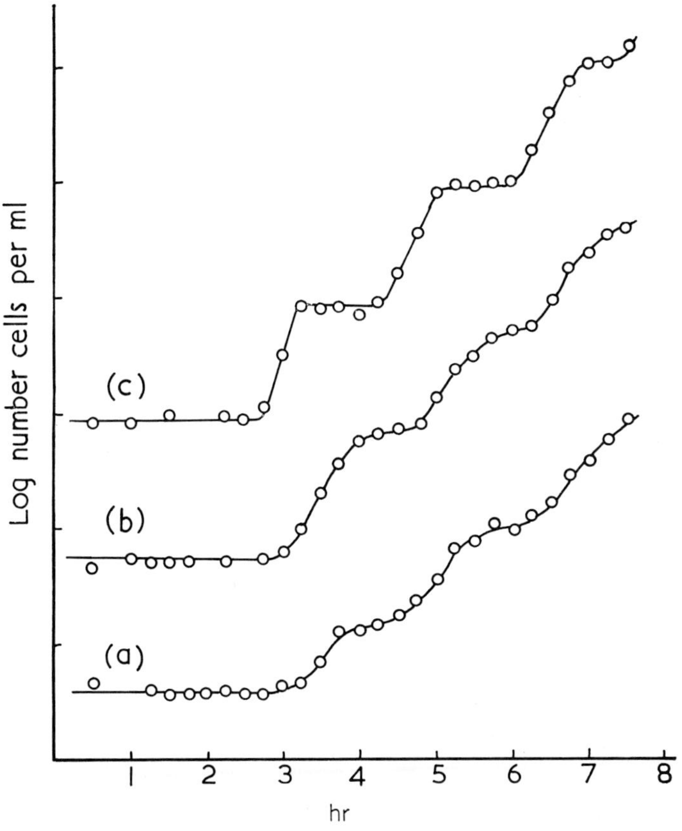

FIG. 5. Growth curves of *Saccharomyces cerevisiae* (N.C.Y.C., strain 239) incubated with aeration at 25° in a defined medium (Williamson and Scopes, 1960) enriched with 0·1 per cent yeast extract. Inoculated with (a) cells from a fully grown culture, comprising 'small' and 'large' individuals (see text); (b) 'large' cells selected by differential centrifugation but otherwise untreated; (c) 'large' cells which had been subjected to a synchronizing routine (Williamson and Scopes, 1962). Each interval on the ordinate corresponds to a doubling in cell numbers; the cultures were inoculated with 1–2 × 10^6 cells/ml. (Previously unpublished data of Williamson and Scopes.)

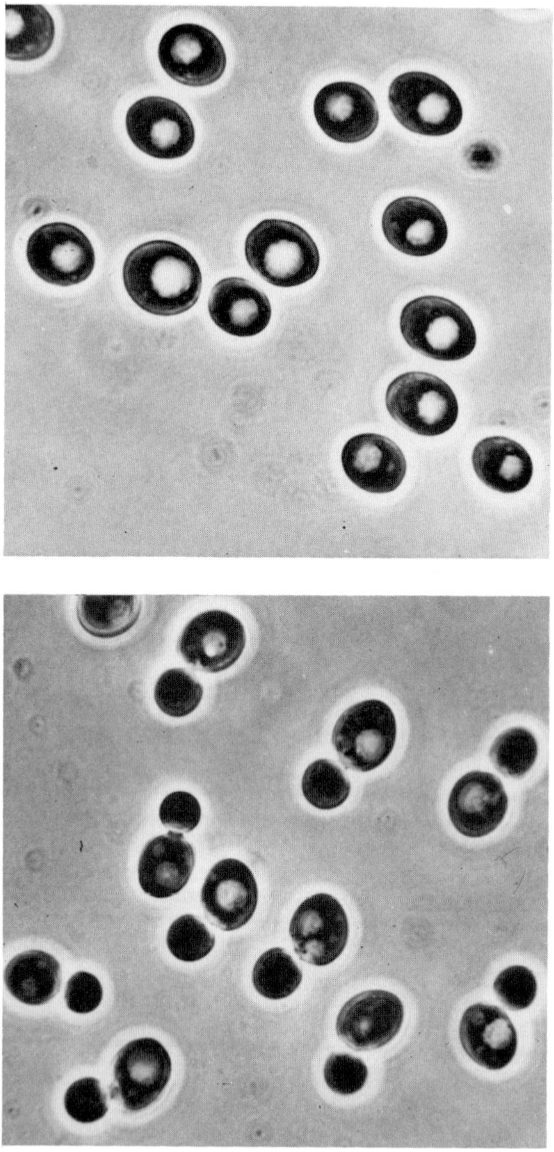

(a)

(c)

FIG. 6. Stages in the growth of a synchronized yeast culture: (a) at inoculation; of the first generation; (d) during the first period of division. (Phase-contrast.)

(b)

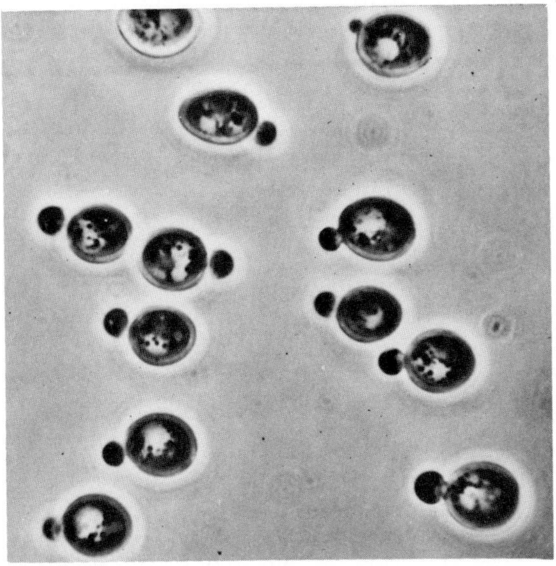

(d)

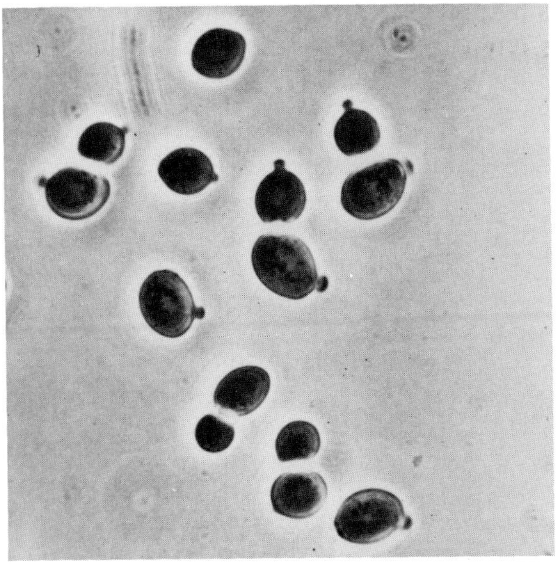

(b) shortly after the appearance of the first generation of buds; (c) nearing the end being such that no new buds were formed, the organisms remaining in an effectively 'resting' state. Accordingly, both the synchronizing processes devised by these authors incorporate the principles of selection and of cyclic environmental changes. Practical details will be found in chapter 27 but it may be noted here that the first procedure (Williamson and Scopes, 1960) involved alternation of temperature as well as of the supply of nutrients. In the second process (Williamson and Scopes, 1962), the temperature changes, and the possible complications arising therefrom, have been eliminated, and on theoretical grounds therefore, the latter process is to be preferred. Nevertheless, both these techniques give rise to a population of uniformly-sized resting cells which exhibit a good degree of synchrony when inoculated into a suitable medium (Figs. 5, 6), and they have been used to study changes in a variety of parameters during the lag phase and growth cycle of the cell.

Finally it should be mentioned that Yanagita and Sando (personal communication) have succeeded in synchronizing division in the fission yeast, *Schizosaccharomyces pombe*, by a method which probably depends to some extent on starvation. Based on a technique developed for *E. coli*, it involves harvesting cells from a log-phase culture and making a dense suspension (approximately 10^9 cells per ml) in a growth medium. After 2 hr incubation at 30°, the culture is diluted 100-fold with fresh medium, whereupon moderately synchronous division occurs. This system has been used to study changes in various biochemical properties during the cell cycle of this organism.

2. *Synchronous Division Induced by X-rays*

It has long been known that the application of suitable doses of ionizing radiation to yeast cells may prevent them from dividing, while allowing growth (in terms of cell mass) to proceed more or less normally, at any rate for a limited period (Spoerl *et al.*, 1954). This gives rise to cells having a 'figure-eight' configuration, for the bud reaches a size equal to that of the parental unit, but the two components fail to separate or produce new buds. Spoerl and Looney (1959) found that by subjecting a growing culture to a carefully controlled period of irradiation with X-rays, the entire culture could be brought to this division-blocked stage. Moreover, on stopping the irradiation, about 60 per cent of the organisms proved to be viable and underwent at least two cycles of growth and division with a striking degree of synchrony (Fig. 7). Greater interest attaches to the mechanism leading to this synchrony than to the behaviour of the synchronized population itself, for analyses of samples taken from the latter are inevitably encumbered by the presence of a large (though diminishing) proportion of dead cells. Nevertheless, the authors were able to follow changes in the incorporation of ^{32}P into various phosphate fractions, and these were related to the progress of the cells through their division cycle.

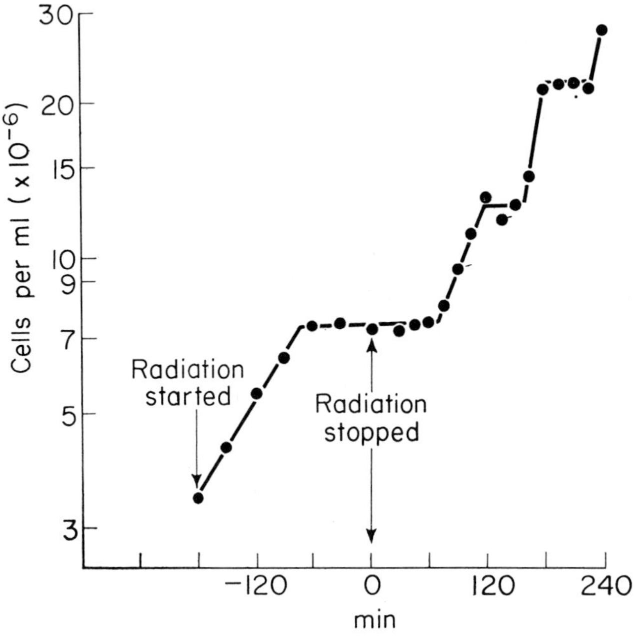

FIG. 7. Synchronization of *Saccharomyces cerevisiae* induced by X-rays. (Reproduced from Spoerl and Looney, 1959.)

It should be pointed out that Spoerl and Looney adopted the convention of counting a bud as a whole cell as soon as it reached about half the diameter of the parental unit. It has already been suggested that this convention may be misleading, and it would have been helpful if, in addition to the growth curves, detailed counts had been presented showing the changes in the proportion of cells bearing young buds. In this connection, the authors had to contend with the problem that buds produced subsequent to irradiation failed to separate from their parents, and this led to the formation of short chains. This was clearly an after-effect of the radiation, for such chains were absent from the original culture. Since the formation of buds continued at regular intervals, the authors concluded that the separate processes of division and bud formation had been affected to different extents, and had thus become dissociated. This has the important implication that the formation of a second generation bud does not necessarily depend on the prior completion of division. It should be pointed out however that, as used by Spoerl and Looney, the term 'division' referred only to the *physical separation* of parent and daughter cells; no evidence was provided that the deposition of cell wall between them had been inhibited. Thus despite their failure to separate, the parent and daughter units might yet have divided in the sense of having achieved physiological independence. This important point deserves further examination.

V. The Mechanisms Underlying Synchronizing Processes

The fact that exposure of a randomly dividing cell population to certain environmental stimuli tends to synchronize division implies that the response of different individuals varies according to the progress each has made through the cell cycle at the time the stimulus is applied. Accordingly, studies of the way in which such stimuli bring about synchrony should enhance our understanding of the mechanisms controlling cellular growth and division, and of the interaction between these mechanisms and the environment. Moreover, knowledge of the biochemical changes induced by a synchronizing process may help in assessing the 'normality' or otherwise of the behaviour of the treated population.

However, elucidation of the factors underlying synchronization is hindered by the fact that most processes have been developed in an arbitrary fashion, and the precise environmental changes involved are often poorly defined. In many cases it is not even possible to decide what the effective stimulus is, let alone the mechanism by which it exerts its effects.

In this respect the procedure described by Spoerl and Looney (1959) is an exception, for the stimulus in this process could only have been the temporary exposure to X-rays. The effects of ionizing radiations on living cells are highly complex and it is unlikely that the ensuing synchronization was mediated by a single, simple process. However, it is at least apparent that the radiation imposed a reversible block on the mechanisms involved in division and bud-formation. Since, as far as can be judged, the blocked cells grew more or less normally up to the stage of incipient division, the block also appeared to be highly specific, and possibly involved quantitatively minor components of the cell, which might be difficult to detect. However, it would perhaps be fruitful to follow changes in the DNA content of the cells during the treatment, for in terms of quantity, this material is certainly a minor component of the yeast cell (Ogur *et al.*, 1952), and the effects of radiation are frequently ascribed to nuclear or chromosomal damage. It is in fact conceivable that the radiation-induced inhibition of division might have been a secondary effect of damage to the mechanism responsible for replicating DNA, rather than a direct consequences of damage to the actual enzymes which carry out the division process. Further discussion is beyond the scope of this article, but it is to be hoped that this system will receive renewed attention, for it promises to be a fruitful source of information about the control of cell division.

The only synchronizing process for yeasts that has been studied with a view to elucidating its mode of action is that devised by Williamson and Scopes (1960). It will be recalled that, essentially, this process involved alternately feeding and starving a resting population, matters being so arranged that no new buds were formed and the cells were effectively maintained in an artificially extended lag phase. This being so, the synchronizing process presumably brought about its effect primarily by acting on those cellular activities concerned with guiding the cell through the lag phase rather than the cell cycle. In a preliminary study of some of the biochemical changes resulting from the process (Williamson and Scopes, 1961b), the authors were unable to correlate the attainment of synchrony with any significant alterations in the cell's mass or gross chemical composition. This led them to suggest that synchronization was mediated by changes in the intracellular concentration of hypothetical 'trigger' compounds which were essential to the emergence of the bud and were normally present only in small (and, therefore, not easily detectable) amounts. It was postulated that as the cells were alternately fed and starved, these compounds were alternately synthesized and destroyed in such a way that eventually every individual in the population would come to contain the same amount.* Once this stage was reached the capacity of all the cells to form a bud on transfer to fresh medium would be the same, and buds would therefore appear in synchrony. Although the existence of such trigger compounds is problematical, their existence would certainly account for some of the observed changes in the kinetics of the appearance of the first generation of buds.

* This is reminiscent of the 'set-back', so prominent in the synchronization of *Tetrahymena* (see chapter 4).

The authors pointed out that there was no evidence that the temperature changes entailed in their process played an essential part in bringing about synchrony, and this would appear to be borne out by the fact that in a recent version of the original process, these changes have been eliminated (Williamson and Scopes, 1962). It does not, of course, follow that both these processes depend on the same underlying principle. However, preliminary studies (Williamson and Scopes, unpublished observations) have indicated that essentially the two procedures induce similar changes (or lack of changes) in the composition of the cells, and the hypothetical models put forward to explain the first process are equally applicable to its successor.

VI. Physiological Behaviour of the Synchronized Yeast Cell

In discussing the results that have been obtained by examining synchronously dividing cultures of *Saccharomyces cerevisiae*, it is convenient to consider separately the lag phase, as defined earlier, and the normal cell cycle, i.e. the interval between successive divisions.

1. *The Lag Phase*

One might have supposed that the lag phase, during which the cell presumably organizes itself in preparation for the emergence of the first bud, would be a period of intense biochemical activity. Although this may be so in some respects, the overall picture one obtains by piecing together the few available observations is one of comparative quiescence. The early indications that the synthesis of macromolecules in general, and DNA in particular, is minimal during this period (Beam *et al.*, 1954; Lindegren and Haddad, 1953; Ogur *et al.*, 1953) have been substantially confirmed. Observations of artificially synchronized populations have in fact shown that there are generally no striking changes in the mass or volume of the cell, or in its content of nucleic acids, protein, glucan, mannan or total nitrogen (Hayashi and Shichiji, 1959; Scopes and Williamson, 1964; Sylvén *et al.*, 1959; Williamson and Scopes, 1960, 1961a).

However, this inactivity is not shared by all the cell's metabolic processes. For instance, respiration continues throughout the lag phase, presumably mainly at the expense of the external energy source, for an adequate supply of sugar seems to be essential to the emergence of the bud (Hayashi and Shichiji, 1959; Scopes and Williamson, 1964; Williamson and Scopes, 1961b). Moreover, Hayashi and Shichiji (1959), using synchronized cultures of baker's yeast, found that the total carbohydrate content of the cell dropped abruptly during the lag phase, mainly because of a sudden loss of trehalose and glycogen.

The significance of these changes is far from being understood, but they imply that the main activities of the cell during this period are directed towards the mobilization of available energy sources. It is relevant here to mention that Swann (1954) has suggested that cell division may be triggered off by the accumulation of a reservoir of energy, possibly in the form of energy-rich compounds such as adenosine triphosphate, and Plesner (1958) has in fact observed such a build up of the latter compound prior to division in *Tetrahymena pyriformis* (see chapter 7). It is possible that a comparable situation holds good for the lag phase yeast cell, the energy store in this case being required to trigger off the appearance of the bud.

In any event, the biochemistry of the lag phase in yeasts should repay further attention. The possibility exists that the substances (or events?) which trigger off the formation of a new bud on completion of the cell cycle are the same as those which induce the appearance of the bud at the end of the lag phase. Should this be so, their detection might be accomplished more readily in the lag phase cells, for examination of these is not complicated by the simultaneous occurrence of active macromolecular synthesis.

2. *The Cell Cycle*

A. Growth in volume and mass

The volume growth curve of the individual cell of *Saccharomyces cerevisiae* was established many years ago by Bayne-Jones and Adolph (1932), who measured the changes in volume of single growing cells by time-lapse photography. They concluded that the curve was essentially sigmoidal in shape, a maximal steady rate of increase being achieved and maintained only in the middle part of the cell cycle, growth ceasing temporarily at division. The only direct measurements of mass changes in individual cells were provided by Mitchison (1958) who used interference microscopy to obtain his results. He found that the mass of the cell increased at a steady rate throughout the cell cycle, and when division occurred, the total rate of mass increase of mother plus daughter abruptly doubled. Mitchison also determined the volume growth curve by photographic measurements, and substantially confirmed the sigmoidal pattern reported by Bayne-Jones and Adolph. He therefore concluded that the density of the cell (strictly speaking, the dry mass per unit volume) fluctuated in a cyclic way during the cell cycle, and increased just before division.

Measurements of mass and volume changes in synchronously dividing cultures of *Saccharomyces cerevisiae* have recently been made by Scopes and Williamson (1964). They used an electronic particle counter to follow the volume changes and their results (Fig. 8)

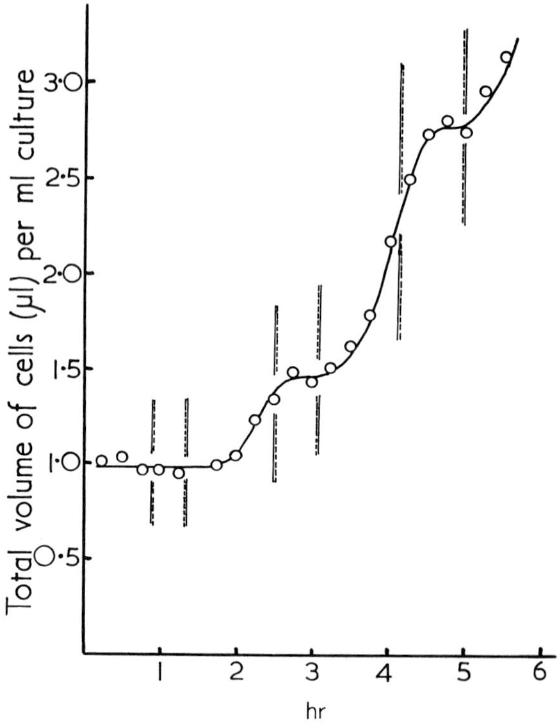

FIG. 8. Changes in volume of a synchronously dividing culture of *Saccharomyces cerevisiae* (Scopes and Williamson, 1964). The pairs of vertical lines indicate, respectively (from left to right), the period occupied by the emergence of the first generation of buds and the first and second periods of cell division.

confirmed the sigmoidal curve reported by the earlier authors. Their measurements of mass changes did not permit precise resolution of the shape of the mass growth curve, but the general impression was gained that the mass of the cell increased more or less steadily throughout the cell cycle, and little or no slowing down was noticeable at division. Thus these studies of synchronously dividing cultures provide general support for the previously described observations of single organisms.

B. Nucleic acids

It will be recalled that observations of naturally synchronous cultures of *Saccharomyces cerevisiae* indicated that the DNA content of the individual cell increased rapidly with the start of active growth (Ogur *et al.*, 1953). By examining artificially syncrhonized populations, Williamson and Scopes (1960) were able to confirm and extend this finding. They reported that the DNA content of

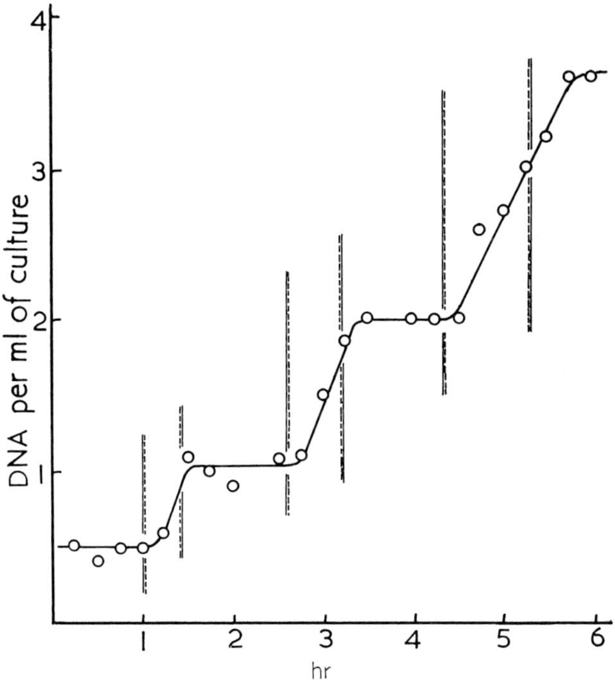

FIG. 9. Changes in the total DNA content of a synchronously dividing culture of *Saccharomyces cerevisiae*. DNA is recorded in arbitrary units. The pairs of vertical lines indicate, respectively (from left to right), the period occupied by the emergence of the first generation of buds and the first and second periods of cell division.

their cultures increased in a series of sudden doublings, each occurring at the beginning of a growth cycle. Since the initial publication of their findings, improved techniques have permitted better resolution of the shape of the DNA synthesis curve, and an example is shown in Fig. 9. The data suggest that DNA synthesis started as soon as, or just after, division was completed, and was thus coincident with the emergence of the bud. The fact that the intervals in which the culture was engaged in synthesis of DNA were little greater than those occupied by the periods of division implies that the individual cell synthesizes all the DNA it requires for division in a very short period. In this respect the yeast cell bears a general resemblance to many plant and animal cells, though in these, the synthetic period is usually located at a later stage of the cell cycle (Stanners and Till, 1960; Yamada and Puck, 1961; chapter 4, this book).

The position with regard to synthesis of ribonucleic acid (RNA) is less certain. In synchronized cultures of *Saccharomyces cerevisiae* this material was synthesized throughout most of the cellcycle, but its formation ceased as the cells passed through the phase of division (Williamson and Scopes, 1960). The only comparable observations of unsynchronized individuals are those of Mitchison and Walker (1959) using the fission yeast, *Schizosaccharomyces pombe*. These authors found an approximately steady rate of synthesis throughout the cell cycle, but there was little indication of a cessation at division. Whether or not these observations indicate an intrinsic difference between the budding and fission yeasts is uncertain. It is, of course, conceivable that the behaviour of the synchronized cultures had been distorted by the synchronizing process, and resolution of this point must await further developments.

C. Protein synthesis and total nitrogen content

A most interesting aspect of the growth of the yeast cell became apparent when Sylvén *et al.* (1959) found that the total nitrogen content of synchronously dividing cultures increased in a series of abrupt steps, one such step occurring in each growth cycle. This was later confirmed by Williamson and Scopes (1961a) who noted that although some variability was apparent in the timing of the steps, they generally occurred soon after the emergence of a new generation of buds, and occupied less than half the cell cycle. In addition, these authors used a specific colorimetric technique to follow changes in the protein content of their cultures. They observed that, in contrast to total nitrogen, the amount of protein in the cells increased more or less steadily throughout the growth cycle, and no reduction in the rate of synthesis was apparent at division (Fig. 10). Unfortunately the precise shape of the synthesis curve was uncertain, the data being too scattered to permit distinction between the likely alternatives of, on the one hand, continuous exponential formation and, on the other, linear synthesis with abrupt increases in rate at division. Even so, it was clearly apparent that the protein and total nitrogen contents of the cell did not change in parallel. The abrupt increases in nitrogen content were interpreted as net uptakes from the medium, and it thus appears that the growing cell absorbs all the nitrogen that will be required for the development of the bud, soon after the latter appears. The form in which this nitrogen is stored prior to its incorporation into macromolecules is uncertain, as are the mechanisms which control its uptake, and this aspect of the growth and division of the yeast cell would profit by further research.

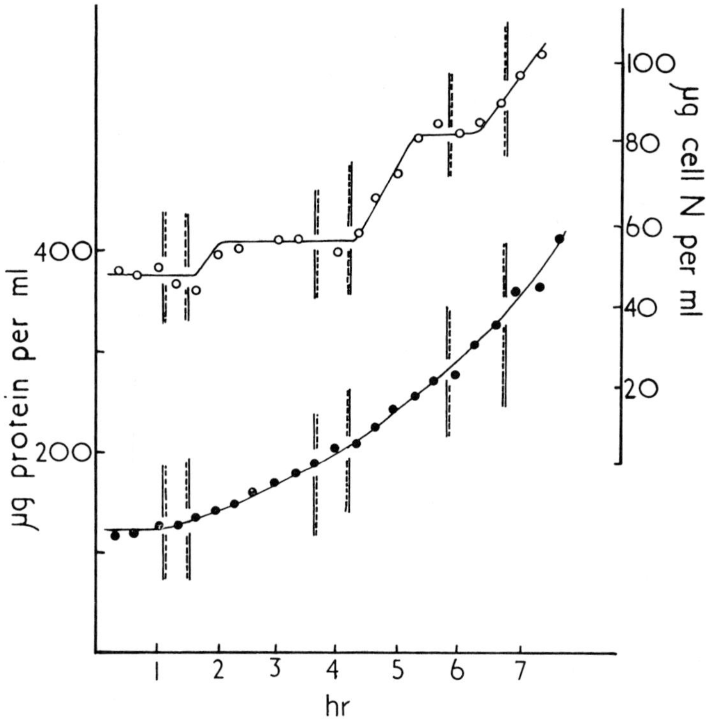

FIG. 10. Changes in the total cell nitrogen (○) and protein (●) contents of a synchronously dividing culture of *Saccharomyces cerevisiae* (Williamson and Scopes, 1961a). The pairs of vertical lines denote, respectively (from left to right), the period occupied by the emergence of the first generation of buds and the first and second periods of cell division.

D. *Oxygen consumption*

In common with other essential components, the enzymes and structures that comprise the respiratory mechanism of the cell must be duplicated during its cycle of growth and division. It might be expected that these enzymes would be synthesized at a rate comparable with that of other cell proteins, and one might therefore expect the potential respiratory capacity of the cell to increase roughly in parallel with cellular mass. If the newly assembled respiratory pathways became fully active as soon as they were formed, the total rate of oxygen consumption of the cell would thus also increase in parallel with the mass of the cell, and in the case of the protozoan *Tetrahymena pyriformis*, this expectation is fulfilled (Zeuthen, 1953; Zeuthen and Hamburger, 1960). However, the growing yeast cell exhibits an altogether different pattern of respiratory changes. Using a polarographic technique, Scopes and Williamson (1964) found that the total rate of oxygen consumption of synchronously dividing aerated cultures of *Saccharomyces cerevisiae* increased in a series of abrupt steps, each occupying only a small fraction of the generation time of the cell (Fig. 11). Although there was some variability in the timing of the increases, in general there

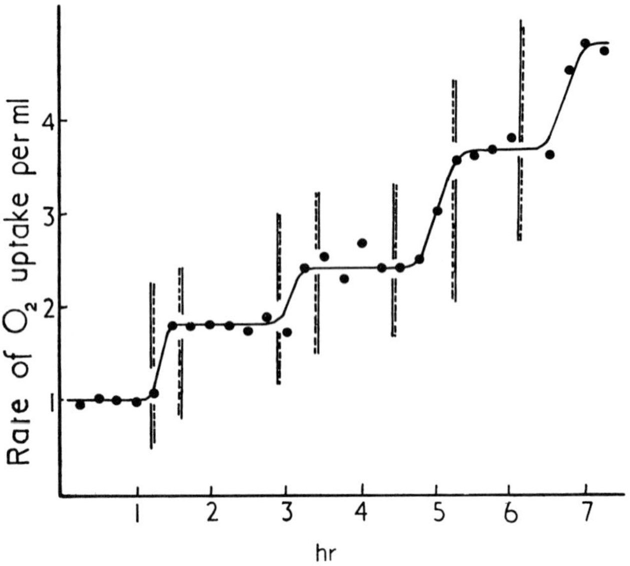

FIG. 11. Changes in the rate of O_2 uptake by a synchronously dividing culture of *Saccharomyces cerevisiae* incubated with vigorous aeration at 25° in defined medium (Williamson and Scopes, 1960) enriched with 0·1 per cent yeast extract. The pairs of vertical lines denote, respectively (from left to right), the period occupied by the emergence of the first generation of buds, the first and second periods of cell division and the start of the third. The arbitrary units on the ordinate denote the rate of uptake per unit volume of culture. (Data by courtesy of A. W. Scopes.)

was only one in each division cycle, and it appeared soon after the formation of the bud. In contrast, the mass of the organisms increased more or less steadily throughout the cell cycle, and there were no discontinuities that might be correlated with the observed changes in rate of respiration. The authors discussed several possible explanations for this situation. One, which appeals by virtue of its simplicity, was that while most of the components of the respiratory chain were formed throughout the cell cycle, a particular rate-limiting element was synthesized in a stepwise fashion, once in each generation. Being governed by the cell content of the latter material, the activity of the system would thus increase in the observed discontinuous manner.

Whatever the true explanation, these observations revealed the existence of a hitherto unsuspected mechanism controlling the respiration of the growing yeast cell, and further study is clearly desirable.

E. Cell size and generation time

On examining the publications referred to above, it emerges that many of the cultures employed were growing in an unbalanced way. That is to say, certain of the observed parameters (e.g. total nitrogen, Fig. 10) failed to double exactly in the course of every growth cycle. Although such behaviour may be an artefact resulting from synchronization, there are grounds for believing this is not the case (Williamson and Scopes, 1961a, b), and it seems more likely that unbalanced growth is a normal (and presumably transient) feature of the early growth of initially resting cells on transference to a fresh medium.

Whether or not this view is correct, the fact remains that many of the cultures described were markedly unbalanced in certain respects and, as shown in Fig. 8, this was reflected in the absence of a doubling in cell size. Inspection of the results obtained with this particular culture reveals some interesting features. In the first place it will be seen that during the first generation, the volume of the cells increased by only about 40 per cent and, bearing in mind that the parental part of the yeast cell does not alter in size during the growth of the bud, it is inferred that the daughter cells of this generation were less than half the size of their parents. However, despite this difference in size, both parent and daughter organisms took about the same time to complete their next cell cycle, for the synchrony of the second division was only slightly poorer than that of the first. It may be concluded therefore that the time taken for a bud to grow to maturity and divide from its parent is independent of the size of the latter. It also transpires that division is not mandatorily linked to the rate of growth of the cell. This deduction stems from the fact that the period occupied by the growth of the second generation was about the same as that occupied by its predecessor, and yet in the course of this second cell cycle, the volume of the culture almost exactly doubled. This implies that at some stage during this cycle, possibly right at its start, the cells acquired an enhanced ability to grow. The cause of this change is not known, but it is at any rate apparent that the mechanism controlling the act of division was not affected in the same way, and to this extent was independent of the factors controlling the overall growth rate of the cell.

These considerations lend support to the general belief that division is not simply a direct consequence of the cell reaching a certain size (see the review by Swann, 1957). Instead, in common with other authors (Scherbaum, 1957; Swann, 1954) we should consider the possibility that the generation time of the cell is governed by the rate at which it can accumulate hypothetical substances which specifically trigger off division. In the case of the yeast cell, the ability to synthesize these substances appears to be partitioned equally between cells of different sizes, and is to some extent independent of the cell's capacity to grow. Undoubtedly any such hypothesis must be an over-simplification, but it is only by erecting and testing such simple models that we can hope to unravel the true complexities of the situation.

VII. Conclusions

From the above survey, it will be apparent that although techniques for phasing division in cultures of *Saccharomyces cerevisiae* are a comparatively recent development, they have already revealed much of interest about the biochemical changes underlying the cell cycle of this organism, and an attempt to summarize some of this new knowledge is presented in Fig. 12. At the same time, it is only too obvious that more questions have been raised than answered, and there is abundant room for further study. However, certain salient features begin to emerge.

Perhaps the outstanding impression is the apparent intricacy of the control exerted by the cell over its various activities. It goes without saying that all aspects of the growth and division of the cell must be highly co-ordinated, or chaos would result. Nevertheless, it would have seemed reasonable to this author to have supposed that much of this integration would have stemmed from the continuous interplay between sequences of linked reactions poised in equilibrium. One might have anticipated, for instance, that nitrogen would have been taken up from the medium at a rate governed primarily by its speed of incorporation into protein. In this case, uptake would have continued more or less steadily throughout the cell cycle. Similarly, one might have expected the rate of oxygen uptake at any moment to reflect the protein content and thus, indirectly, the mass of the cell. But these are both instances in which we find evidence for the existence of a more subtle controlling mechanism. This is also true of the act of division itself. As pointed out earlier, the behaviour of synchronized yeast cultures displaying unbalanced growth supports the well established belief that division is not dependent on the cell having attained a particular size. Instead, it seems to be triggered off by a mechanism which, in a given environment, operates at regular intervals, more or less independently of the size or growth rate of the cell.

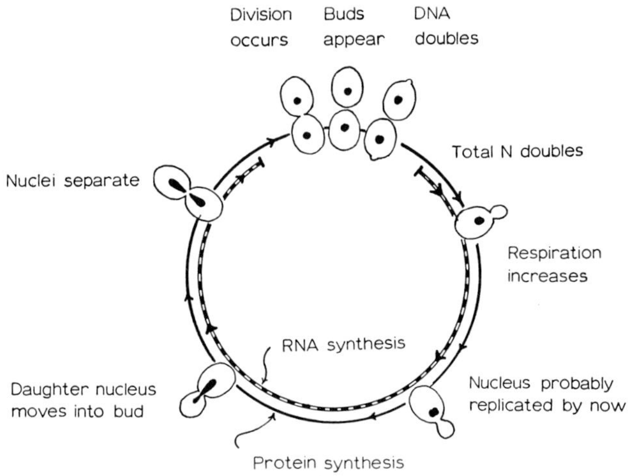

FIG. 12. Diagrammatic summary of events in the yeast cell cycle.

All these observations point to the existence of highly complex systems for co-ordinating the activities involved in cellular growth and division. This is not the place in which to speculate about the nature of these controls. However, our understanding of the gross biochemistry of the cell cycle of *Saccharomyces cerevisiae* has perhaps reached the stage at which experimental study could profitably be turned in this direction.

REFERENCES

BAYNE-JONES, S. and E. F. ADOLPH (1932). Growth in size of microorganisms measured from motion pictures. I. Yeast, *Saccharomyces cerevisiae. J. Cellular Comp. Physiol.* **1,** 387.

BEAM, C. A., R. K. MORTIMER, R. G. WOLFE and C. A. TOBIAS (1954). The relation of radio-resistance to budding in *Saccharomyces cerevisiae. Arch. Biochem. Biophys.* **49,** 110.

BRADLEY, D. E. (1955). A carbon replica technique for microbiological specimens applied to the study of the division of *Saccharomyces cerevisiae. J. Roy. Microscop. Soc.* **75,** 254.

BURNS, V. W. (1954). Physical studies of cell division. Ph.D. Thesis, University of California, Berkeley, Calif.

BURNS, V. W. (1956). Temporal studies of cell division. I. The influence of ploidy and temperature on cell division in *Saccharomyces cerevisiae. J. Cellular Comp. Physiol.* **47,** 357.

CAMPBELL, A. (1957). Division synchronization in a respiratory deficient yeast. *J. Bacteriol.* **74,** 559.

DRABBLE, E., H. DRABBLE and D. G. SCOTT (1907). On the size of the cells of *Pleurococcus* and *Saccharomyces* in solutions of a neutral salt. *Biochem. J.* **2,** 221.

HAYASHI, M. and S. SHICHIJI (1959). Effects of 2:4-dinitrophenol on endogenous respiration of yeast harvested during the first budding cycle. *Bull. Agr. Chem. Soc. Japan* **23,** 159.

LINDEGREN, C. C. and S. A. HADDAD (1953). The control of nuclear and cytoplasmic synthesis by the nucleocytoplasmic ratio in *Saccharomyces. Exptl. Cell Research* **5,** 549.

LINDEGREN, C. C. and S. A. HADDAD (1954). Growth rates of individual yeast cells. *Genetica* **27,** 45.

MITCHISON, J. M. (1958). The growth of single cells. II. *Saccharomyces cerevisiae. Exptl. Cell Research* **15,** 214.

MITCHISON, J. M. and P. M. B. WALKER (1959). RNA synthesis during the cell life cycle of a fission yeast *Schizosaccharomyces pombe. Exptl. Cell Research* **16,** 49.

MORTIMER, R. K. and J. R. JOHNSTON (1959). Life span of individual yeast cells. *Nature* **183,** 1751.

NOVICK, A. and L. SZILARD (1950). Experiments with the chemostat on spontaneous mutations of bacteria. *Proc. Natl. Acad. Sci., U.S.* **36,** 708.

OGUR, M., S. MINCKLER, G. LINDEGREN and C. C. LINDEGREN (1952). The nucleic acids in a polyploid series of *Saccharomyces. Arch. Biochem. Biophys.* **40,** 175.

OGUR, M., S. MINCKLER and D. O. MCCLARY (1953). Desoxyribonucleic acid and the budding cycle in the yeasts. *J. Bacteriol.* **66,** 642.

PLESNER, P. E. (1958). The nucleoside triphosphate content of *Tetrahymena pyriformis* during the division cycle in synchronously dividing mass cultures. *Biochim. et Biophys. Acta* **29,** 462.

SCHERBAUM, O. (1957). Studies on the mechanism of synchronous cell division in *Tetrahymena pyriformis. Exptl. Cell Research* **13,** 11.

SCOPES, A. W. and D. H. WILLIAMSON (1964). The growth and oxygen uptake of synchronously dividing cultures of *Saccharomyces cerevisiae*; in press.

SPOERL, E. and D. LOONEY (1959). Synchronized budding of yeast cells following X-irradiation. *Exptl. Cell Research* **17,** 320.

SPOERL, E., L. E. LOVELESS, T. H. WEISMAN and R. J. BALSKE (1954). Cell division. II. X-radiation as a division-inhibiting agent. *J. Bacteriol.* **67,** 394.

STANNERS, C. P. and J. E. TILL (1960). DNA synthesis in individual L-strain mouse cells. *Biochim. et Biophys. Acta* **37,** 406.

SWANN, M. M. (1954). The control of cell division. In *Recent Developments in Cell Physiology* (J. A. Kitching, ed.) Butterworths, London, pp. 185–196.

SWANN, M. M. (1957). The control of cell division: A review. I. General mechanisms. *Cancer Research* **17,** 727.

SYLVÉN, B., C. A. TOBIAS, H. MALMGREN, R. OTTOSON and B. THORELL (1959). Cyclic variations in the peptidase and catheptic activities of yeast cultures synchronized with respect to cell multiplication. *Exptl. Cell Research* **16,** 75.

WILLIAMSON, D. H. and A. W. SCOPES (1960). The behavior of nucleic acids in synchronously dividing cultures of *Saccharomyces cerevisiae. Exptl. Cell Research* **20,** 338.

WILLIAMSON, D. H. and A. W. SCOPES (1961a). Protein synthesis and nitrogen uptake in synchronously dividing cultures of *Saccharomyces cerevisiae. J. Inst. Brewing* **67,** 39.

WILLIAMSON, D. H. and A. W. SCOPES (1961b). Synchronization of division in cultures of *Saccharomyces cerevisiae* by control of the environment. In *Microbial Reaction to Environment* (G. Meynell and H. Gooder, ed.) (11*th Symp. Soc. gen. Microbiol.*). University Press, Cambridge, p. 217.

WILLIAMSON, D. H. and A. W. SCOPES (1961c). The distribution of nucleic acids and protein between different sized yeast cells. *Exptl. Cell Research* **24,** 151.

WILLIAMSON, D. H. and A. W. SCOPES (1962). A rapid method for synchronizing division in the yeast, *Saccharomyces cerevisiae. Nature* **193,** 256.

YAMADA, M. and T. T. PUCK (1961). Action of radiation on mammalian cells. IV. Reversible mitotic lag in the S3 HeLa cell produced by low doses of X-rays. *Proc. Natl. Acad. Sci., U.S.* **47,** 1181.

ZEUTHEN, E. (1953). Growth as related to the cell cycle in single-cell cultures of *Tetrahymena pyriformis. J. Embryol. Exptl. Morphol.* **1,** 239.

ZEUTHEN, E. and K. HAMBURGER (1960). Some characteristics of growth in normal and synchronized populations of *Tetrahymena pyriformis. Compt. Rend. Trav. Lab. Carlsberg* **32,** 1.

13

Enzymes in Synchronized Yeast Populations

B. SYLVÉN and B. THORELL, *The Cancer Research Division of Radiumhemmet, Karolinska Institute, Stockholm, 60, Sweden*

CONTENTS

I Introduction 381
II The Synchronous System 382
III Activity of Peptidases and Proteinases 385
IV A Protein Disulfide Reductase 388
References 389

I. INTRODUCTION

Yeast forms a very rich source of various enzymes closely studied for a long time. Available reviews and hand-books (Eddy, 1958; Harris, 1958) provide information as to the overall patterns of various enzymic activities extracted from very large yeast populations, which are often not too clearly defined as to their biological states of growth and fermentation. With the advent of more specified cytochemical problems relating to enzymic activities and functions during biosynthetic and metabolic processes, attention has to be paid to the control and homogeneity of the cell materials to be investigated. Apart from naturally occurring cell synchrony, such as for instance of fertilized egg cells, we have now easy access to protozoan, yeast or mammalian cells cultivated *in vitro*, which can be adequately synchronized with respect to cell multiplication by controlling the environmental conditions. These materials lend themselves to detailed studies on the biochemical mechanisms of cell division and many dynamic problems relating to cellular growth and metabolism and the role of enzymes.

Previous average data on enzymatic activities in yeast suggest in a general way that considerable differences occur during different cultural conditions favoring either growth or fermentation or else providing a resting stage. It is, however, not the object of this review to compile these bulk data, but instead to focus the attention strictly on data referring to synchronized yeast populations. Since this field has only recently been opened up a very limited number of enzymes have been cursorily studied. The biochemical changes are suitably expressed on the basis of the cell multiplication timescale (Tobias and Thorell, 1958).

For previous data on bulk cultures of yeast, their pertinent growth curves, and their biochemical characteristics with reference to the interesting fermentation and carbohydrate-synthesizing enzymes the readers are referred to the treatise by Cook (Harris, 1958). The reactions leading to the synthesis of nucleic acids and proteins during non-specified yeast growth were reviewed by Harris (1958), and the specific amino acid-activating enzymes have been surveyed by Koningsberger (1961). Most of these results have so far been obtained on crude samples and subcellular fractions of baker's and brewer's yeast. It would seem reasonable to assume that many of the enzymes thus involved will show cyclic variations in activity during synchronized growth.

II. The Synchronous System

The yeast cell cytology and cytochemistry offer some opportunities to evaluate the cell's functional state with regard to growth and division. In the experiments on enzymatic activities during the cell cycle (Sylvén *et al.*, 1959), which were performed on genetically homogeneous strains of *Saccharomyces cerevisiae*, ultraviolet microspectrophotometry (Thorell, 1958) was used to ensure a homogeneous population and to control the state of nucleotide metabolism.

The latter has earlier been shown to undergo marked changes parallel with starvation and growth (Caspersson and Brandt, 1941). In some experiments, also the 'dry mass' of single cells was judged by X-ray microradiography (Engström, 1956). Cell counting, nitrogen determination and spectrophotometry of the medium completed the analytical setup to secure information about the growth conditions of the yeast cell population.

The essential features of the procedure for obtaining synchronized divisions are summarized in Table I. The yeast cells were initially prepared by an intense growth period for 24 hr in a complete nutrient medium. Immediately after this growth period they were exposed for 3–5 hr to a starvation medium (potassium succinate) at pH 5. The cells must not be allowed to start any fermentation activity. When the yeast cells were starving under such conditions a release of u.v. absorbing substances from the cells into the medium occurred; but the microradiographs indicated no corresponding decrease in dry mass (Tobias, Sondhaus and Thorell, unpublished

TABLE I. Cultural conditions for synchronized growth of *S. cerevisiae*. Temperature 30°C

Mineral requirements: $CaCl_2 \cdot 6H_2O$ 5·0 g, $MgSO_4 \cdot 7H_2O$ 25·0 g, NaCl 5·0 g, H_3BO_3 0·025 g, $CuSO_4 \cdot 5H_2O$ 0·002 g, KI 0·010 g, $FeCl_3 \cdot 6H_2O$ 0·010 g, $Na_2MoO_4 \cdot 2H_2O$ 0·010 g, $ZnSO_4 \cdot 7H_2O$ 0·020 g, $MnSO_4 \cdot 1H_2O$ 0·020 g, water 500 ml. Use 1 ml per 100 ml medium or buffer.

Phase	Medium		Density of cells	Time (hr)
Pre-starvation growth:	Yeast extract (Difco) Dextrose Phosphate buffer pH 6·5 (+ minerals)	0·05% 1%	~ 10^6/ml, final density not exceeding 4×10^7/ml	24–48
centrifuge 700*g*, 5 min, wash twice				
Starvation:	Succinic acid Potassium hydroxide per 1000 ml water (+ minerals)	7·84 g 5·06 g	~ 4×10^7/ml	3–5 (Survival 75–85%)
centrifuge as above				
Synchronized growth:	Same as pre-starvation growth		~ 4×10^7/ml	2–6

experiments), and no significant decrease in the amount of total nitrogen per cell was noted. After the u.v. absorption of the cells had fallen to a minimum value, the cells were washed and again put into a complete nutrient medium consisting of yeast extract, dextrose and the necessary trace elements. After about 2 hr the first divisions occurred.

To obtain more than 75 per cent division within a reasonable short time period it was essential to estimate growth and starvation morphologically and to check the release of nucleotides from the cells. The cells should, at the end of the starvation period, still present a rather uniform size, with a small vacuole and a cytoplasm almost depleted of u.v. absorbing material, except for minute cytoplasmic granules and the somewhat larger nucleus. No budding should be present during starvation. The survival, estimated from plating experiments, should exceed 70 per cent. Fig. 1A shows an example of the cytological picture at the end of the starvation period.

The growth period in the new, complete medium was initiated by a lag phase, during which the cellular u.v. absorption, measured as log I, increased 4–6 times on an average (Fig. 1B). The cellular dry mass and total nitrogen content did not, however, increase correspondingly. Therefore, the cytochemical data could be interpreted as an increase in cytoplasmic nucleotides (Tobias *et al.*, unpublished experiments). In general, the first signs of budding occurred after 60 min and after 2 hr the cells divided. The u.v. absorption of the cells reached its maximum well before the first cell division was completed (Fig. 1C).

By quantitative microspectrometric estimations it was found that the rate of uptake of u.v. absorbing substances from the medium was rather independent of the concentration of nutrients. The cells could further take up nutrients from the external medium sufficient for several cell divisions, i.e. if transferred to a starvation medium after the first lag phase or division, the cells continued to divide.

In conclusion, the procedure worked out to obtain a succession of synchronized divisions, which seemed to imply that a fully developed cellular replication mechanism was brought to a temporary stand-still by the starvation. During the preparation for division in the subsequent first lag phase both the energy metabolism and the biosynthetic processes were again activated by the uptake of essential nutrients from the new complete medium.

Since these studies were performed newer and better methods of synchronizing yeast have been published (chapter 12). Repetition of the studies here reported using these methods might be rewarding.

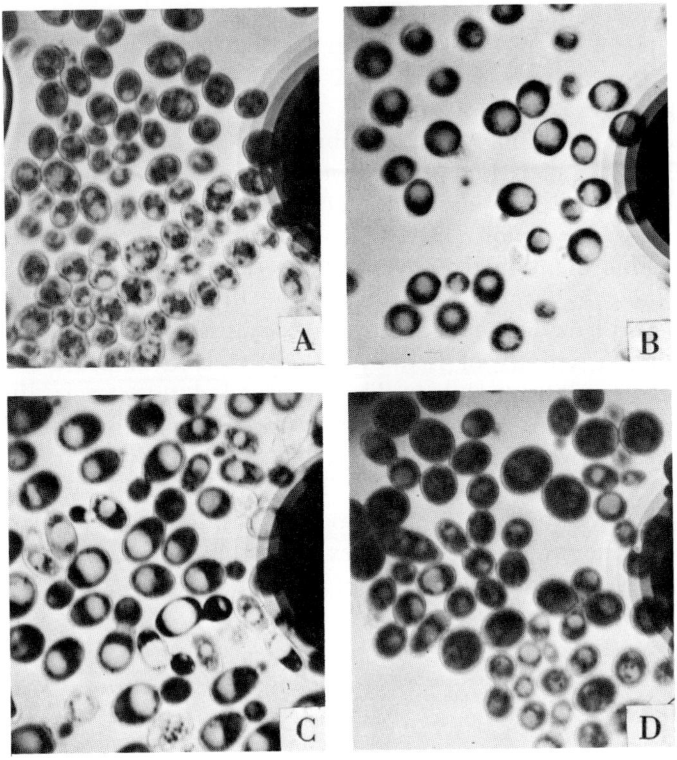

Fig. 1. U.V. microphotographs of intact yeast cells in saline. Image of rotating sector as a reference of optical density (Thorell, 1958).

A. Cells at end of the starvation period.

B. At end of the lag phase in the new complete growth medium.

C. Cells during the first division.

D. Cells irradiated with 15,000 r and then transferred to a complete medium for 5 hr show marked heterogeneity of cell size and u.v. absorption.

Optical data: Wavelength 2570 Å; objective aperture 1·25; condenser aperture 0·80; magnification about 750.

III. Activity of Peptidases and Proteinases

Data collected during the past twenty or thirty years from plant, amphibian and mammalian tissues call attention to the increased dipeptidase and catheptic activities of growing cells as compared with differentiated, 'resting' cells. Peptidases, and to a certain extent also proteinases, have been conceived to play an essential role in connection with growth and protein synthesis (Linderström-Lang, 1952). This does not necessarily refer to a protein synthetic activity, as originally suggested by Wasteneys and Borsook (1930); instead, it seems possible that the hydrolytic activity of these enzymes may have a bearing on the intracellular degradation of peptides and proteins during the growth and normal turnover of cells (Sylvén and Malmgren, 1957).

Now, the commonly used test objects do not allow a distinction as to whether the increased activity of these enzymes is correlated with the cell division or/and with stages in which cellular proteins are synthesized. New information was obtained from the synchronization experiments mentioned above (Sylvén *et al.*, 1959). These experiments have so far only dealt with the dipeptidase activity using alanyl- and leucyl-glycine as substrates and with the 'cathepsin-like' proteinase-A activity of Lenney (1956) assayed at pH 3·8 against acid-denatured hemoglobin. The observed data refer to the total activities extracted after plasmolysis and homogenization of fresh, not autolysed, yeast cells. The activities recorded during the normal growth cycle of this yeast strain (Fig. 2) have been expressed per 10^6 cells as well as on a total nitrogen basis. The variations showed essentially the same course when expressed as total enzyme activities per ml yeast suspension (omitted in Figs. 2 and 3).

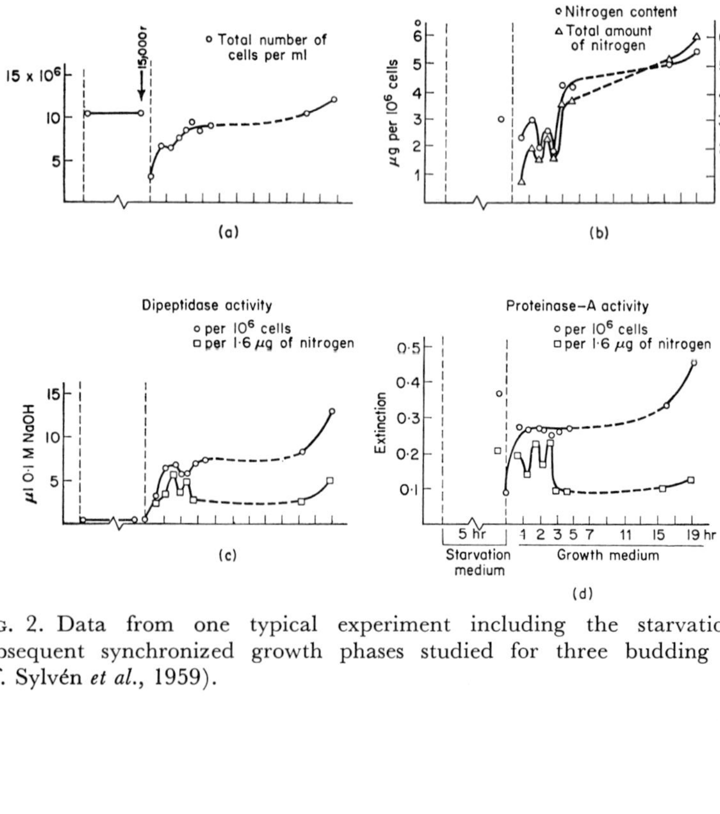

FIG. 2. Data from one typical experiment including the starvation and subsequent synchronized growth phases studied for three budding periods (cf. Sylvén *et al.*, 1959).

The enzymic activities showed marked and regular variations during the growth cycles. Starting from high pre-starvation growth levels, the peptidase activity showed a more marked decrease than the total proteinase-A activity during 3 hr starvation in succinate buffer. Both activities then rapidly increased during the subsequent lag phase. With the onset of the first budding phase very marked drops in activity were observed both per cell and per nitrogen.

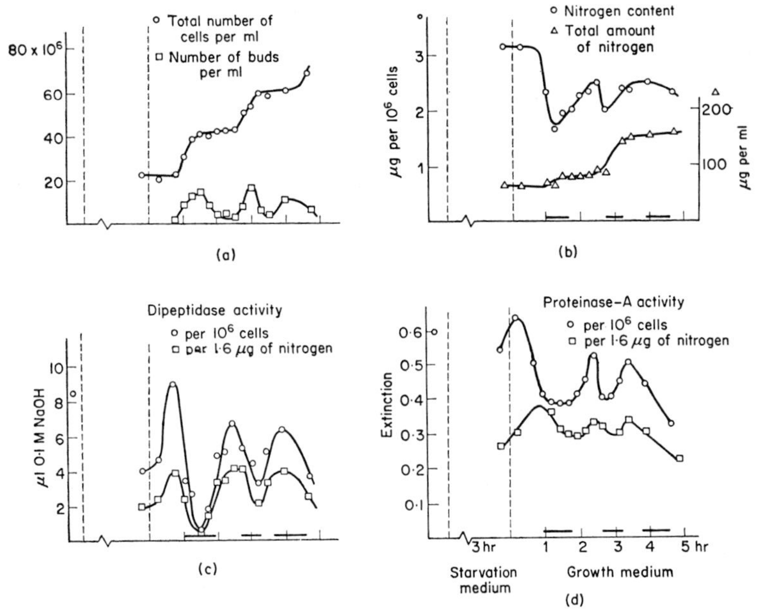

FIG. 3. Following irradiation during the starvation phase, the yeast cells underwent a first complete and a second incomplete division cycle. During the subsequent arrest of cell division for about 14 hr, the synthetic processes continued (cf. Fig. 1D) and the enzyme activities per cell but not per nitrogen remained at a high 'lag-phase' level.

During the subsequent growth phase, the enzymic activities rose to a new maximum, to reach a new minimum in the second budding phase. Such periodic variations could be discerned for three subsequent budding cycles before synchrony faded out. The cyclic variations were more marked in the case of the total dipeptidase activity than in the case of the proteinase-A activity. During phases of decreasing enzyme activities enzymes may wear or become inactivated. In the case of the dipeptidase for which the activity may drop virtually to zero the latter possibility seems the likely one.

It is of interest to mention that the total protein content per ml yeast suspension steadily increased throughout the multiplication cycles (Williamson and Scopes, 1961),* while the total nitrogen content per ml showed a discontinuous increase (Sylvén *et al.*, 1959; Williamson and Scopes, 1961; Figs. 2b and 3b). Periods of accelerated nitrogen uptake from the medium appear roughly to coincide with low peptidase and proteinase-A activities (Sylvén *et al.*, 1959; Williamson and Scopes, 1961; and Fig. 2).

In cultures irradiated during starvation cell growth continued slowly but in most cells was not followed by division. Consequently many supernormal single cells appeared (Fig. 1D). Per cell such arrested cells showed high peptidase and proteinase-A activities (Fig. 3). Per unit nitrogen they showed somewhat lower enzymic activities than did non-irradiated lag-phase cells. In terms of enzyme activity the post-irradiation blockage may thus be regarded as an unusually prolonged lag phase.

The cyclic variations in enzyme activity justify some speculations concerning the biological role of these enzymes during yeast growth. The observations broadly suggest that degrading reactions operating on proteins and peptides are essential *preparatory steps for cell division.* Accelerated uptake from the medium of nitrogenous compounds took place at times when the enzyme activities were low. The maximum of peptidase and proteinase-A activities was reached some 30–60 min later at a time when the nitrogen per cell was constant. It was assumed (Sylvén *et al.*, 1959) that during the lag phases the proteolytic enzymes could operate in feed-back mechanisms necessary to replenish the amino acid pool (Davies and Harris, 1960; Halvorson, 1958a, b, 1960; Halvorson *et al.*, 1955; Spiegelman *et al.*, 1955). Available data do not support the idea of a correlation between these enzyme activities and the synthesis of bulk proteins. Protein synthesis runs at a smoothly accelerating rate during the division cycles (Williamson and Scopes, 1961). More data are needed.

IV. A Protein Disulfide Reductase

Nickerson and Falcone (1956, 1959) have studied a protein disulfide reductase which reduces the disulfide bonds of the glucomannan-pseudokeratin material which is responsible for the rigidity of the

* At this place an error should be corrected in our previous publication (Sylvén *et al.*, 1959). The protein content in Figs. 3–5 was not assayed separately but calculated from the total nitrogen data, which erroneously suggested a discontinuous course of protein synthesis. The error is corrected in the present graphs.

cell wall. The activity of this enzyme may explain the local plasticity of yeast cell walls necessary for the formation of buds. Although direct information is lacking as to the occurrence of the disulfide reductase during the growth cycle of yeast, these authors have shown that normally dividing strains of *C. albicans* carry this enzyme, while a non-dividing filamentous mutant is largely without activity.

This observation invites the closing statement that it would be useful if we had more data for many other enzyme activities during synchronous growth of yeast cells. For *Tetrahymena pyriformis* some are now available (chapters 4 and 6).

REFERENCES

CASPERSSON, T. and K. BRANDT (1941). Nucleotidumsatz und Wachstum bei Presshefe. *Protoplasma* **35,** 507.

DAVIES, J. W. and G. HARRIS (1960). Dynamic aspects of the peptidyl–nucleotidate pool of brewer's yeast during growth. *Biochim. et Biophys. Acta* **45,** 39.

EDDY, A. A. (1958). Aspects of the chemical composition of yeast. In *The Biochemistry and Biology of Yeasts* (A. H. Cook, ed.). Academic Press, New York, pp. 157–249.

ENGSTRÖM, A. (1956). Historadiography. In *Physical Techniques in Biological Research* (G. Oster and A. W. Pollister, eds.). Vol. III. Academic Press, New York, pp. 489–544.

HALVORSON, H. (1958a). Intracellular protein and nucleic acid turnover in resting yeast cells. *Biochim. et Biophys. Acta* **27,** 255.

HALVORSON, H. (1958b). Studies on protein and nucleic acid turnover in growing cultures of yeast. *Biochim. et Biophys. Acta* **27,** 267.

HALVORSON, H. O. (1960). The induced synthesis of proteins. In *Advances in Enzymology* (F. F. Nord, ed.), Vol. 22. Interscience, New York, pp. 99–156.

HALVORSON, H., W. FRY and D. SCHWEMMIN (1955). A study of the properties of the free amino acid pool and enzyme synthesis in yeast. *J. Gen. Physiol.* **38,** 549.

HARRIS, G. (1958). Nitrogen metabolism. In *The Biochemistry and Biology of Yeasts* (A. H. Cook, ed.). Academic Press, New York, pp. 437–533.

KONINGSBERGER, V. V. (1961). Studies on protein synthesis by yeast. In *Protein Biosynthesis* (R. J. C. Harris, ed.). Academic Press, London and New York, pp. 207–232.

LENNEY, J. F. (1956). A study of two yeast proteinases. *J. Biol. Chem.* **221,** 919.

LINDERSTRÖM-LANG, K. U. (1952). *Proteins and Enzymes.* Stanford University Publications, Stanford University Press, Stanford, California, pp. 3–115.

NICKERSON, W. J. and G. FALCONE (1956). Identification of protein disulfide reductase as a cellular division enzyme in yeasts. *Science* **124,** 722.

NICKERSON, W. J. and G. FALCONE (1959). Function of protein disulfide reductase in cellular division of yeasts, VII. 3. In *Sulfur in Proteins* (R. Benesch *et al.*, eds.). Proceedings of a Symposium held at Falmouth, Mass., May 1958. Academic Press, New York, pp. 409–424.

Spiegelman, S., H. Halvorson and R. Ben-Ishai (1955). Free amino acids and the enzyme-forming mechanism. In *A Symposium on Amino Acid Metabolism* (W. D. McElroy *et al.*, eds.). Johns Hopkins, Baltimore, pp. 124–170.

Sylvén, B. and H. Malmgren (1957). The histological distribution of proteinase and peptidase activity in solid tumor transplants. A histochemical study on the enzymic characteristics of the different tumor cell types. *Acta Radiol. Suppl.* **154,** 5.

Sylvén, B., C. A. Tobias, H. Malmgren, R. Ottoson and B. Thorell (1959). Cyclic variations in the peptidase and catheptic activities of yeast cultures synchronized with respect to cell multiplication. *Exptl. Cell Research* **16,** 75.

Thorell, B. (1958). Cell studies with microspectrography. In *Advances in Biol. and Med. Phys.*, Vol. VI (C. A. Tobias *et al.*, eds.). Academic Press, New York, pp. 95–119.

Tobias, C. A., C. A. Sondhaus and B. Thorell; unpublished experiments.

Tobias, C. A. and B. Thorell. Some factors affecting division of yeast cells. The Biophysical Society, 1958 Meeting, Cambridge, Mass., p. 25.

Wasteneys, H. and H. Borsook (1930). The enzymatic synthesis of protein. *Physiol. Revs.* **10,** 110.

Williamson, D. H. and A. W. Scopes (1961). Protein synthesis and nitrogen uptake in synchronously dividing cultures of *Saccharomyces cerevisiae*. *J. Inst. Brewing* **67,** 39.

14

Germinating Conidiospores of *Aspergillus niger*

TOMOMICHI YANAGITA, *Institute of Applied Microbiology, University of Tokyo, Tokyo, Japan*

CONTENTS

I	Introduction	391
II	Physiological Aspects of Spore Germination	392
	1. Nutritional requirements	392
	2. Morphological change	393
	3. Respiratory activity	394
	4. Susceptibility to physical and chemical agents	395
III	Biochemical Aspects of Spore Germination	397
	1. Change in amount of cellular substances	397
	2. Protein synthesis and sulfur metabolism	399
	3. Phosphorus metabolism and nucleic acid synthesis	402
	4. Carbohydrate metabolism	408
	5. Role of alanine and CO_2	411
IV	General Conclusion	413
	References	416

I. INTRODUCTION

The spore as a durable cell form in a microbial life cycle is characterized by its dormancy with respect to metabolic activities. When it germinates, it turns into a metabolically active vegetative cell form. In the asexual life cycle of a microorganism, one generation of vegetative cells follows another in an endless chain. Therefore, if we use such vegetative cells for the study of the cellular life cycle, it may be difficult to know the cause-and-result relationship among the cellular events occurring during one generation. The process of spore germination, in contrast, starts from the breaking of dormancy; this is followed by a vegetative phase. Thus, the investigation of germination may elucidate important aspects of the metabolic shifts from dormancy to the vegetative phase, as well as of the very first step of vegetative cell growth, as long as the germination process proceeds synchronously.

Among various microbial spores, bacterial endospores have been subjected to extensive investigation. The difference in resistance to environmental factors between these spores and their vegetative cells is striking. When they germinate, their optical density and resistance to heat drop instantly, while their stainability and rate of respiration rise in the initial step. Then they swell, emerge, elongate, and finally enter into cell division (Mandels *et al.*, 1956). In the course of such a process of germination of bacterial spores, the initial short period, which is accompanied by the conspicuous changes mentioned above, is generally defined by bacteriologists as 'germination'. In fungi, spores are generally not endowed with high resistance to drastic changes in environmental factors. They seem, however, to exhibit a high degree of biochemical differentiation during germination. In this case no definite criteria of 'germination' have yet been presented. The term 'germination' merely represents two phases of the growth phenomenon—swelling and sprouting—as will be discussed later in detail.

There are a great many varieties of fungal spores, of which the asexual conidia of the *Fungi imperfecti* are considered to be rather simple with respect to their morphology and physiology. We have chosen the conidia of *Aspergillus niger* 1617 (supplied by Dr. J. W. Foster of the University of Texas) for the study of germination. This organism satisfied nearly all the following specifications for the selection of an appropriate fungus, among which abundant spore formation, high percentage of spore germination and high synchrony in germination were especially taken into account. The spore sample was collected by a dry procedure and stored in a refrigerator (Yanagita and Yamagishi, 1958), since fungal spores collected wet, as normally done, undergo physiological changes prior to germination and cannot be stored for a long period.

II. Physiological Aspects of Spore Germination

1. *Nutritional Requirements*

Various fungi have been known to grow abundantly on inorganic media containing glucose as sole carbon source, such as the Czapek medium. It was found, however, that when the spores of *A. niger* were inoculated on this medium, only a small percentage of the spores germinated, although once the spores had germinated, they grew well in the simple medium. If the spores were inoculated on a complex medium, such as one of peptone, yeast extract and glucose, more than ninety per cent of them germinated (Yanagita, 1957). This indicates that the spores have a more rigorous nutritional requirement than do the hyphae, the vegetative cells of the same organism. Such a difference in nutritional requirements between spores and vegetative cells is also known in bacteria. These findings led us to search for a synthetic medium which would allow almost all the fungal spores to germinate, and finally the following composition was found to be ideal for supporting germination: DL-alanine, 0·3 g; L-proline, 0·2 g; guanine, 0·01 g; glucose, 10 g; KH_2PO_4, 2·3 g; $Na_2HPO_4 \cdot 12H_2O$, 1·8 g; water to 1 l. (pH 6·0).

There are two points worth noticing with respect to the composition of the medium: (1) Alanine is an essential requirement for complete germination. It is of interest to recall that spores of several bacilli also require alanine for germination (Halvorson and Church, 1957). Of the two isomers of alanine, L-alanine is more effective in supporting spore germination in this fungus, although D-alanine is also usable, with an effectiveness one-tenth that of L-alanine. (2) This medium does not contain any source of sulfur. There was no apparent effect of the absence of sulfate on the germination process. This is readily explained by the fact that the spores contain a sulfur reservoir which provides sulfur during germination as will be discussed later.

Although CO_2 is not positively added to the medium, it is also one of the essential factors for the initiation of germination (Rippel and Bortels, 1927). When CO_2 was eliminated from the environment, even the swelling of the spores failed to occur in the germination medium. The optimum concentration of gaseous CO_2 for germination of *A. niger* spores has been found to be about 0·1 per cent (Yanagita, 1957).

2. *Morphological Change*

The dormant spores of *A. niger* have an average diameter of about 3 μ. When they were inoculated into the germination medium, their packed volume and dry weight changed almost exponentially, after a lag period of about 3 hr. The sprouting of germ tubes began at the sixth hour and was completed at the eighth hour when the size was about 8 μ in diameter (Fig. 1). This is one indication of the degree of synchrony at the end of germination. In general, the time of germination seems to be normally distributed in fungi (Ryan, 1948; Wellman and McCallan, 1958).

Nuclear staining, with iron-hematoxylin-eosin or with Feulgen reagent, of the germinating spores revealed that the dormant spores were mononucleate and the first nuclear division occurred at around the fifth to sixth hour. Sakaguchi and Ishitani (1952) reported that in the spore of *A. oryzae*, nuclear division occurs prior to sprouting and one of the nuclei moves toward the sprouted germ tube while the other remains in the host spore. This seems to be true also in *A. niger* spores.

3. *Respiratory Activity*

Estimation of the respiratory activity of germinating spores under cultivation should be carried out under an appropriate partial pressure of CO_2, using a CO_2 buffer, such as that of Krebs (1951). The constant elimination of CO_2 with KOH as usual results in the retardation of germination. As shown in Fig. 1, the rate of O_2 uptake by germinating spores of *A. niger* increases abruptly when the germ tubes sprout (Yanagita, 1957), as previously observed in *A. oryzae* spores (Terui and Mochizuki, 1955). Contrary to these observations, no such break in the rate of respiration was observed in the germinating spores of *Myrothecium verrucaria*, although measurement of the respiratory activity was carried out conventionally, using KOH solution as a CO_2 absorbent (Mandels *et al.*, 1956).

In general, the presence of O_2 is considered to be a prerequisite for fungal spore germination. The respiration of fungal spores is

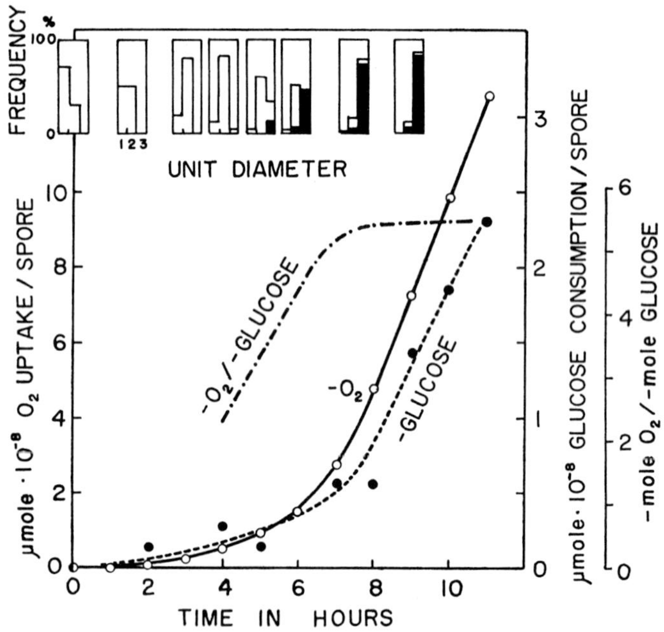

FIG. 1. Course of O_2 uptake and glucose consumption by germinating spores of *Aspergillus niger*. Inserted histograms show shifts in size distribution of spores during germination. Unit diameters 1, 2, and 3 represent the diameters of spores less than 3·2 μ, between 3·2 and 4·8 μ, and more than 4·8 μ, respectively. Black bars represent percentages of spores that sprouted germ tubes (Yanagita, 1957).

characterized by its high activity of endogenous respiration. The substrate of this endogenous respiration is not fully understood. In the spores of *A. niger*, it is highly probable that D-mannitol contained in the dormant spores in a large amount (Sumi, 1928; Takebe, 1960) serves as a substrate, while in the macroconidia of *Neurospora sitophila*, it was suggested that fat might be the substrate because of its high content and the low respiratory quotient (0·7) of the spores (Owens, 1958). In *A. oryzae* spores, a high respiratory quotient, even greater than 1·7, was observed and the high level fell abruptly in the initial phase of germination, followed by a slight increase to more than 1 during the later course of germination (Terui and Mochizuki, 1955).

Although the glucose respiration of *A. niger* spores will be discussed in more detail, the change in the ratio of O_2 uptake to glucose consumed during the course of germination seems to reflect the highly anabolic nature of the germinating spore metabolism. The ratio was found to increase as germination proceeded. Assuming that the O_2 was used only for the oxidation of glucose, incomplete oxidation of glucose occurred during the early phase of germination and almost complete oxidation in the phase of sprouting. Thus an active assimilation of glucose would be expected during early germination (Yanagita, 1962).

4. *Susceptibility to Physical and Chemical Agents*

A. niger spores do not respond to the 'heat activation' which is generally used to stimulate germination in bacterial and some fungal spores. Quantitative comparison of the heat sensitivity of dormant spores and of vegetative mycelia of *A. niger* revealed that the temperatures of 50 per cent survival were 46·5° and 51·5° for mycelia and spores, respectively. Thus the spores showed only slightly higher resistance than did the mycelia, yet the loss of heat resistance of germinating spores in the early phase was demonstrable. As shown in Fig. 2, an abrupt loss of heat resistance occurred within the first 50 min (Yanagita and Yamagishi, 1958). This is exactly comparable to the loss of heat resistance in germinating bacterial spores.

Initial loss of resistance on the part of the germinating spores to a chemical agent, pentachlorophenolate, is also shown in Fig. 2. Recent unpublished observations made by Maruyama of this laboratory show that the resistance to ultraviolet irradiation of the dormant spores also shows an initial decrease. All of these observations suggest that some physiological change may be occurring in the early phase of germination, although no apparent morphological change occurs in this fungal spore. In bacterial spores, the loss of heat resistance is usually accompanied by a marked change in optical density of the spore suspension, caused by the loss of refractility of the spores (Hachisuka *et al.*, 1954). Thus it seems reasonable to assume that the loss of resistance to physical and chemical agents in *A. niger* spores is one of the indications of the initiation of germination. Some experiments carried out along this line reveal that, among various constituents of the germination medium (see section II-1), L-alanine is an essential substrate for the initiation of germination. In the presence of L-alanine, a loss of heat resistance on the part of the spores comparable to that found in the complete medium was observed, while in the presence of each of the other constituents, only a partial or a slight loss occurred (Takebe and Shimizu, 1959). It should, however, be noted that the alanine is continuously required, not only for the initiation but for the completion of germination (Yanagita, 1957).

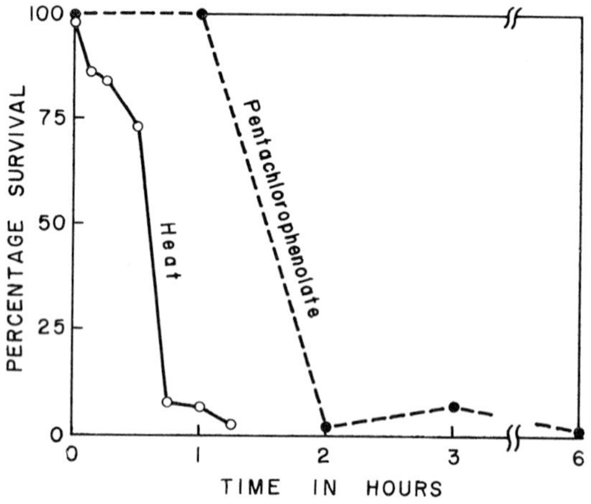

FIG. 2. Loss of heat and germicidal resistance in germinating spores of *Aspergillus niger* (Yanagita and Yamagishi, 1958). Heat treatment: 49° for 3 min; germicidal treatment: $2 \times 10^{-4} M$ pentachlorophenolate for 20 min at 30°.

As for the change in sensitivity to chemical agents of germinating spores, the finding of Terui and his collaborators (1960) is worthy of notice. Spores of *A. oryzae* were found to germinate and grow even in the presence of a considerable concentration of respiratory inhibitors, such as arsenite, arsenate, fluoride and azide, while those initially cultured in the absence of these inhibitors failed to germinate when the inhibitors were later added. They found that this phenomenon of acquisition of resistance to the inhibitor during germination is caused not by the selection of resistant mutants but by the adaptation of each germinating spore to these inhibitors.

III. Biochemical Aspects of Spore Germination

1. *Change in Amount of Cellular Substances*

In the germinating spores of *A. niger*, the total nitrogen content increased almost exponentially, while the protein content also increased, but after a longer lag period than the former. The percentage of nitrogen content per dry weight of spores decreased from the initial high value of 7·3, to 4·2 per cent at the end of germination (Yanagita, 1957). This result is in agreement with the earlier observation that the spores contain a higher amount of nitrogen than the mycelia (Shepherd, 1957; Sumi, 1928), probably reflecting a lower polysaccharide (cell wall substance) content in the spores than in the mycelia. The cell wall polysaccharide of these fungal spores as well as of the mycelia is composed of glucose, glucosamine and small amounts of a few amino acids (Yanagita and Shimizu, unpublished data). Although the net amount of protein did not apparently increase during the first half of germination, actual synthesis of protein was found to take place even in the early phase of germination as revealed by ^{35}S-sulfate and ^{14}C-L-alanine incorporation experiments, which will be discussed later. The active synthesis of protein, preceded by the accumulation of free amino acids, occurred at the time of germ tube sprouting.

An increase in spore nucleic acid occurred earlier in the germination period than the increase in any other macromolecular substances so far examined. In this case also, however, a lag period of less than 2 hr usually preceded the nucleic acid synthesis. As indicated in Figs. 3 and 5 ribonucleic acid (RNA) is shown to increase first, followed by the synthesis of deoxyribonucleic acid (DNA). This sequence of nucleic acid synthesis is also demonstrated in germinating bacterial spores (Fitz-James, 1955; Woese and Forro, 1960). Another point of agreement in the syntheses of nucleic acids in our fungal spores and in the spores of *B. subtilis* (Woese and Forro, 1960) is that the rate of RNA synthesis showed a temporary diminution when the total cell RNA approximately doubled, whereas the rate of DNA synthesis did not change so distinctively. Our experience has shown that the synthesis of DNA in this fungal spore sometimes proceeded in a rather stepwise fashion. The appearance of such steps in nucleic acid syntheses seems to be dependent upon the synchrony of spore germination. Contrary to these observations, the continuous synthesis of DNA was demonstrated in the synchronously germinating spores of *B. cereus*, although that of RNA proceeded discontinuously (Fitz-James, 1955; Young and Fitz-James, 1959). The discontinuity (Maruyama, 1956; Maruyama and Lark, 1959) and the continuity (Abbo and Pardee, 1960; McFall and Stent, 1959; Schaechter *et al.*, 1959) of DNA synthesis in bacterial cells are matters of controversy for the present, but the discontinuity in our fungal spores seems to be a definite fact. Since the synchronization of these cells is not programmed artificially,

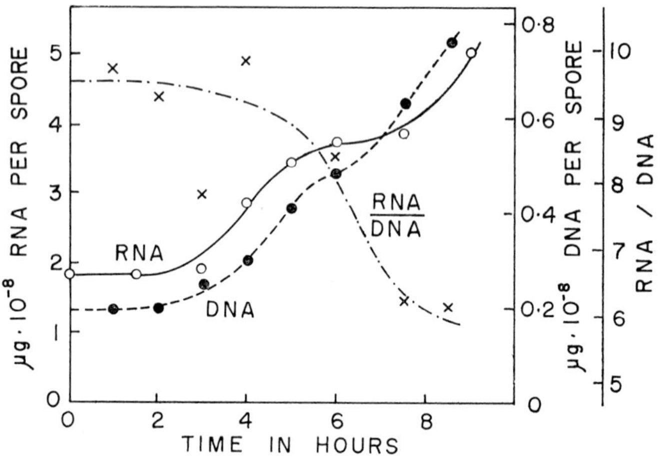

FIG. 3. Change in amount of nucleic acid during germination of *Aspergillus niger* spores (Yanagita, 1957).

the argument that the method of synchronization forced the cells to synthesize DNA discontinuously is precluded. The first nuclear division occurring at the fifth to sixth hour of cultivation must be related to the stepwise increase of DNA. Huebschman (1952) found that the conidia of *Neurospora crassa* have a varying number of nuclei and the average number of nuclei per conidium differs according to the conditions of culture. If such fungal spores are employed as test organisms, the DNA will increase continuously during germination. Thus the continuity or the discontinuity of DNA synthesis is considered to be dependent upon the degree of synchrony of germination, on the one hand, and upon variations in the number of nuclei originally present in the dormant spores, on the other hand. The RNA to DNA ratio in *A. niger* spores decreases from more than 9 to as few as 6 as germination proceeds (Fig. 3). In *B. subtilis* spores, the RNA–DNA correlation is quite the reverse. The ratio rose from 4 to the maximum value of 9 at the time of the first cell division (Woese and Forro, 1960).

2. *Protein Synthesis and Sulfur Metabolism*

As described in the preceding section, a free amino acid pool appeared prior to the active protein synthesis, and the size of the pool diminished as germ tubes sprouted in *A. niger* spores. Qualitative analysis of the pool disclosed the presence of alanine and glutamic acid for the major part and aspartic acid, serine, glycine, threonine, histidine, citrullin, and tryptophan in lesser amounts (Yanagita, 1957). In spores labeled with ^{35}S, the presence of cystine was also demonstrated in this free amino acid pool (Takebe and Yanagita, 1959). Attention was then focused on the origin of these amino acids.

There are two amino acids, alanine and proline, in the present medium. Therefore, the presence of alanine in the pool was supposed to derive from the exogenous alanine. It was, however, shown that after the incorporation of ^{14}C-alanine into germinating spores, no free ^{14}C-alanine could be detected in the spores, at least in the early phase of germination, as will be discussed in section III-5. Alanine seems to be instantly metabolized as soon as it is incorporated.

Another possible explanation of the appearance of the free amino acids is by way of transamination from alanine. The amounts so far detected of the two keto acids, pyruvate and α-ketoglutarate, changed by a course similar to that of the free amino acid, except that they also increased after accumulating at the time of sprouting. Among various transaminase activities demonstrated in spores at 6·5 hr of cultivation, when free amino acids and keto acids accumulated markedly, aspartic–glutamic and alanine–glutamic transaminases were shown to be the strongest. The course of change of these transaminase activities was therefore followed during germination. As shown in Table I, the activity of aspartic–glutamic transaminase was almost constant throughout the course of germination, while that of alanine–glutamic transaminase appeared later, at the time of free amino acid accumulation. It is worth mentioning that vegetative cells of *B. megaterium* contain both aspartic–glutamic and alanine–glutamic transaminases, whereas the spores of this organism contain only aspartic–glutamic transaminase (Levinson and Sevag, 1954). These observations may imply that the latter enzyme plays a role in glutamic acid accumulation in these fungal spores.

TABLE I. Development of activities of aspartic–glutamic and alanine–glutamic transaminases in germinating spores of *Aspergillus niger* (Takebe and Yanagita, 1959)

Time of cultivation (hr)	Aspartic–glutamic transaminase	Alanine–glutamic transaminase
	Qtransam[a]	Qtransam[a]
0	16·6	0
3	20·3	0
6·5	12·3	2·7
10	12·5	2·2

[a] μmoles of glutamate formed per hr per 1 mg of protein nitrogen of enzyme preparation.

One more possible explanation of the accumulation of the free amino acid pool is the decomposition of a reservoir protein in order to reconstruct a new protein. In bacterial spores, the spore coat and some part of the spore proteins are broken down in the initial phase of germination (Levinson, 1957; Powell, 1953), although it is not yet known whether the products of these substances are utilized for the reconstruction of a new protein. In this fungal spore, no such initial decrease in the amount of spore protein was observed (III-1). It remains, however, still unknown whether or not the decomposition of a minor part of such spore protein may occur. Radioactive spores previously labeled with ^{35}S were used to elucidate this problem. In these radioactive spores, the radioactivity was specifically concentrated in the hot-water-soluble fraction (59 per cent) and in the protein fraction (37 per cent), practically none being detected in the acid-soluble and lipid fractions. When such spores were cultured as usual, the radioactivity of each fraction changed as indicated in Fig. 4. The fact that no initial decrease occurred in the protein fraction suggests that the spore protein does not decompose in the initial phase of germination, so far as the sulfur protein is concerned. The proposed possibility that the free amino acid pool is derived from reservoir protein was thus shown to be far from the case.

Figure 4 also shows that the radioactivity of the hot water extract of the spores gradually decreased during germination, together with the reciprocal change in that of the protein fraction. This fact suggests that some water-soluble sulfur compounds in the spores may furnish the sulfur moiety of the newly synthesized protein. Qualitative analysis of the water-soluble sulfur compounds disclosed the presence of choline sulfate (ester) as a major component (Takebe,

1960). The dormant spores of the Aspergilli were found to contain this substance specifically in a high amount, as high as 1·2-1·5 per cent dry weight of spores (Takebe, 1960) as against only 0·2–0·3 per cent dry weight of mycelia (Stevens and Vohra, 1955; Woolley and Peterson, 1937) in *A. niger*. Our recent studies showed radio-autographically that in the growing colony of this fungus, ^{35}S (given as sulfate) once incorporated by a pulse experiment is translocated vertically within the colony into the fruiting bodies and accumulates mainly in the spores (Yanagita and Kogané, 1962).

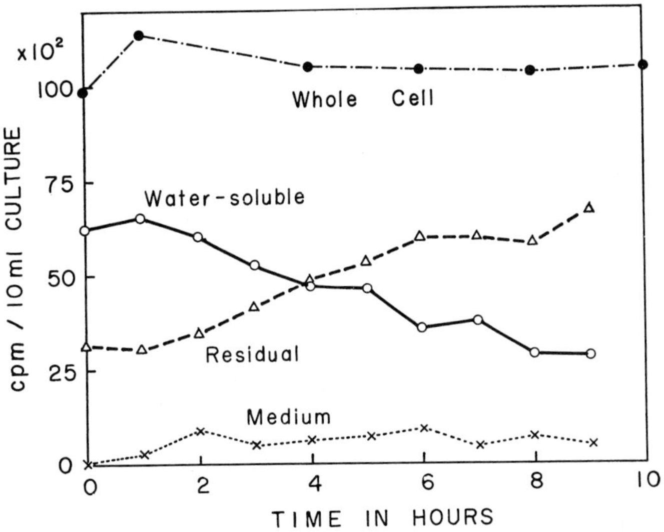

FIG. 4. Change in radioactivity of cellular fractions during germination of ^{35}S-labeled *Aspergillus niger* spores (Takebe and Yanagita, 1959).

Although the pathway of choline sulfate biosynthesis has been studied by some workers (Kaji and Gregory, 1959; Kaji and McElroy, 1958; Spencer and Harada, 1959), that of its utilization in the cell has not yet been elucidated. Takebe found no enzymatic activity of sulfatase, which hydrolyzes choline sulfate in this spore. It may be possible to conjecture that the reverse reaction of a part of the choline sulfate biosynthesis is involved in the construction of sulfur amino acids via 3′-phosphoadenosine-5′-phosphosulfate in the germinating spores, although none of the probable intermediates could be detected in this system during germination. The role of the choline moiety of the sulfate ester in germinating spores still remains unexplained. This substance probably acts as a methyl donor.

It should be recalled that the germination of these fungal spores can proceed unaffected by the absence or presence of sulfate in the medium (II-1). It may thus be concluded that choline sulfate reserved in the dormant spores plays the role of sole sulfur source during germination under the conditions of culture. It is, however, noticed that the incorporation of choline sulfate-sulfur into the protein synthesized *de novo* is retarded to some extent, if other sulfur compounds are added to the medium. Exogenous thiosulfate, sulfite, cysteine and glutathione seem to inhibit incorporation completely, while sulfate and methionine inhibit it only partially (Takebe and Yanagita, 1959). Indeed, ^{35}S-sulfate added to the medium was found to be incorporated into germinating spores after a rather long lag period. It is therefore possible that some of the exogenously added sulfur compounds can be substituted for choline sulfate-sulfur in the spores.

3. *Phosphorus Metabolism and Nucleic Acid Synthesis*

As already shown in Fig. 3, there is a lag period of a few hours before the onset of RNA synthesis in the germination of *A. niger* spores, suggesting the occurrence in the initial phase of some events relative to the nucleic acid synthesis. The ^{32}P-phosphate incorporation experiment showed that active ^{32}P-uptake began to occur after a lag of about 1–2 hr, and the radioactivity appeared first in the acid-soluble fraction, then in the other fractions, including the RNA fraction. Estimation of the phosphorus content of various fractions during the course of germination by the method of Schmidt–Thannhauser–Schneider (Volkin and Cohn, 1954), with slight modifications, disclosed that the amount of polyphosphate and phospholipid decreased rapidly in the initial 1 hr of germination. Thus the polyphosphate and phospholipid are considered to be important phosphagens which provide energy and phosphorus, at least in part, for the synthesis of nucleic acid in the early phase of germination. A similar observation has been made in *A. nidulans* spores (Shepherd, 1957). The most pronounced changes occurring in the early germination of this fungal spore were a decrease in the amount of polyphosphate, organically bound phosphate, and total lipid, with some other changes. Other workers have found that *A. niger* spores contain relatively large amounts of acid-soluble and insoluble polyphosphates (Bajaj, 1954), although this is incompatible in some respects with the observation that our spores contain mainly acid-insoluble polyphosphate. Our recent cytochemical studies demonstrated that in the growing colony of *A. niger* and of *Penicillium urticae*, highly metachromatic granules, possibly polyphosphate in nature, appeared in the fruiting bodies and translocated into spores through sterigmata. The metachromatic substance thus delivered to the spores became diffuse in appearance (Yanagita and Kogané, 1962). The chemical nature of the phospholipid in *A. niger* spores has not yet been determined. Phospholipids identified in various fungi are lecithin and cephalin in *A. sydowi* (Woolley *et al.*, 1935), inositol phosphate in *Penicillium chrysogenum* (Imai, 1950), and inositol phospholipid in *Neurospora crassa* (Fuller and Tatum, 1956).

Although the metabolic role of polyphosphate in microorganisms has been studied (Chayen *et al.*, 1955; Mudd *et al.*, 1958; Sall *et al.*, 1956; Winder and Denneny, 1957; Yoshida and Yamataka, 1953), the significance of this substance for cellular growth has received comparatively little attention (Sall *et al.*, 1958). The use of radioactive spores of *A. niger* labeled with ^{32}P led us to an understanding of the way in which these phosphorus reserves are utilized for nucleic acid synthesis (Nishi, 1959, 1960, 1961).

In Fig. 5 are shown the changes in radioactivity and phosphorus content of various fractions of germinating *A. niger* spores. Among these fractions, that of polyphosphate contains acid-insoluble polyphosphate extractable from spores with a borate buffer of pH 9·0 (at 30° for 1 hr); the 'sugar phosphate' and 'nucleotide' fractions denote, respectively, the acid-soluble phosphates not adsorbable and adsorbable on Norit (Threlfall, 1957). According to this definition the sugar phosphate fraction should contain inorganic and organic phosphates in addition to the real sugar phosphates described later. The radioactivity and phosphorus content of polyphosphate and phospholipid dropped rapidly during the first 1 to 2 hr period. In response to this drop, the radioactivity and phosphorus content of the acid-soluble phosphates increased markedly at a similar rate during this short period, and a gradual decrease in the radioactivity of the acid-soluble fraction ensued. The radioactivity of RNA began to increase a little later than the increase in that of the acid-soluble phosphates. It should be noted that no increase in the radioactivity of DNA occurred, in spite of the definite increase in the amount of DNA at the fourth hour of cultivation. These observations suggest that the phosphorus of polyphosphate and phospholipid can be utilized for the synthesis of RNA but not for that of DNA. The total decrease in radioactivity of these phosphate reserves did not necessarily balance the increase in that of acid-soluble and RNA fractions, but about 50 per cent of the radioactivity was liberated into the medium. The phosphate thus liberated seemed to be orthophosphate.

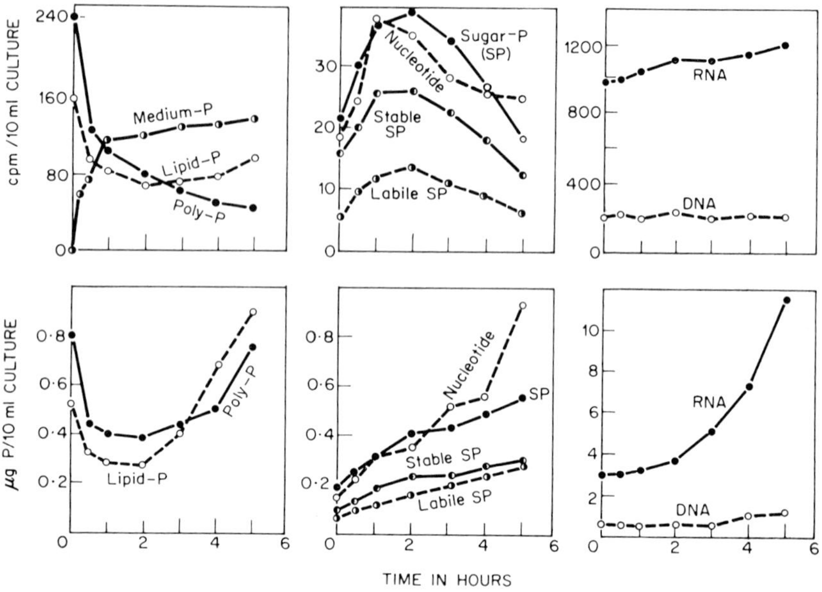

FIG. 5. Change in radioactivity and phosphorus content of various cellular fractions during germination of ^{32}P-labeled *Aspergillus niger* spores (Nishi, 1961).

The above experiments clearly showed that the phosphate reserves furnished first the acid-soluble fraction. The various phosphate entities in the nucleotide and sugar phosphate fractions in the dormant and second-hour spores were therefore determined by radioautography. Generally speaking, a larger number of more distinctive spots were observed in the second-hour spores than in the dormant ones. The radioactive spots identified are as follows:

Sugar phosphate fraction:—glycerophosphate, glyceraldehyde 3-phosphate (GAP), ribose 5-phosphate (R-5-P), glucose 1-phosphate, glucose 6-phosphate (G-6-P), *phytate*, *metaphosphate*.

Nucleotide fraction:—adenosine triphosphate (ATP), *guanosine triphosphate*, *adenosine diphosphate* (*ADP*), adenosine monophosphate.

Among these, the italicized substances were identified only in the second-hour spores. Of these spots, the size and intensity were highest in ATP, GAP, R-5-P, and glycerophosphate in the second-hour spores in this order. These observations suggest that the key intermediates driving the metabolic activities of the initial phase of germination are glycerophosphate and ATP. It thus seems possible that the phosphorus of glycerophosphate and ATP is derived from phospholipid and polyphosphate, respectively.

The above consideration was further confirmed by enzymatic studies. Cell-free extracts of the germinating spores were tested for the activities of pyrophosphatase, phosphomonoesterase, apyrase, polyphosphatase and polyphosphate-kinase. Prior to the discussion of these enzyme activities, the enzyme activities of bacterial spores should be referred to. In these spores there are enzymes which become active when the spores are 'activated', for example, by heat shock. These enzymes are defined as 'dormant enzymes' (Halvorson and Church, 1957). Among all of the enzymes studied with the fungal spores, no enzyme which is inactive in dormant spores and activated by germination was found. It was, however, evident that the rate of increase in enzyme activity, as expressed by the reaction rate per nitrogen content, varied from one enzyme to another. The activities of phosphomonoesterase (β-glycerophosphate as the substrate) and apyrase remained almost unchanged throughout the course of germination, although the activity of the latter was very slight.

The activities of the other enzymes more or less increased as germination proceeded. Pyrophosphatase, being activated by magnesium ions, increased at the time of sprouting to an extent six times that of the dormant spores. This enzyme was also found in the spores of *B. megaterium* by Levinson (1957). He was led to suppose that the physiological significance of pyrophosphatase in bacterial spores may be that of eliminating pyrophosphate, to stimulate the various important biochemical reactions producing pyrophosphate, such as those concerned with the synthesis of coenzyme A and with transacetylation (Lipmann, 1954).

Polyphosphatase in these *Aspergillus* spores mainly depolymerizes polyphosphate, producing orthophosphate to only a slight degree. It may, therefore, be better to call it a polyphosphate-depolymerase. The activity of this enzyme increases rapidly during the initial 2 hr and levels off gradually in the latter part of germination. The enzyme which transfers polyphosphate phosphorus to ADP, yielding ATP, has been studied by many workers (Hoffmann-Ostenhof *et al.*, 1954; Kornberg, 1957; Winder and Denneny, 1955; Yoshida, 1953). This enzyme is considered to be a key enzyme in the utilization of polyphosphate in the early part of germination. An experiment using ^{32}P-labeled polyphosphate as phosphate donor gave the result shown in Table II. This enzyme undergoes a definite increase in activity in the early phase. It should, however, be mentioned that this enzyme activity was rather weak. On recalling the fact that in intact spores a large amount of ATP accumulated at the probable expense of reserve polyphosphate, the low activity of the polyphosphate-kinase seems, on the one hand, to be due to its partial inactivation during the preparation of the cell-free extract. Its weak activity may, alternatively, be attributed to the chemical nature of the synthetic polyphosphate used as the substrate, although the identity of the natural and synthetic polyphosphates has not yet been established. This line of consideration brings the role of polyphosphate-depolymerase into focus, suggesting that the natural polyphosphate contained in dormant spores is in a highly polymerized state which is depolymerized to a substance of appropriate molecular weight; this in turn is utilized for the formation of ATP, with polyphosphate-kinase as catalyst. The molecular weight of the polyphosphate isolated from the mycelia of *A. niger* was reported to be 6000–7000 (Ingelman and Malmgren, 1950).

TABLE II. Enzymatic synthesis of ^{32}P-ATP from ^{32}P-polyphosphate and ADP by cell-free extracts of germinating spores of *Aspergillus niger* (Nishi, 1960)

Spore extract	Phosphorus transferred to ADP after 60 min incubation		
	per tube[a] cpm	per tube[a] μmole	per μg protein nitrogen, μmole
Dormant spores	160	0·7	0·013
3-hr spores	280	1·2	0·030

[a] Reaction mixture contained $3{\cdot}3 \times 10^{-2}M$ Tris buffer (pH 7·0), $2{\cdot}2 \times 10^{-3}M$ sodium^{32} P-polyphosphate (232 cpm/μmole P), $2 \times 10^{-3}M$ ADP, $2 \times 10^{-3}M$ $MgCl_2$, and spore extract.

The above discussion has concentrated solely on the phosphorus metabolism in the early phase of germination. The utilization of endogenous phosphate reserves may be restricted within the initial few hours of germination, during which time these phosphagens may serve for the synthesis of RNA (but not for DNA) and for the generation of energy. In all probability, the chemical equilibria between various phosphorus compounds may shift dynamically with each stage of the germination process. After the exhaustion of the phosphagens, exogenous phosphate is utilized for subsequent cellular growth. It was indeed observed that exogenous phosphate uptake begins after a lag of 1–2 hr. It is significant that polyphosphate and phospholipid are resynthesized after their initial diminution. Furthermore, an active turnover of these phosphates was demonstrated before their net increase could be observed at the third to fourth hour of incubation. Thus these phosphates seem to play an essential role in the generation of energy and in the synthesis of nucleic acid even at a later stage of germination. In contrast to these phosphates, RNA and DNA, once synthesized, were shown to be stable during the course of germination, since no appreciable turnover of these nucleic acids was demonstrated. This finding confirms the view that there is little or no intracellular turnover of RNA in growing cells (Halvorson, 1958).

Further observations on the nucleic acid synthesis are concerned with the qualitative change in RNA during germination. The first finding along this line is that the base composition of RNA definitely changes in the early phase. The RNA isolated by the Schmidt–Thannhauser method was hydrolyzed with *N* KOH and the resultant nucleotides were subjected to column chromatography. The nucleotide composition thus estimated in spores of various ages is shown in Table III (Nishi, unpublished data). It is clearly indicated

TABLE III. Nucleotide composition of RNA isolated from germinating spores of *Aspergillus niger* cultured for various times. (Numbers indicated are relative values taking those for adenylic acid as unity)

Time of cultivation (hr)	Adenylic acid	Guanylic acid	Uridylic acid	Cytidylic acid	Purine / Pyrimidine	6-Amino / 6-Keto
0	1·00	1·22	0·94	0·91	1·20	0·88
2	1·00	1·14	1·10	0·94	1·05	0·87
4	1·00	1·28	1·21	1·18	0·96	0·88
6	1·00	1·17	1·12	0·97	1·02	0·87

that the purine–pyrimidine ratio is highest in the dormant spores and falls to nearly 1 after 2 hr of cultivation. The ratio of adenylic plus cytidylic acids to guanylic plus uridylic acids (i.e., 6-amino to 6-keto), on the other hand, remains almost unchanged. This finding was partly supported by the following observation. After the spores were cultured for 2 hr in a medium containing ^{32}P-phosphate, the RNA was isolated, hydrolyzed to nucleotides, and subjected to paper chromatography and then to radioautography. This procedure gave only one distinctive spot, which was identified as cytidylic acid. Although the hydrolysis of RNA was performed with *N* KOH, which is known to form nucleoside 3′-phosphate, this experiment suggests that the newly synthesized RNA is, in all probability, furnished initially with cytidylic acid, a pyrimidine nucleotide (Yanagita, unpublished data). Further confirmation of this consideration will be discussed later (section III-5). Thus the initial decrease in the purine–pyrimidine ratio of RNA seems to depend on the fact that more of the pyrimidines than the purines are synthesized in the early phase of germination.

4. *Carbohydrate Metabolism*

In bacterial spores, the absence of a functional glycolytic system in the activated spores, or in spore extracts, seems evident from the observations of many workers (except for those of Amaha and Nakahara, 1959), and the lack of hexokinase activity also implies that the hexose monophosphate (HMP) pathway is not functioning in this system (Church and Halvorson, 1955; Hachisuka *et al.*, 1956; Halvorson, 1957). It should, however, be mentioned that the presence in vegetative cells of the Bacillaceae of a glycolytic system as well as of an HMP system (DeLey, 1955; Dedonder, 1952; Keynam *et al.*, 1954) means that these as well as other enzyme systems are actually developing during germination.

By contrast, the spores of *A. niger* seem to be furnished with Embden–Meyerhof–Parnas (EMP) and HMP routes of glucose dissimilation. Maruyama (1957) found that dormant spores dried instantly *in vacuo* with phosphorus pentoxide could oxidize glucose, G-6-P, hexose diphosphate (HDP), 6-phosphogluconate (6-PG), and R-5-P, but not gluconate, although intact and acetone-dried spores could not oxidize these substrates. As mentioned in II-3, it was found that spores of the early germination period oxidize glucose and other substrates incompletely, while sprouted spores oxidize them almost completely. The course of glucose oxidation by the spores of a 3-hr culture transferred to phosphate buffer containing 5×10^{-3} *M* arsenite is shown in Fig. 6. It was the usual course of events throughout germination that O_2 uptake and CO_2 production in such a system were rapid in the initial period and sloped off rather abruptly when the added glucose was completely consumed, as indicated in the figure. Shortly after the glucose had been exhausted, pyruvate, acetaldehyde and ethanol as well as minor amounts of lactate, acetate and α-ketoglutarate accumulated in the medium. The amount of pyruvate thus accumulated remained unchanged, and that of acetaldehyde dropped abruptly in a later period. The cold trifluoroacetic acid extract of such spores at the time of glucose exhaustion was found to contain orthophosphate, GAP and R-5-P, but no other phosphate esters have so far been observed; the amount of R-5-P and GAP was highest shortly after glucose exhaustion, and gradually decreased thereafter. In this case no synthesis of RNA occurred.

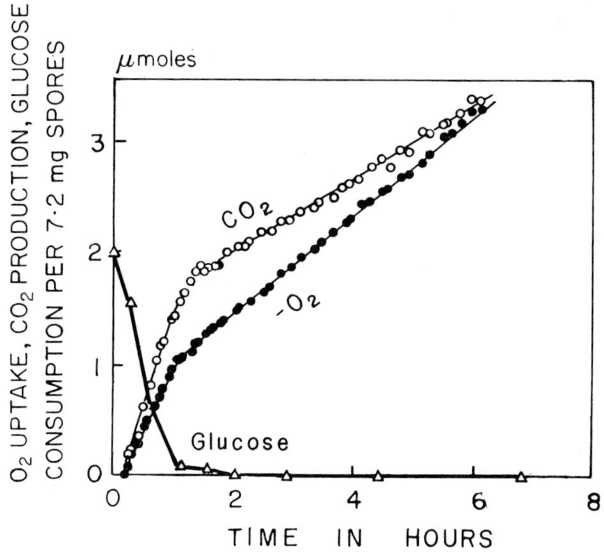

FIG. 6. Course of O_2 uptake, CO_2 production, and glucose consumption by 3-hr spores of *Aspergillus niger* (Maruyama, 1957). Spores are suspended in 0·05 *M* phosphate buffer (pH 6·0) containing glucose (2 μmoles/flask) and arsenite ($5 \times 10^{-3} M$).

The study of carbohydrate metabolism by microorganisms has revealed a number of diverse pathways by which pyruvate is formed from glucose (Gunsalus *et al.*, 1955). Among these routes, the oxidative path of gluconate without phosphorylation was proved to be absent in these fungal spores at least during germination. Although the formation of pyruvate from 6-PG via 2-keto-3-deoxy-6-phosphogluconate, which was found only in species of *Acetobacter*, *Pseudomonas* and *Azotobacter*, was not tested, experiments on the oxidation of R-5-P by these spores revealed that they form GAP, but not sedoheptulose-7-phosphate. These observations, together with the others mentioned above, lead to the assumption that the oxidative pathways functioning in the germinating spores of *A. niger* are the EMP pathway and at least a part of the pentose cycle (HMP shunt pathway). Moreover, in the above experiments, the facts that (1) the amount of pyruvate appearing in the medium leveled off after the time of glucose exhaustion and (2) R-5-P and GAP, accumulated shortly after the glucose had been depleted, gradually disappeared from the system, support the following assumptions: (1) O_2 uptake and CO_2 production occurring during the period of glucose consumption are dependent on the functioning of the HMP shunt pathway together with that of pyruvate oxidation, and (2) those occurring after the time of glucose exhaustion occur principally by way of pyruvate oxidation. In this case, the oxidation of pyruvate may be considered to result in the formation of acetaldehyde and ethanol, as demonstrated above, since the addition of arsenite to this system seemed to inhibit almost completely the functioning of the tricarboxylic acid cycle; in the absence of this inhibitor a fairly large amount of α-ketoglutarate was produced.

Assuming that the rate of O_2 uptake (or CO_2 production) through the HMP shunt pathway represents the metabolic rate of this path, the following relationships may be formulated:

$$V = g/\tau,$$
$$V = V_E + V_H,$$
$$v_1 = V_H + v_2,$$

in which the following notations are used:

V: overall rate of glucose dissimilation,
g: amount of glucose initially added,
τ: period of glucose consumption,
V_E: rate of glucose dissimilation through the EMP pathway,
V_H: rate of glucose dissimilation through the HMP shunt pathway,
v_1: overall rate of O_2 uptake (or CO_2 production) during the period of initial rapid glucose oxidation,
v_2: rate of O_2 uptake (or CO_2 production) through routes other than the HMP pathway. This may be replaced by the rate of O_2 uptake (or CO_2 production) observed after the time of glucose exhaustion.

Using these relationships, values V and V_H were calculated for the spores of various germination stages, employing the CO_2 production–time course curves. Table IV shows the results of these experiments. As already shown in Fig. 3, RNA is synthesized in germinating spores through a course characterized by a stepwise increase. The percentage of the contribution of the HMP shunt pathway in glucose dissimilation (V_H/V) thus calculated is shown to be high just before the times when active RNA synthesis begins. This result seems to bear out the hypothesis that ribose as a building block of RNA originates in the direct oxidative pathway of glucose (Cohen, 1951).

TABLE IV. Change in rate of contribution of HMP pathway in CO_2 production from glucose by germinating spores of *Aspergillus niger* (Maruyama, 1957). (For calculation, see text)

Time of cultivation (hr)	V_H/V
0	0·23
1·5	0·59
3	0·31
5	0·96
9	0·19

V_H = rate of glucose dissimilation through HMP pathway.
V = overall rate of glucose dissimilation.

5. *Role of Alanine and* CO_2

Alanine was shown to be an important exogenous substrate for the initiation and development of germination in this fungal spore (II-1, 4). The metabolism of this amino acid has also received special attention in the studies of bacterial spore germination (Church and Halvorson, 1955). Our approach to elucidating the problem of the mechanism of its role as a trigger in the initiation of germination is directed towards its contribution to the nucleic acid and protein syntheses, which occur early in spore germination (Hoshino and Yanagita, 1960).

When *A. niger* spores are cultured in a germination medium supplemented with [U-^{14}C]-L-alanine, they incorporate this amino acid after a short lag of about 30 min. The incorporation is dependent upon the presence of energy sources, such as glucose and phosphate. Among the Schmidt–Thannhauser fractions of spores, the radioactivity was found to be higher in the residual fraction, containing most of the protein and polysaccharide, than in the nucleic acid fractions; higher in the RNA than in the DNA fraction. Radioautographic analysis of the hydrolysate of nucleic acid obtained from the 2·5-hr spores revealed that pyrimidine bases, such as uracil and thymine, are highly labeled with ^{14}C. This finding confirms the previous observation that the purine–pyrimidine ratio of RNA falls markedly in the early germination stage (see Table III). It should also be noticed that radioactive thymine is present in DNA molecules even at this early stage, when no net increase in the amount of DNA is yet observed. As will be pointed out later in the discussion of CO_2 incorporation, this finding suggests that some minor but important modification in the DNA molecules is occurring in the initial step of germination. This problem is now being subjected to further investigation.

Analysis of the acid-soluble iraction extracted from spores which had incorporated ^{14}C-alanine disclosed the presence of several radioactive amino acids, such as glycine and serine, but not of alanine. This indicates that the alanine is degraded, primarily by deamination, into other compounds as soon as it is taken up. Extract of disintegrated spores deaminates alanine very quickly with the formation of ammonia and pyruvate, as found in bacterial spores (O'Connor and Halvorson, 1959, 1960). This reaction seems to be catalyzed by diphosphopyridine nucleotide-linked L-alanine dehydrogenase described by Pièrard and Wiame (1960). The pyruvate thus formed is further decarboxylated oxidatively with the formation of CO_2, as also found in bacterial spores (O'Connor and Halvorson, 1959).

The CO_2 produced by this degradation of alanine seems to contribute to the synthesis of macromolecular substances in germinating spores. The use of [1-^{14}C]-DL-alanine as a tracer in the incorporation experiment disclosed that the nucleic acid and protein of germinating spores are highly labeled with ^{14}C. This observation led us to recall that these fungal spores can germinate only in the presence of CO_2 (II-1). $^{14}CO_2$ incorporation experiments were then performed (Yanagita, 1961). As shown in Fig. 7, the CO_2 uptake proceeds quite actively, without any lag, from the very beginning of germination. Moreover, a clear cessation of CO_2 incorporation occurred temporarily during germination, almost coincident with the time of slow-down in the rate of RNA synthesis. Among various exogenous substrates tested, such a characteristic course of incorporation by germinating spores is demonstrated only with CO_2. In agreement with these observations, our recent investigation using synchronous cultures of the fission yeast, *Schizosaccharomyces pombe*, showed that active $^{14}CO_2$ uptake occurred subsequent to cell division, followed by RNA synthesis (Sando and Yanagita, 1961).

Upon incorporation of $^{14}CO_2$, the nucleic acid and protein of the spores are labeled with ^{14}C. Nucleic acid isolated from spores which had been cultured for 1 hr in the presence of $^{14}CO_2$ was hydrolyzed and subjected to paper chromatography and then to radioautography. It was thus demonstrated that all the bases are labeled with

^{14}C; among them, the pyrimidine bases such as thymine, uracil and cytosine are highly radioactive, as is the case in ^{14}C-alanine incorporation experiments. A similar experiment using spores cultured for only 15 min showed that radioactive thymine and uracil appear in the nucleic acids. This finding suggests that germination is initiated by a trigger mechanism, which may involve some modification of the RNA or DNA molecules, set off by the incorporated CO_2 or alanine. The modification of the nucleic acids may possibly be characterized by a decrease in the purine–pyrimidine ratio. In this connection, ammonia, produced by the degradation of alanine, and CO_2 may be considered to provide building blocks for the synthesis of pyrimidine bases. Moreover, one is tempted to suppose that the endogenous phosphorus supplied by polyphosphates and phospholipids in the early germination stage may be accepted by some compounds, which are derived from CO_2 and alanine, resulting in the formation of nucleotides.

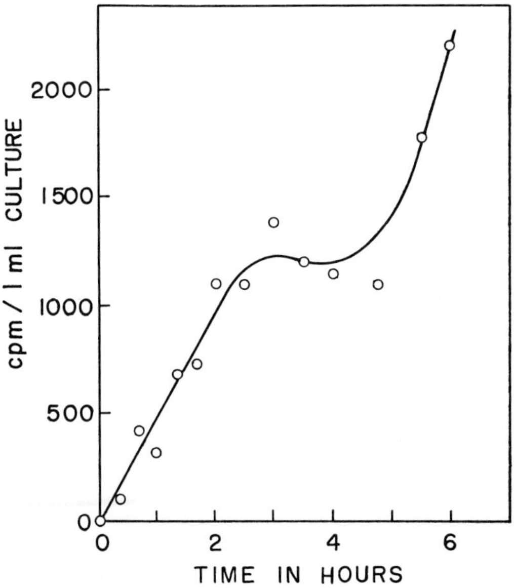

Fig. 7. Course of $^{14}CO_2$ incorporation by germinating spores of *Aspergillus niger* (Yanagita, 1961).

IV. General Conclusion

Physiological changes occurring during germination of *A. niger* spores may be summarized as shown in Fig. 8, in which morphological changes occurring in germinating bacterial spores are also indicated for comparison. It may be apt to define the process of 'germination' as an initial short period, during which many conspicuous physiological changes, comparable to those found in germinating bacterial spores, are occurring (Mandels *et al.*, 1956). This initial period is characterized by events specific to the spores *per se*. After the completion of these events, the metabolic pattern of the spores becomes shifted toward one characteristic of vegetative cells in the phase of 'swelling', which is followed by vegetative growth. The above interpretation of the growth of fungal spores is, of course, based principally on ideas obtained by bacteriologists from bacterial spore germination, although these spores differ in many characteristics. Such a generalization may, however, further the attempt to envisage the mechanisms involved in spore germination of microorganisms.

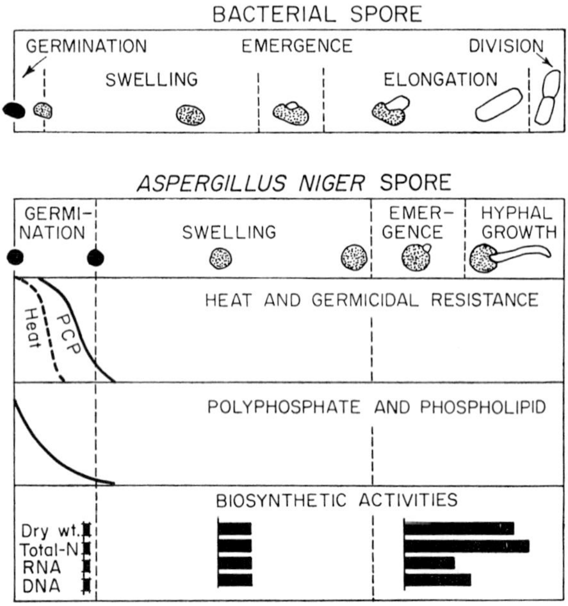

FIG. 8. Schematic representation of germination process of *Aspergillus niger* spores with special reference to that of bacterial spores. Biosynthetic activities represent relative rates of syntheses taking those found in the swelling phase for standard.

A recapitulation of the metabolic pattern occurring during the phase of 'germination' can be schematized as shown in Fig. 9. The exogenous substrates, considered to be the triggers which initiate germination, may be CO_2 and L-alanine. When germination begins, various reservoirs in the dormant spores, such as polyphosphate, phospholipid, choline sulfate, mannitol and others, are utilized in combination with compounds derived from CO_2 and alanine. The primary substance thus synthesized may be ATP, which provides energy and materials for the synthesis of RNA. The important role of the routes leading to pyruvate formation (i.e., alanine degradation and glucose dissimilation) as well as a cycle for triose oxidation should not be neglected in the phase of 'germination'. These metabolic reactions may provide a basis for the synthesis of amino acids (Roberts *et al.*, 1955), which lead to the synthesis of various enzymes as well as of nucleic acid. After the completion of these biochemical processes during 'germination', RNA is then synthesized at the expense, at least in part, of the phosphorus derived from phosphorus reserves; the synthesis of DNA then follows. It is, of course, highly possible that the RNA initially synthesized may be charged with the

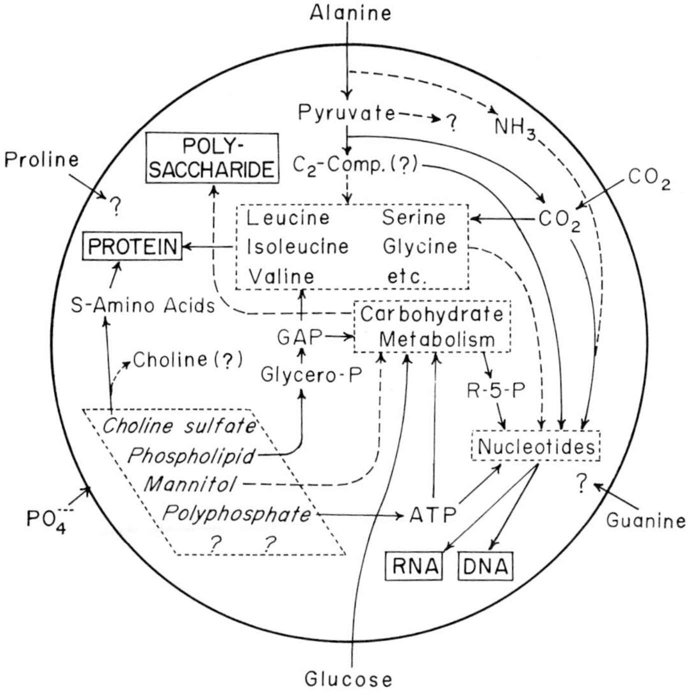

FIG. 9. Schematic diagram of metabolic pattern occurring in the early phase of germination of *Aspergillus niger* spores. Circle represents a spore cell; substances shown outside the circle are exogenous substrates. Italicized substances represent reservoirs in dormant spore. Solid and broken arrows indicate reactions which are demonstrated to proceed and are considered to be probable, respectively.

genetic information carried by DNA in the dormant spores. It is still uncertain at present whether the 'dormant DNA' itself is biologically active. It may be speculated that the 'dormant DNA' may become active only after it undergoes some modification in its molecule. A similar line of speculation may also be applicable to 'dormant RNA' if such an entity is present. It is a temptation to conclude that the characteristic of spore dormancy depends upon the possession of such inactive nucleic acids. However, for the present, this conjecture must await the accumulation of more data to establish its validity.

As the consequence of such initial reactions, the metabolic pattern of the spores is switched to one characteristic of vegetative cells during the phase of 'swelling', as revealed by changes in the rate of O_2 uptake, O_2 uptake–glucose consumption ratio, etc. In this phase, the chemical composition of the cells also undergoes conspicuous changes, as demonstrated by the ratios of total nitrogen–dry weight and RNA–DNA. Although the relationship between subcellular structure and metabolic activities in these spores has not yet been investigated, the finding that an oxidative system, appearing in the supernatant of the cell-free extract of bacterial spores, becomes concentrated in the pellet of the extract as germination proceeded (Matsushiro *et al.*, 1955), suggests an attractive approach to our problem of the growth of fungal spores.

Acknowledgements—The present studies have been performed with the collaboration of Drs. Y. Maruyama, A. Nishi, I. Takebe, J. Hoshino, N. Sando and Miss F. Kogané of this Institute, to whom the author's thanks are due.

REFERENCES

Abbo, F. E. and A. B. Pardee (1960). Synthesis of macromolecules in synchronously dividing bacteria. *Biochim. et Biophys. Acta* **39,** 478.

Amaha, M. and T. Nakahara (1959). Role of the glycolytic enzyme system in the germination of spores of *Bacillus coagulans* var. *thermoacidurans*. *Nature* **184,** 1255.

Bajaj, V., S. P. Damle and P. S. Krishnan (1954). Phosphate metabolism of mold spores. I. Phosphate uptake by the spores of *Aspergillus niger*. *Arch. Biochem. Biophys.* **50,** 451.

Chayen, R., S. Chayen and E. R. Roberts (1955). Observations on nucleic acid and polyphosphate in *Torulopsis utilis*. *Biochim. et Biophys. Acta* **16,** 117.

Church, B. D. and H. Halvorson (1955). Glucose metabolism by resting spores of aerobic bacilli. *Bacteriol. Proc.*, p. 41.

Cohen, S. S. (1951). Utilization of gluconate and glucose in growing and virus-infected *Escherichia coli*. *Nature* **168,** 746.

DEDONDER, R. (1952). Quelques aspects de la glycolyse chez *Bacillus subtilis* et *Bacillus megatherium*. *2nd Intern. Congr. Biochem., Paris*, p. 77.

DELEY, J. (1955). The hexosemonophosphate oxidative route in micro-organisms. *3rd Intern. Congr. Biochem., Brussels*, p. 182.

FITZ-JAMES, P. C. (1955). The phosphorus fractions of *Bacillus cereus* and *Bacillus megaterium*. II. A correlation of the chemical with the cytological changes occurring during spore germination. *Can. J. Microbiol.* **1,** 525.

FULLER, R. C. and E. L. TATUM (1956). Inositol-phospholipid in Neurospora and its relationship to morphology. *Am. J. Botany* **43,** 361.

GUNSALUS, I. C., B. L. HORECKER and W. A. WOOD (1955). Pathways of carbohydrate metabolism in microorganisms. *Bacteriol. Rev.* **19,** 79.

HACHISUKA, Y., N. ASANO, M. KANEKO and T. KANBE (1956). Evolution of respiratory enzyme system during germination of *Bacillus subtilis*. *Science* **124,** 174.

HACHISUKA, Y., N. ASANO, N. KATO and T. KUNO (1954). Decrease of optical density of spore suspensions (*B. subtilis*) and its relation to spore germination. *Nagoya J. Med. Sci.* **17,** 403.

HACHISUKA, Y. and K. SUAGI (1959). Studies on spore germination. IV. Relationship between germination and appearance of glucose dehydrogenase activity in *B. subtilis* spore. *Japan. J. Microbiol.* **3,** 211.

HALVORSON, H. (1957). Oxidative enzymes of bacterial spore extracts. In *Spores* (H. O. Halvorson, ed.). American Institute of Biological Sciences, Washington, D.C.

HALVORSON, H. (1958). Intracellular protein and nucleic acid turnover in resting yeast cells. *Biochim. et Biophys. Acta* **27,** 255.

HALVORSON, H. and B. CHURCH. (1957). Biochemistry of spores of aerobic bacilli with special reference to germination. *Bacteriol. Rev.* **21,** 112.

HOFFMANN-OSTENHOF, O., J. KENEDY, K. KECK, O. GABRIEL and H. W. SCHÖNFELLINGER (1954). Ein neues Phosphatübertragendes Ferment aus Hefe. *Biochim. et Biophys. Acta* **14,** 285.

HOSHINO, J. and T. YANAGITA (1960). Alanine metabolism of *Aspergillus niger* spores during early germination (Abstract in Japanese). *J. Japan. Biochem. Soc.* **32,** 585.

HUEBSCHMAN, C. (1952). A method for varying the average number of nuclei in the conidia of *Neurospora crassa*. *Mycologia* **44,** 599.

IMAI, Y. (1950). Studies on microbial lipid. I. Effect of temperature on the lipid formation by mold mycelia. *J. Japan. Biochem. Soc.* **22,** 192.

INGELMAN, B. and H. MALMGREN (1950). Investigations on metaphosphate of high molecular weight isolated from *Aspergillus niger*. *Acta Chem. Scand.* **4,** 478.

KAJI, A. and J. D. GREGORY (1959). Mechanism of sulfurylation of choline. *J. Biol. Chem.* **234,** 3007.

KAJI, A. and W. D. MCELROY (1958). Enzymic formation of choline sulfate. *Biochim. et Biophys. Acta* **30,** 190.

KEYNAM, A., H. J. STRECKER and H. WALSCH (1954). Glutamine, glutamic acid, and glycolysis in *Bacillus subtilis*. *J. Biol. Chem.* **211,** 883.

KORNBERG, S. R. (1957). Adenosine triphosphate synthesis from polyphosphate by an enzyme from *Escherichia coli*. *Biochim. et Biophys. Acta* **26,** 294.

KREBS, H. A. (1951). The use of 'CO_2 buffers' in manometric measurements of cell metabolism. *Biochem. J.* **48,** 349.

Levinson, H. S. (1957). Non-oxidative enzymes of spore extracts. In *Spores* (H. O. Halvorson, ed.). American Institute of Biological Sciences, Washington, D.C.

Levinson, H. S. and M. G. Sevag (1954). The glutamic–aspartic transaminase of extracts of vegetative cells and of spores of *Bacillus megatherium. Arch. Biochem. Biophys.* **50,** 507.

Lipmann, F. (1954). On the mechanism of some ATP-linked reactions and certain aspects of protein synthesis. In *Mechanism of Enzyme Action* (W. D. McElroy and B. Glass, eds.). Johns Hopkins Press, Baltimore, Md., pp. 599–603.

McFall, E. and G. S. Stent (1959). Continuous synthesis of deoxyribonucleic acid in *Escherichia coli. Biochim. et Biophys. Acta* **34,** 580.

Mandels, G. R., H. S. Levinson and M. T. Hyatt (1956). Analysis of respiration during germination and enlargement of spores of *Bacillus megaterium* and of the fungus *Myrothecium verrucaria. J. Gen. Physiol.* **39,** 301.

Maruyama, Y. (1956). Biochemical aspects of the cell growth of *Escherichia coli* as studied by the method of synchronous culture. *J. Bacteriol.* **72,** 821.

Maruyama, Y. (1957). Alternative changes in the pathways of glucose dissimilation during the course of germination of *Aspergillus niger* spore. *Ann. Rept. Inst. Food Microbiol., Chiba Univ.* **10,** 33.

Maruyama, Y. and K. G. Lark (1959). Periodic synthesis of RNA in synchronized cultures of *Alcaligenes fecalis. Exptl. Cell Research* **18,** 389.

Matsushiro, A., T. Miwatani and D. Nakada (1955). Growth of aerobic spore-forming bacilli. *Proc. Symp. Biol. Sci., Tokyo*, p. 7.

Mudd, S., A. Yoshida and M. Koike (1958). Polyphosphate as accumulator of phosphorus and energy. *J. Bacteriol.* **75,** 224.

Nishi, A. (1959). Identification of acid soluble phosphorus compounds in germinating spores of *Aspergillus niger. Ann. Rept. Inst. Food Microbiol., Chiba Univ.* **12,** 32.

Nishi, A. (1960). Enzymatic studies on the phosphorus metabolism in germinating spores of *Aspergillus niger. J. Biochem.* (*Tokyo*) **48,** 758.

Nishi, A. (1961). Role of polyphosphate and phospholipid in germinating spores of *Aspergillus niger. J. Bacteriol.* **81,** 10.

O'Connor, R. and H. Halvorson (1959). Intermediate metabolism of aerobic spores. IV. Alanine deamination during the germination of spores of *Bacillus cereus. J. Bacteriol.* **78,** 844.

O'Connor, R. J. and H. Halvorson (1960). L-Alanine dehydrogenase: the possible initial binding site for L-alanine in aerobic spores. *Bacteriol. Proc.*, p. 66.

Owens, R. G. (1955). Metabolism of fungus spores. I. Oxidation and accumulation of organic acids by conidia of *Neurospora sitophila. Contrib. Boyce Thompson Inst.* **18,** 125; through Cochrane, V. W., *Physiology of Fungi*, John Wiley, New York, 1958.

Piérard, A. and J. M. Wiame (1960). Propriétés de la L(+)-alanine-déshydrogénase. *Biochim. et Biophys. Acta* **37,** 490.

Powell, J. F. (1953). Isolation of dipicolinic acid (pyridine-2:6-dicarboxylic acid) from spores of *Bacillus megatherium. Biochem. J.* **54,** 210.

Rippel, A. and H. Bortels (1927). Vorläufige Versuche über die allgemeine Bedeutung der Kohlensäure für die Pflanzenzelle (Versuche an *Aspergillus niger*). *Biochem. Z.* **184,** 237.

ROBERTS, R. B., D. B. COWIE, P. H. ABELSON, E. T. BOLTON and R. J. BRITTEN (1955). *Studies of biosynthesis in Escherichia coli.* Carnegie Inst. of Washington, Publication 607, Washington, D.C.

RYAN, F. J. (1948). The germination of conidia from biochemical mutants of *Neurospora. Am. J. Botany* **35,** 497.

SAKAGUCHI, K. and C. ISHITANI (1952). Studies on the natural variation of Koji-molds (*Asp. oryzae* and *Asp. sojae*). I. Cytological studies on the natural variation of Koji-molds. *J. Agr. Chem. Soc. Japan* **26,** 85.

SALL, T., S. MUDD and J. C. DAVIS (1956). Factors conditioning the accumulation and disappearance of metaphosphate in cells of *Corynebacterium diphtheriae. Arch. Biochem. Biophys.* **60,** 130.

SALL, T., S. MUDD and A. TAKAGI (1958). Polyphosphate accumulation and utilization as related to synchronized cell division of *Corynebacterium diphtheriae. J. Bacteriol.* **76,** 640.

SANDO, N. and T. YANAGITA. (1961) Synchronization of cell division in *Schizosaccharomyces pombe.* Monthly Meeting, Japan Botan. Soc., Oct., held in Tokyo.

SCHAECHTER, M., M. W. BENTZON and O. MAALØE (1959). Synthesis of deoxyribonucleic acid during the division cycle of bacteria. *Nature* **183,** 1207.

SHEPHERD, C. J. (1957). Changes occurring in the composition of *Aspergillus nidulans* conidia during germination. *J. Gen. Microbiol.* **16,** 1.

SPENCER, B. and T. HARADA (1959). The role of choline sulphate in the sulphur metabolism of fungi. *Biochem. J.* **73,** 34P.

STEVENS, C. M. and P. VOHRA (1955). Occurrence of choline sulfate in *Penicillium Chrysogenum. J. Am. Chem. Soc.* **77,** 4935.

SUMI, M. (1928). Über die chemischen Bestandteile der Sporen von *Aspergillus oryzae. Biochem. Z.* **195,** 161.

TAKEBE, I. (1960). Choline sulfate as a major soluble sulfur component of conidiospores of *Aspergillus niger. J. Gen. Appl. Microbiol.* (*Japan*) **6,** 83.

TAKEBE, I. and N. SHIMIZU (1959). Some aspects on the nutritional requirement of conidiospores of *Aspergillus niger* for the initiation of germination. *Ann. Rept. Inst. Food Microbiol., Chiba Univ.* **12,** 29.

TAKEBE, I. and T. YANAGITA (1959). Origin of amino acids constituting cellular protein in germinating conidiospores of *Aspergillus niger. Plant and Cell Physiol.* (*Tokyo*) **1,** 17.

TERUI, G. and T. MOCHIZUKI (1955). Studies on the metabolism of mold spores in relation to germination. I. *Technol. Rept. Osaka Univ.* **5,** 219.

TERUI, G., T. MOCHIZUKI, R. IRIE and M. TAKANO (1960). Mass adaptation of mold spores to acquire drug resistance in the course of germination. *Technol. Rept. Osaka Univ.* **10,** 279.

THRELFALL, C. J. (1957). An analytical procedure for the acid-soluble phosphorus compounds in rat skeletal muscle. *Biochem. J.* **65,** 694.

VOLKIN, E. and W. E. COHN (1954). Estimation of nucleic acids. In *Methods of Biochemical Analysis* (D. Glick, ed.), Vol. 1. Interscience, New York, p. 287.

WELLMAN, R. H. and S. E. MCCALLAN (1942). An analysis of factors causing variation in spore germination tests of fungicides. IV. Time and temperature. *Contrib. Boyce Thompson Inst.* **12,** 431; through W. V. Cochrane, *Physiology of Fungi,* John Wiley, New York, 1958.

WINDER, F. G. and J. M. DENNENY (1955). Utilization of metaphosphate for phosphorylation by cell-free extracts of *Mycobacterium smegmatis*. *Nature* **175,** 636.

WINDER, F. G. and J. M. DENNENY (1957). The metabolism of inorganic polyphosphate in mycobacteria. *J. Gen. Microbiol.* **17,** 573.

WOESE, C. R. and J. R. FORRO (1960). Correlations between ribonucleic acid and deoxyribonucleic acid metabolism during spore germination. *J. Bacteriol.* **80,** 811.

WOOLLEY, D. W. and W. H. PETERSON (1937). The chemistry of mold tissue. XIV. Isolation of cycle choline sulfate from *Aspergillus sydowi*. *J. Biol. Chem.* **122,** 213.

WOOLLEY, D. W., F. M. STRONG, W. H. PETERSON and E. A. PRILL (1935). The chemistry of mold tissue. X. The phospholipides of *Aspergillus sydowi*. *J. Am. Chem. Soc.* **57,** 2589.

YANAGITA, T. (1957). Biochemical aspects on the germination of conidiospores of *Aspergillus niger*. *Arch. Mikrobiol.* **26,** 329.

YANAGITA, T. (1961). Physiology of spore germination of *Aspergillus niger*. *Protein, Nucleic Acid and Enzyme* (*Tokyo*) **6,** 140.

YANAGITA, T. and F. KOGANÉ (1962). Histochemical structure of mold colonies and pellets. (Abstract in Japanese). *J. Japan. Soc. Plant Physiol.* **2,** 72.

YANAGITA, T. and S. YAMAGISHI (1958). Comparative and quantitative studies of fungitoxicity against fungal spores and mycelia. *Appl. Microbiol.* **6,** 375.

YOSHIDA, A. (1953). Biochemical studies on metaphosphate. *Sci. Papers Coll. Gen. Educ., Univ. Tokyo* **3,** 151.

YOSHIDA, A. and A. YAMATAKA (1953). On the metaphosphate of yeast. I. *J. Biochem.* (*Tokyo*) **40,** 85.

YOUNG, I. E. and P. C. FITZ-JAMES (1959). Pattern of synthesis of deoxyribonucleic acid in *Bacillus cereus* growing synchronously out of spores. *Nature* **183,** 372.

15

Synchrony of Bacterial Populations as Established by Filtration Techniques

YOICHI MARUYAMA, *The Institute of Food Microbiology, Chiba University, Okubo, Narashino, Chiba-ken, Japan*

CONTENTS

I Introduction 421
II General Considerations Relating to the Filtration Technique . 422
1. Synchrony 422
2. Normality 424
III Changes in Cell Constituents and Cell Functions during Synchronous Growth 425
1. Synthesis of cellular constituents 425
2. Phage reproduction 426
3. Effect of chloramphenicol 427
4. Nucleotide pool 428
5. Respiratory activity 428
6. Sequence of cellular events 430
References 431

I. INTRODUCTION

Methods for obtaining synchronous culture have been developed in various ways which depend on the microbial species and on the working objective (Campbell, 1957; Scherbaum, 1960; Zeuthen, 1958). These techniques have been designed, in principle, along two lines. In the first of these, all the constituent cells of a culture are brought into one stage in the division cycle. By such methods, a homogeneous population can be obtained by shifting the conditions of culture from the normal so that the growth process of the cells is temporarily and preferentially blocked at some stage in the division cycle. It is highly probable, however, that such treatment may affect the succeeding growth stages, resulting in a biased metabolism, regardless of whether the cells have been treated repeatedly or just once under a given condition of culture.

The second line of approach is to select a fraction consisting of uniformly-aged cells from a mixed population of the exponential growth phase, and leave them to grow under normal conditions of culture. The principle of the filtration technique designed along these lines is to pick out the cells at the pre-divisional or post-divisional stage by a filtration procedure which takes advantage of the discontinuous change in cell size at the divisional stage. This method was adopted in our laboratory with the aim of obtaining an unbiased synchronous culture.

In the present paper the degree of synchrony, and problems connected with the question of the metabolic normality of synchronous cultures obtained by the filtration procedure (abbreviated hereafter to 'filtered culture'), are discussed. Studies on changes in cell constituents and in cell functions in synchronous cultures of *Escherichia coli* (strain B) and *Alcaligenes fecalis* (strain LB) are also summarized.

II. General Considerations Relating to the Filtration Technique

1. *Synchrony*

It has been found theoretically, and also empirically, that in an exponentially growing bacterial culture, the cells at any given stage of growth (i.e., cell age) occur in a constant proportion to the total population (Maruyama, 1955; Powell, 1956; Takebe and Yanagita, 1957). Figure 1a shows the pattern of cell age distribution (solid line) derived mathematically on the basis of the assumptions that (a) the cells divide by binary fission and (b) that the distribution of generation times plotted against cell age (broken line) is normal. As indicated in this figure, the range of a steep decrease in the cell age distribution curve around the mean generation time (τ_0) corresponds to the variation in generation time of normal cells. This indicates that even if, for example, all the cells are completely synchronous at cell age zero and are left to grow freely, the following division will occur during the period corresponding to the range of the steep slope. This relationship was clearly shown by Campbell (1957). By contrast, if the cells are synchronous at a stage other than that of cell age zero, the width of the distribution of division time will become much narrower. Thus, in a synchronous culture, there is inevitably a deviation from true synchrony after one generation, caused by variation of the generation time.

In the case of a filtered culture, there is another factor affecting synchrony which should be taken into account. As may be seen in chapter 28, since cells are selected in this method by cell size instead of cell age, there is a possibility that the distribution of cell age at a certain cell size will make the synchrony of the following cell division less clear than would be expected from the distribution of the generation time. The relationship between cell size and cell age in exponentially growing cells of *E. coli* was studied by Takebe and Yanagita (1957), using the longitudinal dimension as the measure of size (Fig. 1b). The histogram of the cell length is rather similar to the cell age distribution curve shown in Fig. 1a. Since, in the case of *E. coli*, the cell grows only in length and not in width, the length of a cell is considered to be proportional to the cell volume (or to the surface area). Thus a rough linear relationship between cell age and cell volume may be assumed. In the case of *A. fecalis* no such linear relationship seems to exist, i.e., the average volume of filtered pre-divisional cells increases rapidly after cell division and

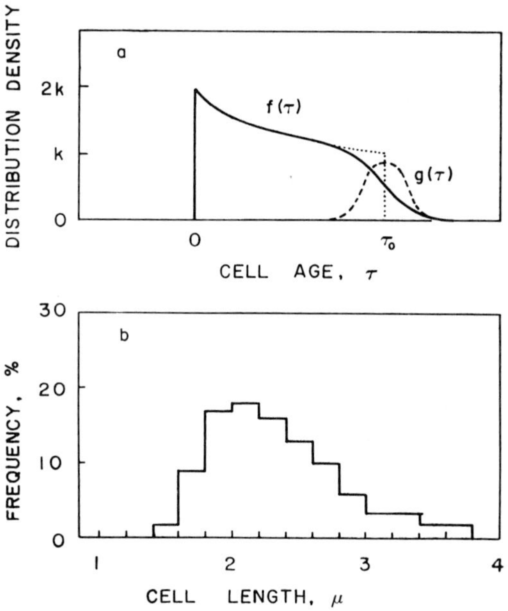

FIG. 1. (a) Schematic diagram showing the relationship between age distribution and generation time distribution. In the figure, $g(\tau)$ denotes distribution density of generation time; and $f(\tau)$, cell age distribution density, which is represented by the following formula (Maruyama, 1955):

$$f(\tau)\ d\tau = 2k \exp(-k\tau)\ G(\tau)\ d\tau,$$

where k is the rate constant of cell multiplication, or $\ln 2/\tau$; τ is the doubling time of the culture; and $G(\tau) = \int_{\tau}^{\infty} g(\tau)\ d\tau$. Dotted line shows a hypothetical distribution when there is no variation in generation time. (b) Frequency–cell length distribution histogram observed during the exponential growth of *E. coli* B (Takebe and Yanagita, 1957).

then remains nearly constant during the remainder of the division cycle (Lark and Lark, 1960).

These considerations on the synchrony of filtered cultures suggest that the deviation from synchrony is caused by various factors, such as the inherent variation of generation time, the variation of cell age at a given cell size, and, in some cases, a non-linear relationship between cell age and cell size. It should, however, be emphasized that the duration of cell division so far studied in filtered *E. coli* and *A. fecalis* is approximately 15 min for a doubling time of 50 min and 18 min for 40 min (or a ratio of 0·3 and 0·5 respectively). The values of such ratios are not necessarily high in comparison with the values for the ratio of standard deviation of generation time to mean generation time which represents the inherent variation of generation time. Such values observed for *E. coli* and *Streptococcus fecalis* have been found to be about 0·3 (Powell, 1955, 1956). The synchrony of the filtered culture is therefore rather higher than expected. This suggests that unknown factors which affect synchrony either positively or negatively cancel each other, resulting in good synchrony.

2. *Normality*

Since supposedly mild treatment, such as centrifugation for harvesting cells and filtration through a paper pile, would be expected to give only a slight stress to the physiological conditions of cells, the filtered cells should recover a fully normal state shortly after they are transferred to a fresh medium. There has been, however, some controversy with respect to the normality of filtered cells. First, the ratio of the number of divided cells to that of starting cells of *E. coli* is 1·6 instead of 2; i.e., only 60 per cent of the starting cells divide at the first step (see Fig. 3 in chapter 28). This may be explained on the basis of the following consideration: In the exponential phase of an *E. coli* culture, there is a fairly large proportion of doubled or chained cells not separated after the division of the member cells. Therefore, if the filtered cells of, for example, the post-divisional stage are left to grow, some of them may divide and separate and others may divide without separating. Thus the ratio of the number of 'separated cells' to that of starting cells would become less than 2. The validity of this consideration has been proved by the observation that the cyclic synthesis of cell materials (such as DNA) in the filtered culture is repeated for several generations (3–4), while the stepwise increase in cell number is not observed after two generations.

It is, however, commonly believed that such discontinuous synthesis of DNA may not reflect a normal growth process, since evidence has been accumulated indicating that in an exponentially growing bacterial culture all the constituent cells are engaged in more or less continuous synthesis of DNA (McFall and Stent, 1959; Schaechter *et al.*, 1959). Although the reason why the stepwise synthesis of DNA readily occurs in our synchronous growth system is not yet clear, the fact that a continuous synthesis takes place in the modified filtered culture used by Abbo and Pardee (1960) is suggestive for answering this problem. These investigators omit the centrifugation procedure in harvesting the cells, in order to reduce the possibility of introducing abnormal conditions.

III. Changes in Cell Constituents and Cell Functions during Synchronous Growth

1. *Synthesis of Cellular Constituents*

As described previously, the DNA content remains constant during the first major part of the bacterial life span, and doubles later within a short period covering less than 20 per cent of the generation time (Fig. 2). In the filtered *E. coli* system, this process appears to be closely related to the change in nuclear pattern during the division cycle (Yanagita *et al.*, 1958). The dinuclear stage in this bacterium covers more than 50 per cent of the generation time, and, prior to the onset of DNA synthesis, these nuclei divide to form quadrinuclear cells, which in turn revert in a short time to the dinuclear condition by cell fission.

This concept of the relationship between the synthesis of DNA and the nuclear pattern seems to be borne out by the multi-target analysis (Atwood and Norman, 1949) of the survival–u.v.-dose curves. A logarithmic plot of the values for survival ratio versus the time of u.v. irradiation showed a linear decrease with a plateau at low u.v.-dosage for different cell ages. The length of the flat portion of these curves differed for different cell ages, whereas the slope was always the same. The identity of the slope of the curve during the life span indicates the invariability of the cross-sectional area of the targets during the division cycle. The target number, as determined from the point of intersection with the ordinate of the extended linear portion of the survival curve, changes with cell age, showing 4·4 for the pre-divisional stage and 1·6 for the post-divisional stage. From such observations we are tempted to assume that the target defined in the multi-target theory corresponds to the cell nucleus. These observations concerning changes in the amount of DNA and in u.v. sensitivity seem to indicate that the DNA content of a nucleus remains almost constant during the greater part of the division cycle. Reference should, however, be made to the finding of Hanawalt *et al.* (1961), that the 'DNA units' per nucleus of *E. coli* show a continuous increase during most of the division cycle.

Lark (1960) has suggested a possible control mechanism in connection with periodic DNA synthesis. He studied the effect of deoxyribosides on synchronously growing *A. fecalis*. First, the addition of four deoxynucleosides destroys periodic stepwise DNA synthesis only when addition is at the stage of DNA synthesis. Secondly, a high concentration of deoxyadenosine selectively blocks DNA synthesis, but this effect can be reversed by the addition of adenosine. The resumption of DNA synthesis by the addition of adenosine occurs at the time of the second step of DNA synthesis observed in the control culture, independent of the time of addition of adenosine. He concluded that DNA synthesis occurs and is interfered with by deoxynucleosides only during a certain 'permitted' portion of the division cycle.

RNA synthesis in *E. coli* cells begins immediately after cell division and then levels off, followed by the occurrence of a gross protein synthesis. The increase in the total nitrogen content, on the other hand, proceeds nearly exponentially. The amount of low molecular weight nitrogen compounds, as estimated by subtracting protein and nucleic acid nitrogen from the total nitrogen, appears to be high during the early stage of cellular growth. This indicates that the building blocks of protein and nucleic acid are accumulating in the cell at this early stage.

A similar result was obtained in *A. fecalis* cells with respect to the sequence of RNA–DNA synthesis. Gross RNA synthesis in these bacteria occurs throughout the division cycle, but most rapidly during the period following the active DNA synthesis. It was disclosed that among various RNA fractions, a fraction, sedimentable in the presence of Mg^{2+} ($3{\cdot}6 \times 10^{-4}$ *M*), was synthesized in one step at the period of rapid gross RNA synthesis, while the other non-sedimentable fractions were synthesized continuously (Maruyama and Lark, 1959). In *A. fecalis*, the gross synthesis of protein, as well as that of cell wall, proceeded continuously (Lark and Lark, 1960).

2. *Phage Reproduction*

Yanagita *et al.* (1958) have studied the change in the infectivity of *T*2 phage on *E. coli* B during the division cycle. First, the phage-adsorbing capacity of the cells was found to increase as growth proceeds, attaining twice the initial value just before cell division. Since, as described previously, the surface area of the bacillary cell is proportional to the cell length, which becomes twice the initial value at the end of a division cycle, it seems most probable that the phage-adsorbing capacity of a cell depends simply on its surface area. Secondly, the phage-reproducing capacity, as determined by the burst size defined in a one-step growth experiment (Adams, 1948), was revealed to be highest before cell division, when gross RNA synthesis had ceased and nuclear division had already started. The latent period of phage reproduction remained constant during the division cycle. These results appear to indicate that the capacity for phage reproduction of the cell depends highly on the physiological state of the nuclei in the cell.

3. *Effect of Chloramphenicol*

Temporary treatment of the synchronously growing cells of *E. coli* at various stages of the division cycle with chloramphenicol at a concentration of 2×10^{-2} *M*, which causes 50 per cent inhibition of the growth rate of exponentially growing cells, showed that it caused the most marked delay of cell division at the post-divisional stage, when active protein synthesis was about to be initiated (Yanagita *et al.*, 1958). In relation to this observation, Maruyama and Lark (1961) studied the effect of chloramphenicol at a higher concentration (500 μg/ml, or 1·55 m*M*) on a synchronized culture of *A. fecalis*. Under this condition, cells at the post-divisional stage cease to grow for about one generation and then rapidly lose their viability, while those treated just before cell division can divide once before they begin to lose viability. Since the protein synthesis is blocked almost completely in the presence of such a high concentration of chloramphenicol, the above observations suggest that the process of cell division itself is independent of protein synthesis. This antibiotic seems also to inhibit some part of the nucleic acid synthesis in *A. fecalis*. When it was added to the synchronous culture at the concentration of 10–500 μg/ml, the rate of RNA synthesis was reduced but the course of RNA synthesis became discontinuous, while that of DNA synthesis retained its regular periodicity. Moreover, it is worthy of note that the periodic syntheses of nucleic acids are repeated for at least four cycles, keeping a regular generation time, irrespective of the dramatically reduced protein synthesis. 10 μg/ml ($3{\cdot}1 \times 10^{-2}$ m*M*) of chloramphenicol inhibits gross protein synthesis to the extent of 70 per cent, whereas 500 μg/ml inhibits it almost completely. In the above experiment, the use of the higher concentration of this antibiotic resulted in the accentuation of the characteristic repetition of discontinuous (or periodic) nucleic acid syntheses. This indicates that the mechanism of either RNA or DNA synthesis includes a step which depends on the *de novo* synthesis of some protein. This step will be greatly slowed by chloramphenicol, and this results in the remarkable stepwise syntheses of nucleic acids. Especially in the case of RNA, at least two mechanisms of its synthesis may be assumed: one is resistant and the other is sensitive to chloramphenicol; the former proceeds in a stepwise fashion and the latter continuously. The sedimentable RNA in normally growing cells seems to be closely related to this chloramphenicol-resistant RNA, although the identity of the two RNAs awaits further analysis.

4. *Nucleotide Pool*

The size of the nucleotide pool of *A. fecalis*, as measured by the estimation of the ribose phosphorus in the cold-acid-soluble fraction (Norit-adsorbable), exhibits periodic variations during the division cycle (Maruyama and Lark, 1962). Since the main constituents of the pool were identified as ribosenucleotides and a few deoxyribose-nucleotides (Lark, 1961), the pool may be assumed to be functioning as a common precursor of both RNA and DNA. The pool size, as determined by the radioactivity of cells after 20-min incorporation of ^{32}P-phosphate, increases rapidly during the period when RNA synthesis is most rapid, and then levels off during the remainder of the division cycle (Fig. 2). The synthetic rate of the pool, as estimated by the increase in radioactivity after 5-min pre-incubation with ^{32}P-phosphate, on the other hand, showed a peak during the period of rapid increase in pool size. It was also disclosed that the amount of phosphate incorporated into the pool during a division cycle is sufficient and not excessive to supply material for nucleic acid synthesis. These characteristics of the pool indicate that at the time of active synthesis of RNA there is an increased rate of phosphorus incorporation, resulting in an equilibrium at the succeeding stage of cell growth, in which the nucleotide pool synthesis continues at a rate balancing the rate at which phosphorus is supplied for nucleic acid synthesis.

5. *Respiratory Activity*

O_2 consumption and CO_2 production of *E. coli* cells show a continuous increase during the division cycle (Maruyama, 1956). However, the rate of O_2 consumption per total nitrogen remains constant. It is worth referring to the observation of Hershey and Bronfenbrenner (1938) that the rate of O_2 consumption per nitrogen is almost constant during the whole course of bacterial culture. The rate of CO_2 production per total nitrogen in the filtered culture of *E. coli* shows two peaks during the division cycle: higher and lower peaks appear after and just before cell division, respectively. A related

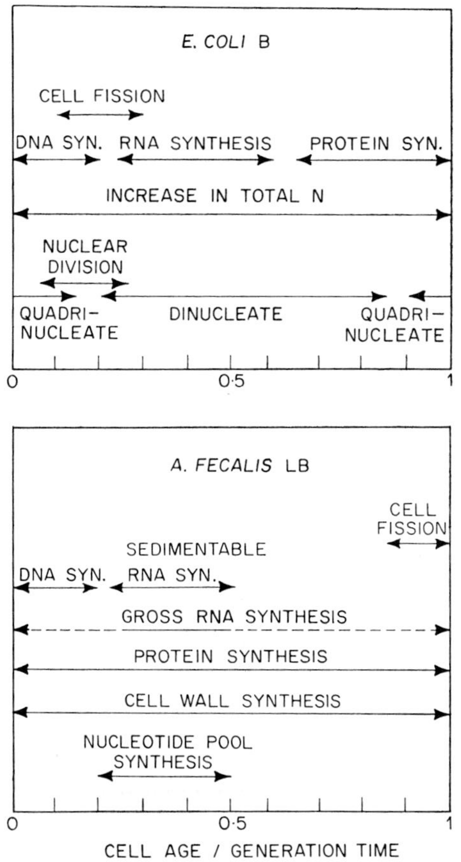

FIG. 2. Sequences of cellular events in filtered cultures of *E. coli* B and *A. fecalis* LB. (For illustrations, see text.)

result was obtained for the consumption of glucose in the growth medium (Maruyama, unpublished data); i.e., its rate becomes a maximum just after cell division when the rate of CO_2 production becomes highest. These results are considered to reflect changes in the pathways of glucose dissimilation during the division cycle.

The effect of cyanide added at different growth stages is of interest in this connection (Yanagita *et al.*, 1958). When cells are put in contact with this reagent for 15 min (30 per cent of a generation time) after cell division, the onset of the next cell division is delayed by 25 min, while no such retarding effect is observed when it is added at the pre-divisional stage. This would indicate that the terminal respiratory system is not playing an essential role in the growth process of such cells, which are already prepared for cell division (cf. also chapter 4).

6. *Sequence of Cellular Events*

The sequence of cellular syntheses in filtered cultures of *E. coli* and *A. fecalis* are summarized in Fig. 2. In this figure the onset of DNA synthesis rather than the onset of cell division (strictly speaking, cell separation) is taken as the starting point of cellular growth. It was observed quite frequently in our system that the time of cell division varies, presumably because of the failure of divided cells to separate even after the cell division is otherwise complete. Therefore the time course of their separation process is not consistent in the various experiments (Lark and Lark, 1960). Cell separation as measured by viable counting or by a particle counter occurs directly before (*A. fecalis*) or after (*E. coli*) the period of DNA synthesis.

The period of DNA synthesis is immediately followed by RNA synthesis in *E. coli*. In *A. fecalis*, the RNA synthesis takes place rather gradually throughout the course of a division cycle, but it occurs most rapidly after the completion of DNA synthesis; at this period the syntheses of chloramphenicol-resistant, sedimentable RNA proceed most actively. Such a temporal relationship between the RNA and DNA syntheses has been pointed out by Lark (1960). He stated that the rate of RNA synthesis depends upon the amount of DNA present at any time. This view was further confirmed by the experiment on the effect of adding chloramphenicol at different periods of the division cycle (Maruyama and Lark, 1961). The RNA-synthetic capacity, as expressed by the ratio: RNA content 2 hr after the addition of chloramphenicol/RNA content at the time of its addition, shows the highest value when chloramphenicol is added just after the completion of DNA synthesis.

The gross protein synthesis follows the RNA synthesis in *E. coli* cells. In *A. fecalis*, however, it proceeds continuously over the whole division cycle.

Acknowledgements—The major parts of the studies published by the author have been performed in the laboratories of Dr. T. Yanagita (the Institute of Food Microbiology, Chiba University) and of Dr. K. G. Lark (St. Louis University School of Medicine), to whom his thanks are due.

REFERENCES

ABBO, F. E. and A. B. PARDEE (1960). Synthesis of macromolecules in synchronously dividing bacteria. *Biochim. et Biophys. Acta* **39,** 478.

ADAMS, M. H. (1948). Methods of study of bacterial viruses. In *Methods in Medical Research,* Vol. II (J. H. Comroe, Jr., ed.). Year Book, Chicago.

ATWOOD, K. C. and A. NORMAN (1949). On the interpretation of multi-hit survival curves. *Proc. Natl. Acad. Sci., U.S.* **35,** 696.

CAMPBELL, A. (1957). Synchronization of cell division. *Bacteriol. Rev.* **21,** 263.

HANAWALT, P. C., O. MAALØE, D. J. CUMMINGS and M. SCHAECHTER (1961). The normal DNA replication cycle. II. *J. Mol. Biol.* **3,** 156.

HERSHEY, A. D. and J. BRONFENBRENNER (1938). Factors limiting bacterial growth. III. Cell size and 'physiologic youth' in *Bacterium coli* cultures. *J. Gen. Physiol.* **21,** 721.

LARK, K. G. (1960). Studies on the mechanism regulating periodic DNA synthesis in synchronized cultures of *Alcaligenes fecalis. Biochim. et Biophys. Acta* **45,** 121.

LARK, K. G. (1961). Variation in bacterial acid-soluble deoxyribotides during discontinuous deoxyribonucleic acid synthesis. *Biochim. et Biophys. Acta* **51,** 107.

LARK, K. G. and C. LARK (1960). Changes during the division cycle in bacterial cell wall synthesis, volume, and ability to concentrate free amino acids. *Biochim. et Biophys. Acta* **43,** 520.

MCFALL, E. and G. S. STENT (1959). Continuous synthesis of deoxyribonucleic acid in *Escherichia coli. Biochim. et Biophys. Acta* **34,** 580.

MARUYAMA, Y. (1955). The age distribution of cells in a microbial population at the logarithmic growth phase, statistical consideration. *Ann. Rept. Inst. Food Microbiol., Chiba Univ.* **8,** 75.

MARUYAMA, Y. (1956). Biochemical aspects of the cell growth of *Escherichia coli* as studied by the method of synchronous culture. *J. Bacteriol.* **72,** 821.

MARUYAMA, Y. and K. G. LARK (1959). Periodic synthesis of RNA in synchronized culture of *Alcaligenes fecalis. Exptl. Cell Research* **18,** 389.

MARUYAMA, Y. and K. G. LARK (1961). Periodic synthesis of bacterial nucleic acids in the absence of protein synthesis. *Exptl. Cell Research* **25,** 161.

MARUYAMA, Y. and K. G. LARK (1962). Periodic nucleotide synthesis in synchronous cultures of bacteria. *Exptl. Cell Research* **26,** 382.

POWELL, E. O. (1955). Some features of the generation times of individual bacteria. *Biometrika* **42,** 16.

POWELL, E. O. (1956). Growth rate and generation time of bacteria, with special reference to continuous culture. *J. Gen. Microbiol.* **15,** 492.

SCHAECHTER, M., M. W. BENTZON and O. MAALØE (1959). Synthesis of deoxyribonucleic acid during the division cycle of bacteria. *Nature* **183,** 1207.

SCHERBAUM, O. H. (1960). Synchronous division of microorganisms. In *Annual Review of Microbiology,* Vol. XIV (C. E. Clifton, ed.). Annual Reviews, Palo Alto, p. 283.

TAKEBE, I. and T. YANAGITA (1957). Changes in cell size distribution during the course of cultural development of *Escherichia coli*. *Ann. Rept. Inst. Food Microbiol., Chiba Univ.* **10,** 27.

YANAGITA, T., Y. MARUYAMA and I. TAKEBE (1958). Cellular response to deleterious agents during the course of synchronous growth of *Escherichia coli*. *J. Bacteriol.* **75,** 523.

ZEUTHEN, E. (1958). Artificial and induced periodicity in living cells. *Advanc. in Biol. and Med. Phys.* **4,** 37.

16

Synchronization of Division in Bacteria by Nutritional Changes

VICTOR W. BURNS, *Biophysics Laboratory, Stanford University, Stanford, California*

CONTENTS

I Introduction 433
II Specific Nutritional Synchronization 434
III General Nutritional Synchronization 437
References 438

I. INTRODUCTION

Cell division takes place only after numerous metabolic processes have been carried out within the cell. Some of these are known for certain types of cells, others can only be guessed. Some are specific in the sense that the process is not immediately essential for most cellular activities but is essential for cell division—for example, DNA synthesis. Others are general in the sense that the process is immediately essential for many cellular activities including cell division—for example, energy yielding processes such as glycolysis, Krebs cycle function, etc. Correspondingly, synchronizations by nutritional changes can be classified as specific if the changes affect primarily a known specific process, or general if the changes affect many processes in such a way that synchronization results.

The special advantage of any specific nutritional synchronization method lies in the fact that the primary effect of nutrient manipulation on the cell can be described in some detail with reference to biosynthetic pathways, and the relation of the specific process to other metabolic processes and to cell division is readily accessible to investigation at the biochemical level. General synchronization has an apparent advantage, not yet exploited, in that it is applicable in principle to wild-type prototrophs whereas specific synchronization has required the use of auxotrophs so far.

Synchronization by nutritional changes involves inducing all of the individuals of a population to get in step or come into the same stage or phase with respect to cell division. The process of induction alters the pattern of preparations for division with respect to an uninduced control population. The information now available concerning patterns of preparation for division indicates that many different patterns are possible for any one cell line. Labeling any one pattern as 'normal' is an arbitrary decision which in no way invalidates knowledge of division dynamics gained from the study of other patterns. It is important to realize, however, that an event which may appear decisive, a 'trigger' event let us say, in one pattern cannot be assumed to play the same role in another pattern without corroborative evidence. The effects obtained by any particular method of synchronization are the result of the interaction of the induced intracellular metabolic changes with the system of processes involved in division. It is of advantage to know what the induced metabolic changes are so that the response of the system can be distinguished from them.

An hypothesis to explain how synchronization by nutritional means may come about can be based on the ideas of Zeuthen (1958). An interphase cell is thought to prepare for division along many more or less independent channels, not through one chain of events in which various syntheses and structural developments are linked up in a fixed sequence. Synchronization depends on the establishment of conditions which will reduce the rate of preparation along one or a few channels primarily, the change in rate depending on stage of division such that individuals closest to division are most retarded, those furthest away are least retarded. If preparation along these channels is stopped completely this blockage must be restricted to a particular stage of the division cycle, so that cells in other stages initially can continue until they enter the blocked stage. The cells are given time to mature for division along all other channels and then synchronous division can be triggered by release of the inhibitory influence on the special channel or channels.

II. Specific Nutritional Synchronization

The ideal cell for specific nutritional synchronization would possess a well characterized channel involved only in cell division which could be blocked specifically in a single stage without interfering with other channels. As our knowledge of biochemical pathways, specific inhibitors and deficiency mutation blocks grows it is becoming increasingly easy to manipulate specific channels. The requirement for no interference with other channels, which implies that the specific channel is independent of other channels, has never been satisfied rigorously. However at least one channel has been found, DNA synthesis, which can be blocked over a period corresponding to one generation time, more or less, with apparently little effect on many other channels important in cell division. Choosing such a channel requires foreknowledge of processes relevant to cell division and a good guess, followed by experimental determination of the independence of the channel. There may be other channels as good or better than DNA synthesis; if an inhibitor or mutation specific for nucleoprotein synthesis should become available, this would be a promising channel.

Barner and Cohen (1956) have employed a DNA inhibition method, using an *E. coli* 15 T-auxotroph requiring thymine. Cells were incubated in medium lacking thymine until on the verge of death (about 30 min). If thymine was then restored, division began again after a lag of 35 min. The published data suggest that the ensuing divisions were partly phased. Synchronization was obtained only if thymine was added during a very brief period immediately preceding death of the cells. Brock and Brock (1960) and Maaløe and Hanawalt (1961) present evidence that death is due to DNA turnover or attempted DNA synthesis respectively. Apparently then, blocking DNA synthesis by thymine deprivation causes degenerative changes in the genetic material which quickly become irreversible.

Many studies support the view that DNA synthesis is largely independent of protein and/or RNA synthesis in *E. coli* during the greater part of the division cycle. Barner and Cohen (1958) investigated a thymineless and uracilless auxotroph. This auxotroph synthesized protein, RNA and the inducible enzyme xylose isomerase at nearly the same rates with or without concurrent DNA synthesis. Inhibition of DNA synthesis in *E. coli* by a variety of chemical and physical methods gives similar results. (See discussion by Frisch *et al.*, 1960.) Maaløe and Hanawalt (1961) show that protein and/or RNA synthesis is required to initiate but not to sustain each round of DNA replication.

Burns (1959, 1961) has reported synchronization of a deoxyriboside auxotroph of *Lactobacillus acidophilus*, R26 (Fig. 1). The nature of DNA synthesis in this strain has been elucidated by Okazaki and collaborators in a series of papers (Okazaki and Okazaki, 1959) and by Hoff-Jørgensen (1952). Division is synchronized by growing the cells for 2 or 3 hr in the absence of deoxyriboside and then adding a deoxyriboside. After a time lag there are two or three synchronized divisions, the first of which shows about 86 per cent phasing under optimal conditions. In contrast to thymineless *E. coli* the *Lactobacillus* R26 strain can be starved for deoxyriboside for up to 6 hr without loss of viability. The basis of this longevity may lie in the different mechanisms of inhibition of DNA synthesis utilized for thymineless *E. coli* and *Lactobacillus* R26. The latter is thymine sufficient and not just one but none of the deoxynucleotides necessary for DNA synthesis can be produced during deoxyriboside starvation. Since R26 can be synchronized by restoration of deoxyriboside many hours before viability would decrease, the method does not require rescue of cells from the brief stage preceding death. During the first 2 to 3 hr of starvation RNA and protein synthesis continue at approximately the same rate found in the presence of deoxyriboside. Maturation for division along the protein and RNA channels presumably then takes place in the absence of DNA synthesis.

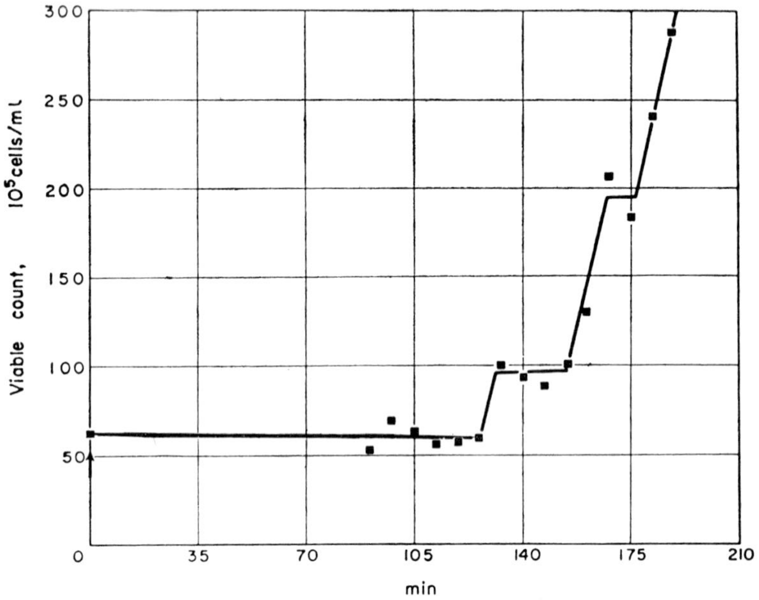

Fig. 1. Synchronization of *Lactobacillus acidophilus* cells deprived of deoxyriboside for 3 hr; restoration of deoxyriboside was at time zero in the figure (Burns 1959).

After restoration of deoxyriboside to cells which were in late lag phase at the start of starvation, DNA synthesis begins immediately and DNA increases from four- to eight-fold before the first division. During this period the DNA synthetic rate shifts from slow to fast and back again several times, possibly reflecting waves of nuclear division occurring without cell division. During cell division there is simultaneous cessation of DNA synthesis and abrupt increase in rate of protein synthesis; under some conditions RNA rate also increases. Since RNA and protein synthetic rates remain high during deoxyriboside starvation and the inhibition of division resulting therefrom, there is an abnormally high content of RNA and protein per cell when deoxyriboside is restored. By limiting the amount of deoxyriboside supplied it was possible to limit DNA increase to two-fold or less. It was then possible to show that the ratio of RNA or protein to DNA at the time of division varies over a wide range so that the cell is apparently not constrained to restore the balance of DNA to other macromolecular constituents. Initiation or partial replication of DNA was not sufficient to allow division; doubling was necessary.

Stevenson (1961) describes division synchronization of a vitamin B12 arthrobacter. The cells do not divide in the absence of vitamin B12 and when it is restored they divide synchronously after a lag. Synchronization was also found on transfer of lag-phase cells growing in medium containing B12 to fresh medium also containing B12, so manipulation of vitamin B12 may not be the most critical feature of this synchronization.

III. General Nutritional Synchronization

This approach has received relatively little attention in comparison to the equally general method of synchronization by means of temperature changes. Early observations of synchronous *growth* in bacteria have been reviewed by Zeuthen (1958). Scott, McNair and Chu (1958) have described synchronization of *E. coli* by means of a combination of nutritional change and temperature shock. *E. coli* B cells were transferred from a glucose-containing medium at 37°C to a glucose-free medium at 6°C for 45 min and then back to the medium at 37°C, to which glucose was added quickly. If the cells were initially in lag phase, division occurred 60′ after the final transfer; if they had exhausted glucose at least 4 hr before the first transfer they divided 120′ after the final transfer. Cells exhausting glucose less than 4 hr before first transfer showed intermediate behavior and poor synchrony. Thus the physiological state of the cells with reference to glucose determined their response to synchronization by temperature shock. The relative importance of the glucose shifts in inducing synchrony is not clear however. Scott *et al.* (1959) used this system to investigate metabolic events correlated with division; evidence was presented that acid-soluble ribose compounds wax and wane before the first division. Perry

(1959) used the method to demonstrate that there is no significant change in respiration during division.

Kjeldgård, Maaløe and Schaechter (1958) have investigated the effects of shifting *S. typhimurium* from minimal to rich medium and vice versa. Although synchronous division is not produced by their treatments, the results delineate some of the dynamics of growth rate changes which should be useful in attempting synchronization by means of general nutritional changes. The sequence of events attendant on shift from low growth rate to high growth rate is: (1) a quick increase in rate of RNA and protein synthesis to the value characteristic of rapidly growing cells, (2) the rate of DNA synthesis shifts up, followed by increase in nuclei/cell, (3) after 70′ the division rate shifts up. In line with this, Doudney (1961) and Billen (1960) have presented evidence that synthesis of RNA, protein, or both, is a necessary activation step for subsequent synthesis of new DNA. Maaløe and Hanawalt (1961) report partial synchronization of cells selected for immunity to thymineless death. This selection amounted to selection of cells from a brief period of the division cycle during which no DNA replication occurs. DNA replication can be initiated in these cells only if preceded by protein and/or RNA synthesis. Hence by control of one aspect of the general processes of protein and/or RNA synthesis specific control of DNA synthesis and division was achieved. This general approach is worthy of further investigation.

REFERENCES

Barner, H. D. and S. S. Cohen (1956). Synchronization of division of a thymineless mutant of *Escherichia coli*. *J. Bacteriol.* **72,** 115.

Barner, H. D. and S. S. Cohen (1958). Protein synthesis and RNA turnover in a pyrimidine-deficient bacterium. *Biochim. et Biophys. Acta* **30,** 12.

Billen, D. (1960). Effects of prior alteration in nucleic acid and protein metabolism on subsequent macromolecular synthesis by irradiated bacteria. *J. Bacteriol.* **80,** 86.

Brock, T. and M. Brock (1960). Studies on thymineless death. *Bacteriol. Proc.*, pp. 98, 178.

Burns, V. W. (1959). Synchronized cell division and DNA synthesis in a *Lactobacillus acidophilus* mutant. *Science* **129,** 566.

Burns, V. W. (1961). Relations among DNA and RNA synthesis and synchronized cell division in *L. Acidophilus*. *Exptl. Cell Research* **23,** 582.

Doudney, C. O. (1961). Recovery of deoxyribonucleic acid synthesis in bacteria after amino acid starvation. *Biochem. Biophys. Res. Commun.* **5,** 405.

Frisch, D. M., J. Gregory and D. W. Visser (1960). Effects of 5-bromodeoxycytidine in *Escherichia coli* and bacteriophage. *J. Bacteriol.* **79,** 666.

Hoff-Jørgensen, E. (1952). A microbiological assay of deoxyribonucleosides and deoxyribonucleic acid. *Biochem. J.* **50,** 400.

KJELDGÅRD, N. O., O. MAALØE and M. SCHAECHTER (1958). The transition between different physiological states during balanced growth of *Salmonella typhimurium*. *J. Gen. Microbiol.* **19,** 607.

MAALØE, O. and P. C. HANAWALT (1961). Thymine deficiency on the normal DNA replication cycle. I. *J. Mol. Biol.* **3,** 144.

OKAZAKI, T. and R. OKAZAKI (1959). Studies of deoxyribonucleic acid synthesis and cell growth in the deoxyriboside requiring bacteria, *Lactobacillus acidophilus*. II. Deoxyribonucleic acid synthesis in relation to ribonucleic acid and protein synthesis. *Biochim. et Biophys. Acta* **35,** 434.

PERRY, R. (1959). On some properties of growth in a synchronized population of *Escherichia coli*. *Exptl. Cell Research* **17,** 414.

SCOTT, D., B. MCNAIR and E. CHU (1958). Synchronized division of growing cultures of *Escherichia coli*. *Exptl. Cell Research* **14,** 166.

SCOTT, D., B. MCNAIR and E. CHU (1959). Periodicity in the synthesis of RNA in synchronized cultures of *Escherichia coli* in the presence and absence of 4-mercapto pyrimidine. *Exptl. Cell Research* **18,** 392.

STEVENSON, I. (1961). Natural synchronization of a vitamin B12 Arthrobacter-sp. *Can. J. Microbiol.* **7,** 667.

ZEUTHEN, E. (1958). Artificial and induced periodicity in living cells. *Advances in Biol. and Med. Phys.* **6,** 37.

17

Synchronous Division of Animal Cells in Culture

ALISON A. NEWTON, *Department of Biochemistry, University of Cambridge, England*

CONTENTS

I Introduction 441
II Description of the Synchronous Systems 443
1. Synchronous division induced by application of temperature shock 443
A. Account of the system obtained 443
B. Investigation of conditions giving optimal synchrony . . 445
C. Mitotic rate in synchronized cultures 448
D. The mechanism of induction of synchronous division by chilling 450
2. Induction of synchronous division by variation of nutritional factors 452
A. Account of the system obtained 452
B. Investigation of conditions giving optimal synchrony . . 453
C. Mitotic rate in synchronized cultures 453
D. The mechanism of induction of synchronous division . . 454
3. Synchronization of division by physical separation of dividing cells 455
A. Account of the system obtained 455
B. The mitotic rate in synchronized cultures 456
III The Study of Cell Biology by the Use of Synchronous Cultures . 456
1. Biochemical investigations 457
2. The sensitivity of synchronously-dividing cells to irradiation . 461
3. The sensitivity of synchronously-dividing cells to infection . 461
IV Conclusion 462
References 463

I. INTRODUCTION

Early in the development of tissue culture it was realized that this new technique offered unique advantages for the study of the physiology of cell division. Before the nineteen-twenties most cytological study of dividing animal cells had been limited to description of the static morphology of mitosis, and even these investigations were hampered by the low mitotic rate of most adult tissues. However, with the introduction of methods for the maintenance of cell growth *in vitro* tissue explants exhibiting a relatively high mitotic rate at their proliferating margins could easily be obtained; moreover in such cultures individual cells could be observed throughout the course of one or more division cycles. Cells in tissue culture are usually thinly stretched and hence well suited to microscopic examination; detailed study of dividing cells was made possible by the development of the technique of phase contrast microscopy (Zernicke, 1935). In addition, since conditions in these early tissue culture experiments were at least partially defined, the effects of variation in environment on the course of division could be investigated.

Much of our knowledge of the kinetics of cell division is derived from these early investigations of cells in culture; however even in healthy tissue explants the mitotic index seldom exceeds 3 per cent. In order to increase the proportion of dividing cells tissue culturists have resorted to various experimental manipulations. Of these the most common has been the application of temperature shock to the cultures, brief exposure to a low temperature being more effective for this purpose than incubation at supra-optimal temperatures (Bucciante, 1928; Spear, 1928). The addition of growth promoting substances such as embryo extract has also been used to increase the mitotic rate in cultured cells (Wilmer, 1935). Such devices have indeed been used quite recently to increase the yield of mitotic cells for biochemical investigations (Firket, 1956; Hoffmann-Berling, 1954). Undoubtedly many of these early workers obtained partial synchrony of cell division. However techniques were not then available for the precise estimation of cell numbers in these cultures; determination of growth rates depended on such methods as measurement of the area of the explants or of the increase in dry weight rather than estimation of any property directly associated with cell number (Wilmer, 1935). For this reason results were often expressed as numbers of mitoses present in the cultures at any given time. However mitotic rate is not a reliable indication of the rate of cell division. The frequency of mitosis is likely to vary in different parts of a culture, particularly in primary explants where only peripheral cells are multiplying rapidly. In addition any agent causing a block of one particular mitotic phase will increase the mitotic index whilst actually reducing the rate of cell division; this is particularly true of temperature changes which are known to cause accumulation of certain mitotic figures (Barber and Callan, 1943; Lewis, 1933).

Detection of synchronized cell division in tissue cultures had therefore to wait until the development of techniques for the reliable enumeration of cell populations. During the last decade methods for the estimation of cell number and of viability have been used with increasing frequency by tissue culturists. In addition other variable factors have been eliminated or become more closely defined. Pure lines of cells derived from cloned stock have been maintained under standard conditions for several years; the growth media used may be completely defined, although in many cases serum remains in use as one variable component (Levintow and Eagle, 1961). With this application of microbiological technique to the field of tissue culture it was inevitable that those methods which had been used successfully for the initiation of synchronous division in populations of microorganisms should also be applied to cultures of animal cells. Within the last few years accounts have been published of the successful synchronization of division in animal cells cultures by the application of each of the three main types of agent, namely temperature shock (see chapters 4 and 8), variation in the nutritional state of the culture (see chapters 12 and 16) and physical selection of dividing cells (see chapter 15). These will now be discussed in detail.

II. Description of the Synchronous Systems

1. *Synchronous Division Induced by Application of Temperature Shock*

A. *Account of the system obtained*

Although Gey, Bang and Gey (1954) reported that cultures of human cervical carcinoma cells, strain HeLa . . . 'manifest synchronously dividing cells following a period of cooling' the first detailed account of synchronized cultures was made in 1958 by Wildy and Newton. HeLa cell cultures subjected to one brief period of chilling followed by replacement at the normal growth temperature (37°C) showed little increase in cell number for 17 hr; between 18 and 20 hr after the cold shock up to 90 per cent of the cells divided. In the early investigations the cultures were again exposed to a temperature of 4°C for 1 hr after the first period of division; a second cycle of division was then observed which showed similar characteristics to the first. Later experiments (Newton and Wildy, 1959) indicated that this repetition of the chilling process was unnecessary and as many as three synchronized periods of division have been obtained following one initial period of incubation at 4°C (Fig. 1A).

Certain characteristics of the growth curves which were obtained regularly are worthy of comment. Firstly the increase in cell numbers during the first 17–20 hr following chilling never exceeded 10 per cent—which is of the same order as the accuracy of the cell-counting method. This is in marked contrast to certain bacterial systems where cell division follows almost immediately upon replacement of the cells at the optimal-growth temperature (Hotchkiss, 1954; Maaløe and Lark, 1954). In addition the interval between successive division cycles in synchronized cultures was equal to the initial lag period between chilling and the first increase in cell numbers. Under conditions which were optimal for the growth of control cultures this interval was equal to 17–18 hr, which is in good agreement with the normal interphase period of HeLa cells. Alteration of the growth medium to give slower growth rates in control cultures (e.g. using sera to which the cells were not adapted)

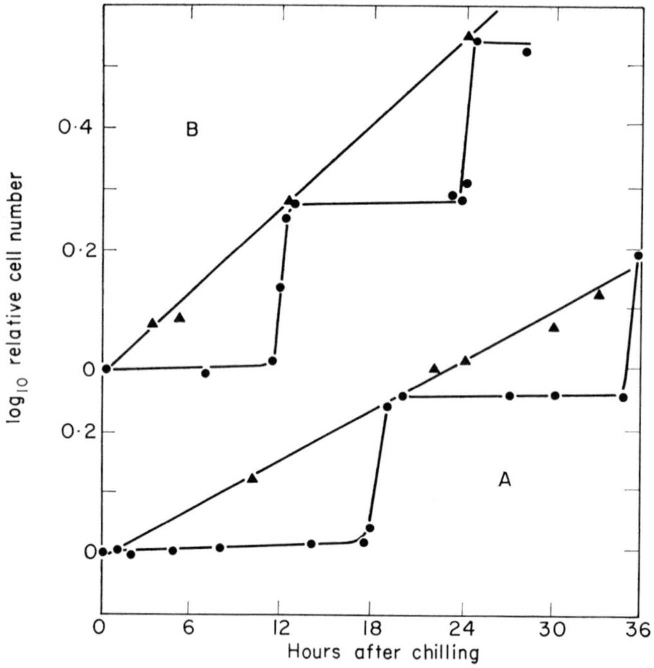

FIG. 1. Synchronous division induced by cold shock in cultures of human cells. Cultures were chilled at 4°C for 1 hr and then replaced at 37°C. ●—● Chilled cultures. ▲—▲ Control cultures.

A. HeLa cell cultures. (Data from Newton and Wildy, 1959.)

B. H Ep.-2 cell cultures (subline C/3). (Data from Rapp, 1960.)

lengthened this time without necessarily affecting the degree of synchronization. At the end of the division period the number of cells in the synchronized cultures was always approximately equal to that in the control unchilled cultures. This suggests that the chilling process delays cell division but does not cause any permanent inhibition in the cells.

The equality of cell number after division between control and chilled cultures was noticeable even though in many cases the synchronized cultures had not doubled at the end of the division period. The results indicate that in all the HeLa cell cultures examined a proportion of the cells did not undergo division; this idea is substantiated by the fact that during the second division cycle in the synchronized cultures the increase in cell numbers was always double that observed at the first cycle. Thus if the degree of synchronization is expressed as the proportion of the population dividing within a certain time interval, the synchronization improved at each division cycle.

Other features of the growth cycle were not constant. The proportion of dividing cells in the cultures varied from as low as 40 per cent to as high as 95 per cent. The reason for this variation is not known, but since the same average growth rates were observed in control as in chilled cultures it seems probable that it reflects the viability of the cells at the time of subculture. Selection of cells for the inoculum was important; if the stock culture was nearing the stationary phase of growth, only a small proportion of the cells divided during the subsequent synchronization experiment. Any undue disturbance of the cells during the subculturing process had the same result. Optimum results were obtained by selecting cells for subculture which were in the early logarithmic phase of growth, and by carrying out all the operations of subculture at 37°C and as quickly as possible.

Variation in the degree of synchronization was also observed. The proportion of the total growth cycle occupied by the division period never exceeded 1/5 and frequently was as little as 1/24. The factors which affect this have not been investigated, but it seems possible that variation in the rate of cooling and reheating of the cultures is responsible. In the normal system used the rate of heat exchange through the thin-walled vessels is rapid, which may account for the success of synchronization in most cases.

B. Investigation of conditions giving optimal synchrony

No detailed study of the cooling conditions necessary for synchronization of division has been made. Earlier workers had shown that chilling of chick cell cultures to 4°C for 10 min had no influence on the subsequent mitotic rate (Spear, 1928). However cooling of the same cells at this temperature for 24 hr resulted in degeneration of mitotic cells (Chèvremont-Comhaire and Chèvremont, 1956). For chick cells 21°C seems to be the critical temperature below which they cannot enter into mitosis, and quite good synchronization of mitosis was achieved by incubating cultures at 16°C for 24 hr before replacing them at 37°C. Cultures of L cells show equally good synchrony if chilled at 4°C or at 12°C, but are not affected by the same period of incubation at 25°C (Newton, unpublished work). Quite mild temperature shocks are sufficient to cause partial synchronization of HeLa cell cultures, which is frequently observed following addition of medium at room temperature to growing cultures.

The effects of incubation at temperatures higher than 37°C have not been carefully investigated. Kemp and Juul (1931) reported that heating of chick cell cultures at temperatures of 46–48°C for 5 min caused a disappearance of mitotic figures for 3 hr after replacement at 37°C; between 3 and 8 hr after the heating there was a rapid increase in the mitotic index of the cultures until this equalled 170 per cent of the control value. Even short periods of heating at temperatures greater than 48°C killed many of the cells, and a few hours of incubation at 42°C also kills several strains of tissue culture cells (Kokott, 1930; Pincus and Fischer, 1931; Scherbaum, 1960). However it may be possible to exploit this aspect of the temperature sensitivity of the division mechanism to yield satisfactory synchronous cultures.

It is not certain that a period of 1 hr at 4°C is necessary for the successful induction of synchronous division. Longer incubation at the low temperature does not improve the synchronization, but shorter periods may be sufficient. It is of interest that the duration of the temperature shock is so short compared with the overall growth cycle of the cells. This is in contrast to many microbial systems where prolonged treatment of the culture is necessary for the induction of synchrony (Scherbaum, 1960).

The population growth phase of the cells at the time of chilling also seems to be an important factor in determing the character of the subsequent synchronized growth curve. With HeLa cells the most successful results have been obtained by chilling cultures just after their entry into the logarithmic phase of growth (about 24 hr after subculture). Chilling during lag phase resulted in poor and inconsistent growth of chilled cultures. Miura and Utakoji (1961) have recently published growth curves obtained by chilling cultures of human amnion cells strain FL (Fogh and Lund, 1957) at various stages of growth; they deduce that similar partially synchronized cultures showing three phases of growth were obtained in each case. However careful examination of their data reveals a lack of consistency in their results. Cells chilled 24 hr after subculture gave a curve reminiscent of the results obtained with HeLa cells when growth was estimated from a photographic record of the culture; cell counting methods however indicated only slight phasing of growth. Cultures chilled 48 hr after subculture exhibited a stepwise increase in cell count, rapid increases occurring 12–16 and 36–40 hr after chilling but with appreciable multiplication taking place between these periods.

Although the population growth phase of the cells is clearly of importance in regard to the overall physiological state of the cells at the time of chilling, cell density is another critical factor. In the experiments of Newton and Wildy (1959) cell division was never observed when the cell number exceeded 10^6 cells/ml even if fresh growth medium was added to the cultures. Thus in order to observe two cycles of division it was necessary to have an initial cell density of $1–2 \times 10^5$/ml. However since cell densities of less than 10^5/ml cannot be counted with efficiency by the methods used variation in size of the inoculum is strictly limited.

The published experiments of Wildy and Newton were all carried out with a 'wild-type' strain of HeLa cells. They have since been repeated with mouse fibroblast cells (strain L) which give essentially similar results having an inter-division period of 17–18 hr, which is comparable to that of HeLa cells. However there are clearly differences in the way in which cell lines react to the chilling process. The S_3-cloned substrain of HeLa cells gave only poor results despite several attempts to induce synchrony by the methods which had proved successful with the wild type. Such a variation between different strains of the same cell line has been carefully investigated by Rapp (1960) using another cell line derived from human carcinoma (H Ep.-2) (Toolan, 1954). One fast growing subline (C/3) which had a mean generation time of 12 hr gave excellent synchronization curves, there being no increase in cell number for 11·5 hr after chilling; in the next half-hour more than 80 per cent of the cells then divided; two division cycles were regularly obtained (Fig. 1B). Attempts to obtain synchrony with other more slowly-growing sublines were unsuccessful; it would have been interesting in these cases to know the phase of growth of the slow-growing cultures at the time of chilling.

C. Mitotic rate in synchronized cultures

Cells in tissue culture possess one unique advantage over other synchronized cell systems, namely that in addition to cell numbers the mitotic rate can also be determined. Thus an independent method for evaluation of the degree of synchronization is available. Although mitotic index is not by itself a reliable index of cell multiplication, an ideally synchronized system should exhibit synchronization of mitosis in all its stages, and of division. Such a system has not yet been achieved, although partial synchronization of the mitotic process has been recorded (Fig. 2). Spear (1928) found that chick cells which had been chilled at 0·5°C for 4 hr and then replaced at 37°C showed a gradual decline in the mitotic index in the first 5 hr after temperature shock. Similar results were obtained by Newton and Wildy (1959) in chilled cultures of HeLa cells, when the prophase index in particular declined, suggesting that no new cells entered mitosis although a small increase in cell number was observed. At some time after replacement at 37°C (6 hr for chick cells, 13 hr for HeLa cells) there was an increase in the mitotic index which became much greater than that observed in control cultures. In the case of the HeLa cells this burst of mitotic activity preceded the increase in cell numbers by about 2 hr, and has been shown to be composed of successive waves in the prophase, metaphase and anaphase activity. Thus the observed results were indeed due to a partial synchronization of mitosis and were not due to accumulation of any one phase. However there existed a discrepancy between the number of mitotic cells and the actual increase in cell number in any one experiment. In no experiment was a mitotic index of more than 10 per cent recorded although up to 90 per cent of the cells were known to have divided in the same system. It is clear that the mitotic processes of these cells had not been closely synchronized. Although this lack of synchronization of the phases of mitosis is partially due to fluctuations in the mitotic period of HeLa cells (Moorhead and Hsu, 1956) the recorded mitotic index is insufficient to account for the increase in cell numbers. Thus in one experiment calculations, based on a prophase time of 8 min determined from control cultures, indicated that 32 per cent of the cells entered mitosis before the period of cell division. However in this period 78 per cent of the cells were known to have divided. There seem to be two possible explanations for this discrepancy. Possibly the method of estimation of mitotic cells is at fault—thus cells in mitosis are known to be attached less firmly to the glass than during interphase (Axelrad and McCulloch, 1958) and so might be selectively lost during the fixation procedure. This

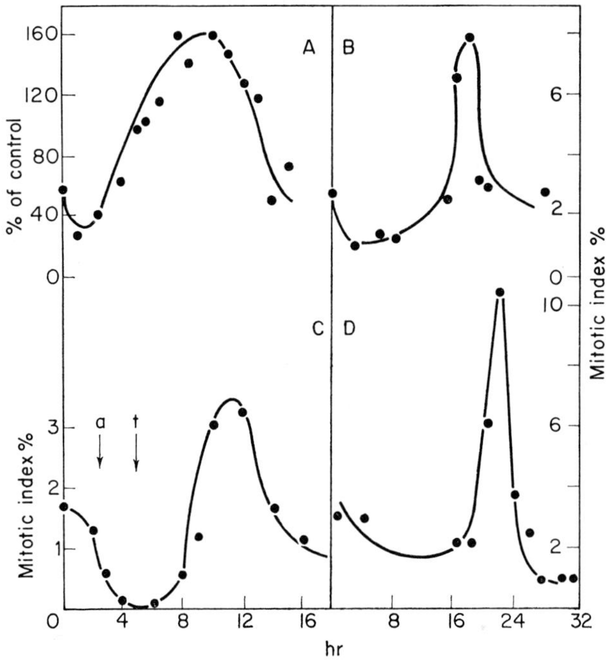

FIG. 2. Mitotic rate in synchronized cell cultures.

A. Chick fibroblast cells chilled at 0·5°C for 4 hr and replaced at 37°C at time 0. (Data from Spear, 1928.)

B. HeLa cells chilled at 4°C for 1 hr and then replaced at 37°C at time 0. (Data from Newton and Wildy, 1959.)

C. Mouse mastocytoma cells treated with amethopterin and thymidine. (a) addition of $5 \times 10^{-7} M$ amethopterin. (t) addition of $3 \times 10^{-5} M$ thymidine. (Data from Schindler, 1960.)

D. HeLa S_3-9IV cells showing synchronous division following selection of mitotic cells from a randomly-dividing culture. (Data from Terasima and Tolmach, 1961.)

may indeed account for some of the discrepancy, since larger numbers of mitotic cells were observed by examination under phase contrast than were subsequently found upon fixation; this point clearly deserves further investigation. However the fact that the same lack of correlation between cell numbers and mitotic index is observed in other synchronized systems suggests that some more fundamental factor is involved. It has been reported that tissue culture cells may undergo amitotic division although the occurrence of such events is rare in normal cell cultures (Hughes, 1952). If the synchronization process stimulated such division then the result would be explained.

In addition cells maintained in tissue culture are known to exhibit a certain proportion of multinucleate and multipolar mitotic cells. Division of these cells would clearly affect results based on cell numbers. Newton and Wildy (1959) studied the appearance of these cells in synchronized cultures and found not only that the proportion of these cells was not changed by the chilling process but that this proportion was less than 0·05 per cent of the cells, making it unlikely that their contribution to the results was significant.

D. The mechanism of induction of synchronous division by chilling

No conclusions have yet been reached as to the mode of action of the temperature shock in inducing synchronous division in tissue culture cells. However two characteristics of the system may provide some clue as to the nature of the physiological changes produced. Firstly the lag period following replacement at the optimal growth temperature, during which little cell division is observed, is approximately equal to the normal interphase period of the cell cycle. Secondly only one period of incubation at the lower temperature is required and this period is short compared to the overall growth cycle of the cell. These facts suggest that during incubation in the cold some change occurs which places most cells in a state comparable to late telophase. No chemical change has so far been detected in the cells during or immediately following incubation at 4°C, although it is known that more prolonged incubation at this temperature causes considerable loss of potassium from chick embryo cells (Wesson *et al.*, 1949). From the results of experiments in which synchronized cultures were exposed to a further low temperature shock it was concluded (Newton and Wildy, 1959) that cells chilled at least 3 hr before expected division had once again to complete the full interphase period of 17–18 hr before division could occur. Cells chilled 2–3 hr before division were however little affected by the incubation at 4°C and proceeded to divide with a delay equal only to the period of chilling. It is probably not a coincidence that in these cells deoxyribonucleic acid (DNA) synthesis (p. 458) also occurs in these 3 hr before division.

It seems likely that cells which have synthesized some of the DNA in preparation for mitosis are unaffected by the incubation at low temperatures. All other cells will be affected by incubation at 4°C in such a way that some components necessary for division must again be built up to the critical concentration. These may be components of one or of several distinct systems. However one attractive hypothesis is based on the properties of the system responsible for synthesis of DNA. It has been suggested recently that the system responsible for synthesis of thymidine triphosphate (TTP) in animal cells may be one of the controlling factors in DNA synthesis (Bollum and Potter, 1959; Ord and Stocken, 1960) and that one or all of the enzymes of this system may have to be regenerated during each division cycle of the cell in order for DNA synthesis to occur (Bianchi *et al.*, 1961; Weissman *et al.*, 1960). One may speculate that this enzyme system is labile at low temperatures; thus cells which had been chilled would be comparable to late telophase cells in having to renew this TTP synthesizing system before DNA synthesis and subsequent division could occur. There is as yet no experimental evidence in support of this explanation of the synchronizing effect of the low temperature shock and of course many other trigger mechanisms may be involved.

Miura and Utakoji (1961) have pointed out that if cells chilled 2–3 hr before division are only little affected by the temperature shock, one might expect that chilling of a randomly-dividing culture having a doubling time of 18 hr would not affect 10–20 per cent of the cells. In practice the increase in cell numbers observed after chilling was usually less than 10 per cent. However, in routine experiments cultures were chilled just after the end of the lag phase, so that a smaller proportion of the population would be in the temperature stable condition. In the experiments described by Miura and Utakoji (1961) cultures were chilled later in the logarithmic phase of growth and their figures show that 12–17 per cent of the cells divided after the chilling period, which agrees more closely with the above calculations. If the above hypothesis is correct the success of the synchronization will depend upon the number of cells which have synthesized DNA at the time of chilling. If DNA synthesis occurs early in interphase synchronization may be poor. If however, as with HeLa cells (Painter and Drew, 1959; Smith *et al.*, 1959), DNA synthesis takes place shortly before prophase good synchronization may result from chilling a randomly-dividing culture.

2. *Induction of Synchronous Division by Variation of Nutritional Factors*

A. Account of the system obtained

The initiation of 'waves of mitosis' in cultures of animal cells by the addition of nutrients to the medium is a well-known phenomenon. The most effective of these mitotic stimulators is chick embryo extract, which is known to contain nucleoprotein as one active principle (Waymouth, 1952). The action of this embryo extract has been studied in some detail by Jacoby, Trowell and Willmer (1937). They found that cells grown in a medium containing only plasma and Tyrode solution showed no signs of cell division but responded to the addition of embryo extract by an increase in the mitotic index to 3–4 per cent some 10–20 hr later. The same effect was achieved even if the embryo extract was in contact with the cells for a few hours only. A further period of mitotic activity could be induced only if embryo extract was present during or after cell division. These results suggest that some essential factor could be rapidly absorbed from the embryo extract and that this factor was probably only effective in cells which had completed a division cycle.

This partial synchronization of mitosis was brought about by the addition of a complex substance to cells which had been starved

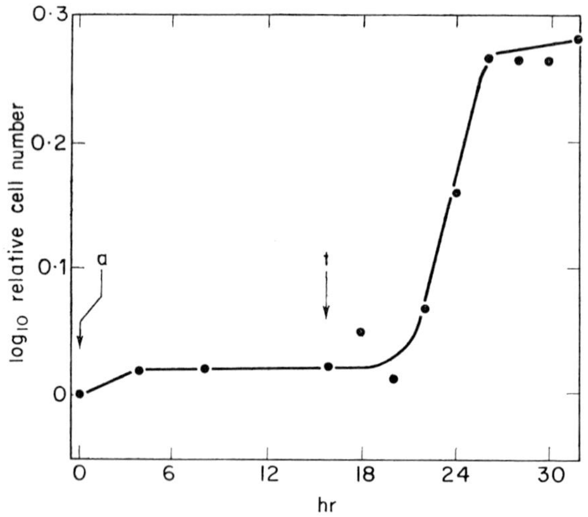

FIG. 3. Synchronous division in HeLa cell cultures treated with amethopterin and thymidine. (Data from Rueckert and Mueller, 1960.) (a) $10^{-6}M$ amethopterin added. (t) $10^{-5}M$ thymidine added.

of most essential growth factors. A high degree of synchronization may be achieved by the relief of the specific starvation caused by the use of competitive inhibitors. Rueckert and Mueller (1960) extending the observations of Hakala and Taylor (1959) found that the addition of amethopterin or of 5 fluoro-2-deoxyuridine (FUDR) to cultures of HeLa cells resulted in a rapid cessation of cell division. Cell numbers increased by only 4–10 per cent following addition of the inhibitor (Fig. 3). If the drug was present for longer than 20 hr the viability of the cells decreased rapidly; however for periods up to 16–20 hr the effects of the inhibitors could be reversed by the addition of excess thymidine to the culture medium. The addition of thymidine to cultures 'on the brink of imminent death' produced a burst of cell division 6–8 hr later. In typical experiments 70–95 per cent of the cell population divided in a period of less than 4 hr. It is not clear whether the population remained synchronous after the division period, but in one experiment no further cell division was observed over the following 15 hr.

B. *Investigation of conditions giving optimal synchrony*

Detailed investigation of conditions giving rise to synchronous division following the use of analogues has not been reported. However similar results have been obtained using either amethopterin or FUDR as the inhibitor. Reversal was achieved in each case by the addition of $10^{-5}M$ thymidine to the cultures not later than 16 hr after the addition of the drug. Addition of thymidine after this time could not reverse the inhibition in all the cells. Synchronization of division by this method was observed not only in monolayer cultures but also in suspension cultures, which raised the hope of obtaining large numbers of synchronized cells. However cell densities greater than 10^5 cells per ml could sometimes overcome the effects of the inhibitor unless high drug concentrations were used. The most satisfactory results were obtained by the use of initial cell densities of approximately 10^5/ml in conjunction with $10^{-6}M$ amethopterin or FUDR.

C. *Mitotic rate in synchronized cultures*

Unfortunately Rueckert and Mueller (1960) have not described any variations in mitotic rate in their synchronized cultures so that the relationship between cell division and mitosis in this instance cannot be determined. However Schindler (1960) using a somewhat similar system has studied the variation in mitotic index. The addition of 5×10^{-4} mM amethopterin or FUDR to cultures of transplanted mouse mastocytoma cells caused a complete disappearance of mitotic figures within 4 hr (Fig. 2C). Addition of thymidine ($3 \times 10^{-5}M$) 6 hr after the addition of the drug resulted in the reappearance of mitotic figures. A peak in the mitotic index was apparent about 5 hr after the reversal of inhibition, which agrees well with the 6-hr lag period before division observed by Rueckert and Mueller (1960). However the maximum mitotic index observed was only 3 per cent. Either there was little cell division under the conditions used by Schindler or else the same discrepancy between mitotic count and cell count exists in these synchronized cultures as in those synchronized by temperature shock.

D. The mechanism of induction of synchronous division

The comparisons between the system outlined above and that described by Barner and Cohen (1956) following thymidine rescue of unbalanced growth in cultures of a thymidine-less mutant of *E. coli* are clear. In each case optimal synchronization was achieved by the addition of thymidine just before the onset of 'thymidine-less death'. In each case also the addition of thymidine was followed after a short lag period by a renewal of DNA synthesis which preceded cell division. The animal cell cultures also showed an increase in the amount of recoverable protein shortly after the addition of thymidine. Preliminary experiments using puromycin (Rueckert and Mueller, 1960) indicated that protein synthesis might be involved in preparation of the cells for DNA synthesis. It thus seems probable that synchronization is made possible by the preliminary starvation of the cells to the point where no thymidine nucleotides are available for DNA synthesis. Those cells which have already prepared the DNA for division will divide unimpeded and it is significant that in this instance a 4–10 per cent increase in cell numbers was observed after the addition of inhibitor.

Before incorporation into DNA, added nucleotides must first be phosphorylated to give the triphosphate form; thymidine is not the natural precursor of TTP in these cells and it is possible that an adaptive enzyme is involved in this process (Weissman *et al.*, 1960). Thus one might anticipate that all starved cells would show a lag after the addition of thymidine during which time the enzyme system necessary for TTP synthesis would be evolved. Only after this period could DNA synthesis take place, to be followed in due course by cell division. Thus the degree of synchronization of the cultures would depend upon the uniformity of this lag period in the starved cells.

3. *Synchronization of Division by Physical Separation of Dividing Cells*

A. Account of the system obtained

The most satisfactory systems for obtaining synchronous division are those which cause least trauma to the cell and which do not result in unbalanced growth of the culture. If some unique property of dividing cells can be used to separate them without damage from other cells in the culture such a system should be obtained. Dividing tissue culture cells are known to be less firmly attached to the supporting surface than interphase cells (Axelrad and McCulloch, 1958). Terasima and Tolmach (1961) have exploited this fact to obtain suspensions of mitotic HeLa cells by gentle washing of monolayer cultures grown on a plastic surface. These suspensions showed a high plating efficiency on being placed in new culture vessels, and most of the cells which became attached under these conditions divided within 3 hr of collection (Fig. 4). 17–21 hr later a second division cycle occurred in which 70–80 per cent of the cells divided over a period of 8 hr; a third division cycle was also observed after a further 15–17 hr. However the degree of synchronization gradually

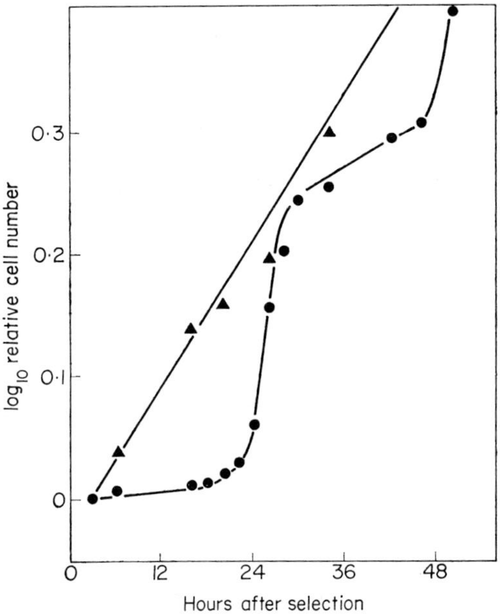

FIG. 4. Synchronous division in cultures of HeLa S_3-9IV cells. Mitotic cells were selected from a randomly-dividing culture at 0 time. (Data from Terasima and Tolmach, 1961.) ▲—▲ Control cultures. ●—● Selected cultures.

lessened during the growth curve, a 10 per cent increase in cell number being observed between the first and second division periods and a 40 per cent increase between the second and third periods. In these experiments as in other systems, it seems that some of the cells present were not dividing although the average growth rate was the same in control and synchronous cultures.

B. *The mitotic rate in synchronized cultures*

Terasima and Tolmach (1961) followed the mitotic rate of the culture synchronized by selection of dividing cells, and observed a rapid increase in the value of the mitotic index 2 hr before the onset of the division cycle (Fig. 2D). The maximum mitotic index observed was 11 per cent. Assuming a duration of mitosis of about 45 min, these figures suggest that in the period 20–28 hr after collection of the cells 60–70 per cent of the cells went through mitosis, whereas estimation of cell number indicates that 70—80 per cent of the cells divided in the same period. The agreement between these figures is much better than in the case of cells synchronized by chilling (see section II-1, C). Nevertheless a maximum value of 11 per cent for the mitotic index suggests that in this system also the events of mitosis are only poorly synchronized.

III. The Study of Cell Biology by the Use of Synchronous Cultures

The methods used for the induction of synchronous division and the properties of the system so obtained can themselves provide a great deal of information about the mechanisms controlling cell division. However it is the further study of these systems that can eventually yield the answers to some of the problems of cell biology. A well synchronized population of cells should provide sufficient material at a known stage of the cell cycle for investigation of chemical and physical properties to be made, previous studies in animal cells having been limited either to embryonic material or to single cells. There is for example the hope that such cultures will be useful for the preparation of the mitotic apparatus and so make possible an extension of the work of Mazia (1958) to other types of cell. Moreover study of the effects of external agents at various stages of growth could provide information about the reproductive processes of the chemical and morphological constituents of the cell. The mode of action of many drugs which are known to interfere with growth processes could well be elucidated by study of such systems. Synchronized cultures can also be of use in the study of cell properties not themselves connected with the process of cell division since synchronization eliminates some of the many variables of a randomly-dividing culture. Such studies of animal cell cultures are only in their infancy but a beginning has been made in the investigation of the biochemistry of the dividing cell. In addition some study of the effects of viruses and of irradiation on synchronously dividing cultures has been made.

1. *Biochemical Investigations*

Disappointingly few biochemical investigations of synchronously dividing tissue culture cells have been reported. One reason for this may be that at the present time it is difficult to obtain the quantity of cells required for the analysis of chemical composition or of enzymic properties. However some study of the changes in nucleic acid content has been made, particular attention being paid to the relationship between the timing of DNA synthesis and cell division. Although synchronous cultures of bacterial cells (Abbo and Pardee, 1959) and of ciliate protozoa (chapter 4) have now been obtained in which no phasing of DNA synthesis is apparent this is unlikely to be the case with animal cells since data from many different systems suggest that DNA synthesis is not a continuous process in these cells but takes place at some definite time during interphase (Vendrely, 1955). Results obtained with synchronized cultures have confirmed these observations and have given a more precise indication of the time at which DNA appears in the cell.

Rueckert and Mueller (1960) found that addition of amethopterin to growing HeLa cells caused eventual inhibition of DNA synthesis although some DNA-like material accumulated in the cells for the first few hours after addition of the inhibitor (Fig. 5A). Induction of synchrony by the addition of thymidine to the system was followed after 2 hr by a rapid synthesis of DNA. This synthesis preceded cell division by 2–4 hr and continued for about 2 hr. Schindler (1960) suggests that the timing of the effects of amethopterin and thymidine on mitosis in cultured mouse mastocytoma cells could best be interpreted if DNA synthesis started 3–5 hr before mitosis and was completed by 1 hr before mitosis.

A second period of DNA synthesis took place in HeLa cells after cell division in the system studied by Rueckert and Mueller (1960); however this post-mitotic synthesis represented a smaller increase in the amount of DNA per cell than that taking place before cell division. Comparable results have been obtained for cultures of HeLa cells where synchrony was induced by chilling (Newton and

Wildy, 1959) (Fig. 5B). The cold shock resulted in inhibition of DNA synthesis for 14 hr, but 3–4 hr before cell division a rapid increase in the DNA content of the cells was observed; the DNA content reached its maximum value about 1 hr before division. Immediately after division the DNA content of the daughter cells was about 60 per cent of this maximum value and in this instance also a small post-mitotic increase in the amount of DNA was observed. There followed a resting period of 14 hr in which little DNA synthesis occurred and the bulk of the DNA for the second division cycle again appeared in the 4 hr which preceded mitosis. In cultures of chick cells partially synchronized by cold shock DNA synthesis precedes prophase by at least 1 hr (Firket, 1956).

These results are in general agreement with others obtained for cultures of animal cells. The quantitative distribution of Feulgen staining material in randomly-dividing HeLa cell cultures suggests that DNA synthesis starts 5 hr before division and occupies about

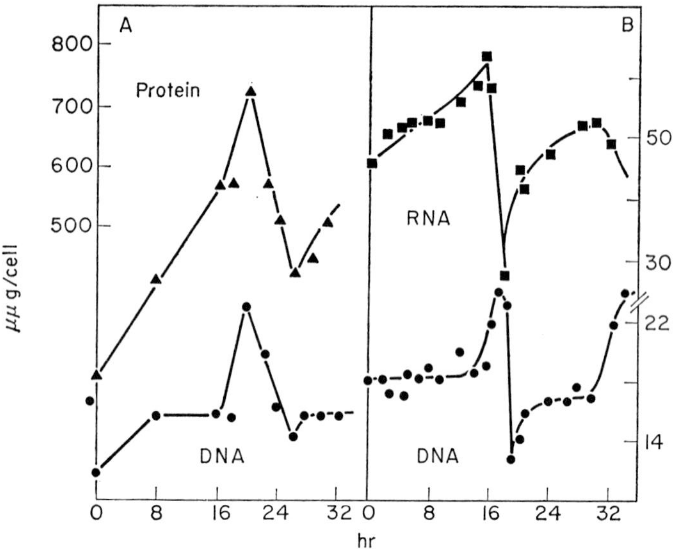

FIG. 5. The nucleic acid and protein content of synchronously dividing HeLa cells.

A. Protein and DNA content of cells synchronized following the addition of amethopterin at 0 time and thymidine at 16 hr. Cell division occurred between 6 and 10 hr after the addition of thymidine. (Data from Rueckert and Mueller, 1960.)

B. RNA and DNA content of cell cultures synchronized by the application of cold shock. The cultures were replaced at 37°C at 0 time. Cell division occurred between 17 and 19 hr after chilling. (Data from Newton and Wildy, 1959.)

4 hr (Smith *et al.*, 1959). Painter and Drew (1959) followed the uptake of tritiated thymidine by cultures of HeLa cells having a rather long mean generation time, and found in this case also a post-mitotic resting phase of 14 hr in which no DNA synthesis occurred; in their system DNA synthesis occupied a period of 5–6 hr and was followed by a resting phase of 2–8 hr. Measurements of the absorption of ultraviolet light by chick cells in culture (Walker and Yates, 1952) indicate that in these cells also DNA synthesis occurs in the latter half of interphase. The variation in DNA content of cells both in whole tissues and such systems as regenerating rat liver suggests that this timing for DNA synthesis is characteristic of animal cells (Vendrely, 1955; Whitmore *et al.*, 1961). The general agreement between these results and those obtained in the synchronous systems is encouraging and supports the belief that the observed fluctuations in DNA content are not merely artefacts arising from the system used to induce synchrony.

However the post-mitotic period of DNA synthesis observed in the two synchronous systems has not been generally observed in other animal cells. It may be that this is indeed a spurious effect produced by the techniques inducing synchrony, although in such a case it seems unlikely that similar artefacts should be observed in systems synchronized by such different procedures as amethopterin treatment and temperature shock. The biphasic appearance of DNA synthesis is not merely a result of a degree of asynchrony in the cells since model systems having a single period of DNA synthesis cannot be constructed to fit the experimental data. Rueckert and Mueller (1960) have shown that this characteristic curve of DNA synthesis is not due to genetic variation in their cultures since cloned sublines gave similar results. One possibility is that material which appears in the cells after division is not nucleic acid but is some other compound reacting in a chemically similar manner; since deoxyribose was measured in both cases the accumulation of polysaccharide material is not excluded. Finally it is possible that real synthesis of DNA does occur in animal cells after division but because of the geometry of the cells at this time it has not been recognized in the systems studied by cytochemical methods. Some earlier workers (Pasteels and Lison, 1950) did indeed suggest that the bulk of DNA synthesis occurs at this stage of the cell cycle but their work has since been criticised. However in kidney cells synthesis of DNA may occur both before prophase and after telophase (Fautrez and Roels, 1954; Pisi and Cavalli, 1955) and Woodard, Rasch and Swift (1961) have found a similar biphasic pattern of DNA accumulation in cells of *Vicia faba*. It should be simple to solve the problem of the reality of this post-telophase synthesis of DNA by study of the uptake of radioactive nucleic acid precursors by the synchronously-dividing system.

Newton and Wildy (1959) also followed the synthesis of ribonucleic acid (RNA) in HeLa cell cultures synchronized by cold shock (Fig. 5B). In contrast to DNA, RNA synthesis was not affected by the chilling, but continued throughout the cell cycle until just before division. At this time, when the DNA content of the cell was still increasing, the amount of RNA present showed a sudden decrease. It is not known whether this loss was due to hydrolysis of the RNA or to a loss into the surrounding medium. A decrease in the total mass of HeLa cells at the time of division has been suggested by other workers (Sandritter *et al.*, 1960) and loss of RNA at division has been observed in amoebae (Rabinovitch and Plaut, 1956). Woodard, Gelber and Swift (1961) and Woodard, Rasch and Swift (1961) have found a loss of both RNA and protein during fission of *Paramecium aurelia* and *Vicia faba* cells, and suggest that this effect is primarily due to a disruption of the normal equilibrium which exists between breakdown and synthesis of these substances. Since it is clear that synthesis of RNA does not continue throughout division in HeLa cells (Feinendegen *et al.*, 1960) and other organisms (Mitchison *et al.*, 1959) this explanation of RNA loss is a feasible one. Redistribution of RNA within the dividing cell has also been observed (Jacobson and Webb, 1952; Swann, 1957) and it would be interesting to know whether the RNA that is associated with the chromosomes at mitosis is subsequently lost to the cell. The change in protein content of cells synchronized by amethopterin and thymidine treatment has been studied by Rueckert and Mueller (1960) (see Fig. 5A). Some protein synthesis continued throughout the period of amethopterin inhibition but addition of thymidine was followed by a rapid increase in the amount of protein recovered from the cells. Some of this protein was lost at cell division but synthesis was again resumed at the same rate once division was completed. Since experiments with ^{14}C-leucine did not confirm the stimulation of protein synthesis following thymidine addition, the authors suggest that some change in state of the protein prior to division might account for the increase in the amount of protein estimated. However it is also possible that the losses of protein at division were due to the same phenomenon as that described above.

No other biochemical studies on synchronously dividing animal cells have been reported. Clearly it would be of the greatest interest to follow any changes that might occur in the concentrations of nucleotides or other 'energy reservoirs'. Moreover the intriguing possibility that some enzymes change in activity during the cell cycle (Weissman *et al.*, 1960) deserves careful study.

2. *The Sensitivity of Synchronously-Dividing Cells to Irradiation*

It is well known that irradiation of cells causes a delay or inhibition of cell division (reviewed by Errera, 1959) and that the sensitivity of the cell to such damage varies according to the stage of growth cycle (Carlson and Harrington, 1954). Although many workers have studied the effects of various radiations on cells grown in culture there is only one account of such investigations on synchronized cultures of cells. Terasima and Tolmach (1961) estimated the survival of cells following irradiation (X-rays, 220 keV) of cultures synchronized by selection of mitotic cells. Cells in mitosis were markedly more sensitive than the randomly-dividing culture. In addition interphase cells showed a maximum sensitivity to irradiation 14 hr after division, which is precisely the time at which DNA synthesis begins in these cells. Cells irradiated earlier or later in interphase were relatively resistant to damage. The extension of these investigations should eventually help in the elucidation of the varied and complex inhibitory effects of high-energy radiation on growth and cell division.

3. *The Sensitivity of Synchronously-Dividing Cells to Infection*

An understanding of the biology of the cell–virus relationship must include a knowledge of the interaction of the virus with the dividing cell. Replication of foreign nucleoprotein particles within a cell would be expected to have a profound effect on the delicately controlled process of cell division. In addition it is quite possible that the mitotic state of the cell at the time of infection may influence the subsequent growth of the virus. Such problems could well be studied in a quantitative manner by infection of synchronously-dividing cells in tissue culture. The effect of *Herpes* virus has been studied in some detail by these techniques (Stoker and Newton, 1959). Although the growth cycle for this virus occupies some 12 hr in HeLa cells infection of synchronously-dividing cultures showed that the virus was able to inhibit division within 1 hr of penetration into the host cell. Since *Herpes* virus contains DNA and multiplies in the cell nucleus it was not surprising that some disturbance of the normal division mechanisms should occur, but the rapidity of the effect was unexpected. Some RNA-containing viruses, on the other hand, have been found to multiply in the dividing cell without interruption of mitosis (Wheelock and Tamm, 1959). It would be of interest to study the effects of certain tumour viruses in such synchronously dividing systems, particularly in view of the recent suggestion that cells infected with a tumour virus are themselves unable to divide but secrete some substance stimulating mitosis in adjacent tissues (Morgan, 1959).

Since the metabolism of the host cell varies with the mitotic phase it seems probable that susceptibility to infection may vary during the cell cycle. Indeed this may account for the well-known asynchrony of infection observed when certain animal viruses are added to randomly dividing populations of cells. Variation in the yield of phage by synchronous cultures of *Escherichia coli* infected at different stages has been reported (Yanagita *et al.*, 1958). However, in the case of *Herpes* virus, it has been shown that the mitotic state at the time of infection does not affect the growth rate or the final yield of the virus (Newton, 1959).*

IV. Conclusion

The ideal system for induction of synchronous division in animal cell cultures has not yet been achieved. The highest degree of synchrony is obtained by subjecting the cultures to a period of incubation at low temperature, but this process can be criticised as it clearly causes some block in a vital, but as yet unknown, pathway of cellular metabolism. In addition, until this method is applied to synchronization of cells grown in suspension culture the amount of material available for investigation is limited. A larger number of synchronously dividing cells can be produced if amethopterin treatment is used to induce the synchrony; this system is however far from ideal for a study of the normal properties of dividing cells since the cultures have to be brought to the point of 'imminent death' before synchronous division can be initiated. Synchronization of division in cultures obtained by selection of mitotic cells from a randomly dividing population is open to the least criticism on theoretical grounds, since cell metabolism should not have been disturbed by the selection process. Unfortunately the number of cells recoverable by this procedure is small and the degree of synchrony in the cultures is low. It is possible that more satisfactory results could be obtained if two of these methods were combined. Thus the application of temperature shock would produce large numbers of mitotic cells which could then be selected as the

*An addendum to this chapter will be found on page 605 of this book.

inoculum for a synchronized culture; on reaching the second division cycle such cells would be sufficiently distant from the original traumatic process to be considered as physiologically normal.

Although ideal synchronization of the division of animal cells in culture has not been achieved the systems are certainly adequate for a detailed investigation of their biological properties to be undertaken. Indeed if one criterion of synchrony is taken as the proportion of the total growth cycle occupied by the period of division animal cell cultures are superior in this respect to many of the microbial cultures which have been the subject of intensive study. Much more information about the division processes in animal cells could be obtained using such systems and it is to be hoped that the scope of these investigations will soon be extended.

REFERENCES

ABBO, F. E. and A. B. PARDEE (1959). Synthesis of protein, enzyme, RNA and DNA in cell-division synchronized bacteria. *Federation Proc.* **18,** 178.

AXELRAD, A. A. and E. A. MCCULLOCH (1958). Obtaining suspensions of animal cells in metaphase from cultures propagated on glass. *Stain Technol.* **33,** 67.

BARBER, H. N. and H. G. CALLAN (1943). The effects of cold and colchicine on mitosis in the newt. *Proc. Roy. Soc. London* **B131,** 258.

BARNER, H. D. and S. S. COHEN (1956). Synchronization of division of a thymineless mutant of *Escherichia coli. J. Bacteriol.* **72,** 115.

BIANCHI, P. A., J. A. V. BUTLER, A. R. CRATHORN and K. V. SHOOTER (1961). The thymidine-phosphorylating kinases. *Biochim. et Biophys. Acta* **48,** 213.

BOLLUM, F. J. and VAN R. POTTER (1959). Nucleic acid metabolism in regenerating rat liver. VI. Soluble enzymes which convert thymidine to thymidine phosphates and DNA. *Cancer Research* **19,** 561.

BRACHET, J. (1957). *Biochemical Cytology.* Academic Press, New York, pp. 178–194.

BUCCIANTE, L. (1928). Ulteriori ricerche sulla velocità della mitosi nelle cellule coltivate *in vitro* in funczione della temperatura. *Arch. Exptl. Zellforsch.* **5,** 1.

CARLSON, J. G. and N. G. HARRINGTON (1954). X-ray-induced 'stickiness' of the chromosomes of the *Chortophaga* neuroblast in relation to dose and mitotic stage at treatment. *Radiation Research* **1,** 491.

CHÈVREMONT-COMHAIRE, S. and M. CHÈVREMONT (1956). Action de temperatures subnormales suivies de réchauffement sur l'activité mitotique en culture de tissus. Contribution a l'étude de la préparation à la mitose. *Compt. Rend. Soc. Biol.* **150,** 1046.

ERRERA, M. (1959). Effects of radiations on cells. In *The Cell,* Vol. I (J. Brachet and A. E. Mirsky, eds.). Academic Press, New York, p. 695.

FAUTREZ, J. and H. ROELS (1954). Mitose et synthèse d'acide désoxyribonucléique. Le cas des mitoses provoquées dans le tube contourné du rein par l'hydronéphrose unilatérale. *Arch. Biol. (Paris)* **65,** 459.

FEINENDEGEN, L. E., V. P. BOND, W. W. SHREEVE and R. B. PAINTER (1960). RNA and DNA metabolism in human tissue cells studied with tritiated cytidine. *Exptl. Cell Research* **19,** 443.

Firket, H. (1956). Mesures cytophotométriques de la synthèse d'acides désoxyribonucléiques dans les cultures de tissus soumises au refroidissement puis réchauffées. *Compt. Rend. Soc. Biol.* **150,** 1050.

Fogh, J. and R. O. Lund (1957). Continuous cultivation of epithelial cell strain (FL) from human amniotic membrane. *Proc. Soc. Exptl. Biol. Med.* **94,** 532.

Gey, G. O., F. B. Bang and M. K. Gey (1954). Responses of a variety of normal and malignant cells to continuous cultivation, and some practical applications of these responses to problems in the biology of disease. *Ann. N.Y. Acad. Sci.* **58,** 976.

Hakala, M. T. and E. Taylor (1959). The ability of purine and thymine derivatives and of glycine to support the growth of mammalian cells in culture. *J. Biol. Chem.* **234,** 126.

Hoffmann-Berling, H. (1954). Die Bedeutung des Adenosintriphosphat für die Zell- und Kernteilungsbewegungen in der Anaphase. *Biochim. et Biophys. Acta* **15,** 226.

Hotchkiss, R. D. (1954). Cyclical behavior in pneumococcal growth and transformability occasioned by environmental changes. *Proc. Natl. Acad. Sci., U.S.* **40,** 49.

Hughes, A. (1952). *The Mitotic Cycle.* Butterworths, London, p. 151.

Jacobson, W. and M. Webb (1952). The two types of nucleoproteins during mitosis. *Exptl. Cell Research* **3,** 163.

Jacoby, F., O. A. Trowell and E. N. Willmer (1937). Studies on the growth of tissues *in vitro.* V. Further observations on the manner in which cell division of chick fibroblasts is affected by embryo tissue juice. *J. Exptl. Biol.* **14,** 255.

Kemp, T. and J. Juul (1931). Der Einfluss der Wärme auf die Zellteilung. *Arch. Exptl. Zellforsch.* **11,** 602.

Kokott, W. (1930). Zur Frage des Einflusses erhöhter Temperatur auf die Mitosen in Gewebskulturen. *Z. Zellforsch. Mikroskop.* **11,** 484.

Lewis, M. R. (1933). Reversible changes in the nature of the mitotic spindle brought about in living cells by means of heat. *Arch. Exptl. Zellforsch.* **14,** 464.

Levintow, L. and H. Eagle (1961). Biochemistry of cultured mammalian cells. *Ann. Rev. Biochem.* **30,** 605.

Maaløe, P. and K. G. Lark (1954). A study of bacterial populations in which nuclear and cellular divisions are induced by means of temperature shifts. In *Recent Developments in Cell Physiology* (J. A. Kitching, ed.). Butterworths, London, p. 159.

Mazia, D. (1958). Cell division. In the *Harvey Lectures,* Series LIII., Academic Press, New York, p. 130.

Mitchison, J. M. and P. M. B. Walker (1959). RNA synthesis during the cell life cycle of a fission yeast *Schizosaccharomyces pombe. Exptl. Cell Research* **16,** 49.

Miura, T. and T. Utakoji (1961). Studies on synchronous division of FL cells by chilling. *Exptl. Cell Research* **23,** 452.

Moorhead, P. S. and T. C. Hsu (1956). Cytologic studies of HeLa, a strain of human cervical carcinoma. III. Durations and characteristics of the mitotic phases. *J. Natl. Cancer Inst.* **16,** 1047.

Morgan, C. (1959). Viral multiplication and cellular hyperplasia. *Nature* **184,** 435.

Newton, A. A. (1959). Infection of synchronously dividing HeLa cells by *Herpes simplex* virus. *Bacteriol. Proc.*, p. 77.

NEWTON, A. A. and P. WILDY (1959). Parasynchronous division of HeLa cells. *Exptl. Cell Research* **16,** 624.

ORD, M. G. and L. A. STOCKEN (1960). Deoxyribonucleic acid synthesis—a serial story. In *The Cell Nucleus* (J. S. Mitchell, ed.). Butterworths, London, p. 151.

PAINTER, R. B. and R. M. DREW (1959). Studies on deoxyribonucleic acid metabolism in human cancer cell cultures (HeLa). The temporal relationships of deoxyribonucleic acid synthesis to mitosis and turnover time. *Lab. Invest.* **8,** 278.

PASTEELS, J. and L. LISON (1950). Recherches histophotométriques sur la teneur en acide désoxyribonucléique au cours de mitoses somatiques. *Arch. Biol. (Paris)* **61,** 445.

PINCUS, G. and A. FISCHER (1931). The growth and death of tissue cultures exposed to supranormal temperatures. *J. Exptl. Med.* **54,** 323.

PISI, E. and G. CAVALLI (1955). Teneur en acide désoxyribonucléique et activite mitotique dans le rein du Rat blanc dans diverses conditions expérimentales. *Arch. Biol. (Paris)* **66,** 439.

RABINOVITCH, M. and W. PLAUT (1956). Cytochemical and autoradiographic observations on nuclear ribonucleic acid in *Amoeba proteus. Exptl. Cell Research* **10,** 120.

RAPP, F. (1960). Observations of measles virus infection of human cells. III. Correlation of properties of clones of H Ep.-2 cells with their susceptibility to infection. *Virology* **10,** 86.

RUECKERT, R. R. and G. C. MUELLER (1960). Studies on unbalanced growth in tissue culture. I. Induction and consequences of thymidine deficiency. *Cancer Research* **20,** 1584.

SANDRITTER, VON W., H. G. SCHIEMER, H. KRAUS and U. DÖRRIEN (1960). Das Wachstum von Einzelzellen in Gewebekultur. Interferenzmikroskopische Untersuchungen an HeLa-Zellen. *J. Biophys. Biochem. Cytol.* **7,** 579.

SCHERBAUM, O. H. (1960). Synchronous division of microorganisms. *Ann. Rev. Microbiol.* **14,** 283.

SCHINDLER, R. (1960). Der zeitliche Ablauf der Desoxyribonukleinsäure-Synthese im Teilungszyklus von Krebszellen in Kultur. *Helv. Physiol. Pharmacol. Acta* **18,** C93.

SELAWRY, O. S., M. N. GOLDSTEIN and T. MCCORMICK (1957). Hyperthermia in tissue-cultured cells of malignant origin. *Cancer Research* **17,** 785.

SMITH, C. L., A. A. NEWTON and P. WILDY (1959). Deoxyribonucleic acid formation in multiplying HeLa cells. *Nature* **184,** 107.

SPEAR, F. G. (1928/29). The effect of low temperature on mitosis *in vitro. Arch. Exptl. Zellforsch.* **7,** 484.

STOKER, M. G. P. and A. NEWTON (1959). The effect of *Herpes* virus on HeLa cells dividing parasynchronously. *Virology* **7,** 438.

SWANN, M. M. (1957). The control of cell division: A review. I. General mechanisms. *Cancer Research* **17,** 727.

TERASIMA, T. and L. J. TOLMACH (1961). Changes in X-ray sensitivity of HeLa cells during the division cycle. *Nature* **190,** 1210.

TOOLAN, H. W. (1954). Transplantable human neoplasms maintained in cortisone-treated laboratory animals. *Cancer Research* **14,** 660.

VENDRELY, R. (1955). The deoxyribonucleic acid content of the nucleus. In *The Nucleic Acids*, Vol. II (E. Chargaff and J. N. Davidson, eds.). Academic Press, New York, p. 155.

WALKER, P. M. B. and H. B. YATES (1952). Nuclear components of dividing cells. *Proc. Roy. Soc. London* **B140,** 274.

WAYMOUTH, C. (1952). In *The Mitotic Cycle* (A. Hughes, ed.). Butterworths, London, pp. 163–183.

WEISSMAN, S. M., R. M. S. SMELLIE and J. PAUL (1960). Studies on the biosynthesis of deoxyribonucleic acid by extracts of mammalian cells. IV. The phosphorylation of thymidine. *Biochim. et Biophys. Acta* **45,** 101.

WESSON, L. G., Jr., W. E. COHN and A. M. BRUES (1949). The effect of temperature on potassium equilibrium in chick embryo muscle. *J. Gen. Physiol.* **32,** 511.

WHEELOCK, E. F. and I. TAMM (1959). Mitosis and division in HeLa cells infected with influenza or Newcastle disease virus. *Virology* **8,** 532.

WHITMORE, G. F., C. P. STANNERS, J. E. TILL and S. GULYAS (1961). Nucleic acid synthesis and the division cycle in X-irradiated L-strain mouse cells. *Biochim. et Biophys. Acta* **47,** 66.

WILDY, P. and A. A. NEWTON (1958). The 'synchronous' division of HeLa cells. *Biochem. J.* **68,** 14P.

WILLMER, E. N. (1935). *Tissue Culture.* Methuens, London, pp. 21–23.

WOODARD, J., B. GELBER and H. SWIFT (1961). Nucleoprotein changes during the mitotic cycle in *Paramecium aurelia. Exptl. Cell Research* **23,** 258.

WOODARD, J., E. RASCH and H. SWIFT (1961). Nucleic acid and protein metabolism during the mitotic cycle in *Vicia faba. J. Biophys. Biochem. Cytol.* **9,** 445.

YANAGITA, T., Y. MARUYAMA and I. TAKEBE (1958). Cellular response to deleterious agents during the course of synchronous growth of *Escherichia coli. J. Bacteriol.* **75,** 523.

ZERNICKE, F. (1935). Das Phasenkontrast verfahren bei der microskopischen Beobachtung. *Z. Tech. Phys.* **16,** 454.

SECTION III

GENERAL CONSIDERATIONS

A.	Mathematical considerations	Chaps. 18–20
B.	Detection of synchrony in amoebae	Chap. 21
C.	Development of techniques for flagellates	Chap. 22
	Outlook	

18

The Theoretical Basis of Synchronization by Shifts in Environmental Conditions

ALLAN CAMPBELL, *Department of Biology, University of Rochester, Rochester, New York.*

CONTENTS

I Introduction 469
II The Cell as a Simple Clock 470
1. Continuous timepieces—almost perfect clocks 470
A. The effect of a single shift of environmental conditions . 470
B. Multiple shifts and periodic alterations of growth conditions 473
2. Continuous timepieces—imperfect clocks 477
3. Discontinuous timepieces—division as a sequence of steps of finite number 478
III A Non-clock Model for Cell Division 482
IV Conclusions 483
References 483

We cannot make 100 uniform bacterial cells divide all at the same time any more than we can make all sucrose molecules of a sugar solution invert at the same moment (Otto Rahn, 1932).

I. INTRODUCTION

In this chapter we shall assume that, with respect to the division cycle, the cell behaves like a simple clock. The consequences of this assumption for the cellular age distribution are the province of another chapter and we shall here merely avail ourselves of the results of this type of analysis. The present chapter will be concerned primarily with how sudden shifts in the growth conditions might have a synchronizing effect on a population of such clocks. The first sections will deal with clocks which keep perfect time, and the treatment will then be extended to those which do not. By simple we mean that there is a single variable on which all the interesting properties of the system depend; that any cell can be assigned a time on the clock, and that the expected behavior of any two clocks which read the same time will be the same regardless of their histories. The only aspect of their behavior with which we shall concern ourselves here is the timing of cell division.

We shall not deal in much detail with the experimental facts. It will be seen that some cases definitely do not fit the theory. This is not surprising, since uncomplicated systems are seldom found by accident. One function of theory is to define the properties of simple systems so that they may be recognized when encountered.

II. The Cell as a Simple Clock

1. *Continuous Timepieces—Almost Perfect Clocks*

A. The effect of a single shift of environmental conditions

Consider a population of simple, almost perfect clocks which have been growing* for many generations in some environment E_1. We assume that either the population density is kept low or that the environment is replenished so that the growth conditions do not change with time. The population will assume a steady state age distribution :

$$\begin{aligned} \varphi_1(a_1) &= 2k_1 e^{-k_1 a_1}, \quad 0 \leqslant a_1 \leqslant (\ln 2)/k_1 \\ \varphi(a_1) &= 0, \quad (\ln 2)/k_1 < a_1 \end{aligned} \tag{1}$$

where k_1 and a_1 are the growth rate of the population and the age of the individual, respectively. As long as a constant environment is maintained, growth will be exponential and the age distribution will remain the same.

Now suppose we were to introduce into the same kind of environment a population of clocks with some arbitrary initial age distribution. In general, growth will not be exponential, and the age distribution will change with time. Since we are dealing with clocks which are almost perfect, the variation with time of the age distribution will be almost periodic.

How do we obtain the second population with a different age distribution? One method is to take a population growing in the steady state in a second environment E_2. These must, of course, follow the distribution

$$\varphi_2(a_2) = 2k_2 e^{-k_2 a_2} \tag{2}$$

Distributions (1) and (2) may appear identical except that the time scale has been altered in the ratio of the two growth rates. However, it is not quite that simple, because the fraction of a generation required to reach the stage denoted by a_1 in E_1 may be quite different from that required to reach the same stage in E_2.

* The clocks of this chapter will increase in numbers by undergoing binary fission whenever they reach 12·00.

The situation is most readily visualized by a diagram such as that of Fig. 1. In E_1, the population has a certain generation time, proportional to the circumference of the circle shown, and during this doubling cycle it passes through a sequence of stages A, B, C, D, E, etc. The second circle represents the same population in E_2.

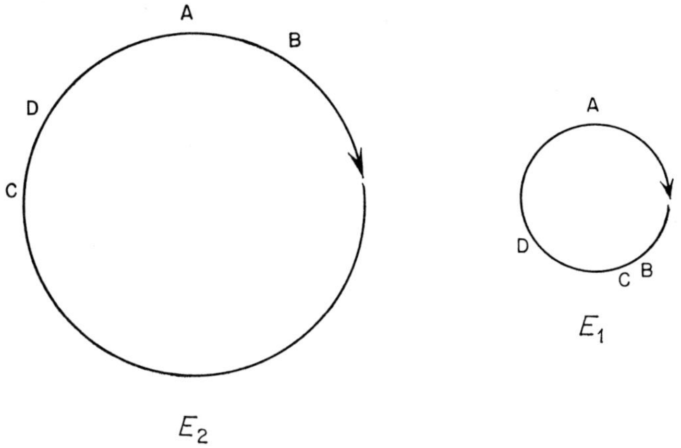

FIG. 1. Diagram of cell division cycle in two different environments, following the simple clock model. (Modified from Campbell, 1957.)

The diagram is drawn with different generation times in the two environments, but that is not the crucial point here. The important thing is that the relative length of time spent in the different stages AB, BC, CD, as represented by the lengths of arc, is different in the two cases. If the variable distinguishing the two environments were the temperature, one would say that some step between B and C had a temperature coefficient much larger than those of the other steps of the cycle.

Two points which are brought out more analytically below are intuitively clear from the diagram. First, if we suddenly place a group of clocks from E_2 into E_1, the population will be somewhat synchronized, because a large proportion will be in stages between B and C, and they will all reach 12·00 at close to the same time. Secondly, there is obviously a mathematical relationship between what happens after such a shift and what would happen if the shift were made in the opposite direction.

In order to calculate $\varphi_2(a_2)$ from $\varphi(a_1)$ we may suppose a mapping function f_{21} such that a clock of age a_1 in E_1 will behave, upon transfer to E_2, just like a clock of age $f_{21}(a_1)$ which had been born in E_2. This supposition is a strong assumption when applied to actual objects such as cells which might not be perfect clocks, and thus allows experimental test of the clock hypothesis. The method is to compare the pattern of cellular increase following a shift from E_1 to E_2 with that following the reverse shift from E_2 to E_1.

If we let $N(t)$ = total number of cells at time t

$$g = \text{doubling time} = (\ln 2)/k$$

$$w = g - t$$

$$\varphi_{12}(\beta) = 2k_1 e^{-k_1 f_{12}(\beta)} f'_{12}(\beta)$$

it can be shown that

$$\varphi_{12}(w_2) = N'(t)/N(0) \tag{3}$$

and

$$f_{12}(w_2) = -\frac{1}{k_1} \ln \left(1 - \frac{\int_0^{w_2} \varphi_{12}(\beta)\, d\beta}{2}\right) \tag{4}$$

We are thus able, given the measurable growth curve immediately following a shift from E_1 to E_2 and knowing the generation time under both conditions, to calculate the function f_{12}. The reverse experiment of shifting from E_2 to E_1 would enable us to calculate

$$f_{21}(w_1) = -\frac{1}{k_2} \ln \left(1 - \frac{\int_0^{w_2} \varphi_{21}(\beta)\, d\beta}{2}\right)$$

For a perfect clock, f_{12} and f_{21} thus calculated would be inverse functions. This experiment has been performed at least once (Bruce and Maaløe, 1956), and the results were negative. Whereas shifts from 25° to 37° did cause synchronization, those from 37° to 25° did not.

The derivation of equations (3) and (4) is straightforward but requires several steps, which we can outline here. The first problem is to calculate the effective age distribution in E_2 of a population which has reached the steady state in E_1. We can call this distribution $\varphi_{12}(a_2)$. It will differ from the steady state distribution $\varphi_2(a_2)$ unless the mapping function f_{12} is linear.

A frequency distribution function such as φ_1 says that the proportion of individuals with ages between a_1 and $(a_1 + da_1)$ is approximately $\varphi_1(a_1)da_1$ for any sufficiently small increment da_1. We wish to calculate the proportion of individuals between a_2 and $(a_2 + da_2)$ as a function of a_2. To do this, we first apply the function φ_1 to the point which corresponds to a_2; in other words, we calculate

$\varphi_1(f_{12}(a_2))$. Then we must remember that $f_{12}(a_2 + da_2) = f_{12}(a_2) + f'_{12}(a_2)da_2$. So $\varphi_{12}(a_2)da_2$ will equal $\varphi_1(f_{12}(a_2))f'_{12}(a_2)da_2$, and hence

$$\varphi_{12}(a_2) = 2k_1 e^{-k_1 f_{12}(a_2)} f'_{12}(a_2)$$

This relation is not useful experimentally because neither a_1 nor a_2 nor f_1 is directly measurable. What we can measure is the growth curve as a function of time following the shift. This depends directly on the function φ_{12}, because the number of cells dN which will double between times t and $(t + dt)$ following the shift is precisely the number with ages between $(g - t)$ and $(g - t - dt)$ at the moment of shifting; i.e., $N(0)\varphi_{12}(w_2)dt$. Therefore

$$dN = N'(t)dt = N(0)\varphi_{12}(w_2)dt$$

from which equation (3) follows directly.

We can thus calculate φ_{12} from the experimental data, and it remains to express f_{12} explicitly in terms of φ_{12}. To do this, we note from the definition of φ_{12} that

$$\int \varphi_{12}(\beta)d\beta = -2e^{-k_1 f_{12}(\beta)}$$

and hence

$$\int_0^{w_2} \varphi_{12}(\beta)d\beta = -2\left(e^{-k_1 f_{12}(w_2)} - e^{-k_1 f_{12}(0)}\right)$$

Remembering that $f_{12}(0) = 0$, we obtain equation (4) by substitution and rearrangement.

B. Multiple shifts and periodic alterations of growth conditions

A method which has been frequently employed with some success in synchronization studies is to submit a cell population to successive shifts between two growth conditions, the period of the shifts being so chosen that the total cycle will correspond to one population doubling. We will examine the effect of such shifts on a population of perfect clocks.

Given the environments E_1 and E_2, we so choose the time unit that $g_2 = 1$. We consider the effect on the age structure of putting the cells in E_1 for a time A_1, and in E_2 for a time A_2, for an indefinite number of cycles. We so choose A_1 and A_2 that

$$A_1 = g_1(1 - A_2) \tag{6}$$

No prediction is possible without knowledge of the function f_{12}. A simple form for this function would be one which deviates from a straight line

$$a_2 g_1 = a_1$$

in only one direction and shows no inflections. This is satisfied by the family

$$f_{12}(a_2) = \frac{K a_2 g_1}{K + g_1 - a_2 g_1} \qquad (7)$$

where K is any positive number.

Consider the fate of a clock which has age 0 at the time of the initial shift from E_1 to E_2. At time A_2 it will be at point A on the curve of Fig. 2, and after time A_1 in E_1 it will be at B—i.e. it will not have completed a whole cycle. (For the inverse function f_{21}, a clock initially of age 0 would have completed more than one full cycle.)

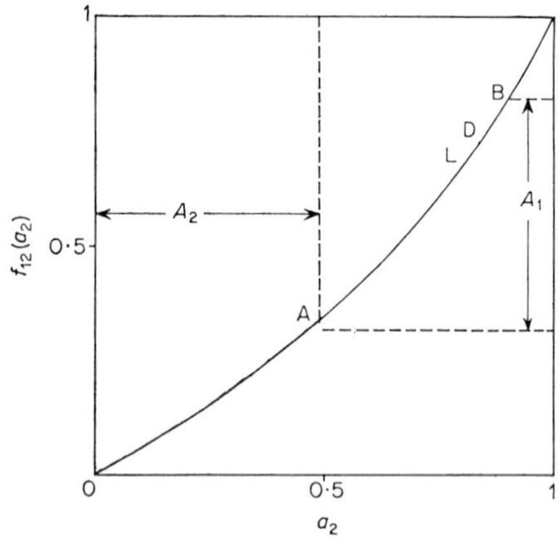

FIG. 2. Plot of mapping function for shifts between E_1 and E_2. Drawn for $K = 1$, $A_2 = 0{\cdot}5$, $g_1 = g_2 = 1$.

At the end of the second cycle they will have regressed further to some point D. The value of a_2 at the end of subsequent cycles will gradually approach some value a, assumed at point L of Fig. 2, at which it is possible to complete exactly one division cycle during one culture cycle. The same considerations apply to a clock of any initial age.

There are two points on the curve which have this property. If, in the region between 0 and 1, one draws the set of chords whose horizontal projection is $1 - A_2$, there will be exactly one such chord which has slope g_1. This is also true for the set of chords of horizontal projection A_2. The upper endpoint of the former chord and the lower endpoint of the latter both satisfy the condition that a clock initially at this point will return to it after one culture cycle. However, only the first-mentioned of these points is stable in the sense that a point slightly displaced from it will tend toward it in subsequent cycles.

Therefore, if a population of clocks is subjected to a series of such cycles, one will eventually tend to a situation wherein all the clocks at the moment of the shift from E_1 to E_2 are of age $\hat{a}$ given by

$$f_{12}(\hat{a}) - f_{12}(\hat{a} + A_2 - 1) = (1 - A_2)g_1 \tag{8}$$

the useful solution of which, for the particular family we are considering, is

$$\hat{a} = \frac{3 - A_2}{2} + \frac{K}{g_1} - \left(\frac{A_2{}^2}{4} - \frac{A_2}{2} + \frac{K^2}{g_1{}^2} + \frac{K}{g_1} + \frac{1}{4}\right)^{\frac{1}{2}} \tag{9}$$

Two limits are of interest

$$\lim_{K \to \infty} \hat{a} = 1 - (A_2/2) \tag{10}$$

$$\lim_{K \to 0} \hat{a} = 1 \tag{11}$$

In other words, if K is large, which corresponds to a small deviation of f_{12} from a linear relationship and therefore a small phasing effect of a single shift, the age of the clocks at the beginning of the cycle depends on the relative proportions of time spent in E_1 and E_2. On the other hand, if a single cycle has a strong phasing effect, the age of the clocks at the beginning of the steady state cycle is independent of these proportions.

The family of functions f_{12} may be chosen differently. Perhaps a more natural choice would be one in which

$$f'_{12}(0) = f'_{12}(g_2)$$

and which has only one inflection point between 0 and g_2. As in the above case, cycling will put all individuals in phase. However, if there is more than one inflection point, there may be more than one stable cycle—so that one would expect the population at any one time to be a mixture of individuals in two or more phases, the relative proportions of which would depend on the function f_{12} and, under the completely deterministic assumptions we have made, upon the initial age distribution.

We can now discuss cycles more complex than those involving shifts between two environments. If more than two environments are used, or even if there is a continuous variation from one to the other and back, the problem can be reduced as follows. Consider an individual of age a_2 at time 0 subjected to periodic changes of period T. At time T it will have undergone n divisions and will be of age a_2^{**}. If we plot the quantity $(a_2^{**} + n - a_2)$ against a_2 we will obtain, for the simple shift cycle already discussed, a curve of the form shown in Fig. 3.

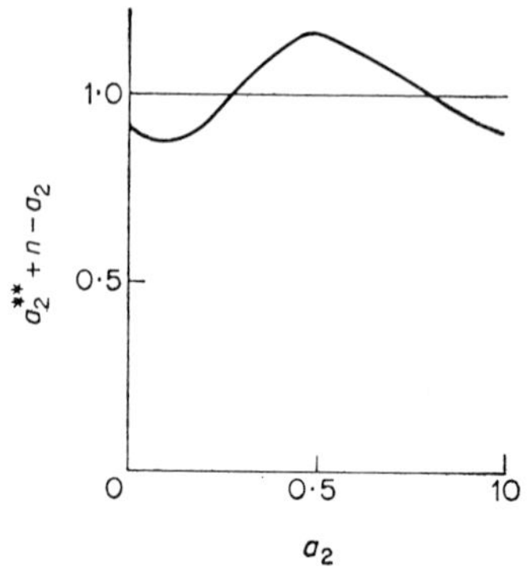

FIG. 3. Plot showing the fraction of a division cycle which a cell initially of age a_2 will traverse during one culture cycle. Drawn for $K = 1$, $A_2 = 0{\cdot}5$, $g_1 = g_2 = 1$.

In the general case, and ignoring a few unlikely exceptions, we can say that whenever the periodic process is so chosen that in such a plot the curve crosses the line

$$a_2^{**} + n - a_2 = 1 \qquad (12)$$

exactly twice, the clocks will accumulate at some point such as L of Fig. 3.

These considerations predict some liberty in the choice of cycles which will produce synchronization. It is not at all necessary that the process correspond precisely to any natural frequency of the clock itself. The only limitation is that, in a plot such as that of Fig. 3, the curve must intersect the straight line described by equation (12). Equation (5) is thus not a necessary condition. It happens to be the only one which will work for all the curves of the family, but inspection will reveal that the smaller the value of K, the larger are the limits within which A_1 can be chosen for a given A_2.

2. *Continuous Time Pieces—Imperfect Clocks*

The observed behavior of cells is not that expected from perfect clocks. Specifically, if one measures the length of the division cycle for a large number of individual cells under steady state conditions, one finds variations from cell to cell which exceed any possible errors of measurement. Protozoa are good, though imperfect, timepieces by this criterion, whereas bacteria are extremely poor; so much so that it is difficult to understand how they can be synchronized at all.*

If all the cells of a species were perfect clocks, division synchrony would occur automatically in any culture derived from a single cell. In order to arrive at the steady state age distribution (equation (1)), one must therefore introduce at least some small degree of randomness into the timing of the cycle. At the other extreme, one can imagine a purely stochastic mode of growth in which the probability per unit time of division is independent of the time elapsed since the previous division. Such a population could never be synchronized.

In the intermediate cases which correspond more closely to reality, we can think of synchronization as a procedure for obtaining a population of clocks all of which are in phase with each other. The best synchronization obtainable is one where all of the clocks are made to read exactly the same time at the same moment. If the clocks are imperfect, however, this maximal synchronization will not lead to perfect division synchrony during subsequent division cycles in a constant environment. Inasmuch as bacteria are very poor clocks, maximal synchronization would still lead to very poor synchrony, much worse than that which has frequently been observed.

Without inquiring into the mechanism of imperfection, let us suppose that a group of individuals initially of age a will, after one culture generation, be normally distributed with a mean of a and a variance σ_a^2. For cyclical processes of the type previously discussed, let us suppose that cells which have age a_2 at a given time will be normally distributed about a_2 * * (a_2) with variance $\sigma_{a_2}^2$ one period later.

Whereas the deterministic (perfect clock) model predicts that any periodic process will ultimately lead to complete synchroniza-

* It has recently been shown (Schaechter *et al.*, 1962) that bacterial cells elongate at a constant rate and divide when they reach a certain average length. The variability about this average length is rather large, but a cell which divides early in one generation will generally divide late in the next. So, although the variability of individual doubling times is quite large, the underlying clock is fairly precise.

tion, this will no longer be so. The cells at the beginning of the time-independent cycle will be distributed, as a first approximation, according to

$$\varphi(a) = \int_0^1 \frac{\varphi(\alpha)e^{-[a_2-a_2^{**}(\alpha)]^2}d\alpha}{2} \tag{13}$$

If we assume σ to be constant, the limiting cases which concern us are

$$[a_2 - a_2^{**}(\alpha)]^2 \ll \sigma^2 \tag{14}$$

and

$$[a_2 - a_2^{**}(\alpha)]^2 \gg \sigma^2 \tag{15}$$

If a_2 belongs to an interval for which inequality (14) holds, σ will be approximately constant throughout the interval. If inequality (15) holds, it will be approximately zero. Inequality (15) can never hold for the point $\hat{a}$.

A crude estimate of the degree of synchronization is therefore the inverse of the width of the interval surrounding a_{2L} for which $[a_2 - a_2^{**}(a_2)]^2 < \sigma^2$. For the family of curves we have considered, the negative of the slope of the curve of Fig. 2 at the point L has at most the value 1 (as $K \longrightarrow 0$), which corresponds to a minimum width of σ for the interval. Whereas for the perfect clock, A_2 could have any value between 0 and 1, if it approaches closely either of the limits in the present case, then

$$a_2^{**}(a_2) \longrightarrow a_2$$

and phasing will be poor. There will be an optimal value of A_2 which will give the best phasing for a given value of K.

3. *Discontinuous Timepieces—Division as a Sequence of Steps of Finite Number**

There is another reasonable approach to the problem of imperfect clocks which may be closer to physical reality than that used above. We imagine a clock whose hands, instead of continuously traversing the whole circumference of a circle, move by discrete jumps from

* A more rigorous mathematical discussion of many of the points developed here is given by Kendall (1948). For a critical discussion of the present status of theory and experimentation in this area, the papers of Powell (1955) and Koch and Schaechter (1962) are highly recommended.

one location to the next and remain for some time at each location. Everyone has seen such clocks. They are so constructed that internal parts continue to move while the hands remain motionless, thereby informing the hands of exactly how long to stay in each position. The reader should try to forget this fact for the analogy made here. We are concerned always with the sequence of primary determinative events in the cycle, and the only case in which a stepwise process is of interest is one in which there is no other clock to tell one how long one has been on a particular step. Passage from one step to the next must then be a completely stochastic process. It is as though we had connected our clock to a Geiger counter and arranged that it would advance one step with each click of the counter.

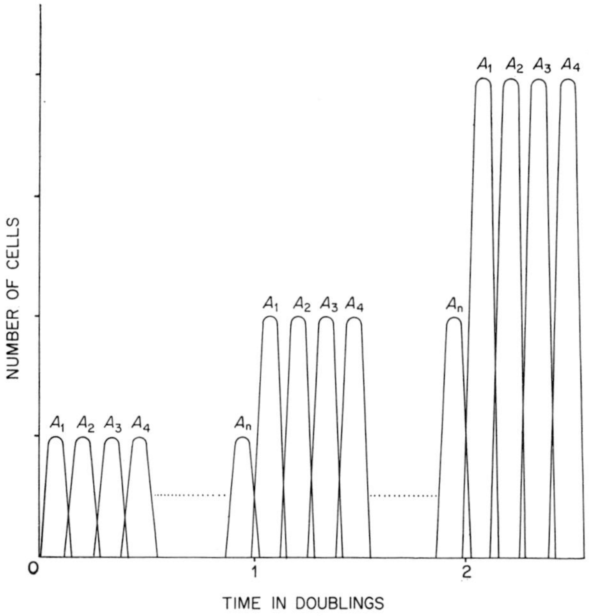

FIG. 4. General expectation for behavior of sequential age classes for very large n in a population initially in complete synchrony.

If the number (n) of steps with similar rates in the cycle is very large, the system will approach a perfect clock in its behavior. If $n = 1$, division will be completely random. The interesting cases are the intermediate ones.

For $n = 2$, the appropriate equations are

$$\frac{d(A_1)}{(A_1) + 2(A_2)} = \frac{d(A_2)}{(A_1) - (A_2)} = k\,dt \tag{23}$$

the solution of which is

$$(A_1) = e^{-kt} \cosh (\sqrt{2}kt)$$

$$(A_2) = \frac{e^{-kt}}{2} \sinh (\sqrt{2}kt)$$

It is clear that this system has little resemblance to that shown in Fig. 4, and that the overall growth curve has really no stepwise character. The first two cycles of growth for $n = 2$ and $n = 4$ are plotted in Figs. 5 and 6.

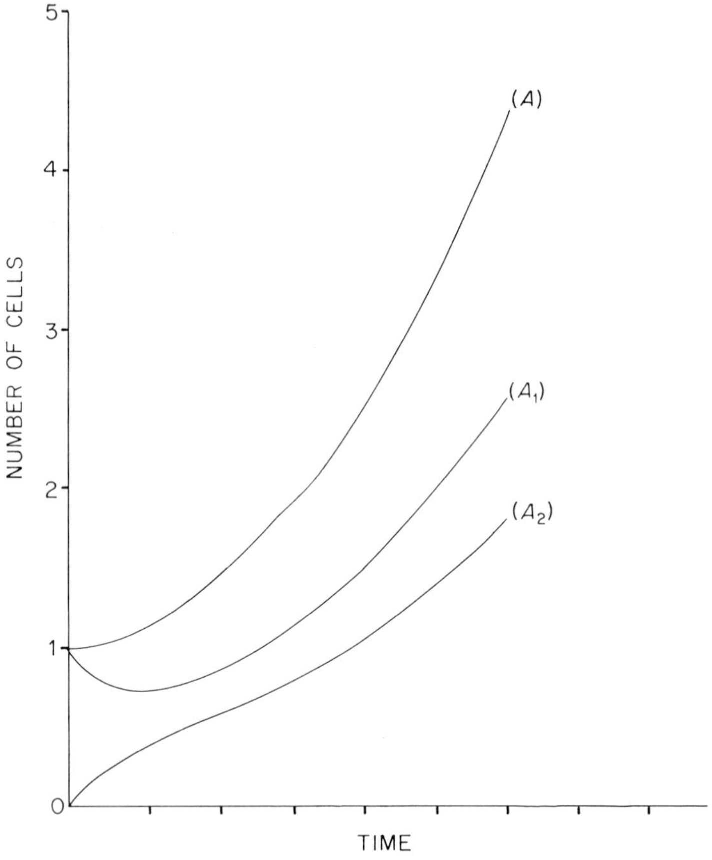

FIG. 5. Calculated curves of increase for individual age classes and total population when $n = 2$.

How large must n be in order that the picture of Fig. 4 be approached? A minimum criterion should be that the growth curve involve at least some periodic terms multiplied by positive exponentials. The solutions of these equations are always derived from the n-complex roots of 2, and a positive exponential appears in a period term only when the real part of the first complex root exceeds one.

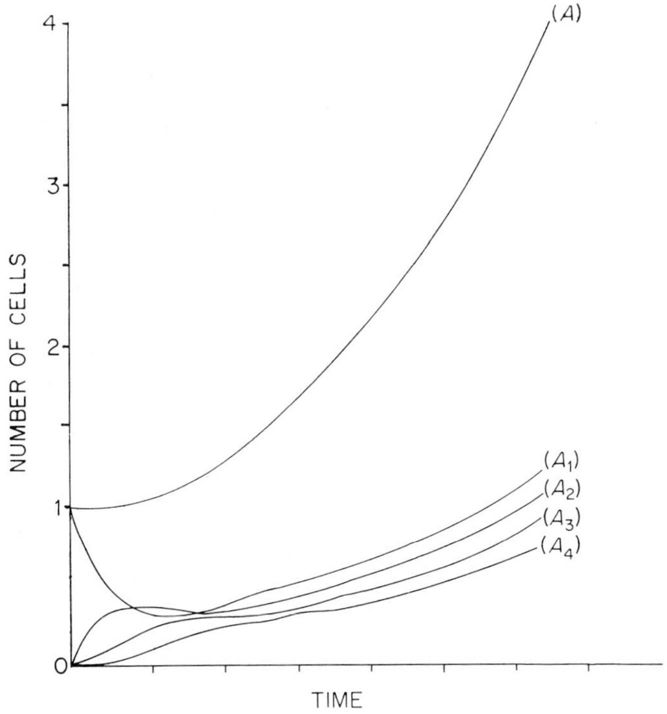

FIG. 6. Calculated curves of increase for individual age classes and total population when $n = 4$.

This happens only for $n \geqslant 29$. For smaller values of n, synchronization would hardly seem possible.

One can obtain a minimal estimate of n for any real population from the measured distribution of individual doubling times. This is most simply calculated from the relationship

$$\frac{\text{standard deviation}}{\text{mean}} = \frac{1}{\sqrt{n}} \qquad (24)$$

Even for the bacteria, the value of n thus derived has a value of about 20, and species which behave as good clocks would require much higher numbers. If the rate constants of the individual steps are unequal rather than equal, it requires a larger number of them to reduce the standard deviation to the same value. This is easily understood by considering a cycle in which one step is much slower than all the rest. The cells will accumulate at the stage immediately preceding the slow step, and the overall process will be distributed as if $n = 1$.

We thus conclude that if this model is correct, the number of steps must be quite large. One is therefore justified for most purposes in ignoring the stepwise nature of the process and approximating it by a continuous function as we have done in the preceding sections of this chapter.

III. A Non-clock Model for Cell Division

One difficulty arises in explaining synchronization if division depends on a sequence of stochastic steps. If the mapping function f_{12} for the continuous clock is to be derived as an approximation from a large but finite sequence of steps, the departure of f_{12} from linearity which is required for synchronization will correspond to a differential effect of the environment on the rate constants of individual steps. However, if we change rate constants at random, we will in general accelerate certain reactions and slow others adjacent to them in the sequence. The net result will be that f_{12} will remain essentially linear, with perhaps a few minor wiggles. In order to create a general concavity of the curve, it is necessary that the differential effect of the environment should gradually change as one proceeds along the sequence; for example, that the reactions with the highest temperature coefficients should be grouped together in the sequence. This is of course possible, but it represents a purely gratuitous assumption.

It is of interest to consider an alternative hypothesis proposed many years ago by Rahn (1932), in which cell division is pictured as requiring the completion of a certain number of steps which are parallel rather than sequential and therefore need not occur in a specified order. If one starts with a group of cells of age 0, the fraction f which have not yet divided at age a will follow a multi-hit curve

$$f = 1 - (1 - e^{-\bar{k}t})^n \tag{25}$$

assuming n steps each of rate $\bar{k}$. Rahn imagined that each of the n steps constituted the replication of a single gene, but this specification is not really necessary.

This model is of interest here for two reasons. First, it alleviates the necessity to assume that the effect of the environment on a particular step should be related to its position in the sequence. If the slowest steps in E_1 are not identical to those in E_2, a shift will have some synchronizing effect, because in either environment the slowest steps will generally be the last to be completed. Second, the cell is no longer a simple clock. Different cells will follow different paths. In any one environment, one could assign, to all cells with an equal expectation of division, the same time, but the cells within this group would in general differ among themselves as to their expectation of division in a second environment.

IV. Conclusions

We have deliberately confined our treatment to a restricted class of models, chosen primarily for their mathematical rather than their physical simplicity. It is obvious that the dividing cell is not simply a clock, and that there are relevant internal changes which will affect the subsequent growth process and which cannot adequately be described merely by advancing the hands of the hypothetical clock for some distance along a unidimensional scale. We believe that the cell division cycle can follow a variety of alternative routes, and that some steps proceed in parallel and others in series.

It is not so clear whether the concept of a simple clock in cell division can be a useful one. In a certain sense, it seems obvious that considerations of this type must contribute to synchronization although they may not explain it entirely. In a purely formal manner, one can always postulate a clock and then divide the totality of the attributes of the actual system into those which are directly tied to the clock and those which are not. Such an approach may or may not be fruitful, depending on whether any meaningful synthesis can be made of the residual properties of the system. At the present time, any attempt at such synthesis would seem premature.

References

Bruce, V. G. and O. Maaløe (1956). The ultraviolet inactivation of synchronously dividing *Salmonella typhimurium*. Post-irradiation effects on viability. *Biochim. et Biophys. Acta* **21,** 227.

Campbell, A. (1957). Synchronization of cell division. *Bacteriol. Rev.* **21,** 263.

Hoffman, J. (1953). *The Size and Growth of Tissue Cells.* Charles C. Thomas, Springfield, Ill.

Kendall, D. G. (1948). On the role of variable generation time in the development of stochastic birth process. *Biometrika* **35,** 316.

Koch, A. L. and M. Schaechter (1962). A model for statistics of the cell division process. *J. Gen. Microbiol.* **29,** 435.

Powell, E. O. (1955). Some features of the generation times of individual bacteria. *Biometrika* **42,** 16.

Rahn, O. (1932). A chemical explanation of the variability of the growth rate. *J. Gen. Physiol.* **15,** 257.

Schaechter, M., J. P. Williamson, J. R. Hood and A. L. Koch (1962). Growth cell and nuclear division in some bacteria. *J. Gen. Microbiol.* **29,** 421.

Of those papers not referred to in the text, the review by Campbell contains a different exposition of some of the ideas developed here, and the book by Hoffman discusses the effect of the mode of division on the age distribution of the population.

19

Age Distribution of Cells in Logarithmically Growing Cell Populations

JAMES R. COOK* and THOMAS W. JAMES,† *Laboratory of Nuclear Medicine and Radiation Biology, University of California, Los Angeles, California, and Department of Zoology, University of California, Los Angeles, California*

CONTENTS

I	Introduction	485
II	Historical	486
III	The Distribution Function	487
IV	The Fission Time	489
V	Age of the Average Cell	490
VI	Variation of Member Cells in Real Populations	493
	References	494

I. INTRODUCTION

While the time course of events occurring over the life cycle of cells can be known with a great deal of certainty when the experimental material is a single growing cell, or a synchronized population, some such information can also be conveniently obtained from instantaneous sampling of exponentially expanding cell populations. Since all ages of cells are represented in a large growing culture, it is possible, at least in theory, to array cells chosen at random into a temporally logical sequence with respect to any variable of interest such as volume, DNA content, and mass. In a frequency histogram constructed from the observed values, the amount per cell axis is proportional to cell age, and the character of the frequency distribution is thus a measure of cellular activity at any age (for example, see Walker and Mitchison, 1957). Other kinds of information can be obtained from logarithmic cultures. For example, the time spent in cell division can be determined with a knowledge of the average generation time and the instantaneous fraction of cells in division at any given time. For all such measurements, it is of obvious importance that experimental conditions be so standardized

* Public Health Research Fellow (CF-9395-C2) of the National Cancer Institute.
† Supported in part by Grant 5666 from the National Science Foundation, Washington, D.C., and in part by Grant 536 from the California Cancer Research Committee.

that the many cellular parameters, for instance the logarithmic growth rate, be rendered highly reproducible. Since many characteristics of the average cell change considerably with culture age, or population density, and this while still in steady logarithmic expansion (Summers, 1960), it is further desirable that this additional variable be known or controlled, as for example by continuous dilution techniques. Beyond these precautions, it is essential to consider the contribution made by the variable distribution of the ages of member cells within the population in any interpretation of data so obtained. If a growing population is sufficiently large in number, the age distribution will be for all practical purposes a continuous function, the number of filial cells being always greater than the number of parent cells, with the frequencies of those fractions having intermediate ages falling off exponentially between these two limits. It is the purpose of this chapter to discuss some of the ways in which this 'age distribution' may be used, and to point out some of its practical limitations. In synchronization procedures, the wide distribution of cell ages (with respect to time of division) normally found in logarithmic cultures is in essence compressed toward a uniform value, and the division 'burst' of synchronized populations—both in shape and degree of dispersion—is in part a function of the extent of this compression (James, 1960). The age distribution in logarithmic populations is thus a factor of some importance in any consideration of synchronized populations.

II. Historical

The distribution of ages in growing populations appears to have been derived first by Sharpe and Lotka in 1911. They further demonstrated that the distribution is a stable one, and that perturbations will in time be smoothed over and the original distribution reestablished. These matters are discussed in Lotka's book (1925). Walker (1954) has considered the age distribution in an analysis of certain interphase processes. The distribution function has also been derived by Powell (1955) and Scherbaum and Rasch (1957), who have in addition treated statistically the role of individual variation in generation times. More recently, Frankel (1960) has made use of the function (derived in this case by F. Forro) to study morphogenesis in *Glaucoma chattoni*; and a similar derivation also appears in a paper by Edwards *et al.* (1960), in conjunction with a study of DNA synthesis in Ehrlich ascites tumor cells. The derivation of the functions describing the age distribution and the fission time given below is essentially that previously given by one of us (James, 1960).

III. The Distribution Function

Except as noted below, the following discussion is confined to those micro-organisms which divide to yield two cells, rather than a larger number. The *parent* cell thus gives rise to two *filial* (or daughter) cells. All the cells in the population are considered to have ideal behavior in terms of their life cycle, i.e., the generation time is invariant. Unfortunately, no cell population is known to be so ideal; complicating effects of individual variation in generation times are briefly discussed below.

In such an ideal population, growing exponentially and with cell divisions completely randomized, the instantaneous number of filial cells is always exactly twice as great as the number of parent cells. The number of cells generated by division at any instant, all of which therefore belong to the same age class, remains constant as this class progresses toward the subsequent division. Since the absolute number of cells entering division is constantly increasing in an exponential manner, each new filial-cell class formed is larger (by an exponential factor) than the one which preceded it. It is thus intuitively apparent that the instantaneous frequency distribution of cell ages will have a shape like that shown in Fig. 1, where age is shown on the abscissa and cell frequency on the ordinate. The whole area under the (negative) exponential curve represents the total number of cells in the population at any time, those cells preparing

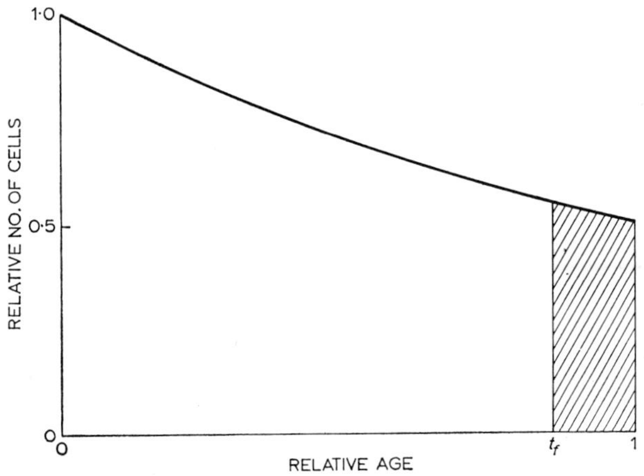

Fig. 1. The ideal distribution of cell ages in an exponentially expanding population. The number of filial cells (age = 0) and the total generation time are both arbitrarily made equal to unity. Fission commences at t_f, and the shaded area represents the instantaneous fraction of cells in fission.

to divide being at the right and those just divided at the left of the distribution (ages, respectively, GT and 0).

To derive the function describing the idealized age distribution, it is necessary to make use of the growth equation,

$$N = N_0 2^{t/GT}, \tag{1}$$

where N is the total number of cells at time t, N_0 the total number at $t = 0$, GT is the generation time, and the time t takes on all values.

If $n_0, n_1, n_2, \ldots n_n$ represents the number of cells in each age class, then

$$N = n_0 + n_1 + \ldots + n_n. \tag{2}$$

If the number of cells in the filial-cell class is n_0, then the frequency of this class at time 0 ($t = t_0$) is

$$f_0 = \frac{n_0}{N} = \frac{n_0}{N_0 2^{t_0/GT}} \tag{3}$$

where here $N = N_0$. As this particular class of cells ages, its relative frequency becomes smaller, since n_0 will remain constant while the exponential term in the denominator of (3) increases in value. The frequency at any age of the class in question may be written as

$$f_i = \frac{n_0}{N} = \frac{n_0}{N_0 2^{t_i/GT}} \tag{4}$$

where t_i is both the time elapsed since $t = 0$ and also the age of the class under consideration.

As the population ages, the daughter-cell class just considered will trace out the distribution curve as it itself ages from $t = 0$ to $t = GT$. This follows from equation (4). Assuming the age distribution to be stable, equation (4) must therefore also describe the frequency of all age classes at the time the daughter-cell class was first considered, namely at $t = 0$.

Consider a class of cells which had its inception at a time $t = t_{-X}$, prior to $t = 0$. If the total number of cells in this class is n_X, then at age X, when $t = 0$, their frequency is

$$f_X = \frac{n_X}{N_0}.$$

When the new class of cells n_0 formed at $t = 0$ has aged X time units, the frequency of this latter class will be

$$f_0 = \frac{n_0}{N_0 2^{t_X/GT}}.$$

If the age gradient is stable, the relative frequency of cells at any given age must be unchanging with time, although the absolute number of cells increases. Thus, f_x at $t = 0$ must be equal to f_0 at $t = t_x$. Since t_x can have any value between $t = 0$ and $t = 0 - GT$, a similar relation holds for cells of all ages at $t = 0$. This is equivalent to saying that the instantaneous frequency of all ages at $t = 0$ is described by equation (4).

Since this generalized frequency function is derived for a population of size N_0, the absolute number of cells in each class is obtained by multiplying each frequency by N_0:

$$n = \frac{N_0 n_0}{N_0 2^{t_i/GT}} \tag{5}$$

or

$$n_i = n_0 2^{-t_i/GT} \tag{6}$$

where n_i is the number of cells in the class of age t_i. The total number of cells is the sum of those in all classes:

$$N = N_0 = \sum_0^{GT} n_i = \sum_0^{GT} n_0 2^{-t_i/GT}, \tag{7}$$

or, as a continuous function,

$$N = n_0 \int_0^{GT} 2^{-t/GT}\, dt. \tag{8}$$

Graphically this can be represented as the area under the age distribution of Fig. 1. When the generation time is unity, this area has a value of $n_0[1/(2 \ln 2)]$.

It should be remembered that in many species—e.g., *Chlorella* and *Chlamydomonas*—more than two filial cells are generated from each parent cell, and the base number of the growth equation (1) is changed accordingly. This correction will of course also be reflected in the function describing the distribution of ages. When the generation time is made unity, the integrated value of equation (8) is $n_0[3/(4 \ln 4)]$ for a yield of four daughter cells, $n_0[7/(8 \ln 8)]$ for eight progeny, etc.

IV. The Fission Time

Equation (8) may be used to determine the fraction of the generation time required for completion of some terminal process, e.g., mitosis or cytokinesis, and consequently the relative age of the cell at the inception of the process. For this determination it is necessary only to know what proportion of the whole population is actively engaged in the process of interest at any instant. Since the generation time generally will be known also, the absolute time required will follow.

As an illustration, the time spent in recognizable fission will be considered. The fission index (the number of cells in fission divided by the total number) is easily measured with the microscope.

If, in Fig. 1, t_f is the age at which fission commences, then the shaded area to the right of this age when divided by the total area under the frequency distribution curve will be equal to the fission index. 'Working back' from the known value of this ratio will yield the numerical value of t_f.

Since the total area is given by equation (8), and the area to the right of t_f (also the number of cells in fission n_f) is

$$n_f = n_0 \int_{t_f}^{GT} 2^{-t/GT}\, dt \tag{9}$$

then

$$FI = \frac{n_0 \int_{t_f}^{GT} 2^{-t/GT}\, dt}{n_0 \int_{0}^{GT} 2^{-t/GT}\, dt} \tag{10}$$

where FI is the fission index. The value of $(GT - t_f)$, which may be called the fission time FT, is given by

$$FT = \frac{\ln(FI + 1)}{\ln 2} GT. \tag{11}$$

Dividing both sides of (11) by GT will give the fraction of the generation time spent in fission.

V. Age of the Average Cell

The protein content, dry weight, volume, etc., of the 'average' single cell in logarithmic growth is commonly determined by random sampling of a large number of cells from the population, performing the measurement on the entire sample, and dividing the value obtained by the number of cells used. This average value will lie somewhere between the values held by the filial and parent cells; specifically, it may be said that a value obtained for the 'average' cell will also be characteristic of a particular class of cells whose age is intermediate between the two limits. If the age of this class were known, it would be possible to determine the comparable values for filial and parent cells (or cells of any age) and with a knowledge of the generation time to calculate absolute synthetic rates over the life cycle. This can be done by use of equation (8), and the additional (and important) assumption that cellular accretion of the variable of interest is continuous and linear over the entire life cycle of the single cell, a circumstance known to be true in many instances. Similar treatments can be made with continuous non-linear rates of synthesis provided that a function for the rate can be described. This need not be a restriction if estimates of the rate function can be approximated and limiting values can be assigned. In general the rate functions will fall between two limiting cases, i.e. concave with an ever-increasing rate, or convex with an early maximum rate which gradually decreases. Assuming a two-to-one relationship in balanced growth between the parent value and the filial value, the random population average can be used to construct an approximate rate function. It must be granted that too much weight should not be placed on a curve derived in such a way, but in situations where no other data or means of obtaining them are available it may provide additional information of some importance; one example would be a comparison between synthetic patterns in logarithmic and synchronized cultures.

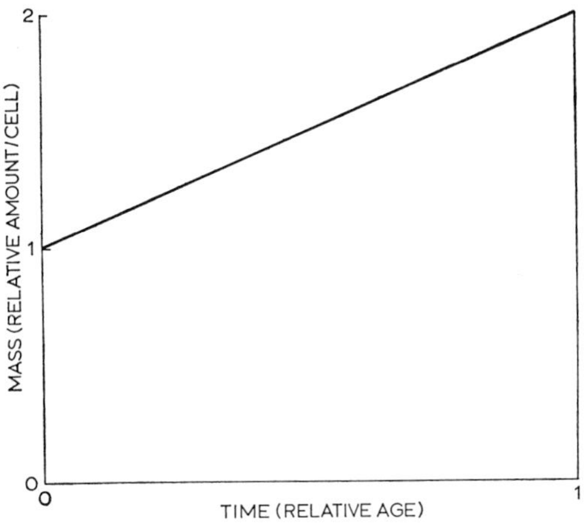

FIG. 2. Hypothetical linear growth curve for a single cell between divisions. The ordinate may represent total cell mass or the amount of any cellular constituent.

For purposes of discussion, cellular mass will be considered. Fig. 2 shows the hypothetical linear mass increase for a single cell over its life cycle, where the ordinate y represents mass. If the generation time is unity, and y is assigned a value of unity at age zero, the mass at any age t_i is

$$y_i = mt_i + 1 \tag{12}$$

where the slope m is equal to 1. Since the single cell of age t_i has a mass y_i, then all of the cells n_i in the same age class will have a total mass of $n_i y_i$. The value of $n_i y_i$ integrated over all ages is the mass of the entire population sampled; dividing this by the total number of cells (i.e., the area under the age distribution curve) gives the mass of the average cell. The age of the class in which the member cells have the same mass as the 'average' cell can readily be found from equation (12).

From equations (6) and (12), we may write

$$\begin{aligned} n_i y_i &= (n_0 2^{-t_i/GT})(t_i + 1) \\ &= n_0 t_i 2^{-t_i/GT} + n_0 2^{-t_i/GT} \end{aligned} \quad (13)$$

The total mass M of the population is

$$M = n_0 \int_0^{GT} (t2^{-t/GT} + 2^{-t/GT})\, dt. \quad (14)$$

The value of this integral is

$$M = n_0 \frac{1}{2(\ln 2)^2}. \quad (15)$$

Since the total number of cells N_0 (or the area under the age distribution curve) is $n_0(1/2 \ln 2)$, then the mass $\bar{y}$ of the average cell is

$$\bar{y} = \frac{M}{\overline{N}_0} = \frac{n_0\ (1/2(\ln 2)^2)}{n_0\ (1/(2 \ln 2)\ )} = 1{\cdot}44. \quad (16)$$

From this and (12), it is seen that cells which have completed 44 per cent of their life cycle have a mass equivalent to that of the average cell. The mass of cells of other ages readily follows.

The age of the average cell changes with the number of progeny at division of the parent, and also with the growth pattern of the constituent of interest. Table I summarizes some of these permutations. Values are given for linear growth of single cells (or any constituent thereof) for yields of two, four, and eight filial cells, and for convex and concave growth patterns for yields of two filial cells. An exponential function ($2^{t/GT}$) and a logarithmic function [$\log_2(2t + 2)$] were used to describe the non-linear growth patterns; it will be obvious that these terms are only representative members of families of curves. It is of interest to note that the age of the average cell lies always to the left of the median age.

TABLE I. Age of the average cell

Type of growth	No. of progeny	Interphase growth equation	Average mass	Average age
Linear	2	$y_i = \left(\frac{t_i}{GT} + 1\right) y_0$	$1{\cdot}44y_0$	$0{\cdot}44GT$
	4	$y_i = \left(\frac{3t_i}{GT} + 1\right) y_0$	$2{\cdot}15y_0$	$0{\cdot}38GT$
	8	$y_i = \left(\frac{7t_i}{GT} + 1\right) y_0$	$3{\cdot}37y_0$	$0{\cdot}34GT$
Exponential	2	$y_i = y_0 2^{t/GT}$	$1{\cdot}38y_0$	$0{\cdot}47GT$
Logarithmic	2	$y_i = \left[\log_2\left(\frac{2t}{GT} + 2\right)\right] y_0$	$1{\cdot}46y_0$	$0{\cdot}36GT$

The age of the 'average' cell is shown for three types of single-cell growth. The average age varies with the number of progeny resulting from each parent cell. Also indicated are the functions used to describe single-cell growth, where y_i is the mass of any cell of age t_i, and y_0 is the mass of the filial cell (age = 0). The mass of the average cell is given in terms of the filial-cell mass.

VI. Variation of Member Cells in Real Populations

All of the above considerations have been based on the view that the single cell cycle is invariant with respect to time from one generation to the next. This assumption is in point of fact probably never strictly true, since all species so far examined for this property exhibit a measure of variation in the generation time of individual cells. This variability was first noted by Kelly and Rahn (1932), and later extended to other species by several workers, notably Powell (1955, 1956). The degree of variation is apparently characteristic of the species, although it may be subject to some modification by culture conditions (Cook and Cook, 1962). The variation is often considerable; in certain species of bacteria, Powell (1955, 1956) has found that the coefficient of variation for individual generation times approaches 0·5, although in other species the coefficient is only slightly greater than 0·1. In the flagellate protozoan *Euglena gracilis* (strain Z) the coefficient of variation is about 0·1 (Cook and Cook,

1962); *Tetrahymena pyriformis* HS, the only other protozoan cell studied in this respect, also has a coefficient of about 0·1 (Prescott, 1959). Since the ideal age distribution is distorted by individual variation in generation times, the degree of confidence one can place in its use must be tempered by a knowledge of the extent of this variation.

It may also be mentioned that any particular phase of the life cycle need not always stand in the same ratio to the total generation time. The fact that individual generation times vary, for example, does not necessarily imply that the time spent in fission varies to the same degree; it is at least conceivable that the fission time could be fairly constant, and the total generation time longer or shorter depending on the time course of preceding events. Furthermore, values obtained for the fission time under one set of culture conditions may not hold for other conditions. This could also be true of the amount of individual variation found. While these latter considerations will be of little importance when compared to the effects of variable generation times, they do emphasize that caution should be exercised in any practical application of the ideal age distribution.

REFERENCES

Cook, J. R. and B. Cook (1962). Effect of nutrients on the variation of individual generation times. *Exptl. Cell Research* **28,** 524.

Edwards, J. L., A. L. Koch, P. Youcis, H. L. Freese, M. B. Laite and J. T. Donalson (1960). Some characteristics of DNA synthesis and the mitotic cycle in Ehrlich ascites tumor cells. *J. Biophys. Biochem. Cytol.* **7,** 273.

Frankel, J. (1960). Morphogenesis in *Glaucoma chattoni*. *J. Protozool.* **7,** 362.

James, T. W. (1960). Controlled division synchrony and growth in protozoan microorganisms. *Ann. N.Y. Acad. Sci.* **90,** 550.

Kelly, C. D. and O. Rahn (1932). The growth rate of individual bacterial cells. *J. Bacteriol.* **23,** 147.

Lotka, A. J. (1925). *Elements of Physical Biology*. Williams and Wilkins, Baltimore.

Powell, E. O. (1955). Some features of the generation times of individual bacteria. *Biometrika* **24,** 16.

Powell, E. O. (1956). Growth rate and generation time of bacteria, with special reference to continuous culture. *J. Gen. Microbiol.* **15,** 492.

Prescott, D. M. (1959). Variations in the individual generation times of *Tetrahymena geleii* HS. *Exptl. Cell Research* **16,** 279.

Scherbaum, O. and G. Rasch (1957). Cell size distribution and single cell growth in *Tetrahymena pyriformis* GL. *Acta Pathol. Microbiol., Scand.* **41,** 161.

Sharpe, F. R. and A. J. Lotka (1911). A problem in age distribution. *Phil. Mag.* (Series 6) **21,** 435; through Lotka, A. J. (1925), *Elements of Physical Biology*, Williams and Wilkins, Baltimore, p. 110.

SUMMERS, L. G. (1960). The physiological variation of cell and nuclear volumes. Ph. D. Thesis, University of California, Los Angeles, California.

WALKER, P. M. B. (1954). The mitotic index and interphase processes. *J. Exptl. Biol.* **31,** 8.

WALKER, P. M. B. and J. M. MITCHISON (1957). DNA synthesis in two ciliates. *Exptl. Cell Research* **13,** 167.

20

Measurement of Degrees of Synchrony in Cell Populations

JOSEPH ENGELBERG,*† *Department of Zoology, University of California, Berkeley, California*

CONTENTS

I	Introduction	497
II	Determination of Degree of Synchronization from Growth Curve Data	498
	1. Measurement of the growth curve	498
	2. The rate of cell division	498
	3. Asynchronous growth	499
	4. Parasynchronous growth	500
	5. Degree of synchronization	500
III	The Decay of Synchronization	501
IV	What Constitutes a Synchronized Culture?	502
V	Estimation of the percentage Synchronization from the Mitotic Index	502
VI	Point Estimates of percentage Synchronization	505
VII	Critique of Different Definitions of percentage Synchronization	506
VIII	Generalization of the Above Concepts	507
	References	507

I. INTRODUCTION

A 'synchronization procedure' consists of a set of experimental operations designed to increase the degree of synchronization with which a cell culture passes through a given point of the division cycle. Ideally one attempts this way to produce a 'synchronous culture', that is, a culture in which every cell passes through the given point at precisely the same time. By contrast a culture in which the times at which the individual cells pass through this point are completely randomized is said to be 'asynchronous'. In practice synchronization procedures yield populations which are 'parasynchronous', i.e., the populations are neither synchronous nor asynchronous; they are partially synchronized. Our discussion will center on the simple question: Following a given synchronization

* Aided by a Postdoctoral Fellowship from the American Cancer Society.

† Present address: Department of Physiology, University of Kentucky Medical College, Lexington, Kentucky.

procedure, how can one determine the degree of synchronization, with respect to cell division, which has been induced?

II. Determination of Degree of Synchronization from Growth Curve Data

1. *Measurement of the Growth Curve*

Consider a cell population in which there are n cells at a given time t. When no cells are cytolyzed n is a number which either remains constant or increases with time. If one is to obtain a precise measure of the degree of synchronization, one must, in principle, know the value of n at each instant of time following the synchronization procedure. For example, if one only knows that the cell number is 200 at one instant, and 280 10 hr later, then one cannot distinguish between the possibilities of cell division having taken place at the rate of eight cells per hour during the entire 10-hr period and, alternatively, all eighty cells having divided during a single 10-min period. It will therefore be assumed that the experimenter has obtained a complete growth curve for his culture, that is, a curve giving the number of cells in the culture as a function of time following the synchronization procedure. How can he deduce from these data a single number to describe the degree of synchronization of the culture?

2. *The Rate of Cell Division*

Consider the following image for cell division in a synchronized population: at first the rate of cell division is small; there follows a burst of mitotic activity during which the rate of cell division assumes some high value, following which the rate becomes small again. These considerations suggest that synchrony is associated with the peaking of the curve of the rate of cell division, dn/dt, vs. time, t: the sharper the peak the greater the degree of synchrony. The variable dn/dt is not, however, ideally suited for quantitating synchrony since its magnitude is a function of cell population size. Thus, were we to subject 1000 cells to some synchronization procedure, and then arbitrarily separate these cells into two groups of 200 and 800 each, although both groups would thereafter exhibit the same growth pattern, the rate of cell division of the second group would always be four times that of the first.

This ambiguity can be avoided by defining another quantity, the 'normalized rate of cell division', R, as

$$R = (dn/dt)/n. \qquad (1)$$

R is the fraction of cells dividing per unit time. If, for example, there are, at a given moment, 1000 cells dividing at the rate of fifty cells per hour, then $R = (50)/(1000) = 0{\cdot}05$ per hour. In the example of the previous paragraph although the rates of cell division for the two sub-populations differ by a factor of four, the normalized rate of cell division is the same for both.

Given a plot of n vs. t based on experimental data one can readily calculate (Engelberg, 1961) the value of R at any given time and thus obtain a plot of R vs. t (Fig. 1). When cell division is at its ebb R will be small; when it proceeds at a high rate R will be large.

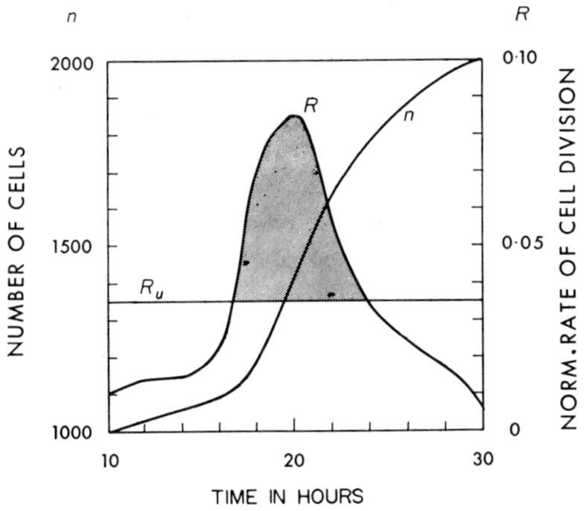

FIG. 1. Growth parameters of a hypothetical cell culture: n, growth curve; R, normalized rate of cell division ($R = (dn/dt)/n$); R_u, normalized rate of cell division for an exponential culture having a generation time of 20 hr ($R_u = (\ln 2)/20\ \text{hr}^{-1}$). The shaded area is the 'overlap area'. Percentage synchronization = 30 per cent. (Reproduced by permission of Academic Press.)

Hence a peak in the R vs. t plot represents a burst of mitotic activity. The R plot also has another important property: Suppose at some given time, t_2, there are exactly twice as many cells as at some previous time, t_1. Then it can be shown (Engelberg, 1961) that the numerical magnitude of the area under the R curve in the interval between t_1 and t_2 is constant for all possible growth curves and equals $\ln 2 = 0{\cdot}693 \ldots$.

3. *Asynchronous Growth*

An asynchronous culture will be defined as one in which the cell number, n, increases exponentially with time, i.e.,

$$n = n_0 2^{t/T_g} \tag{2}$$

where n_0 is the cell number at zero time, and T_g is the *generation time** of the culture. For an asynchronous culture R is constant in time and the R plot is a horizontal line located a distance $(\ln 2)/T_g$ above the time axis (Engelberg, 1961).

4. *Parasynchronous Growth*

Figure 1 gives the growth curve (n vs. t) and the R plot (R vs. t) for a parasynchronous population. In the interval between $t_1 = 10$ hr and $t_2 = 30$ hr the cell number doubles. The horizontal line labeled R_u is the R plot which would have been obtained had the cell number doubled asynchronously between t_1 and t_2. In this time interval the numerical magnitude of the area beneath R is exactly equal to that beneath R_u. Since the two areas are equal in magnitude it follows that if R sometimes lies below R_u, it must at other times lie above R_u. We call the area under R which projects above R_u the *overlap area*. This area is shaded in Fig. 1. The greater the degree of synchronization, the greater the normalized rate of cell division, R, will be at some time, hence the greater the overlap area. This property of the R curve allows us to define a numerical measure of the degree of synchronization of a cell population.

5. *Degree of Synchronization*

We define the *percentage synchronization* of a culture to be

$$\text{Percentage synchronization} = \frac{\text{overlap area}}{\text{total area under } R \text{ between } t_1 \text{ and } t_2} \times 100 \qquad (3)$$

where $t_2 - t_1$ is a time interval during which the population doubles its cell number. A numerical example involving the data exhibited in Fig. 1 and illustrating the use of this equation has been given in a previous publication (Engelberg, 1961). It can be shown that an

* *Definitions:* Any time interval $(t_2 - t_1)$ during which a population of cells doubles its cell number will be said to be a 'doubling period' of the population. In general the length of the doubling period will be different for different time intervals along the growth curve. For the special case of exponential growth the length of the doubling period remains constant in time and is then defined as the *generation period* of the population. The generation period of a population should also be distinguished from the doubling periods of the individual cells of which it is composed. The generation period of a population of cells is not, for example, in general equal to the average doubling period of individual cells (Engelberg, 1961). In view of the perturbations in growth patterns encountered with synchronized cultures it is recommended that the concepts of doubling period and generation period not be used interchangeably, and that the words 'generation period' be restricted to the description of exponential growth alone.

asynchronous population has a percentage synchronization of 0, a perfectly synchronized population a percentage synchronization of 100. The growth pattern of Fig. 1 exhibits a percentage synchronization of 30 per cent.

III. The Decay of Synchronization

Consider a hypothetical population in which every cell follows a division cycle having a duration of precisely T hours. The R curve for this population will be perfectly periodic in time, its pattern remaining the same from doubling period to doubling period. It follows that such a population maintains its level of synchronization indefinitely.

By contrast consider a population in which there is some variability in the doubling times of the individual cells—some cells having longer doubling times than others. Such a situation would arise, for example, when the two daughter cells of a dividing cell have at the end of telophase, due to the limited precision of the cell division mechanism, somewhat different volumes or quantities of metabolic units. Here the differences in doubling times among members of the population are not hereditary; rather the doubling times fluctuate randomly from division to division. We will now show that for this population the degree of synchronization decreases with time until the population becomes asynchronous.

Consider a population for which the mean doubling time of a cell is T, the maximum doubling time is $T + \delta$, and the minimum doubling time is $T - \delta$. Assume further that the population has been synchronized so that it divides with perfect synchrony (percentage synchronization $= 100$) at $t = 0$. From $t = 0$ to $t = T - \delta$ no cell division takes place since the shortest doubling period for any cell is $T - \delta$; hence $R = 0$ during this period (Fig. 2). Cell division will again take place in the interval from $t = T - \delta$ to $t = T + \delta$. Thus during the second doubling period the peak in the R curve has a width in time of 2δ; similarly during the third doubling period its width would be 4δ. By extending this reasoning to subsequent divisions it is clear that the peaks become progressively wider until the peaks of adjacent doubling periods merge to form a continuous curve. Since the area under R must be the same for each doubling period, as R becomes progressively wider it must also progressively decrease in height. The widening and lowering of R however reaches a limit since the area under R during a given doubling period must equal a constant, namely, ln 2. It follows that R approaches a horizontal line of height $(\ln 2)/T_g$, where T_g is the generation time of the culture. An R curve consisting of a horizontal line signifies exponential growth and therefore asynchrony. Hence we see how a synchronized population is converted in time to an asynchronous population due to the variation in the doubling times of individual cells.

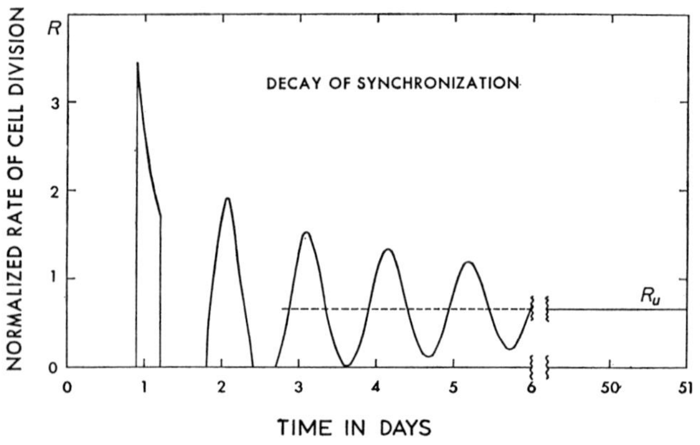

FIG. 2. Normalized rate of cell division, R, for a cell culture in which individual cell doubling times follow a statistical distribution given by a rectangular probability density function (Engelberg, 1961). $T = 1{\cdot}05$ days, $\delta = 0{\cdot}15$ days. The cells were assumed to have divided with perfect synchrony at time $t = 0$ hr. (Reproduced by permission of Academic Press.)

IV. What Constitutes a Synchronized Culture?

From our definition of percentage synchronization it follows that any deviation away from exponential growth partially synchronizes a culture. This follows from the fact that a displacement of the growth pattern from that of exponential growth displaces R below R_u for part of the doubling period, and above R_u for the rest of the period; thus an overlap area results and with it partial synchronization.

V. Estimation of the percentage Synchronization from the Mitotic Index

In the previous paragraphs a method has been described for calculating the percentage synchronization from data on cell number vs. time. It is also possible to estimate percentage synchronization from data on mitotic index vs. time. Such an estimate can be made when (1) the time for mitosis (time from the beginning of prophase to the end of telophase), T_m, is small compared to the doubling period, $t_2 - t_1$, of the culture, and (2) the variation of the mitotic index during any time interval of duration T_m is likewise small.

An average value for T_m can be computed from measured values of mitotic index and generation period obtained for a population in exponential growth as follows: The variation of cell number with time is given for such a population by the equation

$$n(t) = n_0 2^{t/T_g} = n_0 \exp[(t \ln 2)/T_g]. \quad (4)$$

The number of cells, $n_m(t)$, in mitosis at time t equals the number of cells which entered mitosis between $t - T_m$ and t, and hence equals the number of cells dividing between t and $t + T_m$. Thus the mitotic index, I_m, is

$$I_m = [n_m(t)/n(t)] = [n(t+T_m) - n(t)]/n(t). \quad (5)$$

Substituting equation (4) into equation (5) one obtains (Hoffman, 1949)

$$I_m = \exp[(T_m \ln 2)/T_g] - 1, \quad (6)$$

so that

$$T_m = [T_g \ln(1 + I_m)]/\ln 2 \quad (7a)$$

$$= T_g(I_m - I_m^2/2 + \text{higher order terms})/\ln 2. \quad (7b)$$

T_m is often computed using only the first term of equation (7b), thus:

$$T_m = I_m T_g/\ln 2. \quad (8)$$

Equation (8) would give, for example, for a culture with mitotic index 0·028 and generation time 750 min, $T_m = (0{\cdot}028)(750)/(0{\cdot}693) = 30$ min. When I_m is greater than 0·1, T_m should be computed from equation (7a), since the error resulting from the use of equation (8) will be in excess of 5 per cent. Thus for a culture with $I_m = 0{\cdot}25$, and $T_g = 120$ min, $T_m = [(120)\ln(1{\cdot}25)]/(\ln 2) = (120)(0{\cdot}223)/(0{\cdot}693) = 39$ min. (A similar calculation using equation (8) would have yielded $T_m = 43$ min.) Noting that for cultures in exponential growth $R = (\ln 2)/T_g$, it follows from equation (8) that for exponential cultures with small mitotic index

$$R = I_m/T_m . \quad (9)$$

Consider a culture in which there are, at time t, n cells of which n_m cells are in mitosis. The mitotic index, I_m, is defined by

$$I_m = n_m/n . \quad (10)$$

At time t we would see nI_m cells in mitosis. In the time interval from t to $(t + T_m)$ all of these cells divide, and thus nI_m cells are added to the culture. Hence the rate of cell division at time t is approximately

$$dn/dt = nI_m/T_m \tag{11}$$

from which it follows that

$$R = (dn/dt)/n = I_m/T_m. \tag{12}$$

Equation (12) allows us to convert a plot of I_m vs. t to a plot of R vs. t, providing the average time for mitosis, T_m, is known. One simply divides the value for I_m obtained at each time t by the time for mitosis, T_m. Note that when T_m is a constant, R is directly proportional to the mitotic index, I_m.

Although equation (12) allows one to process data in the form of mitotic index vs. time so as to obtain an R plot, the percentage synchronization cannot be calculated until some time interval $(t_2 - t_1)$ during which the cell number doubled is known. This interval can be determined as follows: One selects a time t_1 to mark the beginning of the doubling period in question. Usually t_1 can be set at the time at which the cells were returned to normal environmental conditions following the synchronization procedure. One then guesses a time t_2' at which the population might have just doubled in size. Next one finds the area under R between t_1 and t_2'. This area will either be larger or smaller than $\ln 2 = 0{\cdot}693$. If it is larger one chooses for one's next estimate of t_2 a value which is smaller than t_2', if it is smaller a value which is larger than t_2'. Proceeding in this manner one can establish the doubling period, $t_2 - t_1$, to any degree of accuracy. A planimeter is a useful instrument for expediting this trial and error calculation. Once the doubling period is known one proceeds with the calculation of percentage synchronization as before.

The above procedures for calculating percentage synchronization from the mitotic index can be modified for the case when the time for mitosis, T_m, is large compared to the population doubling time. In this case one selects an observable portion of the mitotic cycle, say anaphase, the duration of which is short compared to the duration of the whole cycle, and at each time of observation, t, records the number of cells, n_a, in anaphase in a microscopic field containing a total of n cells. The *anaphase index*, I_a, can then be defined as

$$I_a = n_a/n\,. \tag{13}$$

If the average duration of anaphase at time t is T_a, and if T_a is small compared with the doubling time, then R will be approximately given by

$$R = I_a/T_a \,. \tag{14}$$

It should be noted that in principle the determination of percentage synchronization from data on mitotic index versus time is inferior to that based on cell number versus time, due to the error in the calculated rate of cell division arising from the finite duration of T_m, and also due to the possible variation of T_m with time.

What is the relationship between the mitotic index and, for example, the anaphase index? When a culture is in exponential growth there is a direct connection between these two quantities: $I_a = (T_a/T_m)I_m$, i.e., the anaphase index is to the mitotic index as the time for anaphase is to the time for mitosis. For a mammalian cell with T_m of the order of 60 min, and T_a of the order of 5 min, the magnitude of the anaphase index will only be $5/60 = 0{\cdot}08$ that of the mitotic index. When the culture is not in exponential growth, however, and one measures I_m at time t one can infer only that I_a is smaller than, or equal to, I_m at time t.

VI. Point Estimates of percentage Synchronization

The definition of percentage synchronization as developed here has made use of the growth pattern of a culture over an entire population doubling period. Alternatively one can obtain computational simplicity, however, at the cost of completeness and accuracy, by the use of point estimates for percentage synchronization, that is estimates in which the states of the culture at two points in its development are compared without making use of the details of the growth pattern in the interval between the two points.

Zeuthen (1958) has suggested two such estimates. The first relates percentage synchronization to the fraction of the generation time for which 50 per cent of the population doubles. Let T_{50} equal the shortest time interval during which 50 per cent of the population doubles, then

$$\text{Percentage synchronization} = (100)\,[T_g/2) - T_{50}]/(T_g/2)\,. \tag{15}$$

The second estimate is based upon the consideration that the term 'synchronization' should be used to convey information as to how much the steady state, represented by the randomized population, has been shifted towards the situation in which all cells divide or go through a defined morphological state in 100 per cent synchrony, and can be written as

$$\text{Percentage synchronization at time } t = (100)(I_m - I_{m0})/(1 - I_{m0}) \,. \tag{16}$$

Here I_m is the mitotic index at some time t after the synchronization procedure, and I_{m0} is the mitotic index for the culture under conditions of exponential growth just prior to the synchronization procedure.

Scherbaum (1959) defined the *synchronization index*, *SI*, by

$$SI = [(n_2 - n_1)/n_1]\,[1 - \triangle t/T_g] \tag{17}$$

where n_2 and n_1 are the population cell numbers just before and after synchronous multiplication, and $\triangle t$ is the time span during which synchronous multiplication takes place.

VII. Critique of Different Definitions of percentage Synchronization

Although equations (3), (15), (16) and (17) are all labeled percentage synchronization, cognizance must be taken of the fact that when all four equations are applied to the same data, in general four different numbers will be obtained for the percentage synchronization. This is due to the fact that the four definitions are related to different aspects of the growth curve data. Thus equation (3) is based upon the time distribution of cell division during a finite time interval, namely, the population doubling period. Equation (16), on the other hand, counts only the cells which are undergoing division at a given point in time, while equation (15) makes use only of the time at which the growth pattern has passed through the point where the cell number has increased by 50 per cent.

The definitions of equations (3), (15), (16) and (17) tell an investigator who has subjected a population to a synchronizing procedure to what extent he has displaced the growth pattern from random (exponential) growth in the direction of synchronized growth. Consider, however, the case of a biochemist who wishes to harvest as many metaphase cells as possible for an isolation procedure. For his purpose the metaphase index (defined analogously to the anaphase and mitotic indices) is a better criterion for synchrony since he will want to harvest the cells at a time when the metaphase index is a maximum.

Any definition is to some extent arbitrary and can only be judged by its usefulness in a context of a given application. Situations in which the definition of equation (3) would lead to misleading results can be imagined. Consider the hypothetical case where a population has been synchronized in such a way that half the population divides with perfect synchrony at one point in time while sometime later the other half of the population also divides with perfect synchrony.* The R plot for this case consists of two δ-functions, and the percentage synchronization is 100 per cent though only half of the population divides simultaneously. This example suggests that the definition of equation (3) is strictly suitable as a measure of synchronization only when the R plot exhibits a single peak during a given cell population doubling period.

VIII. Generalization of the Above Concepts

Above we have for the sake of simplicity discussed procedures for quantitating the degree of synchronization of cell populations with respect to cell division. The same formalism can, however, be applied to other modes of synchrony. A worker may wish, for example, to synchronize a population of cells to encourage them to initiate DNA synthesis at the same time. (A population synchronized with respect to DNA synthesis need not, of course, necessarily be synchronized with respect to the time of cell division.) One can determine the degree of synchrony for this situation from equation (3) by use of the following artifice: To construct the growth curve (n vs. t) of the culture one counts every cell which has passed the point where DNA synthesis has been initiated as two cells, while counting every other cell as one cell. This growth curve can then be converted into an R plot as before and the percentage synchronization determined.

REFERENCES

Engelberg, J. (1961). A method of measuring the degree of synchronization of cell populations. *Exptl. Cell Research* **23,** 218.

Hoffman, J. G. (1949). Theory of the mitotic index and its application to tissue growth measurement. *Bull. Math. Biophys.* **11,** 139.

Scherbaum, O. H. (1959). A comparison of the degree of synchronous multiplication in various microbial systems. *J. Protozool. Suppl.* **6,** 17.

Zeuthen, E. (1958). Artificial and induced periodicity in living cells. *Advances in Biol. and Med. Phys.* **6,** 37.

* I am indebted to Dr. Allan Campbell for this example.

Note Added in Proof

Since this chapter was submitted for publication, the following articles dealing with measures of synchronization have appeared: 1. BLUMENTHAL, L. K. and S. A. ZAHLER (1962). Index for measurement of synchronization of cell populations. *Science* **135,** 724. 2. SPENCER, H. T., R. R. SCHMIDT, C. Y. KRAMER, W. E. C. MOORE, and K. W. KING (1961). Estimation of the degree of synchrony in microbial populations. *Exptl. Cell Research* **25,** 485. 3. BLUM, J. J. and G. M. PADILLA (1962). Studies on synchronized cells: The time course of DNA, RNA and protein synthesis in *Astasia longa*. *Exptl. Cell Research* **28,** 512.

21

Detection of Synchrony Induction in Amoebae; Problems in Cell Counting

ROBERT J. NEFF,* *Vanderbilt University, Nashville, Tennessee*

CONTENTS

I	Introduction	509
II	Division Index by Direct Microscopic Observation	511
III	Determination of Growth Rates by Counting Methods	511
	1. General problems of counting soil amoebae	511
	2. Chamber counts	512
	3. Automated 'counting' with the nephelometer	513
	4. Automated counting with the cell counter	515
IV	Summary	519
	References	519

I. INTRODUCTION

Basically, all methods of detecting synchronous growth and division are counting methods. Arbitrarily we can recognize two major sub-methods for detecting synchrony; by alterations in division rate and growth rate.

In the first, deviations from the random division rate are determined by frequently enumerating the fraction of cells in some recognizable stage of division. Division bursts appear as sharp, normalized peaks spaced along the time axis. In the second method, frequent counts are made of the culture, or of samples from the culture. Growth curves may be constructed by plotting $\log_{10}$ of cell numbers against time. A randomly-growing and dividing culture produces a straight line which increases 0·3 units for a doubling of cell numbers. A synchronous burst of divisions is detected by a sharp rise in the growth curve adjacent to a flatter portion or plateau.

The two methods may be compared. The growth rate method is simpler since only one count is required at each time interval. The division rate method requires two counts at each time interval; a total count and a count of all dividing cells. From data collected

* Unpublished work of the author, summarized in this article, was supported by grants from Public Health Service (E-623, E-1908, and RG-6731), National Institutes of Health, and by small equipment grants from the Natural Sciences Research Committee, Vanderbilt University. Support is gratefully acknowledged.

in the division rate method, one can construct both growth and division index curves. If division index and $\log_{10}$ of cell numbers are plotted on the same time axis it will be seen that division index peaks occur at points of inflexion on the growth curve. Division indices can be calculated from growth rate data if the average duration of a cell division is known. (See other chapters in this section.)

Both methods have common limitations. If anything less than the total population is counted, the accuracy of the count is at the mercy of statistical variation. It follows that the minimal size of a recognizable division burst is limited by counting accuracy. Fortunately, counting errors are easily calculated and confidence levels may be placed on individual counts. Important aspects of handling counting errors are summarized below.

> Errors in counting cells follow a Poisson-type series. This has been shown for the counting of bacteria by plating methods (Fisher *et al.*, 1950), for red blood cells by counting chamber (Berkson *et al.*, 1940), and for blood cells, bacterial cells, and a variety of inanimate particles by an automatic cell counter (Berg, 1958; Brecher *et al.*, 1956; Kubitschek, 1958; Mattern *et al.*, 1957). In the special case of counting cells, the square root of the count represents a close approximation to the standard deviation (Mather, 1943). The counting error becomes smaller as the number of cells counted is increased. The standard deviation for a count of 1600 cells is 2·5 per cent and for a count of 10,000, 1·0 per cent. One can decide the confidence to be placed in a count since, by chance alone, one can expect 67 per cent of all samples of the same size to give counts within $\pm$ 1 standard deviation of the original count.

It is thus clear that a severe practical limitation imposed on growth and division rate studies, as applied to synchrony, is that a large number of cells must be counted at each time interval to assure the detection of small division bursts. In practice, only bursts representing increases considerably larger than the standard deviation are used in assaying the degree of induced synchrony.

Another limitation of rate methods may be noted. The accuracy with which a division burst may be defined (initiation, peak, or peaks if in fact the burst is a series of small bursts, and termination) is limited by the frequency with which counts are made. A special limitation of the division-rate method with a cell type having a small ratio of division time (the duration of recognizable division stages) to generation time will be examined in section II.

This volume of review articles gives ample evidence that synchrony induction can be followed by either major method. Yet, both methods possess a most undesirable feature in common. This is the unavoidable drudgery of counting over long periods. In attempts to lighten this task, several methods of automating the counting procedure have been explored and are described below.

II. Division Index by Direct Microscopic Observation

James (1958), and possibly Roslansky *et al.* (1958), have determined the effect of a temperature treatment on growing cultures of amoebae by recording the number of dividing cells present in the entire small population during and after the treatment. James attempted to avoid continuous observation by making time-lapse photographic recordings at low magnifications during temperature cycling and light cycling. As James has pointed out, the application of this procedure is severely limited by the size of the microscope field and thus only small numbers of cells may be followed. The degree of synchronization resulting from an experimental treatment can be determined only after the film has been developed. However, the film provides a permanent record. The method is useful in analyzing individual cell variability with regard to its generation time and duration of division. It cannot be used with agitated cultures or rapidly motile organisms.

In *Acanthamoeba* sp. a high degree of natural division synchrony was detected by observing division figures (Neff, 1955). However, the division rate method is impractical in synchrony induction experiments with this organism because its division time is at best one-seventieth of its generation time. Dividing cells would be infrequent even in sizable division bursts.

III. Determination of Growth Rates by Counting Methods

Growth rate methods have been applied almost exclusively to studies of synchrony induction in *Acanthamoeba* sp. Much of the data to be discussed originated in the author's laboratory; it has not been published previously. Cell numbers have been determined by chamber counts, plate counts, automated nephelometer readings, and automatic cell counter counts.

1. *General Problems of Counting Soil Amoebae*

Two of the major errors to be overcome in counting cultures of soil amoebae are their tenacious attachment to glass surfaces and associations of two or more cells with each other. The former is largely overcome by coating growth containers with high vacuum silicone grease (Neff *et al.*, 1958) prior to sterilization. In shallow cultures, cells are easily dislodged by tilting the growth flask so that surface tension pulls cells off the bottom. Cells are not ruptured by this procedure. In experiments where agitation is to be avoided, many small replicate flasks are prepared and duplicates taken at each sample time. With these small cultures attachment to surfaces and to each other can be reduced by adding an octyl alcohol solution (3 vols. of octyl alcohol dissolved in 20 vols. of ethyl alcohol) just prior to sampling. One drop per 2 ml of culture medium causes cells to detach and round up. With gentle swirling, cells become uniformly dispersed; they may be counted immediately, or fixed with dilute formalin for subsequent counting. Other uses of octyl alcohol mixtures are described in section III-4.

2. *Chamber Counts*

The average diameter of *Acanthamoeba* sp. is close to 20 μ (Neff, 1957a) but changes with nutritional state and age of the culture. These cells may be counted in standard blood cell counting chambers, 0·1 mm deep. They are more easily and accurately counted in deeper chambers such as the Fuchs–Rosenthal type, which is 0·2 mm deep. If this chamber is obtained with *no* concavity on its undersurface, the phase contrast microscope can be advantageously used in counting. Samples are thoroughly mixed and are rapidly added to the chamber with a large tip pipet. Enough chambers are counted so that 2000 or more cells are enumerated in each sample.

Sampling and counting procedures were tested by Neff (1955) to see if they gave the variability anticipated for a Poisson series. Variations of 5 per cent between duplicates were seldom exceeded. In addition, samples from a large fixed culture containing approximately 100,000 cells per ml were counted repeatedly, a total of about 40,000 cells. The maxium variation from the averaged mean was slightly less than $\pm$ 2·5 per cent when 2000 cells were taken in any order. The observed deviation approximated the theoretical standard deviation.

A plating method for counting amoebae has also been developed (Neff, 1957b). Diluted samples of amoebae are plated with a food bacterium. A single amoeba is trapped in the bacterial carpet where it multiplies and eats bacteria, thereby producing a clear area or plaque. The plaque number is equal to the number of amoebae plated, up to 300 per plate. A detailed account of this method will not be given. It is of limited value in induced synchrony studies because the time of incubation before plaque counts can be made is long, 3 to 5 days. The method is useful in determining survival of variously treated amoebae. The method is similar to that developed for myxamoebae (Sussman, 1951).

3. *Automated 'Counting' with the Nephelometer*

The development of an automated counting instrument capable of instantaneous or very rapid monitoring of cell numbers during growth would greatly accelerate the testing of synchronizing treatments in soil amoebae. We have explored two possible approaches to automated counting. First, (this section) we have constructed and tested on automated nephelometer and second, (section 4) we have examined the feasibility of automating an electronic cell counter.

Light scattering (nephelometry, turbidimetry, or photometry) has been used to approximate bacterial numbers since at least the turn of the century (McFarland, 1907). Recently it has been used with soil amoebae (Band, 1959). A recent review of the wide uses of turbidimetry in bacterial assay methods reveals that little is known of mechanisms of scatter by microorganisms (Gavin, 1957) and by blood cells (Ponder, 1941; Shohl, 1940). A perusal of any of several recent reviews of light scattering of particles suspended in solution (Bender, 1952; Doty and Steiner, 1950; Oster, 1948; Stacey, 1956) indicates that a large number of factors contribute to the amount of light scattered. Some of these variables are particle size, concentration, wavelength of incident light, internal orientation (isotropic or anisotropic), and refractive indices of solvent and particle. Regardless of the complexity of the situation it is apparent that in many cell growth systems the rate of increase of scatter is directly proportional to rate of increase in cell numbers.

There was no *a priori* reasoning that permitted one to decide if scatter from a culture of soil amoeba was primarily a function of cell numbers, cell surface area, or mass. However, it seemed possible that either synchronized growth or division or both could be detected nephelometrically. Accordingly, an automated nephelometer was constructed, and tested with *Acanthamoeba* and *Tetrahymena* cultures.

The automated nephelometer* as constructed measured the 90°

* Equipment and advice for testing the preliminary model of the high sensitivity photomultiplier system were generously provided by Prof. J. E. Gullberg, Zoology Department, University of California, Berkeley, Calif. The well regulated voltage supply and the photomultiplier circuit of the final model were kindly designed by Dr. G. W. Ellis, Division of Molecular Biology, Vanderbilt University.

scatter of white light. Scattered light intensity was directly proportional to cell numbers in the range 5×10^3 to 6×10^5 per ml under standard growth conditions. Above the top concentration shading became important. Cells were agitated continuously by a magnetic stirrer and also by aeration bubbles between readings. Addition of a concentrated suspension of cells to a growing culture indicated that an increase of less than one-twentieth of the cell concentration present was easily detected in the useful range. In general, the machine possessed high sensitivity and reproducibility.

First, the scatter of a growing culture of *Acanthamoeba* sp. was determined, and compared to cell concentration, as determined by chamber counts. Over the log phase of growth, $\log_{10}$ of cell numbers was parallel to the $\log_{10}$ of recorded galvanometer readings, corrected for background scatter, provided the counts and readings were taken at long intervals; 4 hr or more. However, if the counting interval was 1 hr or less the correspondence between cell numbers and light scattering was frequently very poor. Usually the scatter curve increased smoothly even when chamber counts indicated sharp rises (such as one-sixth of the cells dividing in 1 hr) or extended plateaus.

In a second test of the automated nephelometer, cultures of *Tetrahymena* were synchronized by the method of Zeuthen and Scherbaum (1954). In two experiments where the peak division indices for the first post-treatment division were 50 and 65 per cent, the machine showed an increase in slope beginning about 1·5 hr after the peak of divisions, a time when few dividing cells could be found. These results, together with those on *Acanthamoeba* sp., suggested that scatter was proportional to some quantity other than cell number or surface area, perhaps mass.

In a third set of experiments, the relation of light scatter to average cell volume was tested. Aliquots of a concentrated suspension of amoeba cells were diluted in salt solutions of different tonicities. The number of amoebae per unit volume was constant, but the average volume of the cells was an inverse function of tonicity. As is the case with red blood cells (Ponder, 1941), the scatter increased with decreasing cell volume as the tonicity of the medium became greater.

Finally, when anaerobiosis was tested as a synchrony-inducing treatment, a rapid change in scatter occurred. Anaerobiosis, effected by bubbling nitrogen through the growing culture, gave rise to increases in scatter which were usually close to a doubling. This scatter increase was reversible when air or oxygen bubbling was restored. The packed cell volume of oxygen- and nitrogen-treated duplicate samples were equal. The dramatic shift in scatter intensity was probably due to an interesting phenomenon we have called protoplast contraction. Benton (of this laboratory) has observed that cells with an internal pH more acid than 7·0 show a marked packing of visible granules in a ring around the nucleus under anaerobic conditions. This phenomenon is completely reversible by small amounts of oxygen. We have concluded that the increase in the light scattered is a function of *mass distribution* within the cell membrane. It also seems reasonable to assume that less dense portions of the cell, the cell membrane and hyaline cytoplasm, contribute little to scatter. Further, those changes in scatter that occur with changes in tonicity such as observed in *Acanthamoeba*, in red blood cells (Ponder, 1941), and cell particulates (Klein and Neff, 1960), would all appear to be a function of the concentration within a limiting membrane.

An evaluation of the automated nephelometer as a synchrony monitor seems in order. At best, this instrument gives a delayed report of cell division—the increase in light scattering lags behind the increase in cell numbers. Our data suggest that, in soil amoebae, scatter is related to a function of cell mass rather than to cell numbers or total surface area. We must conclude that in this organism, the only legitimate use of an automated nephelometer is for *gross* studies of generation time. Even in this case, caution must be exercised. For example, the growth medium used for *Acanthamoeba* sp. produces a fine precipitate upon aeration, which scatters light. As cell numbers increase, this precipitated material is either ingested by cells or digested by secreted enzymes, since the scatter of the supernatant of a centrifuged culture decreases as cell numbers increase. For cultures in which the cell concentration is low, scatter corrections can be made by subtracting the galvanometer readings of an uninoculated control flask from the readings of the experimental culture.

4. *Automated Counting with the Cell Counter*

The Coulter Counter (Coulter Industrial Sales Co., Chicago, Illinois) Model B, has been used in our studies with *Acanthamoeba* sp. Details of the construction of this instrument (Coulter, 1957), principles of its operation (Berg, 1958), and performance in counting and sizing several cell types (Brecher *et al.*, 1956; Grant *et al.*, 1960; Kubitschek, 1958; Mattern *et al.*, 1957; Richar and Breakell, 1959) have been reviewed. In this account, the usefulness of the Coulter counter for routine counting of soil amoebae and its potential adaptability for automation in constant counting of sterile amoebae cultures will be discussed.

The Coulter counter permits accurate counting of soil amoebae within the limits of accuracy claimed by the manufacturer, but only if the cell suspensions are properly prepared for counting. Several intrinsic sources of error were discovered which were associated primarily with properties of the cells themselves and of their growth medium. These have been circumvented in procedures developed for counting aliquots of large cultures. Unfortunately, it now appears that great difficulties must be overcome to allow correction of the same errors in an automated machine. Sources of error are described in Table I.

Our standard procedure for counting aerated cultures in silicone-coated bottles is summarized below. After a vigorous manual swirling, both the experimental inoculated cultures and the uninoculated controls are sampled at each time interval. The samples are treated identically. Aliquots are diluted with 0·1 *M* KCl, centrifuged, and all but 0·5 ml of the supernatant aspirated off. The cells and/or particles are diluted to a level in the optimal counting range with 0·1 *M* KCl (containing 20 ml diethyl ether, 1 ml of 5:1 ethyl alcohol:octyl alcohol, vol./vol., and 2 mg $CaCl_2$ per liter) and dispersed in a 40 ml conical centrifuge tube by a motor driven spatula.* The resulting cell suspensions are then counted in the Coulter counter and the counts corrected for coincidence. Control counts are subtracted from experimental counts. The lower volume threshold of the counter is set at 800 μ^3, which is below the smallest cell volume encountered under normal or experimental conditions. The 100-μ orifice is used. When this procedure is followed, the corrected counts closely parallel counts made in a blood cell counting chamber. Four cultures can be sampled at the same time, and all counted in approximately the time required for one chamber count

*We have recently found that the rate of rotation of the spatula used to re-suspend cells must be rigidly controlled. In our system, if the rate is below 1000 r.p.m., clumped cells are not completely dispersed. If the rate is greater than 3500 to 4000 r.p.m., cells are pulled in two, but are *not* broken open; that is, as the shear increases (above 4000 r.p.m.) one finds an increase in counted cells which is just equal to the increase in the number of *enucleate* cells. For any given combination of cell container and spatula geometry, the upper and lower limits of rotation rate must be determined, and the motor controlled within these limits for accurate counting. We currently use a motor and controller as supplied by B and B Motor Control Corp., New York, which provides reliable speed control up to 5000 r.p.m.

TABLE I. The control of intrinsic sources of error in counting amoebae with an automatic cell counter

Source of error	Method of control in counting aliquots of cultures	Potential method of control in direct counting of cultures
Cells attach to glass surfaces.	Attachment is eliminated by coating glass with silicone grease and by vigorous swirling before removing the sample.	Coat flasks with silicone grease and shake automatically before sampling.
Cells aggregate.	Cells in each sample are washed free of growth medium, and resuspended in 0·1 *M* KCl solution containing ether and octyl alcohol; cells round up and clumps of cells are dispersed.	None. Ether and octyl alcohol cannot be used in cultures since they do not permit cell growth. Aggregation error is probably small in cultures below 4×10^4 cells per ml.
Non-cellular particles precipitate from growth medium on aeration.	Uninoculated medium is sampled at each interval. Both experimental and control samples are washed centrifugally and cells or particles are diluted to a suitable counting level. The control count is subtracted from the experimental count.	None. The precipitate count approaches the counter limit with time. As cells grow they digest some of the precipitate. These are the most difficult errors to overcome in counter automation.
Coincidence (two or more cells counted as one).	Dilution. After cell numbers reach 90,000 cells per ml, samples are diluted. Correct for coincidence.	The recorded number could be corrected.
Resistivity of the growth medium changes with temperature.	Does not apply here; aliquots are washed and counted in the same KCl solution used to flush the counter, and at constant temperature.	The recorded number could be corrected. The correction would be small.

of 2000 cells—usually 30 to 40 min. Instrument counts are also advantageous since they are numerically large, and their standard deviation correspondingly small.

It is possible, theoretically, to extend the usefulness of the Coulter counter by eliminating the sampling step and counting a growing culture directly. This goal might be reached by introducing an orifice unit of the counter directly into a specially modified sterile growth chamber. It would be hoped that a fully automated machine would be able to monitor the entire growth range of an amoeba culture under standard or experimental conditions; it would produce immediate information, and a permanent record.

In *Acanthamoeba* cultures, the most serious problem to be overcome is the precipitation of material from growth medium during aeration. The amount of precipitate increases with time; the particles count in the cell size range. The magnitude of this factor may be demonstrated by the following experiment.

Standard growth medium (containing 1·5 per cent proteose peptone, 1·5 per cent glucose, added salts and vitamins) was prepared, cleared by filtration or high speed continuous flow centrifugation, dispensed into duplicate culture bottles, and sterilized. They were not inoculated, but were aerated by bubbling at the rate of 2 l. of air per minute per l. of medium. Samples were taken at daily intervals, and counted directly in the cell counter without dilution. The count at zero time was 1000 to 7000 particles per 0·5 ml. It rose to 30,000 to 50,000 particles per 0·5 ml during the first 24 hr. Thereafter, the count continued to rise; in one instance, it reached a maximum of 70,000 particles per 0·5 ml at the end of the fourth day of aeration. Changes in concentration of major constituents of the medium do not materially lessen the amount of precipitate formed.

The practical upper limit of counting is 90,000 particles per 0·5 ml sample volume control. The difference (about 20,000–40,000) between maximum particle count and the upper counting limit does not leave many counts for cells. For this reason, only a small portion of the whole growth curve could be followed accurately. Aerated cultures may reach a concentration of 4 million cells per ml at their maximum stationary phase. Furthermore, the correction for precipitate is complicated by the fact that amoebae may ingest the precipitated particles.

The counting errors due to attachment of cells to surfaces and cell aggregation must also be considered and circumvented in any automated counting instrument. Two other sources of error can be anticipated: those due to coincidence, and those due to a change in resistivity of the culture medium with temperature.

IV. Summary

There is *no* simple and accurate counting method that can be used to detect growth and division synchrony in amoebae which is free of tedium. However, the counting of cells can be greatly expedited by an automatic cell counter.

Similarly, no satisfactory, completely automated method has been developed for following the results of synchronizing treatments. Within its limits, the time lapse film-record method is useful for small cultures. The automated nephelometer has no value in monitoring mass cultures during synchrony studies. While the cell counter is extremely useful in counting samples of a mass culture, the prospects for complete automation, for direct monitoring of experimental cultures, are poor at present. It is suggested that the next step toward complete automation might be the combination of an automatic sampling procedure with the subsequent counting of fixed samples by an automatic counter.

REFERENCES

Band, R. N. (1959). Nutritional and related biological studies on the free-living soil amoeba, *Hartmanella rhysodes*. *J. Gen. Microbiol.* **21,** 80.

Bender, M. (1952). The use of light scattering for determining particle size and molecular weight and shape. *J. Chem. Educ.* **29,** 15.

Berg, R. H. (1958). Electronic size analysis of subsieve particles by flowing through a small liquid resistor. In *Symposium on Particle Size Measurements* (Special Technical Publication, No. 234). American Society for Testing Materials, U.S.A., pp. 245–255.

Berkson, J., T. B. Magath and M. Hurn (1940). The error of estimate of blood cell counts as made with the hemocytometer. *Am. J. Physiol.* **128,** 309.

Brecher, G., M. Schneiderman and G. Z. Williams (1956). Evaluation of electronic red blood cell counter. *Am. J. Clin. Pathol.* **26,** 1439.

Coulter, W. C. (1957). *Theory of the Coulter Counter.* (This pamphlet is obtainable from Coulter Electronics, Chicago, Ill.)

Doty, P. M. and R. F. Steiner (1950). Light scattering and spectrophotometry of colloidal solutions. *J. Chem. Phys.* **18,** 1211.

Fisher, R. A., H. G. Thornton and W. A. MacKenzie (1950). The accuracy of the plating method of estimating the density of bacterial populations. In *Contributions to Mathematical Statistics* (R. A. Fisher, ed.). Wiley, New York, pp. 4·325–4·359.

Gavin, J. J. (1957). Analytical microbiology. III. Turbidometric methods. *Appl. Microbiol.* **5,** 235.

Grant, J. L., M. C. Britton and T. E. Kurtz (1960). Measurement of red blood cell volume with the electronic cell counter. *Am. J. Clin. Pathol.* **33,** 138.

JAMES, T. W. (1958). Synchronization of cell division in amoebae. *Ann. N.Y. Acad. Sci.* **78,** 501.

KLEIN, R. L. and R. J. NEFF (1960). Osmotic properties of mitochondria isolated from *Acanthamoeba* sp. *Exptl. Cell Research* **19,** 133.

KUBITSCHEK, H. E. (1958). Electronic counting and sizing of bacteria. *Nature* **182,** 234.

MCFARLAND, J. (1907). The nephelometer: An instrument for estimating the number of bacteria in suspension used for calculating the opsonic index and for vaccines. *J. Am. Med. Assoc.* **49,** 1176.

MATHER, K. (1943). *Statistical Analysis in Biology.* Methuen, London.

MATTERN, C. F. T., F. S. BRACKETT and B. J. OLSON (1957). Determination of number and size of particles by electronic grating. *J. Appl. Physiol.* **10,** 56.

NEFF, R. J. (1955). Physiology of amoeba. *Terminal Rept. Public Health Research Grant* E-623.

NEFF, R. J. (1957a). Purification, axenic cultivation, and description of a soil amoeba, *Acanthamoeba* sp. *J. Protozool.* **4,** 176.

NEFF, R. J. (1957b). Enumeration of viable *Acanthamoebae* by a plating method. *J. Protozool. Suppl.* **4,** 8.

NEFF, R. J., R. H. NEFF and R. E. TAYLOR (1958). The nutrition and metabolism of soil amoeba, *Acanthamoeba* sp. *Physiol. Zool.* **31,** 73.

OSTER, G. (1948). The scattering of light and its applications to chemistry. *Chem. Rev.* **43,** 319.

PONDER, E. (1941). Red cell counts, percentage volume, and the opacity of suspensions. *Am. J. Physiol.* **134,** 739.

RICHAR, W. J. and E. S. BREAKELL (1959). Evaluation of an electronic particle counter for the counting of white blood cells. *Am. J. Clin. Pathol.* **31,** 384.

ROSLANSKY, J. D., N. H. BRANCHFLOWER and M. S. HUCKMAN (1958). Temperature-induced synchronization of cell division in cultures of trophozoites of *Endamoeba terrapinae. Anat. Record* **132,** 498.

SHOHL, A. T. (1940). The measurement of cell volume by the cell opacity method. *J. Lab. Clin. Med.* **25,** 1325.

STACEY, K. A. (1956). *Light-Scattering in Physical Chemistry.* Butterworths, London.

SUSSMAN, M. (1951). The origin of cellular heterogeneity in the slime molds. *J. Exptl. Zool.* **118,** 407.

ZEUTHEN, E. and O. SCHERBAUM (1954). Synchronous division in mass cultures of the ciliate protozoon *Tetrahymena pyriformis,* as induced by temperature changes. In *Recent Developments in Cell Physiology* (J. A. Kitching, ed.). Academic Press, New York, pp. 141–156.

22

The Development of Techniques for Synchronizing Flagellates

George M. Padilla*† and James R. Cook,‡§
Gerontology Branch, Baltimore City Hospitals, Baltimore, Maryland, and Laboratory of Nuclear Medicine and Radiation Biology, University of California, Los Angeles, California

CONTENTS

I	Introduction	521
II	Development of a Temperature Cycle for *Astasia longa*	522
	1. Long-term responses to incubation temperatures	522
	2. Exploratory studies with temperature cycles	523
	3. Improvement of the temperature program	527
III	Light-Induced Division Synchrony of *Euglena gracilis*	530
	1. Inhibition of growth by visible light	530
	2. Minimal light requirements for optimal growth of *Euglena*	532
	3. Synchronous division on a light–dark cycle	533
	References	534

I. Introduction

Because of their systematic and phylogenetic affinities, the phytoflagellates constitute an important group of organisms (Hutner and Provasoli, 1955). For cell biologists in general, they represent an increasingly studied group of cell types endowed with certain qualities desirable in an experimental organism: a relatively long generation time, photosynthetic ability for the green forms, and chemosynthetic capacity for the colorless forms. In addition to their wide spectrum of temperature viability, most flagellates are easily handled as they possess a tough pellicular covering.

Several flagellates have recently been synchronized: the colorless *Astasia longa*, by means of a repetitive temperature cycle (Padilla and James, 1960), and the chlorophyll-bearing *Euglena gracilis* (Cook and James, 1960) and *Chlamydomonas moewusii* (Bernstein, 1960) with recurrent light–dark cycles. In addition, synchronous cell division

* Public Health Research Fellow (CF–9281–C2) of the National Cancer Institute.
† Present address: Biology Division, Oak Ridge National Laboratory, Oak Ridge, Tennessee, U.S.A.
‡ Public Health Research Fellow (CF–9395–C2) of the National Cancer Institute.
§ Present address: Zoology Dept., University of Maine, Orono, Maine.

has been described for the marine flagellate *Gonyaulax polyedra* (Sweeney and Hastings, 1958; and chapter 10, this book). There are also indications that low accelerative forces (Buetow *et al.*, 1959) and sonic vibrations (Levedahl and Dreher, 1961) may be employed in synchronization studies. In the present paper, we shall deal primarily with recent technical developments in the synchronization of cell division in *Astasia longa* and *Euglena gracilis.*

II. Development of a Temperature Cycle for *Astasia longa*

1. *Long-Term Responses to Incubation Temperatures*

The first series of investigations (with *Astasia longa*) established the long-term (or steady state) response of *Astasia* to incubation temperatures ranging from 15° to 30°C (James and Padilla, 1959). A low temperature (e.g. 15°C) seems to favor protoplasmic growth rather than cell division since the cells grow relatively large. Conversely high temperatures (25° or 30°C) favor cell division. At 25° the cells synthesize twice as much dry matter per unit of oxygen consumed as they do at 15°C. In terms of a cycle, composed of alternating 15° and 25° periods, this implied that the cells would require twice as much time at the low temperature as at the high one in order to produce an equivalent amount of protoplasm, if other effects of the temperature shift are ignored. One would also expect that the cells would tend to divide only at the higher temperature. Thus, a 24-hr cycle must consist of 16 hr at 15°C and 8 hr at 25°C to satisfy these requirements. Before testing this cycle it was necessary to determine the duration of the inhibition of cell division following an abrupt downward shift in temperature. The following experiment was performed: A culture of *Astasia* in logarithmic growth at 25°C (generation time 11·6 hr) was quickly transferred to 15° C. At frequent intervals cell counts were taken and determinations of the dry weight of washed cells were performed. In addition, the division index (expressed as the percentage of cells in a recognizable phase of cytokinesis) was also determined. Fig. 1 shows the results of this experiment.

Cell division is inhibited for about 20 hr following the transfer to 15°. After this time the growth rate is resumed with a generation time characteristic of cells normally grown at 15°C (i.e., about 35 hr). During the time that cell division is inhibited, the dry weight increases from 0·69 to nearly 1·32 mg/10^6 cells. This emphasizes the fact that the primary effect of a shift to a lower temperature is inhibition of cell division.

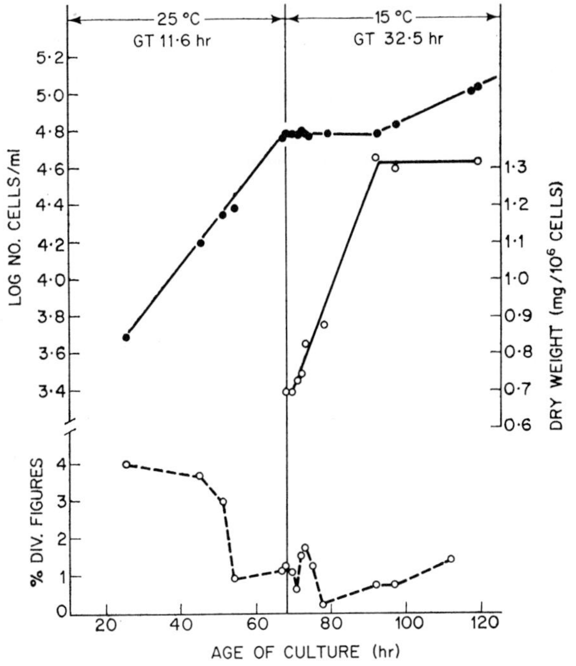

FIG. 1. Changes in *Astasia longa* grown first at 25°C and then transferred (at the time indicated by the vertical line) to 15°C. The filled circles represent the number of cells per ml, and show a temporary inhibition of multiplication before the steady-state generation time of 32·5 hr is assumed at 15°C. The dry mass (open circles, upper) continues to increase; the proportion of cells in recognizable fission (open circles, lower) passes through a minimum.

2. *Exploratory Studies with Temperature Cycles*

The success in attaining a repetitive pattern of synchrony was found to be dependent on at least two factors: (a) the duration of the temperature transitions, i.e., the rapidity with which the temperature is changed and (b) the conditions of growth. Included in the latter category are the nature of the medium, level of aeration, and effect of light (Cook, 1960). We shall consider these factors as they relate to the development of the temperature cycle.

The first cycle which gave a well defined pattern of synchrony consisted of 8 hr at 25°C and 15 hr at 15°C with two 30-min temperature transition periods (Padilla and James, 1960). Cells growing on proteose peptone (1 or 2 per cent) in a 4-1. Erlenmeyer flask regularly divided in the warm period, with over 90 per cent doubling within 3–4 hr. The Cramer–Myers medium was supplemented with

SH-bearing compounds such as cysteine and methionine (Padilla and James, 1960). One of the effects of these compounds on *Astasia* is the reduction of its fission time. Glutathione, ascorbic acid, and thioglycolic acid have a similar property (James, 1960), the latter compound having been substituted for cysteine and methionine in later studies (Blum and Padilla, 1962; Padilla, 1960). With these modifications, the synchronization of *Astasia* on a chemically-defined medium was achieved. However, the usefulness of this technique was limited by the fact that after six or seven generations the medium had become exhausted and the culture had to be discarded. In addition the volume of the culture was limited to 2 l., thus reducing the opportunity of harvesting large quantities of cellular material for biochemical determinations. To solve this problem, semi-continuous culture techniques were instituted by slight modifications in the culture vessel (Padilla, 1960). In essence the culture was regularly diluted with sterile medium, keeping the cell density around 100,000 cells/ml. Although the volume was still limited to 2 l., it became possible to extend the sampling schedule to several cycles.

By using serial dilutions, it was soon discovered that the system was no longer stable, i.e., the phase relationship existing between the onset of the warm period and the onset of the burst of division, previously believed to possess a constant time differential (Padilla and James, 1960), was subject to fluctuations. An attempt was made to remedy the situation by raising the temperature of the warm period to 26·5°C and reducing its duration, and by continuously stirring the cells to insure adequate exchange of gases. In addition the cells were now grown in constant darkness following the demonstration that light has an inhibitory effect on the growth rate of *Astasia* (Cook, 1960). The temperature transition periods remained 30 min long. Fig. 2 illustrates a typical pattern of synchrony under the conditions outlined above.

Except for the first and second cycles seen in Fig. 2, the bursts of division occur well into the warm period, occupying the latter half of it. Yet even in this system, there is some evidence that the bursts of division may gradually move out of phase with the temperature cycle (cf. cycle 6 in Fig. 2). After some time, however, this system becomes quite stable with the cells regularly dividing during the warm period. Unfortunately, this is at the expense of the doubling time (i.e., the time required for 90 per cent of the cells to divide) which is lengthened to approximately 6 hr. Five such consecutive bursts of division are shown in the left-hand portion of Fig. 3.

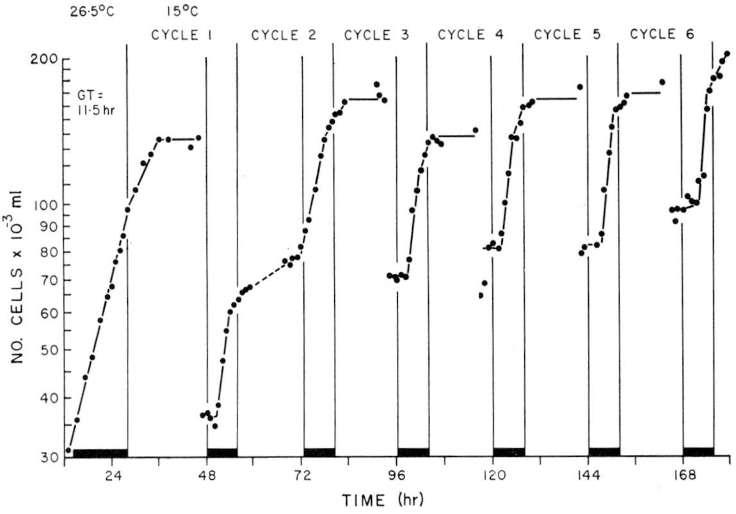

FIG. 2. Synchronous division of *Astasia longa* with periodic dilution, once in each temperature cycle. The areas over the black bars are the warm (26·5°C) periods, the remaining areas the cold (15°C) periods. Prior to Cycle 1, the cells were grown at a constant temperature of 26·5°C.

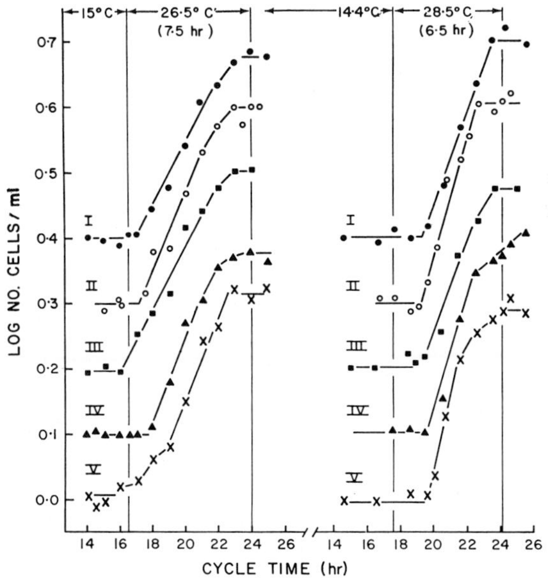

FIG. 3. Comparison of the division behavior of *Astasia longa* in the warm periods of two different temperature programs having the same total cycle time (i.e., 24 hr). (Discussion in text.) Five typical bursts are shown for each program.

The left half of Fig. 4 shows a typical division burst and some biochemical characteristics found with this temperature program.

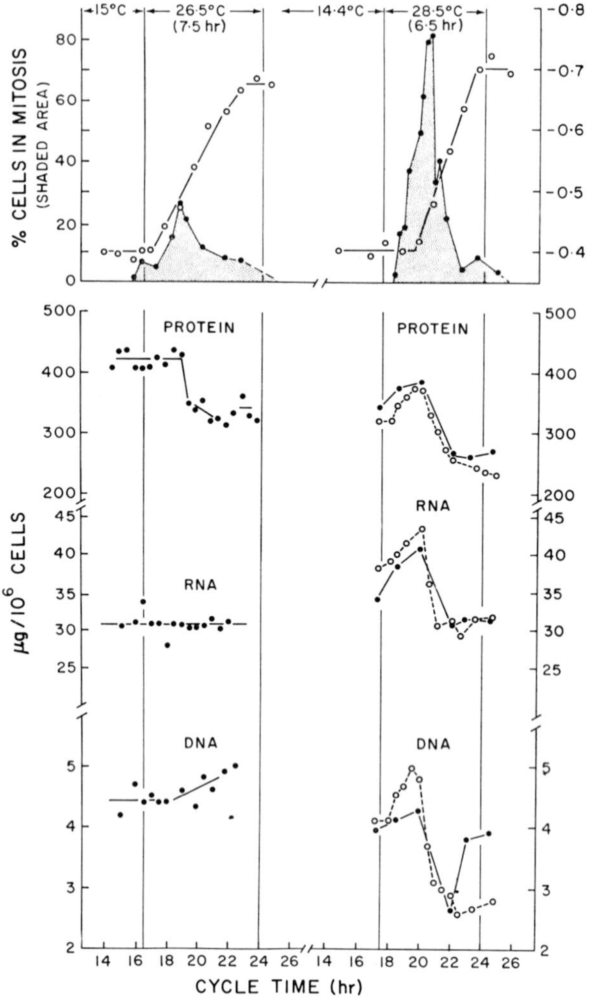

FIG. 4. Cytological and biochemical characteristics of *Astasia longa* during the warm periods of the two temperature programs of Fig. 3. The upper curves show typical synchronous increases in cell number (open circles) and the proportion of cells in mitosis (filled circles). On the same time scale, the lower curves show levels of protein, RNA, and DNA; note that these levels show an increase prior to initiation of the burst in the right-hand curves, but increases are obscured in the other temperature program because of the greater time spread in division activity.

It is quite clear that with a lengthened doubling time, the pattern of synthetic activity is obscured, except for the changes in the protein content of the cells which are correlated with the division process. Although not shown, the dry weight of the cells, reflecting in part the synthesis of paramylum, doubles in the cold period (Padilla, 1960). The mitotic index, measured with ethanol-fixed cells stained with aceto-carmine (Leedale, 1958), has a peak value of 26 per cent but a rather wide distribution. The nucleic acid content of the cells, measured by a modified Schmidt–Thannhauser technique (1945), shows little fluctuation, with the DNA level possibly increasing slightly in the latter half of the warm period. Given the low mitotic frequency and the 10–12 per cent level of experimental error for these determinations, which represent means of samples pooled from several cycles, it is apparent that the nucleic acids are synthesized in the warm period in close association with cell division. These results prompted a re-evaluation of the conditions under which these cells were being synchronized.

3. *Improvement of the Temperature Program*

The primary concern has been the reduction of the two 30-min temperature transition periods. Although this span of time may not be a significant fraction of 24 hr, it is of considerable importance when viewed in terms of the phasing properties of a temperature cycle.

The shift to the lower temperature may be seen as preventing cells from initiating a cellular process, such as mitosis, while at the same time permitting those already engaged in such a process to complete it. Through the cold period these latter cells could advance to the same stage of development but be prevented in turn from initiating the low temperature-inhibited process. A return to the higher temperature would permit all the cells in this particular age class, i.e., late interphase, to proceed as a single unit toward cell division. However, a 30-min temperature transition period may be too long to permit them to do this, particularly when one considers that at 26·5°C *Astasia* has a mitotic time of only 1·42 hr (Blum and Padilla, 1962). Thus, a gradual change in temperature may permit cells to fall out of phase with each other by permitting individual cells to enter into mitosis at different times. The determination of the mitotic time of *Astasia* also showed that a 7·5-hr long warm period could be shortened 1 or 2 hr and still allow the cells ample time to complete their division. With these considerations in mind, the temperature cycle was somewhat modified as discussed below.

A secondary concern dealt with the possibility that a lack of oxygen during the warm period constituted a limiting factor for the cells. Recent studies (Buetow, 1961), as well as our unpublished results, indicate that flagellates have a considerably higher oxygen consumption than had previously been reported, indicating the desirability of aeration.

The above considerations, plus the advantage of increasing the culture volume, led to the adoption of a new culture vessel whose detailed description is given elsewhere (Blum and Padilla, 1962). A schematic drawing of the culture apparatus is shown in Fig. 5. It consists of a 10 in. × 10 in. cylindrical Pyrex glass jar (Corning Co.) fitted with a stainless steel cover. It is provided with an internal stainless steel coil through which water passes at the desired temperature. Several ports in the lid (all of which are not shown in Fig. 5) allow one to stir the culture, insert a temperature sensing element, add sterile medium and sterile air, and remove samples. With this culture vessel, the temperature transition periods have been reduced to 10 min each, even for a 5-l. culture. Moreover, sterile air is pumped in at a minimum rate of 3·5 l./hr, adequately providing the cells with oxygen. All these changes, as well as the adoption of a cycle consisting of 6·5 hr at 28·5°C and 17·5 hr at 14·4°C, have resulted in a highly repetitive pattern of cell division synchrony as shown on the right-hand side of Fig. 3.

Fig. 3 gives, as for the other system, five consecutive bursts, which indicate that the average doubling time has been reduced to approximately 3·5 hr. The phasing of the bursts of division has also been considerably improved, the onsets of division regularly occurring 2 hr after the warm period begins. The reduced doubling time and improved alignment of the division periods are reflected in the cytological and biochemical data (right-hand side of Fig. 4). The mitotic index shows a peak of 81 per cent, as compared to 26 per cent for the previous cycle. The protein and nucleic acid contents of the cells show a definite pattern with a near doubling occurring before the cells enter into division. Two separate experiments are shown and represent means of triplicate samples. As before, the total protein content was measured by Lowry's method (1951). The RNA and DNA determinations, however, were performed by a semi-micro Schmidt–Thannhauser technique developed in Dr. J. J. Blum's laboratory, which considerably reduces the experimental error in comparison with the methods used in the earlier phases of these studies (Padilla, 1960). A detailed description of this technique is given elsewhere (Blum and Padilla, 1962).

Of primary interest is the change in the DNA content of the cells; the maximum DNA level occurs at the peak of mitotic activity, just prior to cell division. By this time, most of the cells have passed

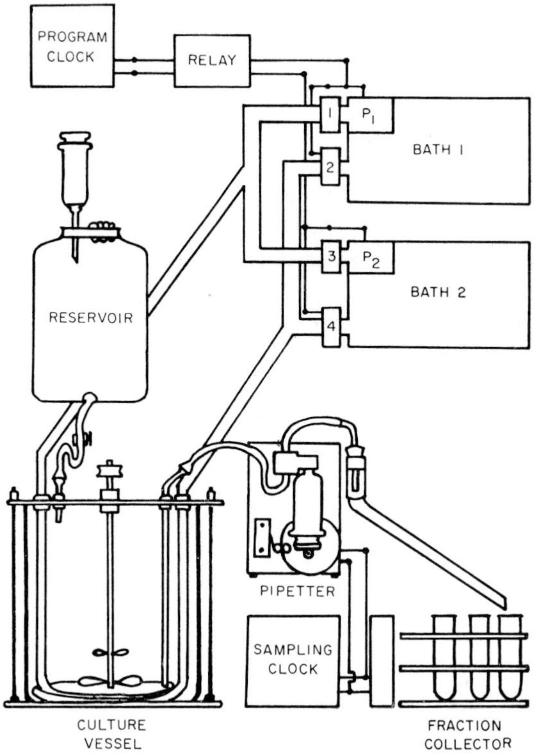

Fig. 5. Schematic of the culture apparatus used in synchronization studies on *Astasia longa*. Bath 1 and Bath 2 are maintained at the two different temperatures. At any given time, water from one of the baths is circulated through coils in the culture vessel; the valves 1, 2, 3, and 4, and the pumps P_1 and P_2, are controlled automatically by the program clock. The automatic pipetter samples the culture periodically. Aeration devices are not shown.

through prophase, suggesting that DNA is being synthesized at this stage (Blum and Padilla, 1962). There is of course some variation, but this is expected as there still exist some fluctuations in the log–linear portion of the burst, particularly towards the end of the warm period (cf. Fig. 3). This may represent a gradual loss of synchrony during the division process. The RNA and protein changes follow a similar pattern, although these compounds appear to be synthesized for a somewhat longer period of time.

III. Light-Induced Division Synchrony of *Euglena gracilis*

1. *Inhibition of Growth by Visible Light*

In like manner, these flagellated cells exhibit a differential response in growth and division to the presence of visible light (Cook, 1960). Light inhibits both growth and division, a fact easily demonstrated by growing the cells in light or dark in a medium supplemented with an accessory carbon source, e.g., sodium acetate. In steady-state logarithmic growth, any one of several species of these flagellates have a shorter generation time when grown in the dark, suggesting that light retards some processes necessary for division; at the same time, the dry mass, protein content, etc., of the average cell are approximately the same whether grown in light or dark, suggesting in turn that synthetic rates are reduced in light, and to about the same degree as the growth rate (see Table I). It may be concluded

Table I. Some characteristics of a light and dark grown bleached *Euglena*

	Dry wt. ($\mu\mu$g/cell)	Vol. (μ^3/cell)	Total prot. ($\mu\mu$g/cell)	RNA ($\mu\mu$g/cell)	DNA ($\mu\mu$g/cell)	G.T. (hr)
Dark-grown	1175 $\pm$ 30	3510	350 $\pm$ 20	30·3 $\pm$ 1·2	5·44 $\pm$ 0·44	31·5 $\pm$ 2·6
Light-grown	1050 $\pm$ 75	3580	335 $\pm$ 20	32·4 $\pm$ 1·5	4·79 $\pm$ 0·99	44·6 $\pm$3·8

Some characteristics of a permanently bleached strain of *Euglena gracilis* grown in light or dark on a salt medium with acetate as the carbon source. The values listed are essentially the same in either case except for the generation times, indicating that visible light inhibits both growth and division. Similar results are found for other euglenoid cells, both green and bleached, at uniform incubation temperatures (Cook, 1960).

that visible light inhibits some process(es) basic to continued optimal multiplication. However, the immediate inhibitory effect is on division rather than incremental growth. This can be seen by first growing the cells in the dark (with acetate as a carbon source) and then transferring the culture, in the middle of the logarithmic phase, to visible light of moderate intensity (300 foot-candles). Such an experiment is shown in Fig. 6. The increase in cell number is abruptly and effectively interrupted for 6 or 7 hr (20 per cent of the normal generation time), after which steady-state growth is resumed at a new and slower rate. Since the temperature remained constant throughout this experiment, it is concluded that light alone affected the growth rate. Upon transfer to light, the proportion of cells in

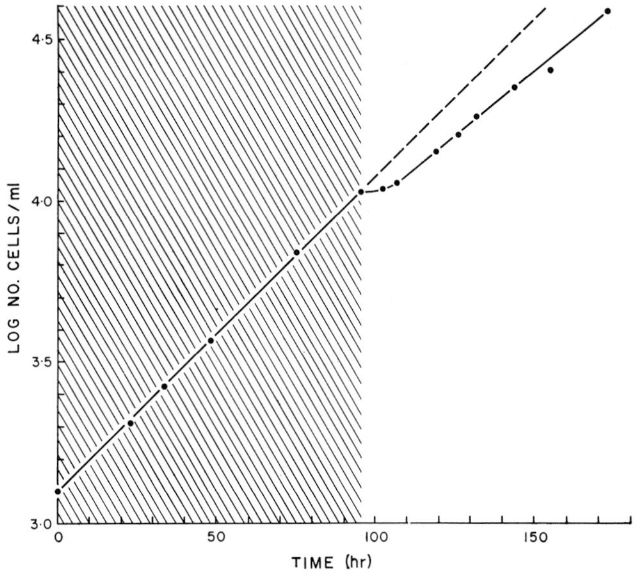

FIG. 6. *Euglena gracilis* grown first in the dark (shaded portion) and then transferred to light of 300 foot-candles intensity. Cramer–Myers salt medium (1952) with acetate as a carbon source; 20°C. (Explanation in text.)

fission falls from a normal value of about 3·5 per cent to less than 0·5 per cent, the lowest level occurring after about 4 hr in the light (Cook, 1960). This result could only arise if those cells already in fission at transfer to light completed division, while no new cells commenced fission. In these cells, fission requires about 30 min; the entire process of mitosis, according to Leedale (1958), lasts 2–3 hr, depending on the species. These time relations imply that light is most inhibitory for division at an age of the cell shortly preceding prophase. Leedale (1959) has more definitive proof of this: when cells are exposed to light after the initiation of prophase, no inhibition is observed; when exposed shortly before prophase, division is delayed. There seems no doubt that visible light has an immediate inhibitory effect on mitosis and division of the flagellates. Inhibition of growth processes, though no less clear, may be rather more delayed; indeed, it has been shown that the RNA content of these cells can be increased to as much as 50 per cent above normal levels by sudden exposure to light. This response is transitory, and RNA levels normally observed are soon re-established (Cook, 1960). Such transitory phenomena, however, suggest the interesting possibility that rates of biosyntheses usually associated with processes of cellular *growth* may instead be immediately affected by those more directly related to mitosis.

Since mitosis of the flagellates is differentially inhibited by visible light, it is not surprising that division of these cells should occur with a greater probability in the dark. When grown on a repetitive light–dark cycle, where the length of each cycle is approximately equal to the generation time, it follows that cell divisions would soon be confined to the dark periods, and the increase in population number would thus be 'rhythmic' or 'phasic'. The large number of references in the literature to phasic division activity of many photosynthetic forms suggests that the inhibitory effects of visible light may extend beyond the flagellate protozoan cells.

2. *Minimal Light Requirements for Optimal Growth of Euglena*

With *Euglena gracilis*, an estimate of the proper light–dark ratio was found for populations growing at 20°C. Axenic cultures of *Euglena* were grown in cylindrical flasks and continuously stirred. The salt medium used was that of Cramer and Myers (1952), identical to the medium used for *Astasia* except that acetate was omitted, so that nutrition was autotrophic. A thin plastic water jacket surrounded the culture flask, and circulating water maintained the temperature constant. The flask was lighted from two sides at a total intensity of 130 foot-candles. (A light intensity of 300 foot-candles does not increase the growth rate, indicating that the intensity chosen was saturating for growth.) The cells were first cultured in continuous light and then transferred, in the middle of the logarithmic growth phase, to complete darkness. Following this, the multiplication rate gradually falls to zero. In the experiment shown in Fig. 7, the last cells to complete division do so about 6·2 hr after transfer to the dark. These particular cells are assumed to have had just sufficient energy and metabolites stored at entry into the dark to complete division. Since the generation time in the light was 19·9 hr, the final group of cells dividing must have spent the first 13·7 hr of their life in the light. Those cells which had received a longer term of light will have divided in the interim; those receiving less will not have divided at all. It is concluded that a recurrent light–dark regimen of 13·7-hr of light and 6·2-hr dark would permit all cells to complete their generation in each light–dark cycle at this temperature;

furthermore, divisions can be expected to be confined for the most part to the dark periods.

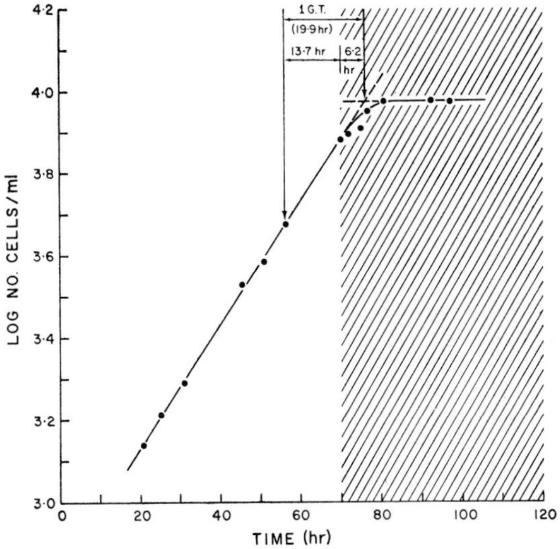

FIG. 7. *Euglena gracilis* grown in constant light (130 foot-candles) and then transferred to the dark (shaded portion), all at 20°C. The generation time in the light was 19·9 hr; multiplication ceased about 6·2 hr after transfer to the dark. (Explanation in text.)

3. *Synchronous Division on a Light–Dark Cycle*

In other studies, the generation time of *Euglena* at 20°C was found to vary between 20–22 hr. With the view that a doubling time of 24 hr does not differ greatly from these, it was decided for convenience to use a diurnal time scale. The light–dark ratio, which is proportional to the one derived above, but covers a 24-hr period, is approximately 16 hr of light to 8 of dark; rounding-off of the values seemed justified by the assumptions used in its derivation. On such a cycle, *Euglena gracilis* was found to divide synchronously in the dark, as expected (Cook and James, 1960). As with *Astasia longa*, addition of small amounts of sulfhydryls (methionine and cysteine) results in a sharper burst, such as is seen in Fig. 8. Cell divisions start about 3 hr after entry into the dark period, and extend about 2 hr into the subsequent light period. Over 95 per cent of the cells divide in each burst at the lower population densities, although this value is reduced as the culture approaches levels of the stationary phase. When diluted with fresh medium once in each cycle, the pattern of synchronous division can be extended for many cycles, much like that of *Astasia longa* in Fig. 2, and that for *Chlorella* in Lorenzen's method (chapter 25, this book). About 6 hr are required for completion of the burst.

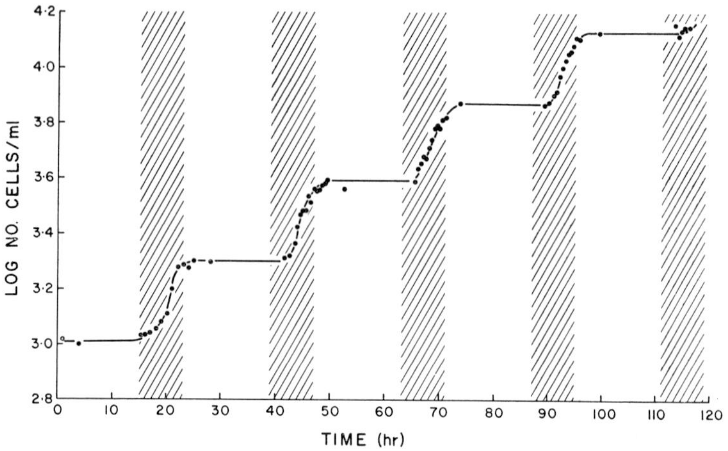

FIG. 8. Synchronous increase in a population of *Euglena gracilis* grown autotrophically at 20°C on a repetitive light–dark cycle. The 8-hr dark periods are indicated by the shaded areas; the light periods are 16-hr long. The ordinate is the logarithm to the base 10 of the cell concentration; at the lower population densities, each burst results in a near doubling of cell number.

While divisions of *Euglena gracilis* are confined to the dark, all measurable mass growth occurs during the light period. This has been described elsewhere (Cook, 1961a, b), and is discussed by James in chapter 11 of this book. With proper attention to the light–dark ratio, it is likely that *Euglena* could be made to divide synchronously at any constant temperature of choice (within the physiological range). A comparison of growth patterns and division behavior of these cells synchronized at several different constant temperatures could provide insight into responsible mechanisms. Comparisons with the closely related *Astasia longa*, synchronized with a different technique but grown on the same salt medium (save for the different carbon sources), also holds promise. These considerations are discussed more fully by James.

REFERENCES

BARNER, H. D. and S. S. COHEN (1955). Synchronization of division of a thymine-less mutant of *Escherichia coli*. *Federation Proc.* **14,** 177.

BERNSTEIN, E. (1960). Synchronous division in *Chlamydomonas moewusii*. *Science* **131,** 1528.

BLUM, J. J. and G. M. PADILLA (1962). Studies on synchronized cells. The time course of DNA, RNA and protein synthesis in *Astasia longa*. *Exptl. Cell Research* **28,** 512.

BUETOW, D. E. (1961). Variation of the respiration of protozoan cells with length of centrifuging. *Anal. Biochem.* **2,** 242.

BUETOW, D. E., G. M. PADILLA and T. W. JAMES (1959). Effects of low accelerative force on cell growth (*Astasia longa* J.). *J. Protozool. Suppl.* **6** (Abstract No. 87).

COOK, J. R. (1960). Light-induced division synchrony of *Euglena gracilis* and the effects of visible light on related cells. Ph. D. Dissertation, Univ. Calif., Los Angeles, Calif.

COOK, J. R. (1961a). *Euglena gracilis* in synchronous division. I. Volume and mass characteristics. *Plant and Cell Physiol.* (*Tokyo*) **2,** 199.

COOK, J. R. (1961b). *Euglena gracilis* in synchronous division. II. Biosynthetic rates over the life cycle. *Biol. Bull.* **121,** 277.

COOK, J. R. and T. W. JAMES (1960). Light-induced division synchrony in *Euglena gracilis* var. *bacillaris*. *Exptl. Cell Research* **21,** 583.

CRAMER, M. and J. MYERS (1952). Growth and photosynthetic characteristics of *Euglena gracilis*. *Arch. Mikrobiol.* **17,** 384.

HUTNER, S. H. and L. PROVASOLI (1955). Comparative biochemistry of flagellates. In *Biochemistry and Physiology of Protozoa*, Vol. II (S. H. Hutner and L. Lwoff, eds.). Academic Press, New York, pp. 17–43.

JAMES, T. W. (1954). Quantitative studies on the ribonucleic acid content in *Amoeba proteus* in relation to the nucleus and the division cycle. Ph. D. Dissertation, Univ. Calif., Berkeley, Calif.

JAMES, T. W. (1960). Controlled division synchrony and growth in protozoan microorganisms. *Ann. N.Y. Acad. Sci.* **90,** 550.

JAMES, T. W. and G. M. PADILLA (1959). Physiological and size changes in protozoan cells in response to incubation temperatures. In *Proc. Natl. Biophys. Conf.* 1*st* 1949 (H. Quastler and H. J. Morowitz, eds.). Yale University Press, New Haven, p. 694.

LEEDALE, G. F. (1958). The time-scale of mitosis in the Euglenineae. *Arch. Mikrobiol.* **32,** 352.

LEEDALE, G. F. (1959). Periodicity of mitosis and cell division in the Euglenineae. *Biol. Bull.* **116,** 162.

LEVEDAHL, B. H. and J. J. DREHER (1961). Effects of sonic vibrations on single cells. *Fifth Annual Biophysical Society Meeting*, St. Louis, Mo. (Abstract TE-3).

LOWRY, O. H., N. J. ROSENBOUGH, A. L. FARR and R. J. RANDALL (1951). Protein measurement with the Folin Phenol reagent. *J. Biol. Chem.* **193,** 265.

PADILLA, G. M. (1960). Studies on the synchronization of cell division in the colorless flagellate *Astasia longa*. Ph. D. Dissertation, Univ. Calif., Los Angeles, Calif.

PADILLA, G. M. and T. W. JAMES (1960). Synchronization of cell division in *Astasia longa* on a chemically defined medium. *Exptl. Cell Research* **20,** 401.

SCHMIDT, G. and S. J. THANNHAUSER (1945). A method for the determination of desoxyribonucleic acid, ribonucleic acid, and phosphoproteins in animal tissues. *J. Biol. Chem.* **161,** 83.

SWEENEY, B. M. and J. W. HASTINGS (1958). Rhythmic cell division in populations of *Gonyaulax polyedra*. *J. Protozool.* **5,** 217.

Outlook

DANIEL MAZIA, *Department of Zoology, University of California, Berkeley, California*

To those future readers of this book who are sensitive to the historic currents in a special field of a science—and after a few years the inevitabilities of progress will see to it that there are no other readers—it will seem that in 1963 a great deal of effort was being spent on the synchronization of cell populations with very limited ends in view. Synchronization will seem to have been purely a tool for obtaining samples of dividing cells, or of cells preparing for division, for the purpose of mass analysis. But such an historically-minded reader—or even the present reader—would be wrong to imagine that the developments he has been reading about were among the routine instances in which the appearance of a tool created a new line of investigation.

Rather, the study of synchrony reflected a genuine shift of focus in Cell Biology during the 1950s. The prior period had been one during which the study of the cell had undergone a renaissance. The fruits of a great period in Biochemistry, the discovery of the *biological* meanings of the nucleic acids, of oxidation and phosphorylation, of structural macromolecules, were being fed back into cytology. Confidence in the new morphology that was being created by the pioneer generation of electron microscopists and other devotees of fine structure was beginning to crystallize. In short, an idea of 'the cell' was coming into being, a more solid, more stalwart, vastly enriched transmutation of Flemming's cell or Wilson's cell. But, dynamic and structured as it was becoming with respect to its composition and metabolism, this cell was rather static in the dimension of time; it existed without birth, death, or history.

When, in 1956, I published a lecture called *The Life History of the Cell*, it still seemed necessary to advocate and defend the necessity of research that would 'define as exactly as possible the events that characterize a given phase of the life history, and especially to discover the events that drive the cell through time, from one phase of its history to the next'. Information was scanty then, and the measurement of the growth-curve of a cell could be presented as a novelty and as a *tour de force*.

By now, we possess a good deal of information about the course of events in the cell cycle, and a major part of this information, though not all, has come from the exploitation of synchronized cell populations. Other methods, notably the ingenious use of pulse-labelling in autoradiographic studies, have led to important findings, but these findings whet our appetite for biochemical detail in proportion to their importance. And where we need biochemical detail, we must rely on synchronized systems. If, then, we were to summarize the present and future of research on the technique of cell synchronization, we could only say, rather platitudinously, that it is justifying itself and that more work toward better techniques on a greater variety of systems is needed. The greatest need is a rather obvious one: a really good system of synchronization for large populations of some plant or animal cells which divide by mitosis. It is frustrating that the best techniques for obtaining synchronized dividing cells have been developed for cell types in which we have the poorest access to chromosomes and nuclear division.

So far, and not surprisingly, most of the applications of synchronized populations have been descriptive, hardly a pejorative assertion when we think of how much more descriptive information we need. Consider an obvious example. We know more about the time relations of DNA synthesis than we do about most aspects of the cell cycle, and can make rather positive statements about the S-period. A good deal is known about the biochemistry of DNA synthesis. But we do not yet possess the facts necessary to decide whether the onset of S is determined by a transformation of the chromosomal DNA to a 'primer' state (in its most general sense), a synthesis or a translocation of polymerase, the appearance of substrates or of the enzymes making substrates, or whether we shall find that the regulatory devices of the cell are so tightly co-ordinated that in fact all of these factors of DNA synthesis go into action essentially at the same time. What little is known of the timing of the appearance of enzymes of DNA synthesis comes from work on synchronous systems, such as the work of H. Stern on the naturally synchronized lily anthers. I use this as an example of the kind of questions we shall soon be asking. Are we to think of a phenomenon such as the Gl period as a pause, waiting for some genetic clock to strike, or shall we find a smooth accumulation of events whose terminal incident we happen to have detected first as the beginning of the S-period? And where we are dealing with a complex synthesis that is well-located in time—one that involves a primer, a polymerase, and a variety of substrates each formed by separate pathways—how closely will the several kinds of biochemical events be regulated to each other?

As we ask such seemingly descriptive questions, and questions which seem to represent mere exploitations of synchronized populations, we see the shape of newer questions that are far from descriptive. These are the questions that have to do with the nature of synchrony and asynchrony itself. The problem of synchronization is given by the fact that cells, even the progeny of a single cell, get out of step in a few generations. We express no surprise at this, for we are accustomed to the idea of variability at every level in biology. The theory of methods of experimental synchronization assumes that the various channels of progress of a cell through its biography are separate to some degree and experimentally separable. We try either to bring them into step (for purposes of true synchrony through the cycle) or to make a controllable process limiting (for the purpose of synchronizing division itself). But we are now also becoming accustomed to the idea of invariability or limited variability. We trace all biosynthetic regulation back to a very stable genome with its sites of structural information and of control. We find feedback controls where we seek them. All told, the present story of cellular controls would tend to predict constancy of generation-time in the progeny of a cell maintained in a constant environment. Shall we find that the inconstancy is a measure of the quantitative limitations of such controls? Shall we discover the sites of 'noise', processes involving unique molecules, which will amplify the consequences of true physical randomness? Shall we find that the control of amounts of enzymes—which is all we know about now—is not an accurate reflection of the control of the process catalyzed by these enzymes, that other, perhaps morphological, variables are less well-regulated?

To learn how the amount of an enzyme is regulated is to ask how the several enzymes of a pathway are regulated to each other. To answer the latter question is to ask how different pathways are regulated to each other, and the inquiry builds up to an inquiry into the whole cell and its biography. And, since it is an inquiry into variability, it addresses itself to the fundamental questions of synchrony. Thus it may well be that the future and deeper study of cell synchronization will out-date the notion of synchronization as a valuable tool for miscellaneous cell-biological research. We are likely to see more and more research on synchronization as a problem in itself, and as a most promising avenue of advance in the study of regulation from the level of the enzyme to the level of the cell in Time.

SECTION IV

TECHNICAL PROCEDURES

A.	Protozoa	Chap. 23
B.	Algae	Chaps. 24–26
C.	Fungi	Chap. 27
D.	Bacteria	Chaps. 28–29
E.	Tissue cells	Chap. 30

23

Techniques Used in the Study of Synchronous *Tetrahymena*

PAUL PLESNER, LEIF RASMUSSEN and ERIK ZEUTHEN,
Biological Institute of the Carlsberg Foundation, Copenhagen, Denmark

CONTENTS

I General Techniques—Rasmussen and Zeuthen 544
1. Media 544
   A. Proteose peptone 544
   B. Inorganic medium 544
2. Stock cultures 544
3. Experimental cultures 545
4. Growth before heat treatment 546
5. 'Synchronizer' (small scale) 548
   A. The periodic temperature bath 548
   B. The shaker 550
6. Transfer of cells to inorganic medium 552
7. Fixing the time point of maximal synchronous division . . 552
8. Methods for cell counting 554
   A. The Sedgewick–Rafter counting chamber in combination with the Whipple micrometer 554
   B. The 'Celloscope' electronic cell counter (Lars Ljungberg, Stockholm, Sweden) equipped with a hole of diameter 125 μ 554
II Techniques Used for Large Cultures—Plesner 555
1. General considerations 555
   A. The medium 555
   B. The density of the cell suspension 556
   C. Centrifugation of cells 556
   D. The true time of observation 556
   E. The number of points of observations 557
2. 'Synchronizer' (large scale) 557
   A. Principle 557
   B. The temperature unit 558
   C. The incubator 559
   D. Procedure 560
   E. Inoculum 560
3. Replacement of medium and collection of cells 561
   A. Outline of procedure 561
   B. Details of procedure 561
References 562

I. General Techniques

1. Media

A. Proteose peptone

Proteose peptone (Difco Laboratories) will support the growth of Tetrahymena in concentrations down to perhaps 0·01 per cent. We have found shortest generation times with concentrations of 1–2 per cent proteose peptone, fortified with liver fraction L (Wilson Laboratories, Chicago 9, Illinois). In routine practice the following media are used: 2 per cent PP, 2 per cent PP plus 0·1 per cent L, and 2 per cent PP plus 0·4 per cent L. The liver fraction extends the logarithmic growth phase and ultimately gives higher yields. At low population densities, however, it does not shorten the generation time appreciably. Salts are added (though without any real justification) as in Kidder and Dewey's synthetic medium A (Kidder and Dewey, 1951) omitting the phosphates. Thus 1 l. of final medium contains 10 ml of each of the following stock solutions, No. 1: 10 g $MgSO_4 \cdot 7H_2O$, 2·5 g $Fe(NH_4)_2(SO_4)_2 \cdot 6H_2O$, 0·05 g $MnCl_2 \cdot 4H_2O$ and 0·005 g $ZnCl_2$ per liter; No. 2: 5 g $CaCl_2 \cdot 2H_2O$, 0·5 g $CuCl_2 \cdot 2H_2O$ and 0·125 g $FeCl_3 \cdot 6H_2O$ per liter. Procedure: The PP, L and salts are mixed in the required volume of water, dissolved by heating to 80°C, filtered and distributed to the culture flasks. These are plugged and then autoclaved at 1·5 atm ($\sim$112°C) for 20 min.

Recipes for synthetic media are given by Kidder and Dewey (1951) and Holz *et al.* (1961).

B. Inorganic medium

The inorganic medium (Hamburger and Zeuthen, 1957) in which synchronized cells will perform division consists of 2·75 g NaCl, 0·25 g $MgSO_4 \cdot 7H_2O$ made up to 1 l. with 5 m*M* potassium phosphate buffer of the desired pH, usually 7·0 (= pH of PPL). Glass distilled water is used. (Copper ions are toxic when present in the inorganic medium.) Since this medium would form a precipitate of magnesium phosphate on heating, for daily purposes it is used non-autoclaved and stored cold.

2. Stock Cultures

Stock cultures are maintained in 5-ml proteose peptone medium in slanted test tubes at 28°C. They are prepared daily by inoculating with 10–20 μl from a one-day old stock. Thus, the cultures can be kept in continuous logarithmic growth. If this is not essential the stocks may be kept in upright test tubes at room temperature and transferred every 3 days.

Permanent stocks are maintained on rat brains—one quarter of a brain per 5 ml tap or distilled water in a test tube. Intestine and probably other organs can also be used. The water is covered with a 10-mm layer of paraffin oil and autoclaved. The culture is inoculated lightly by piercing the oil with the transfer pipette. The preparation keeps for at least a year at room temperature.

3. *Experimental Cultures*

10-*ml medium.* 100-ml Erlenmeyer conical flasks containing 10-ml medium covering the bottom with a layer 4 mm deep have been used. Aeration by shaking is not necessary for logarithmic growth over many generations. Synchronization can also be performed without simultaneous shaking (Scherbaum and Zeuthen, 1954).

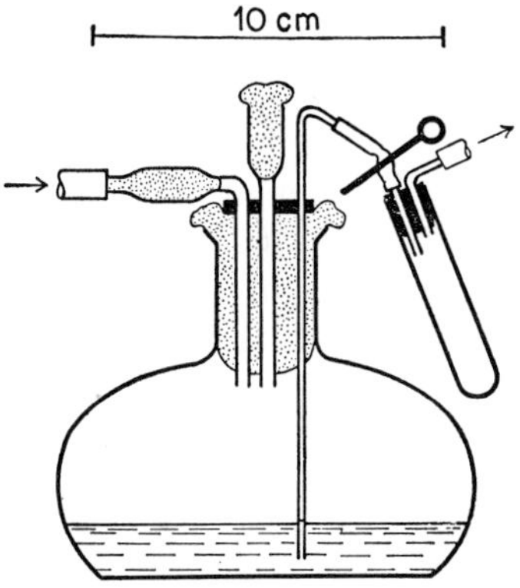

FIG. 1. Culture flask, 450 ml (150 ml medium). From left to right are shown the inlet aeration tube, the inoculation tube and the sampling siphon with small container attached. (From Scherbaum and Zeuthen, 1955.)

150–750-*ml medium.* 450-ml Fernbach culture flasks are used, with 150-ml medium forming an 18-mm layer, area 100 cm^2 (Fig. 1). The inoculated flasks are submerged in a temperature-controlled water bath to a level above that of the culture medium. The flasks are given a horizontal rotatory movement on a shaker (see section 5). The culture medium is thereby kept in constant motion. A stream of humid, sterile air renews the gas phase of the flask at a rate of approximately 500 ml per minute. The neck of the culture flask is equipped with a cotton plug through which are inserted inoculation and aeration tubes and a siphon for removing samples by suction into small test tubes, while preserving the sterility of the culture. The siphon is short (150 mm) and narrow (1·5 mm) in order that the amount of medium (to be discarded) necessary to flush it before sampling should be as small as possible. Similar flasks, 2 l. in capacity, have been used to accommodate larger volumes. In these 500 (750) ml medium makes a layer 30 (42) mm deep, covering 200 cm^2. With the small type of 'synchronizer' (section 5) three 150-ml cultures or two 750-ml cultures can be synchronized simultaneously. When more than 1·5 l. of synchronized population is needed the large type (10 l.) synchronizer can be used.

4. *Growth before Heat Treatment*

In the experiments demonstrated in Fig. 2 (Hamburger and Zeuthen, 1960) we inoculated the flasks with cells from a 4-day old upright, stationary test tube culture (5 ml) grown at 28°C. In the lag phase (2–3 hr) the population density was 200–300 cells per ml. During the ensuing population growth at 28°C, the cell count increased logarithmically through about 10 generations (for 25–30 hr) to about 3×10^5 cells per ml. Then followed a period of negative growth acceleration (6 hr) in which the population increased to a stationary density of 10^6 cells per ml. Cultural growth was followed by cell counts with results similar to what has already been described (Scherbaum, 1957a; Scherbaum and Zeuthen, 1954) and by determinations of dry weight and total phosphorus of cells. In Fig. 2, results from seven experiments are combined. The time scale (abscissa) shows in a generalized manner how the growth phases followed each other: Lag (I), log (II), negative growth acceleration (III) and stationary phase (IV). Curve A (dry weight per cell) shows that the cells in the inoculum are fairly large and tend to grow slightly in mass also during the lag. They become smaller again in the course of the first four to five generations of exponential growth. From now on they keep almost constant dry weight during the remainder of the log phase; however, at the time of the early stationary phase the average cell has again accumulated dry weight. This increase may indeed largely reflect piling of glycogen.

The total phosphorus per cell (curve B) increases during the lag and early log phases. With continued cell division the phosphorus per cell is again reduced, so that—as for the dry weight—a steady level is established after three to four generations of logarithmic growth. By the time of the stationary phase, the content of phosphorus per cell has dropped markedly. In this phase total phosphorus may follow cell volume more closely than cell dry weight. During the lag and the early log phase the phosphorus/dry weight ratio has established a balance which is maintained throughout the remainder of the log phase (curve C).

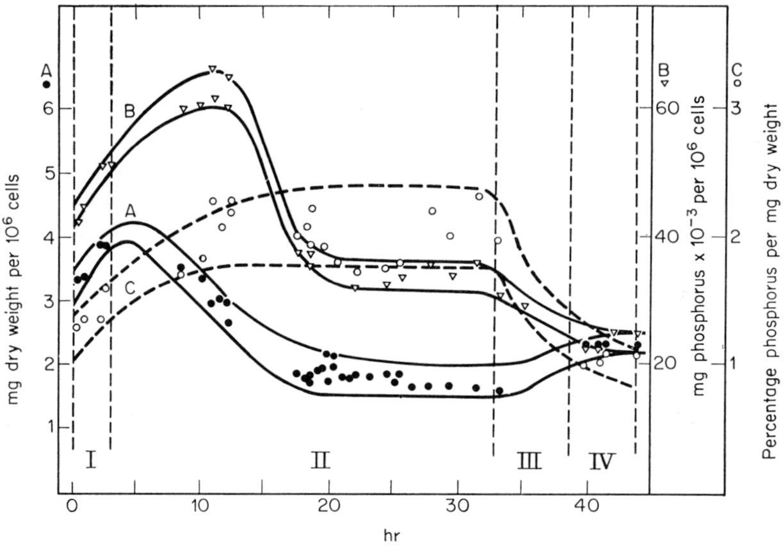

FIG. 2. Cell characteristics during a population growth cycle. I: lag phase, II: logarithmic growth phase ($< 3 \times 10^5$ cells per ml), III: phase of negative growth acceleration (3×10^5 to 10^6 cells per ml), and IV: stationary phase (10^6 cells per ml). (From Hamburger and Zeuthen, 1960.)

If the effects of the synchronizing treatment on chemical and morphological cell parameters are to be studied cleanly (Elliott *et al.*, 1962), it is essential that they are not observed on the background of continuous cellular changes, such as those (demonstrated in the first part of Fig. 2) which occur before the balanced growth situation is established. Except in the study from which Fig. 2 is taken this fact has not been realized in the past. This places doubt on the value of the so-called 'logarithmic controls' often referred to in the literature.

The population density when the synchronization is initiated is only slightly lower than at the time when the first synchronous division begins. In PP, 0·1 per cent L, this density should not much exceed 50,000 cells per ml. With somewhat higher densities (100,000 cells per ml) a good first division may develop, but the second, and especially the third will suffer. When an experiment is planned it is essential that these facts and those observed in Fig. 2 are kept in mind. The decisions to be made concern the duration of the logarithmic growth phase before the initiation of the synchronizing treatment and the population density required in the synchronous division steps. The right inoculum (past history, number of cells) has to be chosen. We consider that after inoculation there is a lag of 2 hr and that henceforth the generation time (at 28°C) is 2·5 hr. The population density 18 hr after inoculation can be predicted to $\pm$ 50 per cent or better. In present routine practice we make inoculations at 1 p.m. using 1-day old *logarithmic* cultures (slanted test tubes) containing about 25,000 cells per ml. Logarithmic growth then proceeds for 12 hr before the temperature shocks begin at 3 a.m. The cycling temperature treatment can be terminated automatically or manually at 8^{20}, 9^{20} or 10^{20} (20′/40′ system) or at 8^{30}, 9^{30} or 10^{30} (30′/30′ system). Synchronous division takes place 80–90 min later.

5. '*Synchronizer*' (*Small Scale*)

In this laboratory we operate two types of synchronizer, small and large scale. In both cases there is an aerated culture vessel (those for the small scale described in section 3), an arrangement which controls and records the temperature of a periodic thermostat and a device for agitating the cell population (not required with 10-ml cultures).

A. The periodic temperature bath

(*a*) Constant flow of cooling tap water—Fig. 3 (Scherbaum and Zeuthen, 1955) is a wiring diagram. The contact thermometer, 7, is set to regulate at the higher temperature. A second contact thermometer, 9, is mounted in the bath and is set to control the temperature at the lower level. A signal watch, 10, and a relay, 11, switch the contact thermometer, 9, in and out of the circuit, regulating the temperature. The contact thermometer, 7, takes control whenever 9 is out of the circuit. 12 represents the heating coil, 13 is an ammeter and 14 is a relay. The heater works against the cooling of a properly adjusted, continuous flow of tap water (temperature 11–13°C throughout the year in Copenhagen) through a coil immersed in the water bath. The water in the bath is well stirred with a motor-driven propeller.

(*b*) Intermittent flow of cooling tap water—In the diagram (Fig. 4), 1 is a relay which controls the temperature of the bath at the levels set by 2, the contact thermometer. 3 is a 1000 W heater,

4 a lamp to indicate 'off' and 'on' of this heater, and 5 a magnetic valve in the cooling tap water coil. The valve, controlled by a 'normally-open' relay, 6, is open when the heater, 3, is 'off'. 7 and 8 indicate a water-saving device (lightly drawn). When water economy

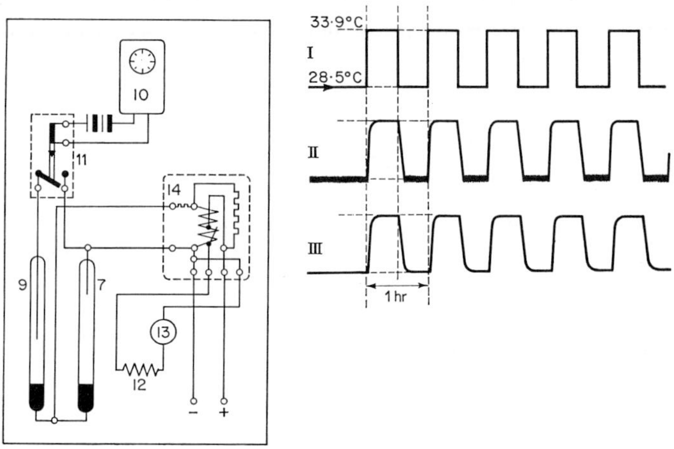

FIG. 3. Left part: Wiring diagram of the synchronizer which is cooled by a constant flow of tap water. Right part: Temperature curves. Curve I: the changing levels set by the signal watch. Curves II and III: Temperature changes recorded in the water bath, and in the culture flask, respectively. (From Scherbaum and Zeuthen, 1955.)

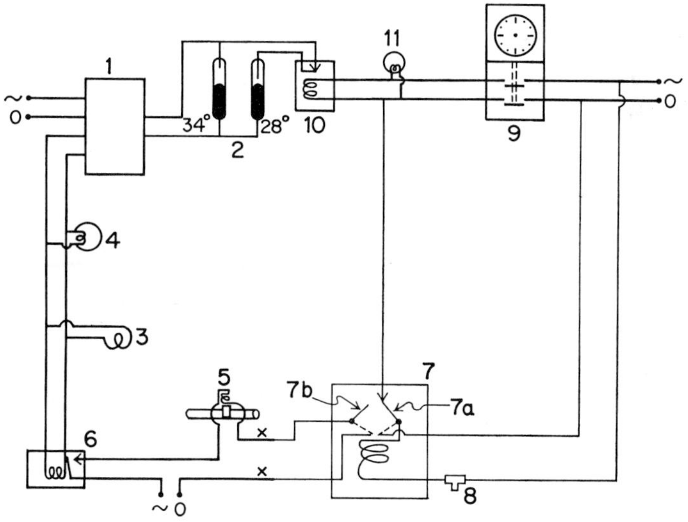

FIG. 4. Wiring diagram of the synchronizer which is cooled with an intermittent flow of tap water. (See text for explanation.)

is not a problem, this part can be eliminated and replaced by a simple connection from x to x. Operation of 7–8 is carried out as follows: When a culture is to be grown at constant 28°C for subsequent synchronization the spring-loaded push-button, 8, is pressed, the switch in the control clock, 9, being in the position drawn. This blocks the flow of tap water at 5. Hours later, when the temperature cycling is initiated, the first signal from 9 activates the double-poled, double-throw relay, 7, with the result that the circuit which controls the magnetic valve becomes established. 9 is an electrical control clock ('Electro-Boy', Firma Hugo Müller, Schwenningen am Neckar, Sturmbüllstrasse 145–149, W. Germany) which connects and disconnects the power supply to the 'normally-open' relay, 10. When 10 is open the temperature is regulated at 28°C. This situation is shown in the diagram. The indicator lamp, 11, is then unlit. Using the 'Electro-Boy' the temperature regimen can be pre-set for a period of 24 hr with smallest intervals of 20 ± 2 min.

(*c*) Cooling using a compressor—When we synchronize by applying cold shocks (7·5–10°C, cf. chapter 4) instead of heat shocks we use the equipment similar to that shown in Fig. 4. However, the magnetic valve, 5, is replaced by an electric water pump which intermittently circulates cooling water through a closed circuit from the cold bath unit of a portable compressor to the water bath of the synchronizer. A compressor has also been used by Lee (1959) (Fig. 5) to obtain cooling in the synchronizer cycling between 28° and 34°. All our baths are supplied with temperature recorders. These can be either thermistor-millivolt recorders ('Varian') or inexpensive mechanical recorders (see Fig. 6).

B. *The shaker*

The mechanical power is supplied by a 50 W single-phase synchronous A.C. motor (type D-73, H. Heidolph, Schwabach B, Nürnberg, Western Germany) with a continuously variable mechanical gear (type RZR, also from H. Heidolph). Inspection of Fig. 6 will give some idea of how the shaking is arranged. The three flasks are given a rotatory movement in the horizontal plane. The shaking rate (90 rotations per min, amplitude 13 mm for 150-ml medium) can be adjusted on the gear (not shown in the figure) and the amplitude at the level where the axis of the motor transfers its power to the shaker.

Further details may be gained from Fig. 6 and its accompanying caption.

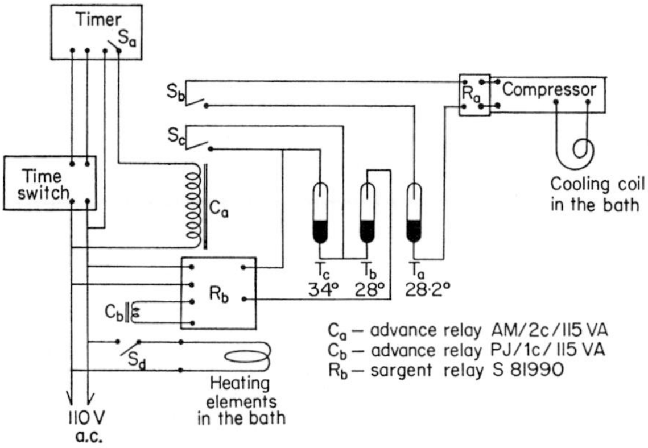

FIG. 5. Diagram of a synchronizer cooled by a compressor. (From Lee, 1959.)

FIG. 6. The small scale synchronizer. In the center, the bath (containing 7 l. of water), the stirring motor, the three flasks with their aeration tubes, the heater and, to the right of it, the thermometers and the cooling coil. To the left of the bath, the shaker and the mechanical temperature recorder. In the upper right corner are the control clock and an indicator lamp.

6. *Transfer of Cells to Inorganic Medium*

When the synchronized population is to be studied in an inorganic medium the cells are transferred by three or four washings in a centrifuge. This is done immediately after a heat shock. When the washing is carried out at room temperature, a hand-centrifuge (4 × 15 ml) is used. For each collection of cells a centrifugal force of 200 × *g* is applied for 20 sec. After removal of the supernatant down to approximately 1 ml the cells are quickly resuspended with a jet of inorganic medium. When necessary the washing can be made quantitative (Hamburger and Zeuthen, 1960), but for this purpose we recommend rapid chilling of the culture (< 500 ml per flask) to below 10°C in an ice bath, followed by use of a refrigerated centrifuge. We employ an M.S.E. Major (8 × 30 ml) and spin for 2 min at 150 × *g*. After removal of the supernatant down to 2 ml we resuspend in cold medium. After the final washing the cells can be concentrated five times in the inorganic relative to the organic medium (i.e. to 500,000 as compared with 100,000 cells per ml) without any loss of synchrony in the first division. The cells are reincubated in the synchronizer and given one or two more temperature shocks (Hamburger and Zeuthen, 1957). The time point at which the last of these shocks is terminated is called EH (End of Heat treatment). This time is defined by the switch of the clock from the 34° to the 28°C contact thermometer. Any harm done to the cells by the washings (delayed divisions by anaerobiosis during the packing of cells or by chilling) is completely corrected for by the extra temperature shock(s).

7. *Fixing the Time Point of Maximal Synchronous Division*

For a great many purposes the developmental stage of the culture is best measured on a time scale which extends from EH to divisions 1 or 2. The exact time of maximal synchronous division can be fixed by frequent (e.g. every 5 min) estimations of the division index (dividers/total number of cells) in the culture. This can be done visually because the process of cytoplasmic cleavage is so conspicuous. The estimates are plotted against time and the position of the 'division maximum' is derived graphically as the time point corresponding to the highest division index (usually 85 to 90 per cent). Proper controls indicate that a single observer fixes division maxima to within ± 3 min. Although several observers may estimate the division indices differently they will all place the division maximum in an almost identical position on the time scale (i.e. within ± 5 min). Many samples (up to 25, each of 1 ml containing approximately 20,000 cells) can be handled simultaneously by one person (Fig. 7). Thus, in one experiment it is possible to study the effects of an agent over a wide concentration range or of many agents in less varied concentrations, or to investigate antagonisms between paired agents varied over wide concentration ranges.

FIG. 7. Recording of the development of division maxima in more than twenty differently-treated samples, removed from the main culture at time EH. The dishes are placed on a dry table thermostatically regulated so that the sample temperature is 28°C.

The effect of an agent is measured in terms of delay of the synchronous division. With long delays the division indices may rise and fall less sharply; at the same time the division maxima consequently become lower. We recommend that conclusions are not based on observed maxima of intermediary values, with which there is the risk that a fraction of the population may fail to divide at all. If necessary, the concentration of the agent can be lowered to sharpen and increase the division maximum, or raised to block division completely.

A recent report (Rasmussen and Zeuthen, 1962) is concerned with the study of cell division, but we want to point out that if the division is used as an indicator of chemical effects, this method, which is fast and reliable, offers great promise as a tool for studies with the emphasis on biochemical antagonisms.

8. *Methods for Cell Counting*

A. The Sedgewick–Rafter counting chamber in combination with the Whipple micrometer

Using the method described in section 3, two parallel samples of 3 ml each were drawn from the vigorously shaking culture, whose density was to be measured (Scherbaum, 1957b). The samples were thoroughly shaken and from each two further samples of 0·7 ml were removed using a wide-mouth Carlsberg constriction pipette. 0·1 ml of 20 per cent formaldehyde was added making a final concentration of 2·5 per cent. The fixed sample was mixed thoroughly, and poured rapidly into the chamber avoiding air bubbles. Counting was carried out a few minutes later when the cells had settled on the bottom of the counting chamber. Approximately 160 cells were counted. The population count was therefore represented by about 4×160 cells. The sampling procedure was meant to minimize errors due to uneven distribution of the cells in the original culture, in the (removed) samples and in the counting chamber. The standard error was within 3·9 to 5·6 per cent.

B. The 'Celloscope' electronic cell counter (Lars Ljungberg, Stockholm, Sweden) equipped with a hole of diameter 125 μ*

For sampling the population we use a pipette with a widebore tip. The sample is withdrawn at a standard rate from the vigorously shaken culture. Fixation solution is added (see Table I) and the mixture allowed to stand for at least 5 min. (Cells can be left overnight in fixation solution, but they must then be resuspended by vigorous shaking before the next step.) Dilution mixture, containing either PPL or inorganic medium according to the composition of the sample is then added (Table I). Before each count the suspension is agitated to distribute the cells uniformly. We perform at least four successful counts, discarding a reading if the counting time is longer than that characteristic of our instrument, or if there is any deviation from the normal image on the oscilloscope screen. The accepted readings are averaged and the background count given by the dilution mixture alone is subtracted. We read the corresponding value of 'cells per ml' from a previously constructed calibration

* The electronic counting procedure has been worked out in collaboration with Dr. Stuart D. Nachtwey.

table and multiply by the dilution factor to obtain the number of cells in the original sample.

TABLE I. Sample dilution scheme

Expected density (cells per ml)	Sample vol. (ml)	Fixation sol. (ml)	Dilution mixt. (ml)	Dilution factor
5000– 25,000	5·0	1·0	4·0	2
15,000– 75,000	2·0	0·4	7·6	5
25,000– 150,000	1·0	0·2	8·8	10
60,000– 300,000	1·0	0·2	18·8	20
150,000–1,000,000	1·0	0·2	48·8	50

Fixation solution: 6 g NaCl + 100 ml (filtered) formalin (40 per cent formaldehyde). Dilution mixtures: (a) 10 g NaCl + 160 ml (filtered) formalin + 840 ml PPL; (b) 10 g NaCl + 160 ml formalin + 840 ml inorganic medium.

II. TECHNIQUES USED FOR LARGE CULTURES

1. *General Considerations*

The amounts of material required for biochemical analysis of many cellular metabolites often exceed the amounts which can be obtained from cultures of the size employed in the standard techniques as described in section I. This is particularly so when studies are to be made of compounds showing a high rate of turnover.

It is believed that the points listed below should be considered before synchronization of divisions in large cultures is attempted.

A. The medium

Synchronized *Tetrahymena* are metabolically very sensitive to the nature of the surrounding medium. This offers an advantage in controlling the metabolic activity of cells dividing in phase but imposes very strict requirements on the investigator in the preparation of the cells for analysis if meaningful data are to be obtained. Thus, Hamburger and Zeuthen (1957) have reported a decrease in respiratory rate of 70 per cent when *Tetrahymena* cells were transferred from nutrient to inorganic medium. The decrease in respiratory rate is steepest immediately after the transfer (authors, personal communication). It is to be expected that generation of ATP by oxidative phosphorylation would decrease concomitantly if the P/O ratio remains constant. A consideration of the central role occupied by ATP in cellular metabolism leads to the assumption that the turnover of a number of low molecular weight metabolites will be affected by a replacement of the medium. Therefore, cells should be allowed to adapt to a change in the quality of the medium for at least 30 min prior to sampling. In addition, the controlled nutritional environment should be maintained during the harvesting procedures by omitting any washing of the cells.

B. The density of the cell suspension

The cell density should not exceed 50–70,000 cells per ml at the time when the synchronizing heat treatment is initiated if the highest possible degree of phasing of cell division is desired (page 547). However, as mentioned on page 552, if cells are transferred to inorganic medium prior to the last one or two heat shocks, the population density may be raised five to tenfold without any loss of division synchrony. The technical advantages of this become apparent when the following points are considered.

C. Centrifugation of cells

When centrifugation is used for harvesting *Tetrahymena* cells for biochemical analysis, the total time spent on centrifugation varies from 6 to 20 min depending on the procedure used (see chapters 6 and 7 of this book).

The cells are undoubtedly affected by the centrifugation. Thus, two consecutive centrifugations of synchronized *Tetrahymena* cells at $120 \times g$ each lasting for 5 min will cause a delay of 40 min in the occurrence of the division maximum (Plesner, unpublished results). Considering the amount of work which has been done on material collected by centrifugation it is unfortunate that so little is known about how the cellular metabolism is affected by centrifugation. We actually do not know if the centrifuged cells are representative of the live cells to which we extrapolate our results. Still, it seems reasonable to assume that the cells suffer from both lack of oxygen and from displacement of subcellular particles when they are packed into a solid pellet. In the present context it is an advantage that the centrifugation necessary to replace the medium can be performed prior to the last heat shock. The cells have time to recover from the centrifugation and the time of the first synchronous division is not delayed.

The reduction in volume when the cell suspension is concentrated 5–10 fold permits that fast-spinning, low-capacity centrifuges can be used for the final harvesting of the cells. It should, finally, be pointed out that the same pattern of changes in the content of nucleoside triphosphates is found whether centrifugation is involved in the analysis or not (see chapter 7).

D. The true time of observation

If the cells are killed at the conclusion of a centrifugal run and if it is assumed that the cells are delayed in their morphogenetic development during the centrifugation, then the death time cannot be the true time of observation. Thus, if a sample from a culture is killed at the conclusion (time t) of a centrifugal run it is not representative of the cells remaining in the culture at the time t but at the best at the time $t - t_e$ where t_e is the developmental delay ($\sim$ time error of observation) caused by the collecting procedure.

The short centrifugal runs which are used when sampling is done from concentrated cultures (see above) obviously reduce the time error to a minimum and makes sampling at short intervals possible; still, only a method which does not include centrifugation will permit a correct correlation in time between changes in biochemistry and morphology (page 199).

E. *The number of points of observations*

Tetrahymena cells pass definable morphogenetic stages of development and cell division rather quickly. If changes in cellular metabolism are related to such fast changes in morphology only several points of observation, at short time intervals, can demonstrate them. The samples for analysis should, furthermore, be obtained from the same experimental culture in order to exclude scatter from culture to culture.

These considerations form the basis for the technique which is used by the author for large scale production of synchronized *Tetrahymena* cells. It has been developed over a period of years in the Institute for Cytophysiology at the University of Copenhagen, The Fibiger Laboratory and The Biological Institute of the Carlsberg Foundation. Dr. Otto Scherbaum, then a guest at the Fibiger Laboratory, made the adaptation of the Lindahl incubator which will be described in section 2, C.

2. '*Synchronizer*' (*Large Scale*)

A. *Principle*

Large volumes of cell suspension are grown in incubators in which the temperature of the cell suspension is regulated by circulating water of a controlled temperature through a suitable radiator system. Water is pumped through the radiator from either of two thermostatic baths; one of them maintains 28°, the other 34°, in the cell suspension when the water from the respective bath is circulating through the radiator of the incubator. Thus, the apparatus appears functionally as two separate units, a temperature unit and an incubator.

B. *The temperature unit*

Fig. 8 illustrates the temperature unit. Two 50-l. thermostatic baths are each fitted with a pump (P1 and P2) and a solenoid valve (S1 and S2). The one-way valves (V1 and V2) prevent refluxes. The solenoid valves, which are wired in parallel to their respective pumps, direct the water back to the proper bath. A siphon (not shown) between the two baths corrects unavoidable small refluxes.

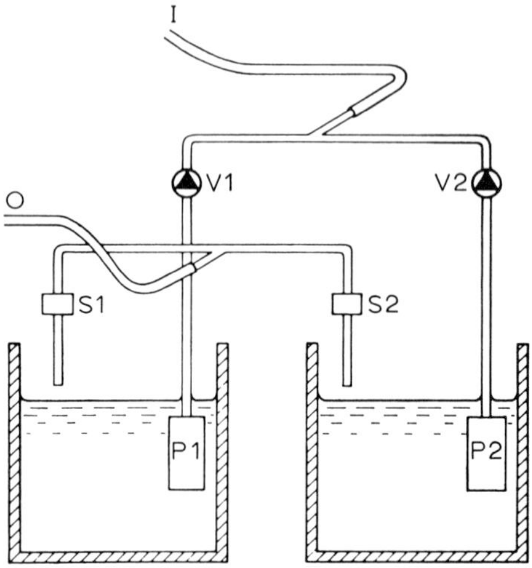

FIG. 8. Temperature unit. Two 50-l. thermostatic baths are each fitted with a pump (P1 and P2) and a solenoid valve (S1 and S2). V1 and V2 are one way valves. O and I are connected to an incubator (see Fig. 10).

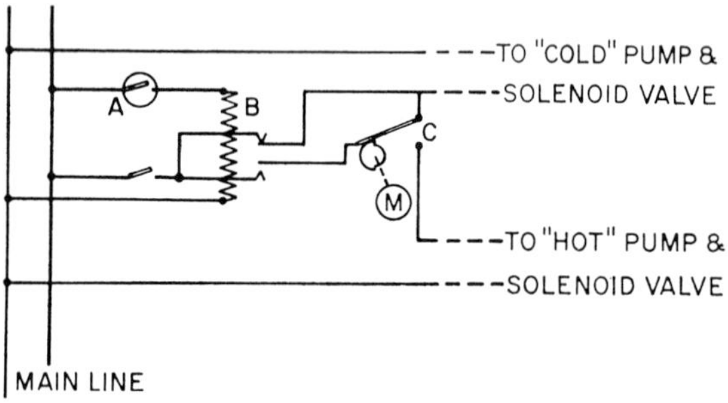

FIG. 9. Wiring diagram for pumps and solenoid valves of temperature unit. A is a pre-set time switch which starts the cycling heat treatment. B is the single pole, double throw relay. C is a single pole, double throw microswitch. The direction of the current through C is changed every 30 min by the synchronous motor M.

The temperature settings of the two baths are determined experimentally so as to compensate for heat losses through piping, etc. In the present example the settings were 28·5 and 34·5 respectively. The heat shocks are applied automatically by letting a microswitch direct the current to either one of the baths, as shown in Fig. 9. The microswitch, C, is opened and closed by a semi-circular disc driven by a synchronous motor (one RPH). The microswitch is shunted out prior to heat treatment by the relay B. At the time of the beginning of heat treatment the pre-set time switch A activates relay B and cycling of the temperature is started.

C. *The incubator*

The incubator described by Lindahl (1936) is used with the modification that the temperature of the cell suspension can be changed so as to make synchronization by heat treatment possible. The incubator is shown in Fig. 10.

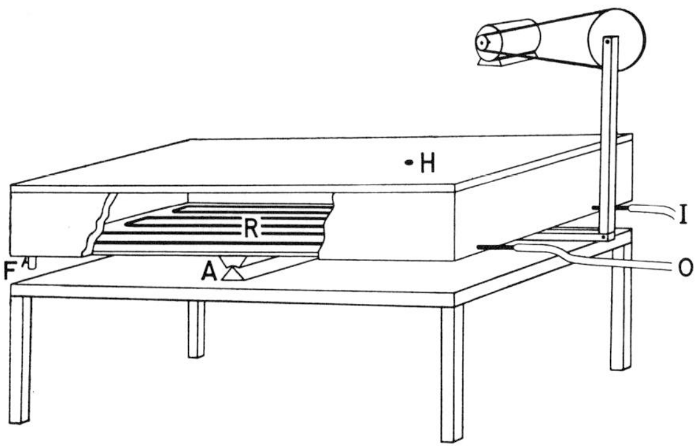

Fig. 10. Tray incubator. Water of controlled temperature is led through the radiator system (R) from I to O. I and O are connected to the inlet and outlet of the temperature unit described in Fig. 1.

A tray (100 × 45 × 10 cm) covered by a lid is supported by the axis A and rocked by a lever fastened eccentrically on a disc which rotates at 20 r.p.m. The amplitude of the oscillations of the tray is 5 cm at the ends. The bottom of the tray is double (2 cm between the two layers). The tray and lid are made of stainless steel. Water from the temperature unit is led from I to O as regulated by a series of incomplete separating walls schematically indicated by the letter R in Fig. 10. Inoculation, aeration, and temperature recording is done through the hole H (actually two holes) and sampling through F.

D. Procedure

The incubator is sterilized by rinsing the tray, radiator system and lid with washings of 70 per cent ethanol followed by boiling water poured directly into the incubator. It is pointed out that the term sterile in this connection is defined as absence of bacterial growth during an experimental run, as estimated by high power phase microscopy. Consequently, the sterilization procedure is modified depending on the duration of the experiment planned: If an experiment is planned to last longer than 18 hr, the incubator is protected by a plastic housing which contains a germicide lamp. We have alternatively, and successfully, used Penicillin (100,000 units/10 l.) which does not interfere with logarithmic growth. The medium is autoclaved in transfusion flasks and transferred to the incubator using commercially available transfusion sets. When the experiments are planned to last less than 18 hr the procedure can be simplified: No plastic housing or germicide lamp or penicillin is used. The medium is heated to 100°, filtered and then sterilized by heating to 112°C for 20 min in pressure cookers. When the medium has cooled to about 95°C it is poured directly into the incubator. The air which is sucked into the incubator during the time when the temperature decreases in the incubator is filtered through sterile filters. The flow of air later used for aeration is filtered in the same way.

E. Inoculum

The volume of the cell suspension in the incubator is usually 5–10 l. An inoculum for a 7-l. culture is grown in a 2-l. Fernbach flask fitted with an aeration inlet and a siphon as described on page 545. The volume of the inoculum is 500 ml with a population density at the time of inoculation of approximately 75,000 cells per ml. For inoculation the inoculum is siphoned into the incubator through a sterilized plastic tubing connected to the siphon of the Fernbach flask. Siphoning is started by applying air pressure to the aeration inlet. After inoculation the cells are allowed to multiply for twelve hours at 28° prior to the beginning of the heat treatment. A cell count is made of the inoculum immediately before the inoculation, to insure that the population density of the experimental culture in the incubator would not exceed 50,000 cells per ml after twelve hours.

3. *Replacement of Medium and Collection of Cells*

A. Outline of procedure

The cell suspension is centrifuged at low speed usually before but sometimes after the last heat shock. The cells are recovered as a loosely packed pellet of undamaged live cells. The ciliates are re-suspended either in fresh proteose peptone or in inorganic medium to make a population density five to ten times that of the original. At suitable times, and usually after an additional heat shock, samples for analysis are removed from this concentrated culture and the cells are collected by high speed centrifugation which yields a firmly packed pellet. The pellet is either treated with acetone and ether and made to an acetone powder or it is frozen immediately. When resuspension of the cells is omitted, the cells which are collected by the low speed centrifugation are immediately submitted to the high speed spinning and the subsequent acetone treatment or freezing.

B. Details of procedure

The centrifuge tubes shown in Fig. 11 are used for collecting the cells by low speed centrifugation. The tubes were made from commercially available centrifuge tubes (Pyrex Glass Sterilizer Bottle for International Serum Centrifuge Cat. no. 606 TOBAC). Each tube has a capacity of 700 ml. The centrifuge used (M. S. E. Major Refrigerator fitted with a 4 × 1 l. head) accommodates four tubes, and 2·8 l. can thus be processed in each centrifugal run. The cells are spun for 5 min at 120 × *g* at room temperature and collected with the pipette shown in Fig. 11. The supernatant is not removed prior to the collection of the cells. The loss of cells during this procedure is usually 3 per cent, and never exceeds 5 per cent.

For the high speed centrifugation, the cells representing a sample are distributed in four 30-ml stainless steel tubes fitting the high speed head of the M.S.E. or International refrigerated centrifuge. They are then centrifuged briefly at room temperature at 7000 × *g*. This procedure yields a firmly packed pellet from which the supernatant can be removed completely by decantation without any loss of cells.

The cells may be frozen in the tubes or made to acetone powder by immediately adding 25 ml of analytical grade acetone, pre-cooled to −20°C, to the precipitated cells in each stainless steel tube. The cells are then washed with a generous volume of cold acetone, either by centrifugation or on filter, and finally treated with 100 ml of pre-cooled peroxide-free ether in the same way. The acetone powder is stored dessicated at −20°C.

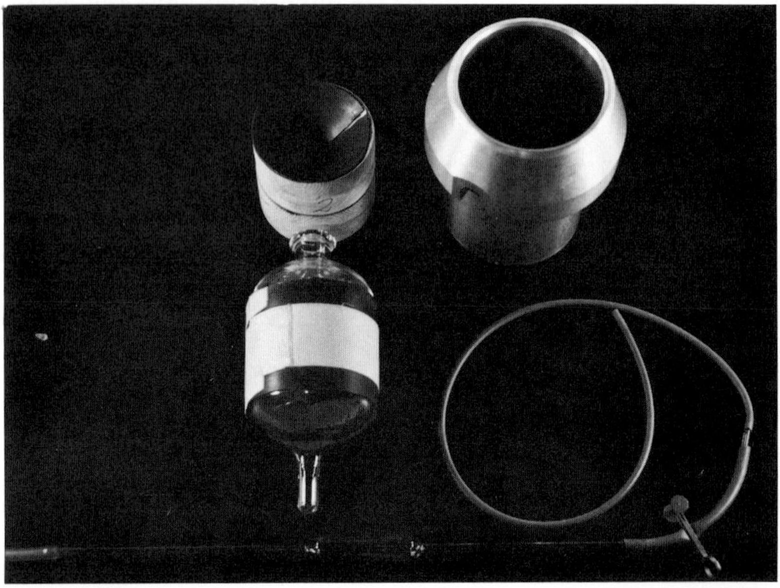

Fig. 11. Centrifuge tube with support and trunnion cup for collecting *Tetrahymena* cells. The pipette is used for aspirating the precipitated cells.

In the described procedure the culture available for biochemical studies is 1 l. in inorganic medium. The total cell weight (acetone powder) is 1–5 g, so using 100 ml samples 0·1–0·5 g aliquots can be obtained. The culture is obtained by concentration of a 10 l. proteose peptone synchronized culture.

REFERENCES

Elliott, A. M., J. R. Kennedy, Jr., and I. J. Bak (1962). Macronuclear events in synchronously dividing *T. pyriformis*. *J. Cell Biol.* **12,** 515.

Hamburger, K. and E. Zeuthen (1957). Synchronous divisions in *T. pyriformis* as studied in inorganic medium. *Exptl. Cell Research* **13,** 443.

Hamburger, K. and E. Zeuthen (1960). Some characteristics of growth in normal and synchronized populations of *T. pyriformis*. *Compt. Rend. Trav. Lab. Carlsberg* **32,** 1.

Holz, G. G., Jr. B. Wagner, J. Erwin, J. J. Britt and K. Block (1961). Sterol requirements of a ciliate *T. corlissi Th–X*. *Comp. Biochem. Physiol.* **2,** 202.

KIDDER, G. W. and V. C. DEWEY (1951). The biochemistry of ciliates in pure culture. In *Biochemistry and Physiology of Protozoa* (A. Lwoff, ed.). Academic Press, New York.

LEE, K. H. (1959). Studies on cell growth and cell division. I. A temperature-controlled system for inducing synchronous cell division. *J. Am. Pharm. Assoc., Sci. Ed.* **48,** 468.

LINDAHL, P. E. (1936). Zur Kenntnis der physiologischen Grundlagen der Determination im Seeigelkeim. *Acta Zool.* **12,** 179.

RASMUSSEN, L. and E. ZEUTHEN (1962). Cell division and protein synthesis in *T. pyriformis* as studied with *p*-fluorophenylalanine. *Compt. Rend. Trav. Lab. Carlsberg* **32,** 333.

SCHERBAUM, O. (1957a). Cell growth in normal and synchronously dividing mass cultures of *T. pyriformis*. *Exptl. Cell Research* **13,** 24.

SCHERBAUM, O. (1957b). The application of a standard counting method in estimation of growth in normal and heat-treated cultures of *T. pyriformis*. *Acta Pathol. Microbiol. Scand.* **40,** 7.

SCHERBAUM, O. and E. ZEUTHEN (1954). Induction of synchronous cell division in mass cultures of *T. pyriformis*. *Exptl. Cell Research* **6,** 221.

SCHERBAUM, O. and E. ZEUTHEN (1955). Temperature induced synchronous division in the ciliate protozoa *T. pyriformis* growing in synthetic and proteose peptone media. *Exptl. Cell Research Suppl.* **3,** 312.

24

Synchronous Cultures of *Chlorella* Starting from a Homogeneous Population*

HIROSHI TAMIYA and YUJI MORIMURA, *The Tokugawa Institute for Biological Research, Mejiro, Tokyo, Japan*

CONTENTS

I	Introduction	565
II	Culture of Inoculum for Pre-culture	567
III	Pre-culture and Separation of D_S Cells	568
IV	Synchronous Culture	568
	References	569

I. INTRODUCTION

As has been described in chapter 9, the synchronous cultures of *Chlorella ellipsoidea* run by Morimura (1959) and Tamiya *et al.* (1961) are started from a homogeneous population of small young cells (referred to as D_S cells) which are separated from less homogeneous populations by differential centrifugation. To obtain the starting D_S cells as uniform in size and properties as possible, their pre-culture has to be effected under constant conditions, using inocula having as constant a pre-history as possible. The culture is, therefore, performed in three steps, namely: (i) culturing of inoculum for the pre-culture, (ii) pre-culturing of D_S cells, and (iii) the synchronous culture of the main experiment. For (i) and (ii), and in normal synchronous cultures also for (iii), a medium of the following composition is used: per liter, 5·0 g KNO_3; 2·5 g $MgSO_4 \cdot 7H_2O$; 1·25 g KH_2PO_4; 0·028 g $FeSO_4 \cdot 7H_2O$; and 1 ml of Arnon's A5-solution (containing B, Mn, Zn, Cu and Mo) (Arnon, 1938).

According to the scale of the experiment, either oblong flat flasks or culture chambers made of lucid acrylate resin are used as containers of the culture. The flasks and chambers are of various sizes, the typical shapes of which are illustrated in Figs. 1 and 2. The features of various types of these containers are given in Tables I and II.

The flask has an opening with a cotton plug carrying a glass tube, the lower end of which almost reaches the bottom of the flask. The

* Information on techniques for the flagellates is to be found in chapter 22 in the section on General Considerations.

upper end of this tube is widened and filled with cotton to serve as an air filter. In the upper opening of the tube is fitted a rubber stopper carrying a glass tube which serves to conduct air containing 5 per cent CO_2 into the culture medium. For the same purpose, the flat acrylate chamber is provided with three glass tubes (or five in the case of the larger type) as shown in Fig. 2. Aeration of CO_2-enriched air is made at a rate of about 50 ml per 100 ml of culture solution

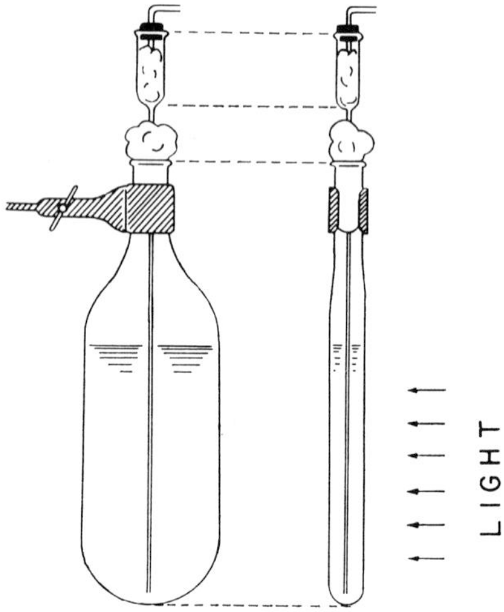

FIG. 1. An example of oblong flat flasks (medium-sized; see Table I).

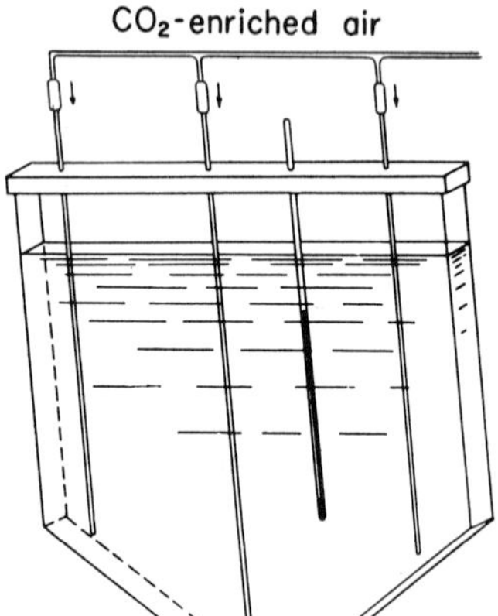

FIG. 2. A culture chamber made of lucid acrylate resin (smaller type; see Table II).

per minute. Before entering the culture solution, the CO_2-enriched air is brought to the same water-vapor pressure as the culture medium by being scrubbed through gas wash bottles containing a solution having the same composition and temperature as the culture medium. The cultures in the flasks are made under aseptic conditions (seeding being made after autoclaving), while those in the chambers are done without the same precautions. Both the flasks and chambers, after being filled with inoculated culture solutions, are placed in a water thermostat with a glass wall on one side, and they are illuminated from outside—in a direction perpendicular to the flat walls of the culture containers—either by a bank of 20–40 Watt daylight fluorescent lamps or of 150–250 Watt reflector flood lamps.

TABLE I. Features of oblong flat flasks of various sizes

Flask	Size of the main body of the flask (cm)			Volume of culture solution (ml)
	Length	Width	Inner thickness	
Small	15	7	2·4	50
Medium	37	12	2·8	500
Large	52	17	3·5	1,000

TABLE II. Features of flat culture chambers made of acrylate resin

Chamber	Size (cm)			Volume of culture solution (ml)
	Height	Width	Inner thickness	
Smaller	47·5–70·0	45·5	2·0	5,000
Larger[a]	66·5–95·5	81·0	2·0	10,000

[a] In the case of the larger chamber, the two plane-parallel walls—at a distance of 2·0 cm—are riveted with eight bolts (also made of lucid acrylate resin, 1·0 cm in diameter) adequately distributed between the plates. The plates of acrylate resin are 0·5 cm in thickness.

II. Culture of Inoculum for Pre-culture

The culture solution is dispensed in 50-ml portions in small, flat oblong flasks, and after being sterilized by autoclaving, inoculated with a pure algal culture taken from a 2- to 4-week old slant culture which has been preserved at 20–25° under illumination with daylight fluorescent lamps, about 5 klux in intensity. The cultures in these flasks are kept at 25° under illumination with daylight fluorescent lamps of 10 klux in intensity and with constant aeration of CO_2-enriched air.

III. Pre-culture and Separation of D_S Cells

After 7–8 days, 25-ml portions of inoculum culture are taken out with sterilized pipettes and poured into medium- or large-sized oblong flasks, each containing 500 or 1000 ml, respectively, of the sterilized culture solution. The cultures are kept at 25° under illumination with incandescent lamps, the light intensity being kept at 10 klux for the first 4–5 days, after which it is reduced to 0·8 klux. Culturing under the lower light intensity lasts for 3–4 days, by which time the algal population density attains the level of about 10 ml packed-cell volume per liter, and more than 90 per cent of the total population are in the form of D cells having diameters ranging from 2·5 to 3·5 μ. The culture is diluted ten- to twenty-fold with fresh culture solution, and then centrifuged at 1200*g* for 5–6 min to discard the larger, heavier cells. The supernatant cell suspension is further centrifuged at 1200*g* for 12–15 min; this time, to remove the cells in the supernatant which are smaller or lighter than the D_S cells. The D_S cells thus separated are somewhat ellipsoidal in shape with longer axes of 2·7–3·3 μ, the maximum frequency (more than 80 per cent) being 3·0 μ.* Although the D_S cells can be obtained quite reproducibly and homogeneously, their yield in the process of differential centrifugation is rather low, being 10–20 per cent of the total population (existing before centrifugation) in terms of packed cell volume.

IV. Synchronous Culture

The D_S cells obtained as above are directly—or after being stored in a refrigerator for not more than 24 hr—used as the starting material of synchronous culture.† According to the purpose of experiment, various types of flat flasks or chambers listed in Tables I and II are

* The D_n cells obtained in the D_SLD and DLD cycles (see chapter 9 in this book) are also somewhat ellipsoidal in shape. They are, however, a little larger than the D_S cells, with longer axes of about 3·0 to 3·8 μ (maximum frequency at 3·6 μ). Both D_S and D_n cells are uninucleate, and the nucleus is situated at one corner of the ellipsoidal cell (Tamiya *et al.*, 1961). The cells at the other stages of development are almost perfectly spherical in shape.

† It has been found that the D_S cells—suspended in the standard medium—can be stored for 24 hr in a refrigerator without any change in the behavior shown in subsequent synchronous cultures.

used for running the synchronous cultures. Most of the experiments are carried out using the medium- or large-sized flat flasks. In experiments such as to make extensive tests with various anti-metabolites relative to their effects upon the algal life cycle, a series of small, flat flasks are used, since, if we only follow the changes of cell number and average cell volume, the quantity of culture solution necessary to be sampled for each measurement is only 0·5 ml in volume. On the other hand, the larger or smaller acrylate chambers have to be used when we intend to follow the changes in quantities of some specific cell substances (such as vitamins) which are contained only in minute amounts in algal cells.

In all these cases, the initial inoculation of D_S cells in the culture medium is made at a concentration of about 7×10^9 cells or 0·15 ml packed cell volume per liter. Maintenance of low cell concentrations during the culture is important for keeping the uniformity of the light intensity available for individual cells. Various types of synchronous cultures thus started from D_S cells have been described, and the merits and demerits of each culture method have been discussed in the section on 'Methodological Considerations' in chapter 9, p. 248.

REFERENCES

Arnon, D. I. (1938). Microelements in culture-solution experiments with higher plants. *Am. J. Botany* **25,** 322.

Morimura, Y. (1959). Synchronous culture of *Chlorella.* I. Kinetic analysis of the life-cycle of *Chlorella ellipsoidea* as affected by changes of temperature and light intensity. *Plant and Cell Physiol.* (*Tokyo*) **1,** 49.

Tamiya, H., Y. Morimura, M. Yokota and R. Kunieda (1961). Mode of nuclear division in synchronous cultures of *Chlorella*: Comparison of various methods of synchronization. *Plant and Cell Physiol.* (*Tokyo*) **2,** 383.

25

Synchronization of *Chlorella* with Light-Dark Changes and Periodical Dilution of the Populations to a Standard Cell Number

HARALD LORENZEN, *Pflanzenphysiologisches Institut der Universität Göttingen, Bundesrepublik Deutschland*

CONTENTS

I Introduction 571
II Method 571
III Comments 576
References 577

I. INTRODUCTION

The present method for the synchronization of *Chlorella pyrenoidosa* was developed at Marburg (Lorenzen, 1955, 1957, 1958 a, b; Lorenzen and Ruppel, 1958; Pirson and Lorenzen, 1958 a, b) and later improved at Göttingen* (Lorenzen, 1959, 1960, 1961; Lorenzen and Ruppel, 1960; Metzner and Lorenzen, 1960; Pirson, 1961; Pirson and Kowallik, 1960; Pirson, Lorenzen and Koepper, 1959; Pirson and Ruppel, 1962; Pirson and Senger, 1961; Senger, 1961).

II. METHOD

Chlorella pyrenoidosa (strain 211-8b of the Algal collection, Institute of Plant Physiology, University of Göttingen) is grown at 30 ± 0·1°C under illumination with light of 10,000 lux and with constant supply of air containing 1·5 vol per cent CO_2. The culture medium contains per litre: 1011 mg ($1 \times 10^{-2}M$) KNO_3; 621 mg ($4{\cdot}5 \times 10^{-3}M$) $NaH_2PO_4{\cdot}1H_2O$; 89 mg ($0{\cdot}5 \times 10^{-3}M$) $Na_2HPO_4{\cdot}2H_2O$; 246·5 mg ($1 \times 10^{-3}M$) $MgSO_4{\cdot}7H_2O$; 14·7 mg ($1 \times 10^{-4}M$) $CaCl_2{\cdot}2H_2O$; 6·95 mg ($2{\cdot}5 \times 10^{-5}M$) $FeSO_4{\cdot}7H_2O$†;

* The author is indebted to Prof. Dr. A. Pirson for valuable discussions and suggestions in preparing the manuscript. Original work was supported by the Deutsche Forschungsgemeinschaft.

† EDTA complex; 0·70 g $FeSO_4{\cdot}7H_2O$ and 0·93 g Titriplex III (like all other substances from Merck, Darmstadt). The complex is dissolved under brief heating to 100° in about 80-ml water and then filled up to 100 ml. 1 ml of this mixture contains the quantity of Fe mentioned above.

0·061 mg ($1 \times 10^{-6}M$) H_3BO_3; 0·169 ($1 \times 10^{-6}M$) $MnSO_4 \cdot 1H_2O$; 0·287 mg ($1 \times 10^{-6}M$) $ZnSO_4 \cdot 7H_2O$; 0·0025 mg ($1 \times 10^{-8}M$) $CuSO_4 \cdot 5H_2O$; 0·0124 mg ($1 \times 10^{-8}M$) $(NH_4)_6Mo_7O_{24} \cdot 4H_2O$.

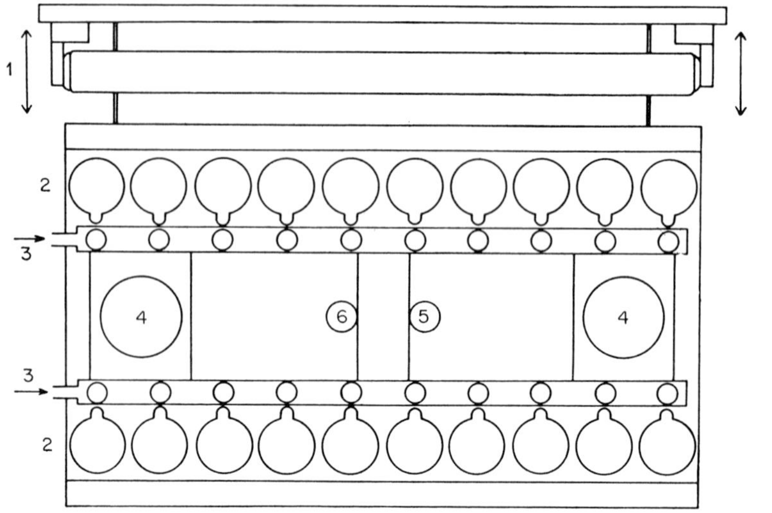

FIG. 1. Culture apparatus.
1, movable illuminating device (painted on one side only);
2, holding device for 10 culture vessels;
3, aeration;
4, pump;
5, heater (cooling coil on the bottom of the bath);
6, contact thermometer.

Figure 1 shows the 120-litre well-mixed thermostat (Lorenzen, 1959) in which twenty culture tubes (Fig. 2) can be held at one time. It has large glass windows on both sides, through each of which the cultures are illuminated with fluorescent lamps (3 Philips TL 40 W 1/32 and 2 Philips TL 40 W 1/55, maximal light intensity 12,000 lux). A heater and a refrigerator unit, both with contact thermometers, regulate the temperature between 4 and 40°C. The culture vessels (Jena glass, Fig. 2a) are fixed in the thermostat, the smaller tubes being connected by rubber tubings to a plastic aeration-tap. Both rows of the culture vessels can be used independently for experiments in light and dark or for experiments with two different light intensities. In dark experiments the glass window is closed with a plastic plate and a light-proof cover is placed over the taps.

To start cultures, cells from pure cultures grown on agar slants are poured into a glass vessel with culture medium to make a volume of 300 ml (Fig. 2a). After two cycles of 14-hr light, 10-hr dark, started immediately, the population density has increased enough for the first dilution to be made to the arbitrary standard cell number of 100 ($= 1{\cdot}56 \times 10^6 \pm 5$ per cent) cells per ml of suspension (counted in the Thoma counting chamber). At this early stage the cell population shows 'complete' synchrony, i.e. at the initiation of the following light period at least 98 per cent of the cells, as counted in the first hour of the preceding dark period, have divided into daughter cells (autospores). (For the other types of synchronization, which we call 'incomplete' synchrony, see Lorenzen, 1957; Pirson and Senger, 1961; Senger, 1961). No further manipulations, such as changes of temperature and light intensity or the use of centrifugal fractionation are necessary to obtain complete synchrony in the present system. After the dilution (with fresh, sterile culture medium) a new light period (14 hr) is initiated immediately. It is

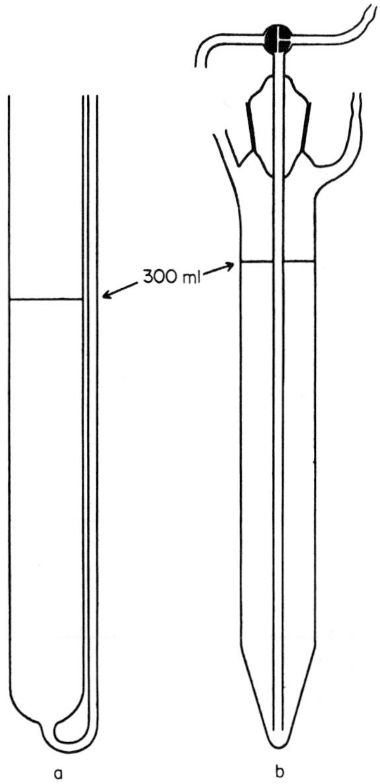

FIG. 2. (a) Culture vessel (Jena glass). Smaller part = gas inlet tube, (b) Culture vessel for sterile conditions.

not necessary to maintain the algae under axenic conditions, since repeated experiments give the same quantitative results. In one case we maintained a synchronous culture for more than thirty subcultures without apparent infection with other organisms. It was diluted regularly every day at the start of illumination and it was repeatedly sampled during the light and dark periods. For special problems (influence of glucose on the growth and in all cases of unfavourable conditions, e.g. the application of shocks of high or low temperature) we grow axenic cultures in special tubes (Fig. 2b).

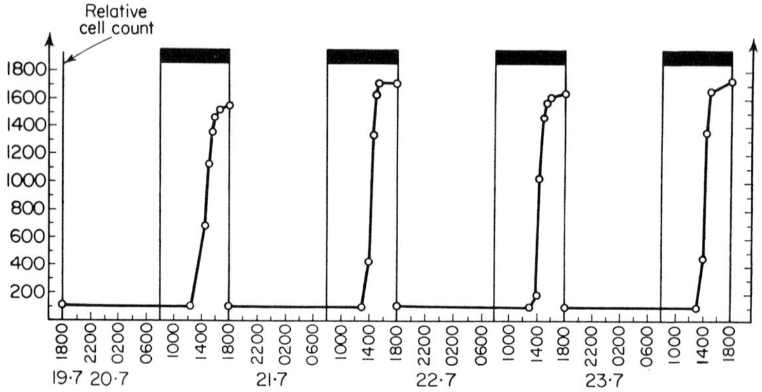

FIG. 3. Cell number (arbitrary unit) in four succeeding periods of the light–dark change of 14:10 hr (30°, 10,000 Lux).

Figure 3 shows the changes in cell numbers in several successive cycles of 14-hr light, 10-hr dark, followed by dilution. There is no cell division during the light periods in which photosynthesis produced larger cells (Fig. 4). After the light is turned off there is at first no increase in cell number, but regularly after 6 hr in the dark the cells begin to divide. The bursting division comes to its maximum during the seventh hour, and after 10 hr in the dark (24 hr after the initiation of the light–dark cycle) no further division is observed. At this stage the suspension is diluted to the standard cell number and the new light period begins. During one light–dark cycle the pH of the medium changes only slightly, from 6·2 to about 6·5.

During the 24-hr cycle, nearly all cells (> 98 per cent) divide, on an average, into sixteen daughters (Fig. 3). The mode of cell division can be checked by the change of frequency curves of cell sizes (undivided, versus autospores) measured before and after the

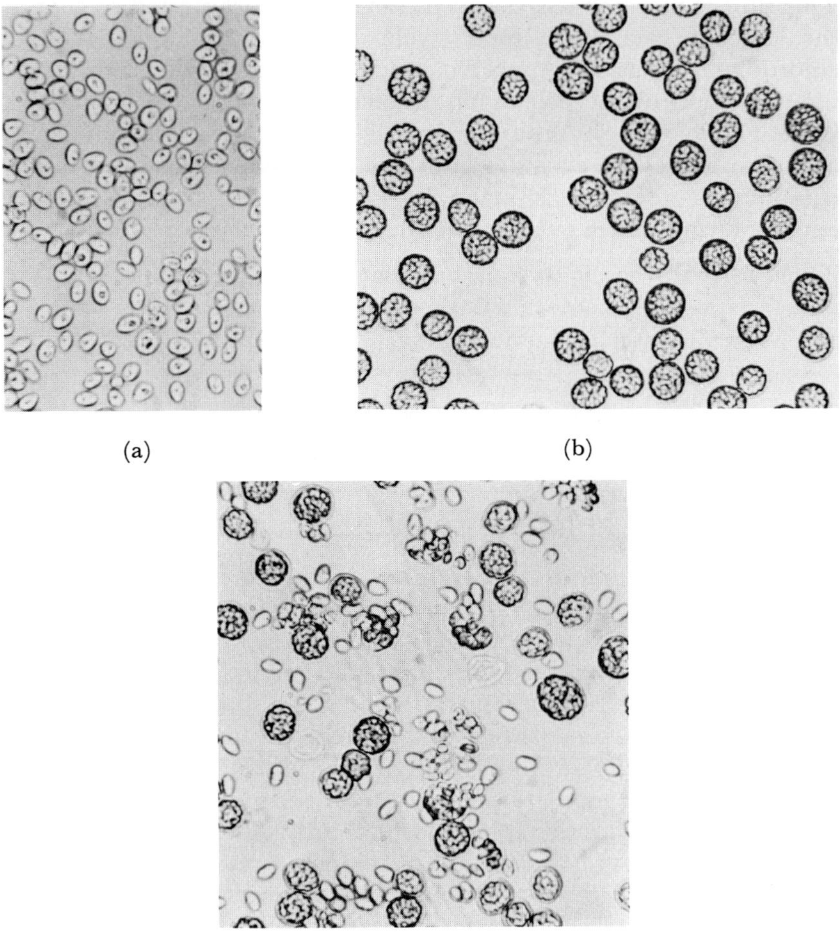

(a) (b) (c)

FIG. 4. Microphotographs of synchronous *Chlorella* cells. (a) Autospores (on average 5 $\mu\varphi$); (b) cells after 14 hr of light (on average 12 $\mu\varphi$); (c) cells during division.

bursting division. In the case of complete synchronization (condition: > 8 autospores per dividing cell) the two frequency curves do not overlap (Lorenzen, 1957; Metzner and Lorenzen, 1960).

During illumination the dry weight per cell increases nearly as much as does the cell number in the following division. During the dark there is only a small respiratory loss in dry weight. During the light there is a considerable shift in the cellular components.

This shift is more or less reversed during the dark period so that at the end of each light–dark cycle each essential component—chlorophyll, DNA, RNA, protein, and carbohydrate—constitutes the same fraction of the cell as at the end of the previous light–dark cycle. For dry weight and chlorophyll Table I shows this conformity over four successive cycles of light–dark dilution. For other cellular components reference is made to Lorenzen and Ruppel (1960), and to Pirson and Kowallik (1960).

TABLE I. Comparison in percentage increase of some periods of the 14 : 10-hr light–dark change

Cell number	Dry weight	Chlorophyll
1440	1460	1400
1500	1450	1520
1460	1490	1460
1530	1510	1480

III. COMMENTS

In our *Chlorella* strain the number of daughter cells produced from one mother cell increases to more than twenty when the temperature is raised from 30 to 33° (Lorenzen, 1961). If a temperature period is superimposed on a light–dark period one cell may give rise to as many as twenty-four autospores. This result was obtained when the normal procedure was changed only in one respect, viz. by heating the culture from 30° (normal) to 37° for the first 10 hr in the light. The high productivity allows samples to be taken at short time intervals for physiological and biochemical tests. At constant 34° and at higher constant temperatures our *Chlorella* stops dividing. Initially the production of cell material increases (at 34°) by more than 20 per cent as compared with a culture kept at a constant 30°. With our *Chlorella* strain a light–dark period of 24 hr is normally chosen to let all the cells complete their life cycle. This light–dark period is a little longer than the shortest possible duration of the life cycle, 20 hr. In earlier papers (Lorenzen, 1960; Lorenzen and Ruppel, 1958, 1960; Pirson and Lorenzen, 1958a) we used a light–dark change of 16:12 hr to obtain complete synchronization. By improving the culture medium, especially by addition of the Fe–EDTA complex, we could later shorten the whole cycle to 24 hr. Complete synchrony was also obtained by the light–dark changes of 13:11 hr and of 12:12 hr. In the last case, the limiting factor for complete synchrony appeared to be the length of the light period. To obtain complete synchrony the light conditions must be such as to allow at least an eight-fold increase in mass permitting later division of one cell into at least eight autospores.

In conclusion, the following points characterize the above described method of synchronizing *Chlorella* cells.

1. The culture conditions have been chosen in such a way that the cells complete their life cycle in the shortest possible time.

2. Almost all cells (more than 98 per cent) participate in each division burst. They divide to increase the population count by 8–24 times according to conditions.

3. At corresponding stages of the life cycle the chemical composition of the cells was constant from one light–dark period to the next (permanent synchronization = 'Dauersynchronisation').

4. While the method described here is confined to the strain of *Chlorella pyrenoidosa* used in our laboratory, we have found it applicable to certain other strains of *Chlorella* and also to other species (*Scenedesmus*, *Ankistrodesmus*).

REFERENCES

KUHL, A. and H. LORENZEN (1964). Handling and culturing of *Chlorella*. In *Methods in Cell Physiology*, Vol. 1 (D. M. Prescott, Ed.). Academic Press, in press.

LORENZEN, H. (1955). Über Wachstum und Stoffwechsel von Grünalgen mit besonderer Berücksichtigung periodischer Erscheinungen. *Diss. philosoph. Fakultät Math.-naturwiss. Abteilung Marburg a.d. Lahn.*

LORENZEN, H. (1957). Synchrone Zellteilungen von *Chlorella* bei verschiedenen Licht-Dunkel-Wechseln. *Flora (Jena)* **144,** 473.

LORENZEN, H. (1958a). Periodizität von Nuklealreaktion und Kernteilung bei *Chlorella. Ber. Deut. Botan. Ges.* **71,** 89.

LORENZEN, H. (1958b). Endogene Rhythmik bei *Chlorella* (Vortragszusammenfassung). *Ber. Deut. Botan. Ges.* **70,** (11).

LORENZEN, H. (1959). Die photosynthetische Sauerstoffproduktion wachsender *Chlorella. Ber. Deut. Botan. Ges.* **71,** 89.

LORENZEN, H. (1960). Über den Einfluss von Glukose auf Synchronkulturen von *Chlorella pyrenoidosa. Naturwissenschaften* **47,** 477.

LORENZEN, H. (1961). Neue Erfahrungen mit synchronisierten *Chlorellen* (Vortragszusammenfassung). *Ber. Deut. Botan. Ges.* **73,** (58).

LORENZEN, H. and H. G. RUPPEL (1958). Periodische Resistenzänderung in einer synchronen Kultur von *Chlorella. Naturwissenschaften* **45,** 553.

LORENZEN, H. and H. G. RUPPEL (1960). Versuche zur Gliederung des Entwicklungsverlaufs der *Chlorella-Zelle. Planta* **54,** 394.

METZNER, H. and H. LORENZEN (1960). Untersuchungen über den Photosynthese-Gaswechsel an vollsynchronen Chlorella-Kulturen. *Ber. Deut. Botan. Ges.* **73,** 410.

PIRSON, A. (1961). Sur l'application de la synchronization à la physiologie cellulaire des algues vertes. In *Colloq. Intern. Centre Natl. Rech. Sci. Dinard, 1960*, 92–109.

PIRSON, A. and W. KOWALLIK (1960). Wirkung des blauen und roten Spektralbereiches auf die Zusammensetzung von *Chlorella* bei Anzucht im Licht-Dunkel-Wechsel. *Naturwissenschaften* **47,** 476.

PIRSON, A. and H. LORENZEN (1958a). Ein endogener Zeitfaktor bei der Teilung von *Chlorella*. *Z. Botan.* **46,** 53.

PIRSON, A. and H. LORENZEN (1958b). Photosynthetische Sauerstoffentwicklung von *Chlorella* nach Synchronisation durch Licht-Dunkel-Wechsel. *Naturwissenschaften* **45,** 497.

PIRSON, A., H. LORENZEN and A. KOEPPER (1959). A sensitive stage in synchronized cultures of *Chlorella*. *Plant Physiol.* **34,** 353.

PIRSON, A. and H. G. RUPPEL (1962). Über die Induktion einer Teilungshemmung in synchronen Kulturen von *Chlorella*. *Arch. Mikrobiol.* **42,** 299.

PIRSON, A. and H. SENGER (1961). Synchronisationstypen bei Chlorella im Licht-Dunkel-Wechsel. *Naturwissenschaften* **48,** 81.

SENGER, H. (1961). Untersuchungen zur Synchronisierung von *Chlorella*-Kulturen. *Arch. Mikrobiol.* **40,** 47.

26

The Use of the Electronic Cell Counter in Studies of Synchronous Cell Division in Dinoflagellates * † ‡

BEATRICE M. SWEENEY and J. WOODLAND HASTINGS,
Biology Department, Yale University, New Haven, Connecticut, and Division of Biochemistry, University of Illinois, Urbana, Illinois

CONTENTS

I Introduction 579
II Making Cell Counts 580
III Determining Cell Size. 584
References 587

I. INTRODUCTION

Studies of synchronous division in populations of any unicellular organism require that the number of cells present be measured frequently, in order that the time of doubling and the degree of synchrony may be determined. To make the required number of cell counts by ordinary visual methods is extremely time-consuming. Furthermore, the accuracy of such methods is often not greater than $\pm$ 10 per cent.

Recently an electronic counter has become available which depends upon the measurement of the change in impedance which occurs when a cell passes through a small pore. The method is relatively rapid, accurate and convenient. Several papers have dealt with the operation and performance of the counter with various types of cells, including red cells, white cells and bacteria (Mattern, Brackett and Olson, 1957; Brecher, Schneiderman and Williams, 1956; Kubitschek, 1958; Grant, Britton and Kurtz, 1960; and Richar and Breakell, 1959). The following is a brief description and evaluation of this counting method as we used it in studies of phased cell division in populations of dinoflagellates.

* The authors wish to acknowledge the support of the National Science Foundation.
† We are grateful to Dr. Marlene Wiesner Karakashian and Evelyn Balbinder for their expert assistance in making this study.
‡ See also chapters 21 and 23 for other uses of cell counters.

II. Making Cell Counts

The electronic cell counter used in these studies, the Coulter Counter*, is based on the principle that cells are relatively poor electrical conductors as compared to the sea water or salt medium in which they are suspended. The counter employs a glass tube which has a minute aperture or pore of known dimensions in its wall, and the impedance across this aperture is measured. Suction is applied to force cells to flow through the aperture from the sample reservoir in which the tube is immersed. As a cell passes through, the impedance across the aperture is transiently increased and counted as a pulse by conventional techniques. The pulses are conveniently displayed on an oscilloscope screen mounted on the instrument panel. One can roughly estimate cell size visually by the pulse height.

The total volume of cell suspension counted is set at a given value (0·5 ml) by means of a mercury column in the suction system which activates and deactivates the scaler as the mercury passes by contacts separated by this volume. The oscilloscope display may be visualized without operating the scaler.

In making cell counts of a population, it is first necessary to determine the appropriate aperture size. A number of different size diameters are available from the manufacturer, ranging approximately from 15 to 300 μ. With large cells, an aperture with a diameter about four times the diameter of the cells to be counted is usually suitable. For counts of *Gonyaulax polyedra* for example, which is about 40 μ in diameter, pores of 140 or 160 μ were most satisfactory. Smaller pores tended to become obstructed, while with larger pores the pulses were too small. The gain control is adjusted so that the pulses are seen almost to fill the oscilloscope screen.

In addition to a gain control, the instrument (Model A) is supplied with a threshold setting which excludes from counting all pulses below a particular height. Since the threshold setting may be varied, a pulse height distribution may be determined by successive counts at increasing thresholds (Fig. 1).

The Model B is equipped with both an upper and a lower threshold setting control and it is possible to count cells within a given range of pulse heights. In making counts, one must select a lower threshold setting which will allow all cells to be counted. With an appropriate setting one may readily eliminate counts due to precipitate or other small debris present in the cell suspension. This is only possible, however, if the pulses arising from noncellular material are signifi-

* In the United States, the Coulter Counter is manufactured by Coulter Electronics, Inc., 590 West 20th Street, Hialeah, Florida, U.S.A.

cantly smaller than those due to cells. This can be determined by counting the suspending medium from which the cells have been removed by decanting, differential centrifugation, or other means.

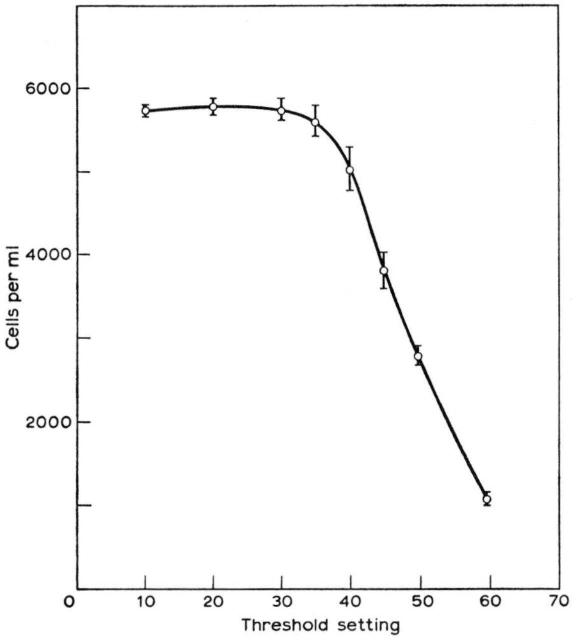

FIG. 1. Cell counts of *Gonyaulax polyedra* made with the Coulter Counter (Model A) at different threshold settings with a pore 140 μ in diameter and gain 3. The standard deviation of each determination, the average of five counts, is shown as a bar at each point.

If two cells pass through the aperture simultaneously, only a single pulse results. Counting errors of this kind (referred to as coincidence errors) will be expected only at relatively high cell densities. The magnitude of this error may be determined by ascertaining whether the machine count decreases linearly when accurately diluted samples of the cells are counted (Fig. 2). At high dilutions (low cell densities) the relationship will always be linear, and with *Gonyaulax* it is clear that coincidence was still negligible in cultures containing 25,000 cells per ml. The accuracy of the determinations, shown in Fig. 2 as the percentage standard deviation, is about $\pm$ 4 per cent when each point is based on five measurements.

It is also important to know how good the absolute accuracy of the Coulter Counter is. This can be roughly checked by comparing machine with visual counts from the same culture. But one cannot be completely sure concerning this point, since an absolute standard is not available. Systematic errors may occur in either or *both* methods, making it impossible to specify absolute values. Systematic errors are very likely in visual counting and the attendant dilutions; it is also easy to imagine such errors in the machine count, as for example by the inclusion of sufficient dead cells and debris.

We have found, with *Gonyaulax polyedra*, that the instrument counts have generally been in very close agreement with the visual counts. In one determination, the machine count gave a value of 5800 cells per ml (± 3 per cent), while the visual count was 5250 cells per ml (± 10·5 per cent). With other species we have actually found large differences between the two methods, so this matter must be investigated with the species being used.

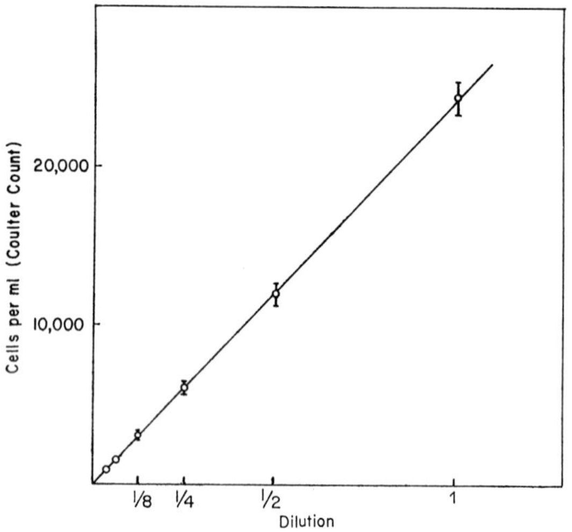

FIG. 2. Test for coincidence in counting a cell suspension of *Gonyaulax polyedra* with the Coulter Counter (Model A, 140 μ diameter pore, gain 3, threshold 10). The standard deviation is shown for each point (the average of five counts). A culture was counted at each of several dilutions, as indicated in the figure.

The relatively large volume of cell suspension necessary for each determination is sometimes a disadvantage of this counting method. Although a volume of only 0·5 ml is used for each determination, a considerable quantity of cell suspension is drawn through the pore between counts. Moreover, it is necessary that there be enough fluid present to immerse the tube to a level above the aperture. When the cell suspension is contained in a small vial (volume 20–25 ml), a minimum of about 15 ml is required for five counts.

By using the method outlined above, it was possible to detect the phasing of cell division in cultures of a number of dinoflagellates (Fig. 3; and Hastings and Sweeney, this volume). The time at which the population divides appears as a step in the growth curve; the steepness and height of each step provides a measure of the degree of synchrony. For example, in the culture of *Scrippsiella* shown in Fig. 3 about 60 per cent of the cells divide each day.

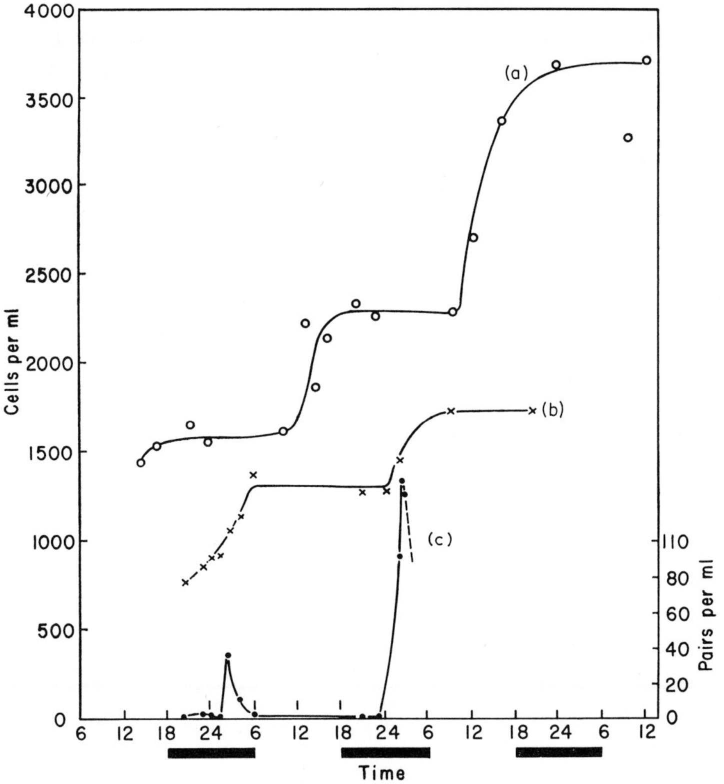

FIG. 3. Growth curves for a culture of *Scrippsiella sweeneyi* (Curve a, O) in natural light, and for *Gonyaulax polyedra* (Curve b, X) in a daily light cycle (12 hr light 0600–1800, and 12 hr dark) showing steps due to the phasing of cell division. Cell counts were made with the Coulter Counter Model A, 140 μ pore, gain 3, threshold 15 for *Scrippsiella*, threshold 20 for *Gonyaulax*. Plotted also are counts of the number of paired cells in the same *Gonyaulax* culture, counted visually (Curve c, ●).

III. Determining Cell Size

The size of the pulse from a cell will depend upon the difference in impedance between the suspending medium and the cell. This difference will be a function of both the size and impedance of the cell, as well as of the impedance of the suspending medium. If the suspending medium in sea water (equivalent to about 3 per cent NaCl), the relative impedance of all cells is very high and the difference in impedance between cells from different species may be neglected. Small differences in the conductivity of the suspending medium, on the other hand, lead to large differences in the pulse height and in pulse height distribution. This is illustrated in Fig. 4, which shows the fact that the pulse height with spores of *Lycopodium* is very different in salt solutions of different concentration.

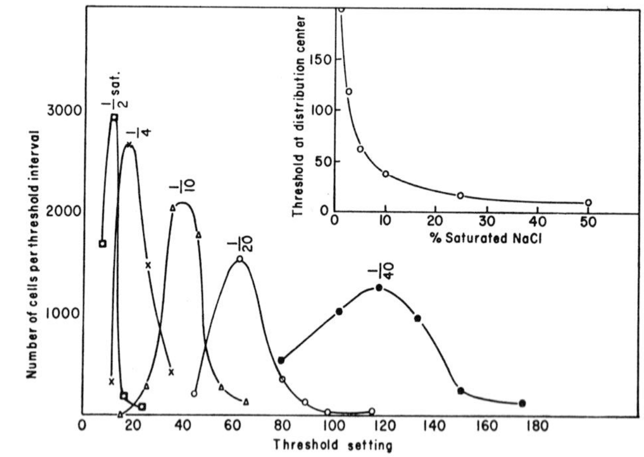

Fig. 4. The apparent size distribution of *Lycopodium* spores suspended in several different concentrations of NaCl as marked on each curve. These figures were obtained by plotting, as differential counts, data such as is shown in Fig. 1. The difference in the number of cells between successive threshold settings (at intervals of 10 units) was calculated and plotted on the graph as being at the centre of the threshold interval. The total number of cells at each salt concentration is the same. The insert shows how the apparent size of *Lycopodium* spores increases with dilution of the suspending salt solution, the ordinate being the threshold value at the centre of the distribution of pulses at each salt concentration. All threshold settings (arbitrary units) correspond to those at gain 4 (Model A Coulter Counter, 140 μ pore).

Since, in a given suspending medium, the pulse height is a function of cell size, it is possible to measure the size distribution of a population by making successive cell counts of the same cell suspension at each of several threshold settings as in Fig. 1, and then differentiating (Fig. 5). In order to interpret the distribution obtained in terms of microns, the threshold scale must be calibrated. This was accomplished by visually measuring (with a microscope and calibrated ocular micrometer) the linear dimensions of cell samples, and plotting the distribution thus obtained as a function of cell size. This curve was then fitted to the distribution curve which was obtained by varying the threshold setting (Fig. 5). Using calibrations

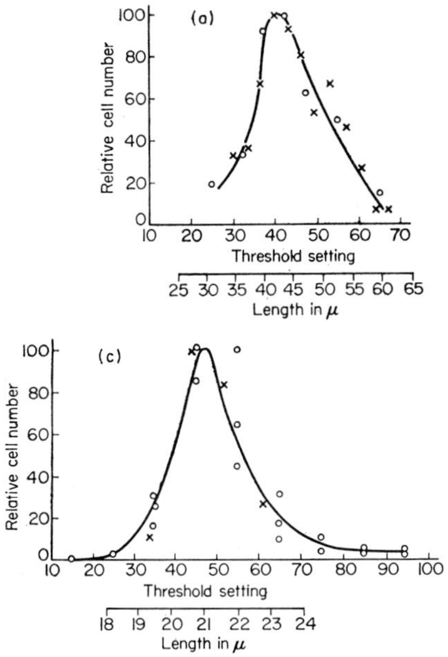

Fig. 5. Calibration of the instrument threshold settings in terms of microns from a comparison of the distribution curves of the cell diameter measured visually (X), and pulse heights determined with the Coulter Counter (O). Curves (a) and (b) were determined with the Model A Counter using a pore 140 μ in diameter; Curves (c) and (d), with the Model B, and 100 μ diameter pore. Curve (a), *Gonyaulax polyedra*, gain 3: Curve (b), *Lycopodium* spores, gain 4: Curve (c), ragweed pollen, gain 1/16, pore current 1/1: Curve (d), pecan pollen, gain 1/32, pore current 1/2. Ordinate; differential counts normalized to a value of 100 for the maximum differential count.

carried out by this method, it was found that one obtained a better fit between the two distribution curves when diameter rather than volume was used as a measure of cell size. This finding is not in agreement with the claim of the manufacturer, who states that the pulse height is proportional to cell volume. Mattern, Brackett and

Olson (1957), studying red blood cells, came to this same conclusion, and suggested that the area of a pulse might be proportional to volume. More recently, Grant, Britton and Kurtz (1960), also using red blood cells, found that threshold settings in their instrument were proportional to volume. The possible reason for the difference was not discussed. Although theory indicates that pulse height should be proportional to volume, it is certainly not clear that this relationship is always obtained in practice.

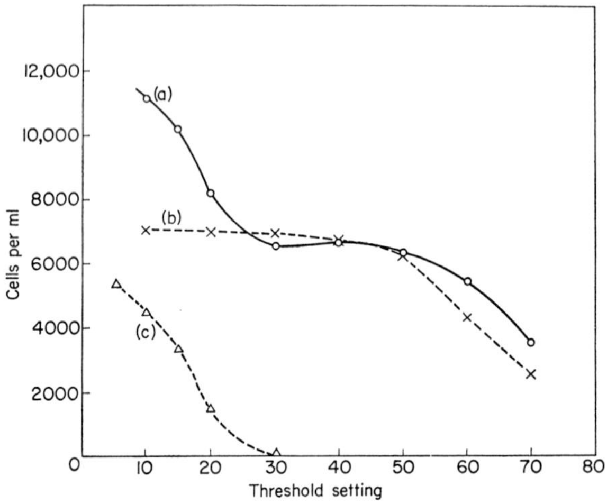

FIG. 6. Cell counts of a 1:1 mixture (Curve a) of cultures of *Scrippsiella sweeneyi* and *Gonyaulax polyedra* (O) compared with counts of each culture separately [*Gonyaulax*-X (Curve b); *Scrippsiella*–△ (Curve c) each diluted 1:1 with sea water] at different threshold settings (Coulter Counter Model A, 140 μ pore, gain 4).

Calibrations made with *Gonyaulax* cultures, fern spores, *Lycopodium* spores and pine pollen showed that the μ dimensions of a threshold unit are the same irrespective of the type of cell used in calibration, provided that the same suspending medium, pore size, and gain setting are used. It is this observation that led us to conclude, as mentioned above, that the impedance of all cells is very great relative to salt solutions and differences between the impedance of different cells are not detectable.

A factor for the interconversion of gain settings is easily obtained by counting cells at two gain settings. Conversely, the measurement in microns of the cell size distribution of a population is readily possible after calibrations have been made.

When cells of sufficiently different sizes are mixed, it is possible to detect the number of cells of each type in the mixture with the electronic counter. This is accomplished by systematically varying the lower threshold setting and plotting the counts, as shown in Fig. 6. A shoulder in the curve distinguishes the larger and smaller species. The cell count at the shoulder gives the number of cells of the larger species, and subtracting this value from the highest counts gives the count for the smaller species.

In *Gonyaulax* the daughter cells adhere for a short while just after mitosis, and enumeration of these 'pairs' may be taken as a measure of a division stage. An attempt was made to determine the presence of paired cells in populations of *Gonyaulax* using the Coulter Counter, but the size difference between paired and non-dividing cells proved too small to be detected by this method. In some organisms, e.g., *Euglena gracilis* (Cook, 1960), the difference in size between undivided and recently divided cells is sufficiently great to allow the enumeration of each from differential counts of mixtures.

Many of the limitations of this counting method may be surmised from the description already given. Although cultures of rather low densities (100 cells per ml) may be counted conveniently, a limit is set by the amount of debris, dead cells or other particulate matter. A medium which causes clogging of the aperture can be particularly bothersome. In general, however, the use of an electronic cell counter greatly facilitates cell counting in algal cultures.

REFERENCES

Brecher, G., M. Schneiderman and G. Z. Williams (1956). Evaluation of an electronic blood cell counter. *Am. J. Clin. Pathol.* **26,** 1439.

Cook, J. R. (1960). Light-induced division synchrony of *Euglena gracilis* var. *bacillaris*. *Exptl. Cell Research* **21,** 583.

Grant, J. L., M. C. Britton and T. E. Kurtz (1960). Measurement of red blood cell volume with the electronic cell counter. *Am. J. Clin. Pathol.* **33,** 138.

Kubitschek, H. E. (1958). Electronic counting and sizing of bacteria. *Nature* **182,** 234.

Mattern, C. F., F. S. Brackett and B. J. Olson (1957). Determination of number and size of particles by electronic gating. I. Blood Cells. *J. Appl. Physiol.* **10,** 56.

Richar, W. J. and E. S. Breakell (1959). Evaluation of an electronic particle counter for the counting of white blood cells. *Am. J. Clin. Pathol.* **31,** 384.

27

Techniques for Synchronizing Yeast Cells

D. H. WILLIAMSON, *The Cell Biology Department, John Innes Institute, Bayfordbury, Hertford, Hertfordshire, England*

CONTENTS

I Introduction 589
II First Method 589
III Second Method 590
References 591

I. INTRODUCTION

Two methods for preparing synchronously dividing yeast cultures have been described by Williamson and Scopes (1960, 1962). They were developed for use with *Saccharomyces cerevisiae* (National Collection of Yeast Cultures, strain 239), but in principle should be applicable to any strain of this species which is free from chain-forming tendencies. The procedures described below are slightly modified from the original processes.

The starting material for both methods is obtained by incubating cultures (with shaking) at 25° in a double-strength version of the malt-extract medium of Wickerham (1951) containing (per litre of medium): glucose, 20·0 g; 'Difco' yeast-extract, 6·0 g; 'Difco' malt-extract, 6·0 g; 'Oxoid' mycological peptone, 10·0 g. Incubation is continued for 10 days, by which time the culture is almost entirely free of cells bearing buds or small daughter cells.

II. FIRST METHOD

The washed cells are suspended at a concentration of 5–10 mg (dry weight) per ml in single strength Wickerham's medium dispensed in 100 ml amounts in 8-oz screw-capped 'medical flat' bottles. The suspension is transferred immediately to the refrigerator (4°) and left there for about 18 hr. At the end of this period the supernatant fluid is carefully decanted and discarded. The sedimented yeast is then rapidly washed several times by centrifugation in distilled water and resuspended in a convenient volume (cell concentration is not critical) of 'starvation medium' containing, per 100 ml, the following: KCl, 0·075 g; $CaCl_2$, 0·026 g; $MgCl_2{\cdot}6H_20$, 0·05 g. This suspension is aerated in a water bath at 25° for about 6 hr and after overnight storage at 4° is subjected to a similar period of aeration on the following day. This is followed by resuspension in the cold malt-extract medium as before, and the whole cycle of events is repeated as often as necessary. The repeated decantations selectively remove the smaller cells in the population, which consequently becomes increasingly homogeneous with respect to cell size as the treatment progresses. Usually a sufficient degree of synchrony is attained after four to seven cycles of feeding and starving.

III. Second Method

This process differs from its predecessor in that the smaller cells in the population are removed before applying the cyclic 'feeding–starving' routine. Moreover, the latter is accomplished without changes in temperature.

The small cells may be removed from the starting population by differential centrifugation in an aqueous solution of mannitol (15 per cent, w/v) (Williamson and Scopes, 1961). The precise details of this step may be varied to meet individual requirements, but in the author's laboratory it has been found convenient to centrifuge at about 600 × *g* for 2–6 min, using 'swing-out' buckets. The suspension should not be too concentrated; about 6 mg (dry weight) of cells per ml is satisfactory. The supernatant fluid, bearing mainly small cells, is discarded, and the deposit repeatedly re-centrifuged until microscopic examination shows it to be essentially free of the smaller individuals.

Although the entire separation can be carried out in this fashion, the labour involved can be greatly reduced by previously subjecting the suspension to a treatment referred to as 'foam flotation'. This is carried out in the following manner. The cells are suspended in 'starvation medium' at a concentration of about 5 mg (dry weight) per ml. The suspension (a total volume of about 250 ml) is transferred to a vertical glass tube about 1·6 m long and 30 mm internal diameter. This is fitted at its lower end with a rubber bung holding a sintered glass filter connected to an air supply. The bung is also pierced by a hypodermic needle connected, by means of flexible tubing, to a syringe loaded with a solution of egg albumen (approximately 1 per cent, w/v). Air is passed through the sinter and, at intervals, small volumes of albumen solution are injected sufficient to cause the formation of a stable foam. This rises to the top of the tube where it is broken by contact with a loose plug of glass wool held in place by a bung pierced by a glass tube. The smaller cells apparently adhere to the denatured protein produced by the foam to a greater extent than the large cells and, being trapped in the glass wool, are selectively removed from the population. Usually aeration for about 1 hr is sufficient to reduce the proportion of small cells to about one third of its initial value, and once this stage is reached, further aeration has no additional effect. The remaining small cells are then removed by a comparatively short period of centrifugation as described above.

The population of 'large' cells resulting from this treatment is suspended in Wickerham's malt-extract medium at a concentration of 5–10 mg (dry weight) per ml, the medium having been previously warmed to 25°. The suspension is held in a water bath at this temperature for 40 min and is then rapidly washed several times by centrifugation in 'starvation medium' previously warmed to 25°. Finally the cells are suspended in the latter medium at about the same cell density as before and are returned to the water bath, where they are vigorously aerated for about 6 hr. The whole cycle of feeding and starving is repeated on the following day, and on subsequent days as often as desired. In general, an adequate degree of potential synchrony is attained after only two cycles of treatment.

In both the above processes, the cells remain in a 'resting' state, and should not at any stage form buds. Their synchronous condition becomes apparent only when they are inoculated into a suitable growth medium.

REFERENCES

WICKERHAM, L. J. (1951). Taxonomy of yeasts. *U.S. Dept. Agr. Tech. Bull.* No. 1029.

WILLIAMSON, D. H. and A. W. SCOPES (1960). The behavior of nucleic acids in synchronously dividing cultures of *Saccharomyces cerevisiae. Exptl. Cell Research* **20,** 338.

WILLIAMSON, D. H. and A. W. SCOPES (1961). The distribution of nucleic acids and protein between different sized yeast cells. *Exptl. Cell Research* **24,** 151.

WILLIAMSON, D. H. and A. W. SCOPES (1962). A rapid method for synchronizing division in the yeast, *Saccharomyces cerevisiae. Nature* **193,** 256.

28

Technique of Synchronizing Bacterial Culture by Filtration

YOICHI MARUYAMA, *The Institute of Food Microbiology, Chiba University, Okubo, Narashino, Chiba-ken, Japan*

CONTENTS

I Principle 593
II Suggested Procedure 594
1. Organism and cultivation 594
2. Filtration 595
III Remarks 597
1. Choice of filter paper 597
2. Scaling up 597
3. Test organism 598
References 598

I. PRINCIPLE

The bulk of a microbial cell population developing exponentially is composed of cells at various growth stages in the division cycle, with a certain distribution pattern (Maruyama, 1955). Since the size of the cells at each growth stage is considered to be relatively constant, it should be possible to obtain a synchronously growing culture by collecting cells of a definite size. In the procedure of synchronization described below, cultures in the exponential phase are fractionated by filtration according to cell size.

When a thick suspension of bacterial cells is placed on top of a paper pile and flushed through it with a constant flow of an eluting medium, the cells migrate at different velocities, depending on their sizes. After a certain amount of eluting medium has passed, the distribution of cells within the paper pile and in the resulting filtrate has two distinct peaks. Fig. 1 shows the results of such a procedure observed with *Escherichia coli*; a fraction of smaller cells is obtained in the filtrate, while the larger cells are retained in the top sheets of the paper pile (Maruyama and Yanagita, 1956).

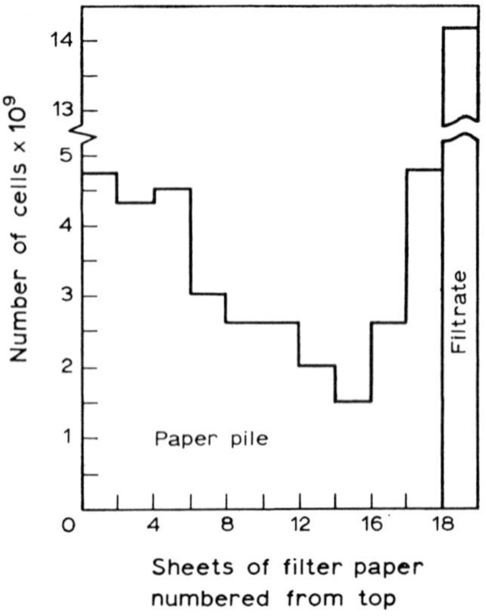

FIG. 1. Retention of cells in paper pile and filtrate. Eighteen No. 126 filter papers were packed together between two No. 1 filter papers. Filtration was stopped after 20 ml of filtrate were obtained. Cells applied: $8{\cdot}4 \times 10^{10}$, and cells recovered: $4{\cdot}7 \times 10^{10}$.

II. SUGGESTED PROCEDURE

1. *Organism and Cultivation*

Since the organism used in the original work was *Escherichia coli* (strain B), the following description will give the procedure adapted to this organism. The medium used for cultivation is as follows: NH_4Cl, 1 g; $Na_2HPO_4{\cdot}12H_2O$, 18 g; KH_2PO_4, 1 g; $MgSO_4{\cdot}7H_2O$, 0·2 g; glucose, 1 g; and Tween 80 (polyoxyethylene solbitan monooleate), 0·01 g in 1 l. of distilled water (pH 7·4). For eluting cells through the paper pile, glucose is omitted from this medium to prevent growth during filtration.

Ten milliliters of an overnight culture are inoculated into 1 l. of fresh medium, and incubated under aeration at 37° in a water bath. The doubling time of this culture is 50 min under these conditions. When the density of the culture reaches 10^8 cells per ml, cells are harvested by centrifugation of 800 × *g* for 20 min at room temperature. In this case the use of a continuous flow centrifuge at a higher speed seems to be desirable to finish the centrifugation more rapidly. The turbid supernatant containing a fraction of the smaller cells is discarded and the sedimented cells are suspended in 2 ml of eluting medium, then subjected immediately to the filtration procedure.

2. *Filtration*

Eighteen sheets of Toyo filter paper No. 126 (rather thick and coarse paper used in filter presses) with two sheets of No. 1 filter paper (thin paper for chemical analysis) on top are cut into 5·5 cm circles, set up in a pile and screwed tightly into a filtration funnel specially designed for this purpose (Fig. 2). The pile is wetted before use by passing 1 l. of water and then 50 ml of the eluting medium through it by suction, and tightened again.

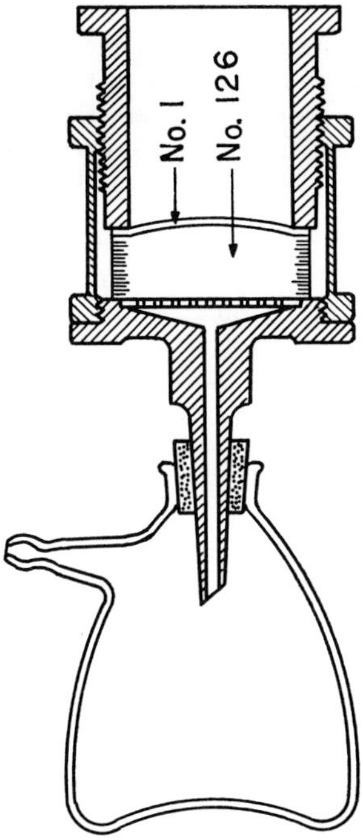

FIG. 2. Paper pile apparatus used for the fractional filtration of cells. Inside diameter, 55 mm.

Two milliliters of the cell suspension prepared as described above, containing approximately 10^{11} cells, are placed uniformly on the top of the paper pile, and sucked into the upper sheets. Vacuum filtration is then carried out, while 50 ml of the eluting medium are gently applied at the surface of the pile. A flow rate of 50 ml per 20 sec is usually maintained during filtration; a lower rate is not favorable for obtaining a suspension of uniformly-sized cells.

Immediately after the completion of filtration, the filtration funnel is unscrewed and the third, fourth and fifth sheets of filter paper from the top are taken out of the apparatus and eluted with culture medium. In this case, the first and second sheets should not be used, since it is highly possible that they contain doubled or chained cells. The average yield of larger cells thus obtained is 5–10 per cent of the initial suspension applied at the top of the paper pile. Table I gives the data for one filtration, showing the yield of larger and smaller cells.

The larger cells thus eluted from the three sheets near the top, when grown at 37° under aeration, showed an immediate

TABLE I. Number of cells in filtrate and in eluate from paper sheets third, fourth and fifth after fractional filtration. Number of cells was counted using Coulter counter (50 μ aperture tube). For experimental conditions, see text

Fraction	Volume (ml)	Density (cells/ml)	Number of cells
Original culture	1,000	$1{\cdot}00 \times 10^8$	$1{\cdot}00 \times 10^{11}$
Filtrate (smaller cells)	50	$0{\cdot}73 \times 10^8$	$0{\cdot}37 \times 10^{10}$
Eluate (larger cells)	100	$0{\cdot}66 \times 10^8$	$0{\cdot}66 \times 10^{10}$

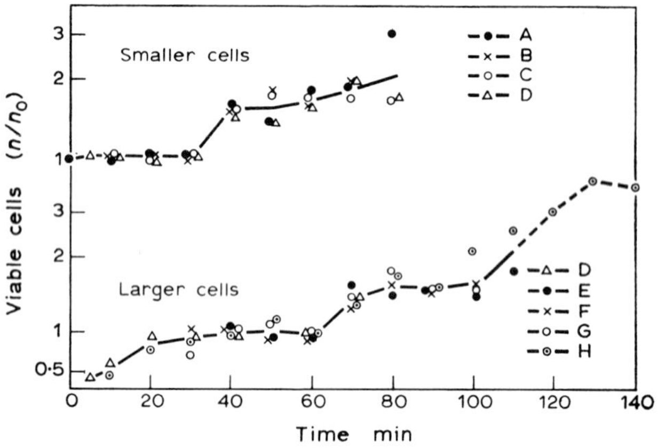

FIG. 3. Growth curves of cells obtained by fractional filtration. Upper and lower curves represent growth curves obtained after inoculations of smaller cells and larger cells, respectively. All experiments from A to H were carried out under the same experimental conditions using different batches of cells. Average numbers (n_0) of cells during the first flat phase in each experiment were as follows: A, 0·65; B, 0·57; C, 0·72; D, 2·33 (larger cells) and 0·83 (smaller cells); E, 1·28; F, 1·00; G, 3·02; and H, 4·57 cells per ml ($\times$ 10^8). In each experiment the ratio of the number (n) of cells to the average number, n/n_0, was plotted logarithmically against time. Curves were drawn through the average points at successive time intervals.

increase and then leveling off in cell number. During the period of constant cell number, each cell grew synchronously in mass until the next division (Fig. 3). On the other hand, cultivation of the smaller cells obtained from the filtrate resulted in an initial period of constant cell number, followed by division.

III. Remarks

1. *Choice of Filter Paper*

A thick and coarse filter paper (Toyo No. 126) was selected for this technique because of its high rate of filtration and high capacity for water retention. Rudner (1960) employed this filter paper for the synchronization of *E. coli*. This No. 126 filter paper is comparable to the Eaton–Dikeman No. 623 or No. 624 filter paper. The use of the latter provided good synchrony (Lark and Lark, 1960). When thin and fine paper, such as Toyo No. 1 (corresponding to Whatman No. 1), was employed instead of the thick paper, 100 sheets were necessary to absorb and fractionate the cell suspension. The employment of such thin paper with the original method was found to be unfavorable for obtaining good synchrony. For the fractionation of smaller number of cells (e.g., 10^{10} cells), however, the use of nine sheets of thin, fine paper (such as Toyo No. 131 adapted for the filtration of $BaSO_4$ precipitates), cut in 3·5 cm circles, resulted in good synchrony. Abbo and Pardee (1960) employed the following Whatman filter papers: from top to bottom, two sheets of No. 1, one of No. 42, twenty sheets of No. 1, and one of No. 42. They obtained successful results by filtering 200 ml of $1–2 \times 10^8$ cells per ml culture of *E. coli* directly through the paper pile.

In a modification of the present method, Anderson and Pettijohn (1960) employed only one sheet of Grade RA Millipore paper (pore size 1·2 μ) for the synchronization of *E. coli* (K 12).

2. *Scaling Up*

The maximum quantity of cells which can be used with the apparatus shown in Fig. 1 is approximately 3×10^{11} cells; above this the resulting culture becomes asynchronous. To scale up the present procedure, it is necessary to enlarge the area of filtration. When the diameter of the paper pile becomes larger, the flow rate of bacterial cells within the paper pile differs considerably from one part of the pile to another, since the cells tend to concentrate more and more at the centers of the sheets toward the bottom of the pile.

The cells eluted from sheets in which they are unevenly distributed include those of heterogeneous sizes, which will therefore grow asynchronously. In such a case, the cells located in the peripheral zone where they are unevenly distributed should be eliminated.

The cell-concentration gradient in the pile can be visualized by the use of ink instead of bacterial cells. After the elution of the ink by the usual procedure, the colored parts of the third, fourth and fifth sheets are checked; when the bacterial cells are filtered, each paper is cut with reference to these colored sheets, leaving the central part that corresponds to the evenly colored area. These discs are then eluted as usual with culture medium. Lark and Lark (1960) have succeeded in obtaining synchronous cultures of *Alcaligenes fecalis* with a paper pile made of nine sheets of 15 cm circles, and $1{\cdot}4 \times 10^{12}$ cells applied at the top.

3. *Test Organism*

This method, with or without modifications, has been tested and proved successful with bacilli, such as several strains of *E. coli* and *A. fecalis*. With the latter organism, which is flagellated and forms clumps in a culture, it is necessary to remove the flagella and break up the clumps before filtration. Lark and Lark (1960) did this by vigorously stirring 15 ml of a thick suspension of this organism in a Waring blendor; they were then filtered as usual. The generation time of such cells, as observed from the synchronous growth, appears to coincide with the doubling time of the original culture.

Although no coccus has yet been subjected to this method, an attempt to synchronize it might be successful. Bacilli which form chains may naturally be unfavorable material for this method.

REFERENCES

Abbo, F. E. and A. B. Pardee (1960). Synthesis of macromolecules in synchronously dividing bacteria. *Biochim. et Biophys. Acta* **39,** 478.

Anderson, P. A. and D. E. Pettijohn (1960). Synchronization of division in *Escherichia coli. Science* **131,** 1098.

Lark, K. G. and C. Lark (1960). Changes during the division cycle in bacterial cell-wall synthesis, volume and ability to concentrate free amino acids. *Biochim. et Biophys. Acta* **43,** 520.

Maruyama, Y. (1955). The age distribution of cells in a microbial population at the logarithmic growth phase, statistical consideration. *Ann. Rept. Inst. Food Microbiol., Chiba Univ.* **8,** 75.

Maruyama, Y. and T. Yanagita (1956). Physical methods for obtaining synchronous culture of *Escherichia coli. J. Bacteriol.* **71,** 542.

Rudner, R. (1960). Mutation as an error in base pairing. *Biochem. Biophys. Res. Commun.* **3,** 275.

29

Techniques Involved in Synchronization of *Lactobacillus acidophilus* R26

VICTOR W. BURNS, *Biophysics Laboratory, Stanford University, Stanford, California*

CONTENTS

I Media 599
II Preparation and Enumeration of Cells 599
References 600

I. MEDIA

Stock cultures were grown and carried in milk media (Løvtrup and Roos, 1957). Two types of synthetic media were employed for preliminary growth and subsequent synchronization. One (LSM) was the same as Hoff-Jørgensen's (1952) but with the addition of 10 mg of uracil and 0·5 mg of pyridoxine per liter, as suggested by Siedler *et al.* (1957); the other (DBM) was the double basal medium described by Løvtrup and Roos (1957). The media differ primarily in that LSM uses hydrolyzed casein to supply amino acids, and DBM employs known quantities of synthetic amino acids instead. Although cell growth seems to be the same in either medium, LSM gave the better synchrony.

II. PREPARATION AND ENUMERATION OF CELLS

Late log phase cells were harvested from cultures grown for 14–16 hr at 37°C in LSM or DBM with 10^{-8} *M*/ml thymidine. Cell concentration at the time of harvest near the end of the log phase was 2×10^8 cells/ml. Middle log phase cells were prepared in a turbidostat. The latter device employs a Sigma pump to sample cell suspension in a 1000 ml flask and pumps it through a continuous recording turbidimeter. The pump does not injure the cells. The turbidimeter controls input of fresh medium and outlet of cell suspension through electric solenoid valves. It can be set to control the optical density of the growing culture at any predetermined level. In practice, the device was set to keep cell concentration at 1–2 $\times 10^7$ cell/ml, and the cells grew for about a dozen generations at this concentration before being harvested for synchronization experiments.

Cells prepared by either of the above methods were harvested, washed, and suspended at about 10^7 cell/ml in synthetic medium lacking thymidine. Two to three hours later thymidine at 10^{-8} M/ml was added to induce synchronized division.

Viable cell counts were made by the indirect method; suitable diluted aliquots were plated on nutrient agar, and the number of viable cells per milliliter was determined from colony counts. Comparison with Petroff–Hausser counting chamber results showed the viability was always above 50 per cent and usually close to 100 per cent.

REFERENCES

HOFF-JØRGENSEN, E. (1952). A microbiological assay of deoxyribonucleosides and deoxyribonucleic acid. *Biochem. J.* **50,** 400.

LØVTRUP, S. and K. ROOS (1957). Microbiological determination of DNA. The growth of *Thermobacterium acidophilum* on a chemically defined medium. *Exptl. Cell Research Suppl.* **4,** 269.

SIEDLER, A. J., F. A. NAYDER and T. B. S. SCHWEIGERT (1957). Studies on improvements in the medium for *Lactobacillus acidophilus* in the assay for deoxyribonucleic acid. *J. Bacteriol.* **73,** 670.

30

Special Methods Relating to Tissue Cells

ALISON A. NEWTON, *Department of Biochemistry, University of Cambridge, England*

CONTENTS

I Synchronization of Cell Division by Temperature Shock . . 601
1. Growth medium 601
2. Cell lines 601
3. Experimental procedure 601
II Synchronization of Cell Division by Amethopterin Treatment . 602
1. Growth medium 602
2. Cell line 603
3. Induction of synchronous division 603
III Synchronization of Cell Division by Selection of Mitotic Cells . 603
References 604

I. SYNCHRONIZATION OF CELL DIVISION BY TEMPERATURE SHOCK

1. *Growth Medium*

All growth media contained 0·002 per cent (w/v) phenol red, 100 units/ml penicillin, 100 μg/ml streptomycin and were based on Gey's balanced salt solution (Gey, 1950) as the electrolyte base. Either 0·5 per cent (w/v) lactalbumin hydrolysate or Eagle's amino acid and vitamin mixture (Eagle, 1955) was added as the source of growth factors. 'Human' growth medium, for cultivation of HeLa cells, was supplemented with 20 per cent whole human serum, and 'bovine' growth medium was supplemented with 10 per cent bovine serum (Newton and Wildy, 1959).

2. *Cell Lines*

Human cervical carcinoma cells (strain HeLa) (Scherer *et al.*, 1953) or mouse fibroblast cells (strain L) (Sanford *et al.*, 1948) were grown routinely as monolayer cultures. Stock cultures were maintained in growth medium on one face of a six-sided glass bottle and were subcultured 24 hr before use as the inoculum for experimental flasks.

3. *Experimental Procedure*

All operations were carried out in a warm room at 37°C and all solutions and glassware were warmed to this temperature before use.

Suspensions of HeLa cells were prepared from the stock cultures after removal from the glass by gentle pipetting in a solution of disodium ethylene-diamine-tetraacetate (1/5000) (EDTA) in phosphate-buffered saline free from calcium and magnesium (Dulbecco and Vogt, 1954). The cells were immediately added to a large volume of warm growth medium to give a final cell density of 5–10 $\times$ 10^4/ml. L cell suspensions were prepared in a similar manner except that cells were removed from the culture vessel by gentle washing with fresh growth medium. Cell suspensions were agitated to prevent sedimentation of the cells.

Aliquots of 2 ml were taken from a cell suspension prepared in this way and placed in thin-walled, screw-capped glass vials of 10 ml capacity; stoppered test tubes could also be used. If cells were to be studied by cytological methods thin glass coverslips were added to some of the vials. The vials were then placed in a horizontal position in metal racks and incubated at 37°C until cells had attached to the glass surface and had entered the logarithmic phase of growth (24 hr for HeLa cells and 12–24 hr for L cells). At this time the cultures were removed from the incubator and placed in their metal racks on the shelf of a refrigerator at 4°C for 1 hr. Test tube cultures were cooled in a water bath at 4°C for the same period. The cultures were then replaced at 37°C and examined at intervals. Cell division occurred 17–18 hr after replacement at the optimal growth temperature.

Determinations of cell numbers and of cell properties were carried out after removal of the cells from the glass in EDTA solution. Replicate cell counts were made using a Fuchs–Rosenthal counting chamber, depth 0·2 mm, and had an accuracy of $\pm$ 5 per cent. For biochemical analysis cells from several replicate cultures were pooled for each determination.

II. Synchronization of Cell Division by Amethopterin Treatment

1. *Growth Medium*

The growth medium consisted of Eagle's HeLa medium (Eagle, 1955) employing Earle's balanced salt solution (Earle, 1943) as electrolyte base, and supplemented with 2 $\times$ 10^{-5} M inositol and 10 per cent whole bovine serum. When dialysed serum was used the medium was further supplemented with 0·1 mM glycine and serine. All cultures were gassed with 5 per cent CO_2 in air. For suspension cultures calcium and magnesium were omitted from the above medium and further additions of 0·1 mM glycine and serine, and 0·1 per cent carboxymethyl cellulose (Type 12HP, Hercules Powder Co, Wilmington, Delaware) were made (Rueckert and Mueller, 1960).

2. *Cell Line*

Human cervical carcinoma cells (strain HeLa) (Scherer *et al.*, 1953) were used in these experiments. Stock monolayer cultures were maintained in growth medium in Roux bottles and were used as the inoculum for experimental flasks 3 days after subculture, thus ensuring that cells in the logarithmic phase of growth were obtained. Replicate monolayer cultures were prepared by pipetting cells from such cultures into 10 ml of growth medium in a 3 oz prescription bottle, to give a final cell density of 5×10^5 cells/bottle. Experiments could also be carried out using suspension cultures grown by the method described by McLimans *et al.* (1957). Addition of 'spinner' medium to the suspension culture maintained a daily initial cell density of $1{\cdot}5 \times 10^5$ cells/ml.

3. *Induction of Synchronous Division*

All manipulations were carried out in a warm room at 37°C and all media were warmed to this temperature to avoid any synchronization due to temperature shock. Experiments were started 24 hr after the preparation of the cultures as described above, by the addition of one-tenth the culture volume of 'amethopterin' solution in growth medium to give the following final concentration:— $10^{-6}M$ amethopterin, $5 \times 10^{-6}M$ adenosine and $10^{-4}M$ glycine and serine. The amethopterin could be replaced by $10^{-6}M$ 5-fluoro-2-deoxyuridine. After a further 16 hr, during which there was little increase in cell number, the inhibition was reversed by the addition of one-hundredth culture volume of $10^{-3}M$ thymidine. Cell division followed shortly after this addition.

III. Synchronization of Cell Division by Selection of Mitotic Cells

The cell line (Terasima and Tolmach, 1961, 1963) used in these experiments was HeLa S_3-9IV (Marcus, 1959) grown as a monolayer culture in growth medium N16HHF (Ham and Puck, 1962). Plastic Petri dishes, 100 mm in diameter, were seeded with about 10^6 cells from these stock cultures, and incubated at 37°C overnight. The medium was then changed in order to remove any dead cells or debris that might have accumulated. Three to 5 hr later the growth medium was removed and the cell sheet was rinsed repeatedly with a firm but gentle stream of 5–7 ml of fresh growth medium played over the whole surface. About 1 per cent of the cells was found in the medium after this procedure and of these 80–90 per cent were in mitosis. These cells were inoculated into new plastic Petri dishes and incubated at 37°C. After a further 3 hr a high proportion of those cells which had attached to the glass completed division. Twenty hours later the next division cycle was observed. Cell numbers were estimated from a photographic record of the cultures.

REFERENCES

DULBECCO, R. and M. VOGT (1954). Plaque formation and isolation of pure lines with poliomyelitis viruses. *J. Exptl. Med.* **99,** 167.

EAGLE, H. (1955). Nutrition needs of mammalian cells in tissue culture. *Science* **122,** 501.

EARLE, W. R. (1943). Product of malignancy *in vitro.* IV. The mouse fibroblast cultures and changes seen in the living cells. *J. Natl. Cancer Inst.* **4,** 165.

GEY, G. O. (1950). *Bull. Tissue Culture Assoc.* March 16th, 1949. Quoted in Parker, R. C., *Methods of Tissue Culture*, 2nd ed., Hoeber, New York, p. 81.

HAM, R. G. and T. T. PUCK (1962). *Methods in Enzymology* (S. P. Colowick and N. O. Kaplan, eds.). Academic Press, New York, p. 90.

MCLIMANS, W. F., E. V. DAVIS, F. L. GLOVER and G. W. RAKE (1957). The submerged culture of mammalian cells: The Spinner culture. *J. Immunol.* **79,** 428.

MARCUS, P. I. (1959). Host–cell interaction of animal viruses. II. Cell-killing particle enumeration: Survival curves at low multiplicities. *Virology* **9,** 546.

NEWTON, A. A. and P. WILDY (1959). Parasynchronous division of HeLa cells. *Exptl. Cell Research* **16,** 624.

RUECKERT, R. R. and G. C. MUELLER (1960). Studies on unbalanced growth in tissue culture. I. Induction and consequences of thymidine deficiency. *Cancer Research* **20,** 1584.

SANFORD, K. K., W. R. EARLE and G. D. LIKELY (1948). The growth *in vitro* of single isolated tissue cells. *J. Natl. Cancer Inst.* **9,** 229.

SCHERER, W. F., J. T. SYVERTON and G. O. GEY (1953). Studies on the propagation *in vitro* of poliomyelitis viruses. IV. Viral multiplication in a stable strain of human malignant epithelial cells (strain HeLa) derived from an epidermoid carcinoma of the cervix. *J. Exptl. Med.* **97,** 695.

TERASIMA, T. and L. J. TOLMACH (1961). Changes in X-ray sensitivity of HeLa cells during the division cycle. *Nature* **190,** 1210.

TERASIMA, T. and L. J. TOLMACH (1963). Growth and nucleic acid synthesis in synchronously dividing populations of HeLa cells. *Exptl. Cell Research* **30,** 344.

Addendum to Chapter 17 'Synchronous Division of Animal Cells in Culture' by Alison A. Newton

The major part of this contribution has dealt with the development of systems by which cultures of animal cells could be induced to divide in a synchronous fashion. The emphasis of research has now shifted from the development of techniques to the application of these techniques to the study of the properties of the dividing cell. Much information has been gained from these studies and a very brief summary of this recent work is included here in an attempt to amplify the description of the properties of the synchronous system.

Most of the results that have been reported describe biochemical investigations of cell cultures synchronized by the induction and reversal of a 'thymidineless' state by treatment with either amethopterin or 5-fluoro-2-deoxyuridine (FUDR) followed by the addition of thymidine. Thus, Littlefield (1962) found a higher degree of synchronization of DNA synthesis in cultures of mouse cells (strain L) synchronized by this method than by subculture of starved cells into fresh growth medium. Two periods of DNA synthesis were observed in these cultures, each of which preceded the mid-point of the period of cell division by about 8 hours.

Mueller *et al.* (1962) have studied some of the changes that occur during treatment of HeLa cells with amethopterin. They have shown that whereas only 20 per cent of the cells in a control culture are capable of incorporating a 'pulse' of thymidine into DNA, the number of such cells increases during amethopterin treatment. Thus 16 hr after the induction of the 'thymidineless' state all cells are ready to initiate DNA synthesis when provided with a source of thymidine. Under these conditions amethopterin does not interfere with the synthesis of protein or RNA and so Mueller (1963) concludes that these processes are not involved in bringing cells to a 'triggered' state. The addition of thymidine to such triggered cells is followed by a wave of DNA synthesis and subsequently by waves of mitosis and of cell division, resulting in a doubling of the cell population. Similar results were obtained when guanine deoxyriboside was used as an inhibitor of DNA synthesis.

Several groups of workers (Paul and Hagiwara, 1962; Mueller, Kajiwara, Stubblefield and Rueckert, 1962; Littlefield, McGovern and Margeson, 1963) have studied the kinetics of DNA synthesis after rescue of the cells from the 'thymidineless' state. In each case synthesis of DNA began slowly in all cells almost immediately after the addition of thymidine. An acceleration of DNA synthesis was also a constant feature of these systems. Mueller *et al.* (1962) and Mueller (1963) found that the addition of an inhibitor of RNA synthesis, actinomycin D, at the same time as thymidine did not affect the initial rate of DNA synthesis but abolished the acceleration of synthesis usually observed. The addition of puromycin, an inhibitor of protein synthesis, also prevented the acceleration of DNA synthesis: when this compound was added to the cells already actively synthesizing DNA it caused a reduction in the final yield of DNA so that cells failed to divide. It appears from these results that the acceleration of DNA synthesis requires the synthesis of both new RNA and new protein. It is of interest that Littlefield, McGovern and Margeson (1963) have found a change in the intracellular distribution of DNA polymerase activity during the period when DNA synthesis is accelerating.

Stubblefield and Mueller (1962) found that the distribution of labelled thymidine amongst the chromosomes varied with the rate of DNA synthesis prevailing at the time of thymidine addition. The thymidine was not randomly distributed either between different chromosomes or along the length of a single chromosome. The authors conclude that the rate of synthesis of particular DNA molecules depends upon the type of association of DNA in the chromosome. They believe that the acceleration of DNA synthesis is a result of the activation of late labelling sites in the chromosomes. A similar variation in the rate of DNA synthesis, in HeLa cells synchronized by selection of mitotic cells, has been observed by Terasima and Tolmach (1963); they also conclude that the asynchrony of DNA replication occurs at the chromosomal level as has been suggested previously by Taylor (1960).

Mueller (1963) also concluded that synthesis of RNA during part of the period of DNA synthesis is necessary if the cell is to undergo division. Similarly, it appears that protein synthesis during and after the DNA synthetic period is mandatory for cell division. Paul and Hagiwara (1963) observed that in cultures of L cells synchronized by inhibition with FUDR, synthesis of nuclear RNA, nuclear protein and cytoplasmic protein was resumed more slowly than DNA synthesis after the addition of thymidine.

Terasima and Tolmach (1963a) have studied the properties of HeLa cell cultures synchronized by selection of mitotic cells. They have shown that the growth parameters of these cultures, such as the duration of the G_1, S and G_2 phases and of mitosis, were the same as for randomly-dividing cultures. Desynchronization of the culture seemed to arise largely through variations in the length of the G_1 phase. Synthesis of RNA continued throughout interphase, with the rate of synthesis increasing towards the end of interphase. No RNA synthesis was observed during mitosis. Determination, by autoradiography, of the rates of RNA synthesis in sister cells showed that there was neither positive nor negative correlation between the amounts of uridine incorporated by the sister cells.

Terasima and Tolmach (1963b) have also studied the sensitivity of the synchronized population of HeLa cells to various insults at different stages of the division cycle. They find that cells are least affected by X-irradiation when treated during the early G_1 phase. Cells irradiated during mitosis complete cell division normally but show a maximal delay in entering the next mitotic cycle. The subsequent rate of DNA synthesis and also the ability of irradiated cells to survive seem to be most severely affected when cells engaged in DNA synthesis are subjected to X-irradiation. As the DNA content of cells increases above a critical value the cells become less sensitive to the damaging effects of irradiation. Similar observations on the sensitivity to ultraviolet and X-rays were made by Erikson and Szybalski (1963) with cultures synchronized by treatment with FUDR. Terasima and Tolmach (quoted by Lark (1963)) have also studied the thermal sensitivity of cells at different stages of the cell cycle. They have shown that cooling of cells during the G_1 or S phase delays mitosis but that the G_2 phase is not temperature sensitive: these results confirm earlier observations made by Newton and Wildy (1958).

REFERENCES

Erikson, R. L. and W. Szybalski (1963). Molecular radiobiology of human cell lines IV. Variations in ultraviolet and X-ray sensitivity during the division cycle. *Radiation Research*, **18,** 200.

Lark, K. G. (1963). Cellular control of DNA synthesis. In *Molecular Genetics*, I (J. H. Taylor, ed.). Academic Press, New York, Chapter IV, p. 153.

Littlefield, J. W. (1962). DNA synthesis in partially synchronized HeLa cells. *Exptl. Cell Research* **26,** 318.

Littlefield, J. W., A. P. McGovern and K. B. Margeson (1963). Changes in distribution of polymerase activity during DNA synthesis in mouse fibroblasts. *Proc. Natl. Acad. Sci., U.S.A.* **49,** 102.

Mueller, G. C. (1963). Molecular events in replication of nuclei. *Exptl. Cell Research Suppl.* **9,** 144.

Mueller, G. C., K. Kajiwara, E. Stubblefield and R. Rueckert (1962). Molecular events in the reproduction of animal cells I. The effect of puromycin on the duplication of DNA. *Cancer Research* **22,** 1084.

NEWTON, A. A. and P. WILDY (1958). Parasynchronous division of HeLa cells. *Exptl. Cell Research* **16,** 624.

PAUL, J. and A. HAGIWARA (1962). A kinetic study of the action of 5 FUDR on synthetic processes in mammalian cells. *Biochim. et Biophys. Acta* **61,** 243.

STUBBLEFIELD, E. and G. C. MUELLER (1962). Molecular events in the reproduction of animal cells II. The focalized synthesis of DNA in the chromosomes of HeLa cells. *Cancer Research* **22,** 1091.

TAYLOR, J. H. (1960). Asynchronous duplication of chromosomes in cultured cells of Chinese Hamster. *J. Biophys. Biochem. Cytol.* **7,** 455.

TERASIMA, T. and L. J. TOLMACH (1963a). Growth and nucleic acid synthesis in synchronously dividing populations of HeLa cells. *Exptl. Cell Research* **30,** 344.

TERASIMA, T. and L. J. TOLMACH (1963b). Variations in several responses of HeLa cells to X-irradiation during the division cycle. *Biophys. J.* **3,** 11.

TOLMACH, L. J. and T. TERASIMA (1962). DNA synthesis in X-irradiated populations of synchronously dividing cells. *Radiation Research* **16,** 559.

Author Index

Authors' and joint authors' names of chapters in this book and the page numbers at which these chapters begin are printed in **heavy type,** page numbers of citations in the text are in ordinary type and those of bibliographical references at the ends of the chapters are in *italics*.

Aach, H. G., *302*
Abbo, F. E., 4, *8*, 342, 347, 398, *416*, 425, 431, 457, 463, 597, *598*
Abelson, P. H., *419*
Adams, M. H., 427, *431*
Adolph, E. F., 76, *91*, 356, 369, *378*
Agrell, I., 39, 41, 43, 44, 45, 46, 47, 48, 50, 52, 53, 54, 56, 58, 59, 60, 61, 62, *63*
Alfert, M., 63, 82, *91*
Amaha, M., *416*
Amano, M., 78, 87, *91*
Anderson, E., *175*
Anderson, N. G., 74, *91*
Anderson, P. A., 342, *347*, 597, *598*
Andrews, E. A., 56, *63*
Angerer, C. A., 235, 236, *244*
Arnon, D. I., *569*
Asano, N., *417*
Astrachan, L., 79, *96*, *320*
Atwood, K. C., 425, *431*
Axelrad, A. A., 449, 455, *463*

Bajaj, V., 402, *416*
Bajer, A., 20, 21, *34*, *35*, *36*
Bak, I. J., *174*, *562*
Balske, R. J., *378*
Band, R. N., 513, *519*
Bang, F. B., *464*
Barber, H. N., *463*
Barka, T., 81, *93*
Barner, H. D., 3, *8*, 81, *91*, 329, *347*, 435, *438*, 442, 454, *463*, *534*
Bayne-Jones, S., 76, *91*, 356, 369, *378*
Beam, C. A., 358, 368, *378*
Beams, H. W., *175*
Beerman, W., 54, *64*
Bender, M. A., 76, 79, 80, 84, 86, 87, 88, 89, 90, *91*, *95*, 510, 513, *519*
Bendich, A., 185, *193*
Ben-Ishai, R., *390*
Bennett, E. L., *305*
Benson, A. A., 293, *302*
Bentzon, M. W., *95*, *419*, *431*
Berg, R. H., 510, 515, *519*
Berkson, J., 510, *519*
Bernstein, E., 319, *320*, 333, 336, *347*, 521, *534*
Bertalanffy, L. von, 346, *349*
Bianchi, P. A., 451, *463*
Bibring, T., 171, *174*
Billen, D., 438, *438*
Bittner, J. J., 309, *320*
Black, C. A., 22, *34*
Blakeslee, A. F., 18, *37*
Bloch, D. P., 82, *91*
Block, K., *562*
Blum, J. J., 330, 339, 343, 345, *347*, *348*, *508*, 524, 527, 528, 529, *535*
Blumenthal, L. K., *508*
Bolton, E. T., *419*
Bollum, F. J., 83, 84, 85, 87, *91*, 451, 463
Bond, V. P., 76, *92*
Bonner, J., *93*
Borgert, A., 317, *320*
Borsook, H., 385, *390*
Bortels, H., 393, *418*
Borum, K., 51, *64*
Boveri, T., *64*
Braarud, T., 308, *320*
Brachet, J., 60, *64*, *156*, *463*
Brachon, S., 160, 161, 171, *173*
Brackett, F. S., *520*, 579, 586, *587*
Bradley, D. E., 353, *378*
Branchflower, N. H., 246, *520*
Brandt, K., 382, *389*
Breakell, E. S., 515, *520*, 579, *587*
Brecher, G., 515, *519*, 579, *587*
Bresler, A., *93*
Bresslau, E., 43, *64*
Briggs, R., 54, *64*
Britt, J. J., *562*
Britten, R. J., *419*
Britton, M. C., *519*, 579, 586, *587*
Brock, M., 435, *438*

Brock, T., 435, *438*
Bronfenbrenner, J., 429, *431*
Bruce, V. G., 313, 320, 472, *483*
Brues, A. M., *466*
Bryan, A. M., *36*
Bucciante, L., 442, *463*
Bücher, T., 192, *193*
Buetow, D. E., 528, *535*
Bullough, W. S., 50, *64*, 90, *91*
Burlingame, L. L., 23, *34*
Burma, D. P., 87, *91*, *94*
Burns, V. W., 73, 74, *91*, 354, 358, *378*, **433,** 436, **599**
Butler, A. V., *463*
Buvat, R., 14, *34*

Callan, G., 442, *463*
Calvin, M., 302, *305*
Cameron, I. L., 75, *91*
Campbell, A., 3, *8*, 74, *91*, 358, *378*, 421, 422, *431*, **469,** 471, *483*
Campbell, D. H., 22, *34*
Cann, J. R., 147
Cardiff, I. D., 23, *34*
Carleton, A., 309, *320*
Carlson, J. G., 461, *463*
Carlson, L., *93*
Carrier, R. F., *95*
Carrière, R., 80, *91*
Casperrson, T., 382, *389*
Casperrson, T. O., *93*
Cavalieri, L. F., 84, *92*
Cavalli, G., 459, *465*
Cerfontaine, P., 41, 46, *64*
Cerroni, R. E., 127, 128, 145, 150, *156*, 210, *211*
Chalkley, H. W., 331, *349*
Chamberlain, C. J., 15, 16, 17, 23, *34*
Chargaff, E., *64*
Chatton, E., 160, 161, 170, 171, 172, *173*, *174*
Chayen, J., *92*
Chayen, R., 403, *416*
Chayen, S., *416*
Chevremont, M., 446, *463*
Chèvremont-Comhaire, S., 446, *463*
Chou, S. C., 188, 189, *193*
Christensson, E., 111, 113, 116, 124, 127, *156*, 178, 179, 180, *193*, 210, *211*, 239, *244*
Chu, E., 437, *439*
Church, B. D., 393, 405, 408, 411, *416*, *417*
Claff, C. L., *194*
Clermont, I., 51, *66*
Cohen, S. S., 3, *8*, 81, *91*, 329, *347*, 411, *416*, 435, *438*, 454, *463*, *534*
Cohn, W. E., 294, *305*, 402, *419*, *466*
Conklin, M. E., 18, *37*, 39, 41, 46, *64*
Cook, B., 493, *494*
Cook, J. R., 326, 332, 335, 343, 344, 345, *347*, 382, **485,** 493, *494*, **521,** 523, 524, 530, 532, 533, 534, 535, 587, *587*
Coulter, J. M., 16, *34*
Coulter, W. C., 515, *519*
Cowie, D. B., *419*
Cramer, M., 326, *348*, 531, 532, *535*
Crathorn, A. R., *463*
Cummings, D. J., *431*
Cutter, V. M., Jr., 19, *34*, *37*
Crocker, T. T., *92*

Daille, J., 58, *66*
Dalco, A., 43, 55, *64*
Dam, L. Van, *194*
Damle, S. P., *416*
Dan, K., 41, 46, *64*, 244, *245*
Danforth, W. F., 329, *347*, *349*
Daniel, H., *302*
Darlington, C. D., 24, *34*
Davidson, D., *92*
Davies, H. G., *92*
Davies, J. W., 388, *389*
Davis, E. V., *604*
Davis, H. S., 51, *64*
Davis, J. C., *419*
Dawson, J. A., 214, *245*
Dedonder, R., 408, *417*
Deeley, E. M., 80, *92*, *95*
DeLey, J., 408, *417*
Denneny, J. M., 403, 405, *419*, *420*
Deutsch, J. F., *92*
Dewey, V. C., 100, 106, *157*, 544, *563*
Diringer, R., *93*
Dixon, H. H., 21, *34*
Döpp, W., 22, *34*
Dörrien, U., 76, *95*, *465*
Dohan, F. C., Jr., *94*

Dolezal-Janisch, R., *96*
Donalson, J. T., *92*, *494*
Doty, P. M., 513, *519*
Doudney, C. O., 438, *438*
Dounce, A. L., *193*
Drabble, E., 355, *378*
Drabble, H., *378*
Dreher, J. J., 522, *535*
Drew, R. M., 80, *94*, 451, 459, *465*
Dubé, L. F., 19, *34*
DuBois, A. M., 43, 54, *64*
Dubow, R., 27, *36*
Ducoff, H. S., 187, *193*
Dürwald, H., 78, 87, *94*
Dulbecco, R., 602, *604*
Durand, M. C., *64*

Eagle, H., 443, 464, 601, *604*
Eames, A. J., 23, *34*
Earle, W. R., *604*
Eddy, A. A., 381, *389*
Edwards, J. L., 80, *92*, 486, *494*
Ehret, C. F., 170, *174*
Eichel, H. J., 190, *193*
Eiler, J. J., 178, *193*
Elliott, A. M., 121, 156, 162, 163, 164, 172, 173, *174*, 547, *562*
Elson, D., 64
Emanuelsson, H., 50, 58, *64*
Engelberg, J., 497, 499, 500, 502, 507
Engström, A., 382, *389*
Erickson, R. O., 11, 25, 26, 28, *34*, *36*, 89
Erikson, R. L., 607, *607*
Errera, M., 78, 87, *94*, *95*, 461, *463*
Erwin, J., *562*

Falcone, G., 388, *389*
Farr, A. L., *535*
Fauré-Fremiet, E., 85, *92*, 160, *174*
Fautrez, J., 459, *463*
Fawcett, D. W., 51, 56, *64*
Feinendegen, L. E., 76, 78, 87, *92*, 460, *463*
Ferguson, M. C., *34*
Firket, H., 80, *92*, 442, 458, *464*
Fischer, A., 446, *465*
Fisher, R. A., 510, *519*
Fitz-James, P. C., 397, 398, *417*, *420*
Flynn, T. T., 44, *64*
Fogh, J., 447, *464*
Forro, J. R., 397, 399, *420*, 486
Foster, A., 15, 30, 31, *35*
Foster, J. W., 392
Foster, T. S., 83, *92*
Frank, G. M., 50, *64*
Frankel, J., 117, 118, 119, 122, 140, 141, 142, 143, 144, 145, *156*, 167, 168, 169, 170, *174*, 210, 211, 486, *494*
Freeman, B., 19, *34*
Freese, H. L., 92, *494*
Fresco, J. R., *193*
Friesner, R. C., 13, 14, *35*
Frisch, D. M., 438
Fry, W., *389*
Frydenberg, O., 80, 89, *92*
Fuller, R. C., 403, *417*
Furgason, W. H., 160, 162, 172, 173, *174*
Furth, J. J., 87, *92*

Gabriel, O., *417*
Gall, J. G., 81, 83, *92*
Gauchery, M., *92*
Gaulden, M. E., 80, 82, *92*
Gavin, J. J., 513, *519*
Gelber, B., 78, *97*, 460, *466*
Gelfant, S., 90, *92*
Gey, G. O., 443, *464*, 601, *604*
Gey, M. K., 443, *464*
Giaque, B. G., *94*
Giese, A. C., 127, 129, 130, *157*, 167, 174, 184, 186, 187, *193*
Giesel, L. O., Jr., 51, *66*
Gifford, E., 15, *35*
Glock, G. E., 192, *193*
Glover, F. L., *604*
Godlewski, E., Jr., 52, *65*
Godman, G. C., 78, 82, *91*
Goldacre, R. J., 239, *245*
Goldmann, M., *92*
Goldschmidt, R., 43, *65*
Goldstein, L., 78, 87, *92*, *93*
Goldstein, M. N., *465*
Gough, L. H., 307, *320*
Goulding, K. J., *304*
Grant, J. L., 515, *519*, 579, 586, *587*

Grant, P., 57, *65*
Gray, J., 56, *65*
Gregory, J., *438*
Gregory, J. D., 401, *417*
Grell, K. G., 317, *320*
Grimstone, A. V., 172, *174*
Gruber, K., 216, *245*
Guilcher, J. M., *35*
Gulyas, S., *466*
Gunsalus, I. C., 409, *417*
Gustafson, T., 50, 62, *65*
Guttes, E., 81, *93*
Guttes, S., 81, *93*, *94*

Hachisuka, Y., 396, 408, *417*
Haddad, S. A., 76, *94*, 355, 356, 358, 368, *378*
Häcker, V., 41, *65*
Hagiwara, A., 605, 606, *608*
Hakala, M. T., 453, *464*
Halberg, F., 50, *65*, 309, *320*
Hale, L. J., 50, *65*
Halmgren, H., 96
Halvorson, H. O., 388, *389*, *390*, 393, 405, 407, 408, 411, 412, *416*, *417*, *418*
Ham, R. G., 603, *604*
Hamburger, K., 110, 123, 124, 125, 127, 137, 138, 139, 145, 149, *156*, 188, 190, *193*, 198, 205, 210, *211*, 237, *245*, 374, *379*, 544, 546, 547, 552, 555, *562*
Hanawalt, P. C., 426, *431*, 435, 438, *439*
Harada, T., 401, *419*
Harrington, N. G., 461, *463*
Harris, G., 381, 382, 388, *389*
Harris, P. J., 80, 84, 171, *174*
Hartman, C. G., 55, *66*
Hase, E., 247, 248, 250, 256, 257, 258, 261, 262, 265, 272, 273, 274, 275, 280, 282, 283, 284, 286, 292, 293, 294, 295, 296, 297, 298, 299, 300, 301, *302*, *303*, *305*, 341, *347*
Hastings, J. W., 307, 309, 313, 316, 317, *320*, *321*, 522, *535*, **579,** 583
Hatschek, B., 41, *65*
Hayashi, M., 358, 368, *378*
Hecht, L. I., 85, *93*
Heilbrunn, L. V., 236, 243, *245*
Hell, A., 78, 87, *94*, *95*
Henneguy, L., 56, *65*
Hershey, A. D., 429, *431*
Hill, J. P., 44, *64*
Hoff-Jørgensen, E., 57, 58, 60, *65*, 435, *438*, 599, *600*
Hoffman, J., *483*
Hoffman, J. G., *507*
Hoffmann-Berling, H., 442, *464*
Hoffmann-Ostenhof, O., 405, *417*
Hogg, J. F., 178, 182, *195*
Holden, C., *36*
Holm, Å., 43, *65*
Holtzer, H., 52, 61, *65*, *67*, 280, *303*
Holz, G. G., 165, 172, 173, *174*, 238, *245*
Holz, G. G., Jr., 114, 116, 121, 122, 140, *156*, *157*, 160, 166, *174*, 198, 210, *211*, 544, *562*
Hood, J. R., *484*
Horecker, B. L., *417*
Horne, R. W., *304*
Hoshino, J., 411, *417*
Hotchkiss, R. D., *8*, 341, *347*, 444, *464*
Hotta, Y., 31, 32, 33, *35*
Howard, A., 71, 80, *93*
Howard, F. L., 56, *65*, 214, *245*
Hsu, T. C., 448, *464*
Huang, R. C., 87, *93*
Huckmann, M. S., 246, *520*
Huebschman, C., 398, *417*
Huettner, A. F., 54, *65*
Hughes, A., 61, *65*, 74, *93*, 450, *464*
Huie, L. H., *65*
Humphrey, H. B., 22, *35*
Humphrey, R. R., 51, *65*
Hurn, M., *519*
Hurwitz, J., 87, *92*, *93*
Huskins, C. L., 24, *35*
Hutchens, J. O., 330, *347*
Hutner, S. H., 521, *535*
Hyatt, M. T., *418*
Hyde, L., 243, *245*

Imai, Y., 403, *417*
Ingelman, B., 400, *417*
Inoué, S., 21, *35*, 238, 243, *245*
Irie, R., *419*
Ishitani, C., 393, *419*
Ito, S., *64*

Iverson, R. M., 127, 129, 130, *157*, 184, 186, 187, *193*
Iwamura, T., 247, 248, 250, 256, 257, 258, 264, 267, 273, 276, 278, 279, 282, 283, 286, 298, *303*, *305*

Jacobson, W., 460, *464*
Jacoby, F., 452, *464*
Jahn, T. L., 111, 121, *158*, *174*, 178, 184, 185, 188, *194*, 210, *212*, *246*, *348*
James, T. W., 111, *158*, 178, *194*, 214, 215, 216, 234, 236, 239, 243, *245*, *246*, **323,** 326, 327, 328, 331, 333, 334, 335, 336, 337, 338, 339, 340, 341, 342, 343, 344, *347*, *348*, **485,** 486, *494*, 511, *520*, 521, 522, 523, 524, 533, 534, *535*
Jensen, W. A., 13, 14, *35*
Jörgensen, E., 307, *320*
Johnson, D. S., 22, *35*
Johnson, F. H., *157*
Johnston, J. R., 355, *378*
Johri, B. M., 18, *35*
Jollos, V., 317, *320*
Juul, J., 446, *464*

Kaji, A., 401, *417*
Kajiwara, K., 605, *607*
Källén, B., 50, *65*
Kamiya, T., 183, 190, 191, *193*
Kanazawa, T., 273, 274, *303*
Kanbe, T., *417*
Kaneko, M., *417*
Kanwisher, J. W., *194*
Kapil, R. N., 18, *35*
Kato, N., *417*
Kaudewitz, F., 77, *93*
Kavanau, J. L., 325, *348*
Kavaljian, L. G., 13, 14, *35*
Kawamura, N., 339, *348*
Kay, E. R. M., 185, *193*
Keck, K., 417
Kelly, C., 493, *494*
Kemp, T., 446, *464*
Kendall, D. G., 478, *483*
Kenedy, J., *417*
Kennedy, J. R., Jr., *174*, *562*
Kessel, R., *175*
Kessler, W. R., *245*
Keynam, A., 408, *417*
Kidder, G. W., 100, 106, *157*, 544, *563*
Kimball, R. F., 73, 77, 78, 79, 80, 81, 82, 83, 85, 86, *93*, *95*
King, K. W., 249, 280, *304*, *508*
King, T. J., 54, *64*
Kinnander, H., 50, *65*
Kinosita, R., 80, *96*
Kirk, P. L., 28, 37, 89, *96*
Kjeldgaard, N. O., 3, *8*, 438, *439*
Klášterská, I., 262, *304*
Klein, B. M., 170, *174*
Klein, R. L., 515, *520*
Klingenberg, M., 192, *193*
Kluss, B., 85, *93*
Knaysi, G., 76, *93*
Koch, A. L., *92*, 478, *484*, *494*
Koepper, A., 278, *304*, 571, *578*
Kogané, F., 401, 403, *420*
Koike, M., *418*
Kokott, W., 446, *464*
Koningsberger, V. V., 382, *389*
Kopsch, F., 41, 42, *65*
Kornberg, S. R., 405, *417*
Korngold, G. C., *193*
Korotneff, A., 42, *65*
Kowallik, W., 571, 576, *578*
Kramer, C. Y., *508*
Kraus, H., 76, 95, *465*
Krauss, W., 249, 255, 301, *305*
Krebs, H. A., 394, *417*
Krishnan, P. S., *416*
Kröger, H., *91*, *94*
Kubitschek, H. E., 1, *8*, 510, 515, *520*, 579, *587*
Kudo, R. R., 214, *245*
Kuhl, A., *577*
Kunieda, R., *305*, *569*
Kuno, T., *417*
Kurtz, T. E., 519, 579, 586, *587*
Kushida, H., 262, *302*

Lackey, J. B., *348*
Laite, M. B., *92*, *494*
Lance, A., 14, *35*
Land, W. J. G., 16, *34*, *35*
Langen, P., *303*

Lark, C., 341, 424, 426, 430, *431*, 597, 598, *598*
Lark, K. G., 81, *94*, 341, *348*, 398, *418*, 424, 426, 427, 428, 430, *431*, 444, *464*, 597, 598, *598*, 607, *607*
Laughlin, H. H., 13, *35*
Leblond, C. D., 51, *66*
Leblond, C. P., 78, 87, *91*
Lederberg, S., 88, *93*
Lee, K. H., 178, *193*, 550, 551, *563*
Leedale, G. F., 324, 326, 338, *348*, 527, 531, *535*
Leloir, L. F., 207
Lenique, P., 62, *65*
Lenney, J. F., 386, *389*
LePage, G. A., 205, *211*
Levedahl, B. H., 522, *535*
Levenbook, L., 58, *66*
Levinson, H. S., 399, 400, 405, *418*
Levintow, L., 443, *464*
Levy, M., 116, *158*, 182, 183, *194*
Lewin, I., *157*
Lewis, M. R., 442, *464*
Lewis, W. H., 55, *66*
Likely, G. D., *604*
Lindahl, P. E., 559, *563*
Lindegren, C. C., 76, *93*, 355, 356, 358, 368, *378*
Lindegren, G., *378*
Linderström-Lang, K. U., 385, *389*
Linskens, H. F., 30, *35*
Lipmann, F., 405, *418*
Lipp, C., 56, *66*
Lison, L., 459, *465*
Liss, E., *303*
Lister, A., 214, *245*
Lister, G., 214, *245*
Littlefield, J. W., 605, 606, *607*
Looney, D., 365, 366, *378*
Lorenzen, H., 248, 249, 254, 255, 256, 261, 265, 266, 270, 276, 277, 278, 282, 286, 298, *303*, *304*, 534, *534*, **571,** 573, 575, 576, *577*, *578*
Lotka, A. J., 486, *494*
Louderback, A. L., 121, *174*, 178, 184, 185, 188, *194*, 210, *212*, *348*
Loveless, L. E., *378*
Løvlie, 111
Løvtrup, S., 57, 60, *66*, 599, *600*
Lowenstein, J. M., *194*
Lowry, O. H., 528, *535*
Ludwig, K. S., 55, *66*
Lund, R. O., 447, *464*
Luoney, D., 364, 365, 366, *378*
Lwoff, A., 100, 146, *157*, 170, 172, *174*
Lwoff, M., 170, 172, *174*
Lyon, F. M., 15, *35*

Maaløe, O., 2, 3, 4, *8*, 81, *94*, *95*, *419*, *431*, 435, 438, 439, 444, *464*, 472, *483*
McCallan, S. E., 393, *419*
McClary, D. O., *378*
McCormick, T., *465*
McCulloch, E. A., 449, 455, *463*
McDonald, B. B., 80, 81, *94*, *157*, 210, *211*, 341, 346, *348*,
McElroy, W. D., 199, 202, *211*, *212*, 401, *417*
McFall, E., 398, *418*, 425, *431*
McFarland, J., 513, *520*
McGovern, A. P., 605, 606, *607*
MacKenzie, W. A., *519*
McLean, P., 192, *193*
McLimans, W. F., 603, *604*
McMaster, R. D., 26, 30, *37*, 46, *66*, 80, 86, *96*
McNair, B., 437, *439*
Magath, T. B., *519*
Mager, J., 150, 151, *157*
Maheshwari, P., 17, *35*, *93*
Malmgren, H., *379*, 386, *390*, 406, *417*
Mandels, G. R., 392, 394, 414, *418*
Manners, D. J., 182, *194*
Mapes, M. O., *37*
Marcus, P. I., 603, *604*
Margeson, K. B., 605, 606, *607*
Markham, R., 207, *211*
Markmann, B., *66*
Marshak, A., 60, *66*
Marshak, C., 60, *66*
Marsland, D., *245*
Martin, F., 43, *66*
Maruyama, Y., 342, *348*, 398, 409, 412, *418*, **421,** 422, 426, 427, 428, 429, 430, *431*, *466*, **593,** *598*
Mather, K., 510, *520*

Mattern, C. F. T., 510, 515, *520*, 579, 585, *587*
Matsushiro, A., 416, *418*
Mazia, D., 53, *66*, 74, 88, 89, *93*, 137, 138, *157*, 237, 238, 243, 244, *245*, *246*, 456, *464*, **537,** *537*
Mead, A. D., 41, *66*
Meffert, M.-E., 249, *303*
Mendelsohn, M. L., 80, *94*
Menkin, V., 52, *66*
Messier, B., *91*
Metz, C. B., 160, 161, 163, *174*
Metzner, H., 571, 575, *578*
Micou, J., 78, 87, *92*
Mihara, S., 256, 261, 273, 276, 280, 284, 293, 294, 295, 296, 297, 298, 299, 300, 301, *302*, *347*
Minckler, S., *378*
Minick, O. T., 163, 164, 172, 173, *174*
Mita, T., 180, 181, 185, *194*
Mitchison, J. M., 76, 77, 78, 79, *94*, 341, 346, *349*, 369, 372, *378*, 460, *464*, 485, *495*
Mitra, J., *37*
Miura, T., 447, 451, *464*
Miwatani, T., *418*
Miyachi, S., 276, 280, 281, *303*, *304*
Mochizuki, T., 394, 395, *419*
Molé-Bajer, J., 20, *34*, *35*
Monod, J. L., 172, *174*
Moore, H. A., Jr., *94*
Moore, W. E. C., *508*
Moorhead, P. S., 448, *464*
Morgan, C., 462, *464*
Morimura, Y., 248, 249, 256, 257, 258, 261, 263, 264, 265, 266, 273, 275, 282, 292, 294, 295, *302*, *303*, *305*, **565,** *569*
Mortimer, R. K., 355, *378*
Moses, M. J., 25, 27, *36*, 86, *94*
Mudd, S., 403, *418*, *419*
Mueller, G. C., 452, 453, 454, 457, 458, 459, 460, *465*, 603, *604*, 605, 606, *607*, *608*
Müller, H.-M., 273, *303*
Murray, F., 43, 49, *67*
Myers, J., 249, 255, 267, 270, 271, 278, 282, *303*, *305*, 326, *348*, 531, 532, *535*

Nachtwey, R., 43, *66*, 103, 167, 170, *174*
Näf, U., 22, *36*
Nakada, D., *418*
Nakahara, T., *416*
Nakamoto, T., 87, *96*
Nasatir, M., 28, 33, *36*, *94*
Nayder, F. A., *600*
Neff, R. H., 213, 214, *520*
Neff, R. J., 213, 245, **509,** 511, 512, 515, *520*
Nevins, B. I., 22, *36*
Newton, A. A., 441, 443, 444, 447, 448, 449, 450, 457, 458, 460, 461, 462, *464*, *465*, *466*, **601,** *604*, 607, *608*
Nickerson, W. J., 388, *389*
Nigon, V., 58, *66*
Nihei, T., 247, 248, 250, 256, 257, 267, 268, 269, 270, 271, 273, 274, 280, 282, 283, 286, *303*, *305*
Nilsson, R., 53, *66*
Nishi, A., 190, 191, *194*, 403, 404, 406, 407, *418*
Norman, A., 425, *431*
Northcote, D. H., 276, *304*
Novick, A., 356, *378*
Nygaard, O. F., 80, 86, *94*

Oberbeek, J. Van, 18, *37*
Ochoa, S., 87, *91*, *94*
O'Connor, R. J., 412, *418*
Ogur, M., 30, 36, *36*, 294, *304*, 355, 358, 368, 371, *378*
O'Hanlon, S. M. E., 22, *36*
Okazaki, R., 435, *439*
Okazaki, T., 435, *439*
Olson, B. J., *520*, 579, 586, *587*
Ono, T., *194*
Ord, M. G., 451, *464*
Oster, G., 513, *520*
Otsuka, H., 256, 273, 280, 281, 283, 284, 292, 293, 294, 295, 298, 299, *302*, *303*, *304*
Ottoson, R., *96*, *379*, *390*
Owens, R. G., 395, *418*

Padilla, G. M., 239, *245*, 326, 332, 335, 339, 343, 344, 345, *347*, *348*, *508*, **521,** 522, 523, 524, 527, 528, 529, *535*

Padmavathi, P. B., 56, *67*
Painter, R. B., 76, 80, 84, *92*, *94*, 451, 459, *463*, *465*
Paladini, A. C., 207, *211*
Pappas, I., 308, *320*
Pardee, A. B., 4, *8*, 342, *347*, 398, *416*, 425, *431*, 457, *463*, 597, *598*
Pasteels, J., 459, *464*
Patterson, J. T., 42, *66*
Paul, J., *466*, 605, 606, *608*
Pelc, R. S., 71, 80, *93*
Perdue, S. W., 79, 81, *93*
Perry, R. P., 78, 87, *94*, *95*, 437, *439*
Persson, H., 52, 58, 59, *63*
Peterson, W. H., *420*
Pettijohn, D. E., 342, *347*, 597, 598, *598*
Piérard, A., 412, *418*
Pincus, G., 446, *465*
Pirson, A., 249, 256, 265, 266, 270, 278, 286, *304*, 571, 573, 576, *577*, *578*
Pisi, E., 459, *465*
Pitelka, D. R., 160, 161, 163, *174*
Pittendrigh, C. S., 313, *320*
Plaut, W., 87, *93*, *95*, 460, *465*
Plesner, P., 90, *95*, 102, 116, 177, **197,** 198, 206, 207, 209, 210, 211, *212*, 369, *378*, **543,** 556
Podolaski, B., 330, *347*
Pollard, J. K., 36, *37*
Polowzow, W., 61, *66*
Ponder, E., 513, 514, 515, *520*
Popham, R. A., 14, *36*
Potter, V. R., 85, *91*, *94*, 451, *463*
Powell, J. F., 400, *418*
Powell, E. O., 8, *8*, 342, 422, 424, *431*, 478, *483*, 486, 493, *494*
Powers, E. L., 170, *174*
Prescott, D. M., 71, 73, 74, 75, 76, 77, 78, 79, 80, 81, 82, 83, 84, 85, 86, 87, 88, 89, 90, *91*, *94*, *95*, 101, 104, 120, 153, *157*, 210, *212*, 214, 216, 242, *245*, *246*, 341, 346, *348*, 494, *494*
Prill, E. A., *420*
Pringsheim, E. G., 326, *348*
Provasoli, L., 521, *535*
Puck, T. T., 372, *379*, 603, *604*
Rabinovitch, M., 87, *95*, 460, *465*
Rahn, O., 482, *484*, 493, *494*
Rake, G. W., *604*
Randall, R. J., *535*
Rapp, F., 444, 447, *465*
Rasch, E., 80, *97*, 459, 460, *466*
Rasch, G., 178, *194*, 346, *348*, 486, *494*
Rasmussen, L., 102, 106, 116, 120, 131, 133, 134, 136, 139, 140, 143, 147, *157*, 167, *174*, **543,** 553, *563*
Read, C. P., 342, *348*
Reich, K., 232, *246*
Rem, L. T., 189, *193*
Rensch, B., 39, *66*
Řetovský, R., 262, *304*
Reynolds, L. G., *36*
Richar, W. J., 515, *520*, 579, *587*
Richards, B. M., 20, *36*, 80, *95*
Rippel, A., 393, *418*
Robert, A., 41, 44, 47, *66*
Roberts, E. R., *416*
Roberts, R. B., 415, *419*
Rodenberg, S. D., 36
Roels, H., 459, *463*
Roos, K., 599, *600*
Roosen-Runge, E. C., 51, *66*
Rosen, G., 294, *304*
Rosen, U. G., *36*
Rosenberg, B. H., 84, *92*
Rosenbough, N. J., *535*
Rosenkranz, H. S., *193*
Roslansky, J. D., 214, 215, *246*, 319, 511, *520*
Roth, L. E., 163, 164, 172, 173, *174*
Rouiller, C., *92*
Rudner, R., 597, *598*
Rueckert, R. R., 452, 453, 454, 457, 458, 459, 460, *465*, 603, *604*, 605, *607*
Ruppel, H.-G., 248, 249, 254, 255, 261, 276, 277, 278, 298, *303*, 571, 576, *577*, *578*
Rusch, H. P., 56, *94*
Ryan, F. J., 393, *419*
Ryley, J. F., 182, *194*, *195*

Sakaguchi, K., 393, *419*
Sall, T., 403, *419*
Sando, N., 364, 412, *419*

Sandritter, W., 76, 77, *95*, 460, *465*
Sanford, K. K., 601, *604*
Sasa, S., 247, 248, 256, 257, 268, 285, *304*
Sasa, T., *304*, *305*
Sato, H., 123, *157*
Savelkoul, R. M. H., 14, *36*
Sax, K. B., *36*
Schaechter, M., 2, 3, *8*, 80, 81, 82, *95*, *419*, 425, *431*, 438, *439*, 477, 478, *484*
Scherbaum, O. H., 3, 4, *8*, 73, *95*, 104, 105, 106, 109, 111, 116, 121, 127, 129, 130, 145, 146, *157*, *158*, 162, 165, 170, 173, *174*, *175*, **177,** 178, 180, 182, 183, 184, 185, 186, 188, 189, 190, 191, *193*, *194*, *195*, 197, 210, *212*, 218, 234, 238, 239, 240, 243, *245*, *246*, 327, 341, 346, *348*, *349*, 376, *378*, 421, *431*, 446, *465*, 486, *494*, 506, *507*, 514, *520*, 545, 546, 548, 549, 554, *563*
Scherer, W. F., 601, 603, *604*
Schiemer, H. G., 76, *95*, *465*
Schindler, R., 449, 453, 454, 457, *465*
Schmidt, G., 527, *535*
Schmidt, R. R., 249, 273, 280, 281, *304*, *508*,
Schmidt, W. J., 243, *246*
Schneider, W. C., 180, *194*
Schneiderman, M., *519*, 579, *587*
Schönfellinger, H. W., *417*
Scholander, P. F., 182, *194*
Schultz, J., *66*
Schweigert, T. B. S., *600*
Schwemmin, D., *389*
Scopes, A. W., 354, 355, 358, 360, 361, 364, 367, 368, 370, 371, 372, 373, 374, 375, *378*, *379*, 388, *390*, 589, *591*
Scott, D., 437, *439*
Scott, D. G., *378*
Seaman, G. R., 197, 199, *212*
Sehl, A., 43, *66*
Seidel, F., 43, 49, *66*, *67*
Selawry, O. S., *465*
Selenka, E., 41, *67*
Senger, H., 571, 573, *578*
Seshachar, B. R., 56, *67*
Sevag, M. G., 399, *418*
Shantz, E. M., 36, *37*
Sharpe, F. R., 486, *494*
Shepherd, C. J., 397, 402, *419*
Shibata, K., 247, 248, 250, 256, 257, 267, 282, 283, 286, *305*
Shichiji, S., 358, 368, *378*
Shimizu, N., 397, *419*
Shohl, A. T., 513, *520*
Shooter, V., *463*
Short, R. B., 56, *67*
Shreeve, W. W., 76, *92*, *463*
Siedler, A. J., 599, *600*
Silberstein, J. K., *245*
Simmons, N. S., *193*
Singer, W., 178, *195*
Sinsheimer, R. L., 84, *96*
Sirakami, K., 41, 46, 57, *67*
Siskin, J. E., 80, 86, *96*
Skoczylas, O., 317, *320*
Slautterback, D., *64*
Smellie, R. M. S., *466*
Smith, C. L., 459, *465*
Smith, D., 309, *320*
Smith, J. D., 207, *211*
Smith, R. W., *36*
Smith, S. G., 23, 24, *35*
Sondhaus, C. A., 382, *390*
Sonnenblick, B. P., 43, 49, 52, *67*
Sorokin, C., 249, 255, 264, 270, 271, 299, 301, *304*, *305*
Sparks, J. T., *158*
Sparrow, A. H., 24, 25, 27, *36*
Sparrow, R. C., 24, *36*
Spear, F. G., 442, 446, 448, 449, *465*
Speese, B. M., 14, *37*
Spencer, B., 401, *419*
Spencer, H. T., *508*
Spiegelman, S., 388, *390*
Spoerl, E., 364, 365, 366, *378*
Stacy, K. A., 513, *520*
Ståfelt, M., 13, *37*, 309, *320*
Stange, 249
Stanners, C. P., 80, *96*, 372, *379*, *466*
Steele, R., 25, *36*
Steiner, R. F., 513, *519*
Stent, G. S., 398, *418*, 425, *431*
Stern, H., 27, 28, 29, 30, 31, 32, 33, *35*, *36*, *37*, 74, 83, 89, *92*, *94*, *96*
Stevens, A., 87, *96*

Stevens, C. M., 401, *419*
Stevenson, I., 437, *439*
Steward, F. C., 20, *36*, *37*
Stockdale, F. E., 61, *67*
Stocken, I. A., 451, *465*
Stoker, M. G. P., 461, *465*
Strecker, H. J., *417*
Strehler, B. L., 199, 202, *212*
Strong, F. M., *420*
Stubblefield, E., 605, 606, *607*, *608*
Suagi, K., *417*
Sugimura, T., *194*
Suhama, M., 160, 161, 163, *175*
Sullivan, W. D., *158*
Sumi, M., 395, 397, *419*
Summers, L. G., 486, *495*
Sussman, M., 513, *520*
Suzuki, K., *304*
Svensson, G., *93*
Swann, M. M., 56, 57, *67*, 74, 90, *96*, 137, 158, 198, *212*, 237, *246*, 369, 376, *379*, 460, *465*
Sweeney, B. M., 307, 309, 312, 313, 316, 317, *320*, *321*, 522, *535*, **579,** 583
Swift, H., 78, 80, *97*, 459, 460, *466*
Sykes, E., 23, *37*
Sylvén, B., 88, *96*, 359, 368, 372, *379*, **381,** 382, 386, 388, *390*
Syverton, J. T., *604*
Szilard, L., 356, *378*
Szybalski, W., 607, *607*

Takagi, A., *419*
Takahashi, T., 190, 191, *193*
Takano, M., *419*
Takats, S., 33, *37*
Takebe, I., 395, 399, 400, 401, 402, *419*, 422, 423, *432*, *466*
Tamiya, H., 247, 248, 249, 250, 251, 252, 253, 255, 256, 257, 258, 259, 260, 261, 264, 265, 267, 273, 275, 276, 278, 280, 281, 282, 283, 284, 286, 288, 291, 293, 294, 295, 296, 297, 298, 299, 300, 301, *302*, *303*, *304*, *305*, *347*, **565,** 568, *569*
Tamm, I., 462, *466*
Tannhauser, S. J., 527, *535*
Tartar, V., 146, *158*, 172, *175*
Tatum, E. L., 403, *417*
Taylor, E., 453, *464*
Taylor, E. R., *520*
Taylor, J. H., 26, 30, 31, *37*, 76, 79, 80, 84, 86, 87, *94*, *96*, 606, *608*
Terasima, T., 449, 455, 456, 461, *465*, 603, *604*, 606, 607, *608*
Terui, G., 394, 395, 396, *419*
Thorell, B., *96*, *379*, **381,** 382, *390*
Thormar, H., 100, 101, 102, 103, 117, 118, 143, 144, 148, *158*, 237, *246*
Thornton, H. G., *519*
Threlfall, C. J., 403, *419*
Tiegs, O. W., 43, 49, *67*
Till, J. E., 80, *96*, 372, *379*, *466*
Tischler, G., 18, 19, 21, *37*
Tobias, C. A., *96*, *378*, *379*, 382, 384, *390*
Tolmach, L. J., 449, 455, 456, 461, *465*, 603, *604*, 606, 607, *608*
Toolan, H. W., 447, *465*
Toyo, 597
Travaglini, E., *66*
Trowell, O. A., 452, *464*
Tschermak-Woess, E., 80, *96*

Umbriet, W. W., 205, *211*
Utakoji, T., 447, 451, *464*

Van Dam, L., *194*
Vendrely, R., 457, 459, *466*
Verly, W. G., 80, *92*
Vialleton, M. L., 42, *67*
Visser, D. W., *438*
Voegtlin, C., 331, *349*
Vogt-Köhne, L., *93*
Vogt, M., 602, *604*
Vohra, P., 401, *419*
Volkin, E., 79, *95*, 294, *305*, 402, *419*

Wagner, B., *562*
Walker, P. M. B., 80, *95*, *96*, 341, 346, *349*, 372, *378*, 459, *464*, *466*, 485, 486, *495*
Walsch, H., *417*
Wanka, F., 249, 266, *305*
Warner, R. C., *91*
Wasteneys, H., 385, *390*
Watase, S., 42, 50, *67*
Waymouth, C., 61, 452, *466*
Webb, M., 460, *464*

Weber, H., 51, *67*
Weill, J. D., *91*, *94*
Weisman, T. H., *378*
Weiss, S. B., 87, *96*
Weissman, S. M., 451, 454, 461, *466*
Welin-Berger, E., 50, *63*
Wellman, R. H., 393, *419*
Welshons, 87
Went, H. A., *246*
Wesson, L. G., Jr., 450, *466*
Westfall, J. A., 160, 163, *174*
Wheelock, E. F., 462, *466*
Wiame, J. M., 305, 412, *418*
Wickerham, L. J., 589, *591*
Wildy, P., 443, 444, 447, 448, 449, 450, 458, 460, *465*, *466*, 601, *604*, 607, *608*
Williams, G. Z., 177, *519*, 579, *587*
Williams, N. E., 121, 122, 123, 145, 146, 152, *157*, *158*, **159,** 161, 163, 165, 170, 171, 173, *174*, *175*, 238, *245*
Williamson, D. H., 351, 354, 355, 358, 360, 361, 364, 367, 368, 370, 371, 372, 373, 374, 375, *378*, *379*, 388, *390*, **589,** *591*
Williamson, J. P., *484*
Willmer, E. N., 442, 452, *464*, *466*
Wilson, B. W., 344, 345, *349*
Wilson, E. B., 238, *246*
Wilson, K. S., 19, 22, *34*, *37*
Wilson, M., *37*
Wimber, D. E., 80, *96*
Winder, F. G., 403, 405, *419*, *420*
Wintermans, J. F. G. M., 280, *305*
Wiser, R., *302*
Withmore, G. F., 459, *466*
Woese, C. R., 397, 399, *420*
Wohlfarth-Bottermann, K. E., 317, *320*
Wolfe, R. G., *378*
Wood, W. A., *417*
Woodard, J. B., 459, 460, *466*
Woodard, J. W., 78, 80, *97*
Woodlands Hastings, J., *see* **Hastings, J. W.**
Woods, P. S., 78, 87, *97*
Woolley, D. W., 325, *349*, 401, *420*
Wu, C., 178, *195*
Wyatt, G. R., 207, *212*
Wylie, A. P., 24, *34*

Yakovlev, M. S., 21, *37*
Yamada, M., 364, 372, *379*
Yamagishi, S., 392, 395, 396, *420*
Yamataka, A., 403, *420*
Yamataka, S., 160, 161, 163, *175*
Yanagita, T., 342, *348*, **391,** 392, 393, 394, 395, 396, 397, 399, 400, 401, 402, 403, 408, 411, 412, 413, *417*, *419*, *420*, 422, 423, 425, 427, 430, *432*, 462, *466*, 593, *598*
Yates, H. B., 80, *96*, 459, *466*
Yoffe, M. D., 21, *37*
Yokota, M., 265, *305*, *569*
Yoshida, A., 403, 405, *418*, *420*
Youcis, P., *92*, *494*
Young, I.E., 398, *420*
Yuhara, T., 275, *305*
Yuzuriha, Y. O., *193*

Zahler, S. A., *508*
Zalokar, M., 78, 87, *97*
Zernicke, F., 442, *466*
Zeuthen, E., 1, 2, 3, 4, *8*, 39, 41, 52, 53, 57, 60, 61, *65*, *67*, 73, 74, 78, 80, 89, *92*, *95*, *97*, **99,** 102, 104, 105, 106, 108, 109, 110, 116, 118, 119, 120, 121, 123, 124, 125, 127, 128, 129, 130, 131, 133, 134, 135, 136, 137, 139, 140, 143, 144, 145, 147, 148, 149, 150, 151, *156*, *158*, 165, 167, *174*, *175*, 177, 180, 184, 188, 190, *193*, *194*, *195*, 197, 198, 205, 210, 211, *211*, *212*, 218, 235, 237, 238, 240, *245*, *246*, 307, *321*, 327, 341, *349*, 374, *379*, 421, *432*, 434, 437, *439*, 505, *507*, 514, *520*, **543,** 544, 545, 546, 547, 548, 549, 552, 553, 555, *562*, *563*
Zimmermann, A. M., 244, *246*

Subject Index

The letters ff indicate that the stated subject is mentioned again on one or both of the next two following pages. Page numbers in **heavy type** denote the beginning of a chapter about the entry, where a list of the chapter subdivisions may be found. These subdivision heads are not all repeated here.

Acalypha, 18
Acanthamoeba, 216 ff, 235 ff, 511 ff
  aerobic–anaerobic treatment for division synchronization, 232
  heat treatment for division synchronization, 225
  nephelometer counting, 513
  temperature effects, 234
Acetate, 323
Acetobacter, 409
Acetone powder, 561 ff
Adaptation to heat, 118
Adenosine triphosphate, 404
Aequorea forskålea, 41
Aesculus, 20
Age distribution of cells, 469
  in logarithmically growing cell populations **485**
Alanine, 399
  required for spore germination, 393, 396, 411 ff
Alcaligenes fecalis, 341, 422 ff, 427 ff, 598
Algae
  induced synchrony in, **247, 307, 323**
  natural phasing in cell division of, 309
  technical procedures for, **565, 571, 579**
Allium, 13
Amaryllidaceae, 20
Amethopterin, 457 ff, 460
  synchronization of cell division by, 602
Amino acids, 177 ff, 399
  as condition for division, 131 ff
  effect on synchrony of *T. pyriformis*, 115
Amitotic splitting, 88
Amoeba, 74, 87
  accommodation and synchrony induction, 240
  carolinensis, 56
  cell counting problems, **509**
  cell damage and synchrony induction, 241
  division synchrony of, 214
  heat treatment results, 216
  improved medium for, 225
  induction of synchronous division in, **213**
  interference microscopy, 216
  length of cycle, 239
  mitotic synchrony of, 214, 237
  proteus, 214 ff, 235
  rate-dependent temperature effects, 235
  structural hypotheses and synchrony induction, 237
  synchronization of division in, 214 ff
  synchronous division, induction in, **213**
  synchrony induction, 222, 234 ff, 239, **509**
    detection of, **509**
  temperature cycles, 217, 226
Amphioxus, 41
Anaerobiosis, 514
Anaphase index, 504
Aneimia, 23
Angiosperms, 17
Animal cells, in culture, synchronous division of, **441**
Ankistrodesmus, 577
Antheridia, 21 ff
Anthers, 24 ff
Antimetabolites, 286
  shock exposures to, 131
Arbacia lixula, 41, 45
Artificially induced synchrony, 358
Ascaris, 54
Ascorbic acid, 331
Aspergillus niger, germinating conidiospores of, **391**

Aspergillus oryzae, 395
Astasia, 239
cytology of, 338
growth rate inhibited by light, 524
Astasia longa, 325 ff, 333 ff, 340, 341 ff, 346, 522
Asynchronous division, 44
in mosaic eggs, divisional pattern, 44
of mammalian egg, 55
Asynchronous state, 3
Asynchrony of DNA synthesis, 83, 127
Autocatalytic growth, 78
Auxotrophs, 433 ff
Axenic stationary cultures, 216 ff
Azaguanine, 345
Azolla, 15
Azotobacter, 409

Bacillus megatherium, 76
Bacteria, 81
as timepieces, 477
synchronization by filtration techniques, **421, 593**
synchronization by nutritional changes, **433**
synchronous growth, 437
Balanced growth, 3
Beta-carotene, 281
Bellevalia, 84
Belone acus, 41
Biochemical mutants, 325
Biochemical parameters, 111
Biochemical storage, 57 ff
Biological clocks, 307, 316
Bioluminescence, 313
Blastomeres, 57
Botrychium, 23
Bowenia serrulata, 16
Bright light inhibits rhythmicity, 311
Bryophytes, 22
Budding yeast, 352
Bufo, 41, 46
Bursting division, 574

^{14}C-histidine, 125, 148
^{14}C-pyruvate, 183
Calandra, 43, 49
Calliphora erytrocephala, 47 ff
L-Canavanine, 140
Carbohydrates, 182
Carbon dioxide fixation, 267
Carbon dioxide for spore germination, 393 ff, 411 ff
Cartesian diver balance, 177
Cell age
and size variation, 424
average, 490
distribution, 422, 469 ff
Cell biology, study of using synchronous cultures, 456
Cell counting
automated
by Coulter counters, 515, 518
by nephelometer, 513
sources of error in, 517
chambers, 512
electronic, of dinoflagellates, **579**
for amoebae, problems in, **509**
methods, 554
Cell cycle
genetic code duplication, 90
normal, **71**
photosynthetic quotient, 268
yeast, 369 ff
Cell density, 447
Cell division, 73
activation energy, 148
clock models for, 470 ff, 477 ff
doubling time, 528
isochronous, 284
kinetics of, 442
non-clock model for, 482
normalized rate of, 498
partial synchrony of, 307
phased, in marine dinoflagellates, **307**
sequence of steps of finite number, 478
Cell growth, 73
before heat treatment, 546 ff
patterns of, 74 ff
periodic alterations in, 473 ff
phases of, 546
stochastic, 477
vegetative phase, 391, 414
Cell populations
age distribution in, **485**
asynchronous, synchronous, parasynchronous, 497

degrees of synchrony of, measurement, **497**
logarithmically growing, 485
variations of real, 493
Cell size
determination, 584
generation time, 375
statistical distribution of, changes in, 252
Cell wall synthesis, 426
Cells, 275
death time, 199
macronuclear events, 210
mouth, development stages, 210
$^{32}PO_4$ uptake rate, 210
respiratory rate, 210
time of observation, 199
total number, 39
Celloscope electronic cell counter, 554
Cellular components, shift in, 576
Cellular division and cellular differentiation, 61, 62
Centrifuge tube, 562
Ceratium, 307
Cervical carcinoma cells, strain HeLa, 443 ff, 446 ff, 451, 455, 458 ff
Chamber counts, amoebae, 512
Chemical agents affecting synchrony, 116 ff
Chick embryo extract, 452
Chick fibroblast cells, mitotic rate of, 449
Chilomonas paramecium, 330
Chinese hamster cells, 84
Chitosan, 276
Chlamydomonas, 319, 345
filial cells, 489
moewusii, 336
Chloramphenicol, 137, 427
Chlorella
antimetabolites, effects due to, 286
assimilation of major elements, 275
combustion quotient, 275
developmental stages, 257
filial cells, 489
growth and cell division of, **247,** 298
growth, dependence on essential mineral nutrients, 283
light-synchronized, 341
nucleic acids content, 276
periodical dilution of populations, **571**
photosynthetic quotient in cell cycle, 268
polyphosphates content, 280
protein, lipid and carbohydrate content, 275
sulfur, role of, 291
synchronization with light–dark changes, **571**
synchronous cultures, starting from homogeneous population, **565**
temperature and light intensity effects, 263 ff
vitamin content, 282
Chlorella ellipsoidea, 248
cell number and volume changes, 253
DLD cycle, 257, 261, 264
DNA and RNA contents, 277 ff
respiratory quotient, 271
Chlorella pyrenoidosa, 248, 262
Chlorella vulgaris, 262
Chlorophyll, 269, 281
Chloroplasts, 261
Choline sulfate, biosynthesis, 401
Chrysanthemum moriflorium, 14
Ciliary apparatus, 122
Cilium, 161
Cleavage cycle, 52 ff
Cleavage synchrony, 41
Cnidoblasts, 51
Cocos nucifera, 18
Coenzyme levels, 189
Cofactor synthesis, 140
Cold shocks, 102, 285
Cold synchronization, 106
Competitive inhibition, 131
Complete synchrony, 573
Conidiospores germinating in *Aspergillus niger*, **391**
Continuous synchrony, 2
Continuous timepieces
almost perfect clocks, 470 ff
imperfect clocks, 477 ff
Cortical differentiation, 170 ff

Cortical morphogenesis, 122
Coulter counter, 580 ff
Cramer–Myers medium, 333 ff
Critical point, 198
Culture tubes, 572
Cultures, 555 ff
Cyanide, 430
Cysteine, 331, 335
Cytochemical characteristics, 53
Cytokinesis, 72

D cells, 258
Datura, 19
Dauersynchronisation, 577
Daughter cells, 487
Deoxyribonucleic acid (DNA), 343
  base ratio of in *Tetrahymena*, 185
  cytoplasmic, 58, 59, 60
  increase of per cell, 129
  inhibition, 435
  of the embryo nuclei, 54
  polymerase levels, 85
  replication, 81
  replication model, 84
  synthesis, 71 ff, 80 ff, 152, 397, 433 ff, 437, 450 ff
    initiation and control, 85
    post-mitotic, 457 ff
    rate of, 210
    relation to cell division, 457
    ultraviolet effect, 425
DNA-histone, 79
DNA–RNA synthesis in *A. fecalis* and *E. coli*. 430
Deoxyribosenucleotides, 428
Deoxyriboside auxotroph, 435
Deoxyribosides, 426
Desynchronization, 118
Determinate cleavage, 55
Differential centrifugation, 590
Digitonin–luciferase method, 199 ff
Dinoflagellates, electronic cell counter studies, **579**
Dioon, 15
Discontinuous timepieces, 478
Diurnal rhythmic mechanism, 307
Dividing clones, 311
Division index, 102
  in amoebae, 511
  of the fertilized egg, 52
Division maximum, 198
Division number, 251
Division protein, 135, 167 ff
Division rate, 509
Division synchrony
  absence of in embryos, 43
  criteria for in flagellates, 332
  during ovogenesis, 51
  during spermogenesis, 50, 51
  induced by temperature shocks, 104 ff, 443
  in egg cocoon, 43
  in turbellarian worms, 43
  in yeasts, **351**
  methods of induction, 131 ff
  of cnidoblasts, 51
  of discoblastic embryos, 41, 42
  of *Euglena gracilis*, light-induced, 530
  of flagellates, induction of, **323**
  of insect embryos, 43
  of mosaic embryos, 41
  of regulative embryos, 40, 41
  of syncytia, 57
Doubling period, definition, 500
Drosophila, 43, 58

Echidna, 44
Echinus esculentus, 41, 45, 58
Electronic cell counter, **579**
Elodea densa, 14
Embden–Meyerhof–Parnas (EPM) route, 408
Embryogenesis, 40
Endamoeba terrapinae, 214 ff
Endogenous rhythm of cell division, 309
Endoplasm, 172
Endoplasmic reticulum, 122, 173
Endosome, 338
Endosperm, 17
Energy metabolism, 89 ff, 140
Enucleate cells, 516
Enucleation experiments, 87
Environmental conditions, shifts in, as theoretical basis of synchronization, **469**
Enzymes, 124 ff, 373
  activity, 189

in synchronized yeast populations, **381**
Ephedra trifurca, 16
Ephestia, 43
Equisetum, 12, 22
Escherichia coli, 85, 422 ff, 428, 435 ff, 594 ff
Ethanol, 323
Euglena gracilis, 325 ff, 335, 342, 346, 493, 530 ff
light-synchronized, 336
Euglenineae, 324
Euplotes, 82, 83
macronucleus of, 85 ff
Excess delay, 140
Experimental cultures, 545

Fern spores, 586
Filial cells, 487
Filtration techniques, for synchronizing bacterial cultures, **421, 593**
Firefly extract, 199
Fission time, 330 ff, 489
Flagellates
division synchrony in
criteria for, 332
induction of, **323**
light effects on mitosis in, 532
synchronized, respiration and photosynthesis of, 344
techniques for synchronizing, **521**
Flowering plants, 17
Foam flotation, 590
for synchronized growth, 382
Fossombronia, 22
Free amino acids, 33, 111
Free-running period, 313
Fritillaria, 18 ff
Fritillaria imperialis, 21
Frontonia leucas, 56
Functional characteristics, 51 ff
Fungi
imperfecti, 392
synchrony in, **351, 381, 391**
technical procedures for synchronizing, **589**

Generation period, definition, 500
Generation time, 118
definition, 500
Genetic code, duplication in cell cycle, 90
Germination, 392
chemical sensitivity, 396
conidiospores in *Aspergillus niger*, **391**
heat resistance, 395
metabolic pattern, 415
Ginkgo, 16 ff, 20, 23
Glucosamine, 276
Glutathione, 331
Glycogen synthesis rate, 182
Glycolysis, 433
Gnetum, 16
Gonyaulax cultures, 313
mitotic cycle, 317
Gonyaulax polyedra, 308 ff, 579 ff
growth rate, 315
Gonyaulax sphaeroidea, 308 ff
Graded mitotic activity, 60 ff
Group synchrony, 266
Growth
and cell division of *Chlorella*, **247**
control, 77
curves of, 444, 498
patterns of single cells, 74 ff
Growth-promoting substances, 442
Growth rate
determination, 442
inhibited by light, 524
Gryllus, 49, 58
Gymnodinium, 312
Gymnodinium splendens, 308 ff
Gymnodinoides, 146
Gymnosperms, 15, 23

Haemanthus katharinae, 20
Heat shocks, 102
Heat synchronization, 106, 118, 165
HeLa cells, mitotic rate, 449
Helminthostachys, 23
Heterosporous plants, 15
Hexose monophosphate (HMP) route, 408
Higher plants, synchronous cell and nuclear division in tissues of, **11**
Histone, 82
synthesis of, 80
Homogeneous population method, 249

Human cervical carcinoma cells, 601 ff
Human leucocytes, 84
Hydra, 51, 56

Incomplete synchrony, 573
Incubation
  at high temperatures, 446
  temperature of, long-term responses to, 522 ff
Induced synchrony, 4
'Information' for protein synthesis, 150
Inhibition of macromolecular syntheses, 126, 131
Inhibitor effects, 345
Inoculum for large-scale synchronizer, 560
Interference microscopy, 77
Interphase nucleus, 78
Introduction, **1**
Ionizing radiation, 27
Iridaceae, 20
Irradiation, 20, 27
Isoetes, 15

Juglans, 20

Keto acids, 399
Ketoglutarate, 399
Kidney cells, 459
Kinetic hypotheses of synchrony induction, 234
Kinetosomes and cell division, 144 ff
Kinetosomes, origin of new, 170
Kinety, kinetosome, kinetodesma, kinetid, 160
Krebs cycle function, 433

Lactobacillus acidophilus, 435 ff
  synchronization techniques, **599**
Lag phase, 359, 368 ff, 387, 547
Lag-phase cells, 437
L cells, 258
Light
  CO_2-fixing capacity induced by, 267
  growth inhibition by, 530
Light–dark changes, synchronization of *Chlorella* by, **571**
Light–dark cycles, 313
Light–dark period, compared with life cycle, 576
Light-and-dark regimen method, 249, 270
Liliaceae, 20
Lilium, 12, 24 ff, 27, 28 ff, 31, 83
Lilium henryi, 30
Lindahl incubator, 559
Logarithmic growth phase, 547
Loligo, 49
Lycopodium, 23, 584
Lytechinus, 46

Macronuclear amitosis, 165
Macronuclear divisions, 121
Macronuclear function, 76
Macronucleus, 83
Mammalian cells, in culture, synchronous division of, **441**
Mammalian egg, 55
Mammalian tissue culture, 76
Mannitol, 590
Marchantia, 22
Marine dinoflagellates, phased cell division in, **307**
Marsilea, 15
Mass synchrony, 3
Maximal synchronous division, time point, 552
Megagametophyte, 15
Meiosis, 12
Mercaptoethanol, 345
Meristems, 13 ff
Mesenchyme blastula, 62
Metabolic activity and dormancy, 391
Metabolic pattern in spore germination, 415
Metaphase index, 506
Metazoan tissues
  mitotic gradients in, **39**
  natural division synchrony in, **39**
Methionine, 331, 335
Microbial systems, 446
Micronuclear divisions, 121
Micronuclei, 165
Microsporangiate strobili, 23 ff
Mitochondria, 173, 261
  per egg, 53

Mitotic activity, 50 ff
  diurnal period, 50
  graded, 60 ff
  normal diurnal, 324
  of micromeres, 46
  renewed, in embryos, 50
Mitotic cells, 455
  synchronizing cell division by selection of, 603
Mitotic counts, 14
Mitotic gradients, 44 ff, 62
  and differentiation, 62
  in anuran egg, 46
  in embryos, 60, 61
  in insect embryos, 47, 49
  in metazoan tissues, **39**
  in mosaic embryos, 46
  in sea urchin embryo, 44, 45, 46
Mitotic index, 13, 528
  percentage synchronization estimated from, 502 ff
Mitotic rate, 442
  in synchronized cultures, 448, 453, 456
Mitotic rhythm
  in energy metabolism, 53
  of germ cells, 49
Mitotic stimulators, 452
Mitotic synchrony of syncytia, 55, 56
Mitotic time in embryos, 52
Mitosis, 39
  in bacteria, 72
  in lily anther, 88
  oxygen consumption during, 89
  partial synchronization of, 448
  rate of protein synthesis during, 87
  static morphology of, 441
  synchrony of, 309
  temperature coefficient, 13
Mnium hornum, 22
Monotremata, 43
Morphogenetic stabilization point, 143
Morphological characteristics of synchronously dividing cell, 51 ff
Mouse fibroblast cells, 601
Mouse mastrocytoma cells, 449
Multiplication rate changes, 100 ff
Musa, 20
Myrothecium verrucaria, 394

Natural period, 313
Natural synchrony, 357 ff
Nematocysts, 51
Nephelometer, 513
Neurospora crassa, 398, 403
Neurospora sitophila, 395
Normal cell cycle, **71**
Nuclear differentiation of embryo, 54
Nuclear division, 71, 393
  temporal relation to DNA synthesis, 82
Nuclear function and synchronous division, 150
Nuclear volume and DNA content, 185
Nucleic acids, 152, 184 ff, 276, 371
  stores of, 127
  synthesis of, 30, 116 ff
    in germination, 402
    in spore germination, 397
Nucleolar period, 121
Nucleolus, 338
Nucleoproteins in embryos, 54
Nucleoside triphosphates (NTP)
  changes in cellular content, 197
  definitions and terminology, 198
  effect of inhibitors on content of, 209
  quantitative estimation of, 205
    results of, 205
  timing in changes of content, 199
Nucleotide metabolism, 382
  in synchronized cell division of *T. pyriformis*, **197**
Nucleus, 172
Nutrition and synchrony induction, 239
Nutritional synchronization in bacteria, **433**

Outlook, **537**

Paeonia, 21
Pantothenic acid, 298, 301
Paper chromatography, 207
Paracentrotus, 46, 61
Paracentrotus lividus, 41, 44 ff, 58
Paramecium, 75, 77, 78

Paramecium aurelia, 460
Parent cell, 487
Partial synchrony, 307
Penicillium chrysogenum, 403
Peptidases, 385
Peptides, 178, 388
Peridinium triquetrum, 307, 308
Permanent synchronization, 577
Permeability, 27
Phage reproduction, 426 ff
Phosphates, 187 ff
Phospholipid in germination, 402
Phosphorus, 123
  metabolism in germination, 402
Photosynthesis, 323
Photosynthetic activity, 267
Photosynthetic capacity, 313
Physarum, 55
Physarum plasmodium, 57
Physiological transition, 131, 136, 147 ff
  time point of, 139 ff, 143, 167
Pilularia, 15
Pine pollen, 586
Pinus, 23
Plagiochila, 22
Plagiostomum girardi, 43
Plant root cells, 80
Platychnemis, 43, 49
Polygonum, 17
Polyphosphatase in *Aspergillus* spores, 405
Polyphosphate in germination, 402
Polystomum, 43
Polytomella agilis, 325
Polyxemic stationary cultures, 214 ff
Population
  growth phase, 446 ff
  temperature shock responses, 101
^{32}P-phosphate, 125
Precursor pools, 83
Primordium resorption and stabilization, 168
Procession index, 13
Proline, 399
Prorocentrum micans, 308 ff
Protein amino acids, 111
Proteinases, 385
Proteins, 124 ff, 177 ff, 388
  in germination, 399
  synthesis, 88 ff, 140, 147 ff, 179, 426, 437, 460
    and levels of enzyme activity, 88 ff
    and total nitrogen content, 372 ff
  synthesis rate, 75
Proteose peptone medium, 544
Protoplast contraction in *Amoeba*, 515
Protozoa
  as timepiece, 477
  consistent cellular design, 323
  induced synchrony in, **99, 159, 177, 197, 213**
  technical procedures for synchronizing, **543**
Psammechinus miliaris, 41, 45, 58
Pseudomonas, 409
Psilotum, 22
Purine, effects on synchrony, 114 ff
Puromycin, 137, 140
Pyrimidine, effects on synchrony, 114
Pyrosoma, 42
Pyruvate, 399

Quantum efficiency of photosynthesis, 269

Radioautographic studies, 125
Ranales, 24
Regnellidium, 15
Reizstoffe, 21
Reorganization band, 81
Replication, 73, 74
  of macronucleus, 81
Reproduction, 73
Reproduction balance, 73
Respiration, 28, 89 ff
Respiratory activity, 267
Respiratory quotient, CO_2/O_2, 271
Ribonucleic acid (RNA), 343
  base composition, 407
  base ratio of in *Tetrahymena*, 185
  in nuclei of embryo, 53
Ribonucleic acid (RNA) synthesis, 75, 78, 397, 437
  cycle variations in, 86
  in *E. coli* cells, 426
  unaffected by chilling, 460

RNA–DNA synthesis in *A. fecalis* cells, 426
Ribosenucleotides, 428
Riccia, 22

Saccharomyces cerevisiae, 76, 352 ff
 growth curves, 361
Salvinia, 15
Scenedesmus, 577
Scenedesmus obliquus, 271 ff
Schizosaccharomyces pombe, 76
Schmidt–Thannhauser technique, 528
Sciara, 43
Scrippsiella sweeneyi, 583
Selaginella, 15
Set-back, 137, 140, 147 ff
 of washed cells, 133
Shift in cellular components, 575
Shock-exposures to antimetabolites, 131
Shocks, morphogenetic, 140
Shoot meristem, 14
Silverline fibers, 161
Single cell isolates, 310
Slime mold, 80
Smilax bona-nox, 14
Soil amoebae, counting, 511 ff
Soluble proteins, 33
Spatangus purpureus, 41, 45
Sphaerocarpos, 22
Spirostomum ambiguum, 56
Sporangia, 22 ff
Spore germination, physiological aspects of, 392 ff
Stage index, 13
Standard cell number, 573
Starvation medium, 590
Sterol synthesis, 140
Stock cultures, 544
Streptococcus faecalis, 76
Structural developments in synchronously dividing *T. pyriformis*, **159**
Sugar phosphate, 403 ff
Sulfhydryl compounds, 29
Sulfolipid, 293
Sulfur metabolism in germination, 399
Sulfur nucleotidic compounds, 294
Synapta digitata, 40
Synchronization
 decay of, 501
 growth in dry matter during, 123
 growth in nitrogen during, 123
 of biochemical events, 123
 of microscopically visible events, 121 ff
 percentage
  definition of, 506
  estimation from mitotic index, 502 ff
  point estimates of, 505
 theoretical basis of, from shifts in environmental conditions, **469**
 variation in degree of, 445
 with *p*-fluorophenylalanine and gaseous nitrogen, 139
Synchronization procedure, 497
Synchronized culture, 502
Synchronized populations, 81
Synchronized *Tetrahymena* systems, 100 ff, **177**
Synchronizer, large scale, 557
Synchronizing processes
 mechanism underlying, 366 ff
 techniques in flagellates, **521**
Synchronous cycle, 333 ff
Synchronous division
 biochemical investigations of, 457
 cell and nuclear, in tissues of higher plants, **11**
 chemosynthetic phase of, 137
 delays of, in *Tetrahymena*, 132 ff
 induced by chilling, 450
 induced by X-rays, 364 ff
 induction by variation of nutritional factors, 452 ff
 induction in tissue cells, 603
 mechanism of induction, 454
 mechano-phase of, 137
 of *Astasia longa*, 525
 on a light–dark cycle, 533 ff
 physical separation of dividing cells, 455
 physiological transition in, 137
 sensitivity to infection, 461
 sensitivity to irradiation, 461
 time point of maximum, 552
Synchronous generation, 108

Synchronous systems, compared with logarithmic, 111 ff
Synchrony, 1
 natural, in metazoan tissues, **39**
 of filtered culture, 422
 of mitosis, 309
 optimal conditions for, 445 ff, 453
 with continued shocking, 111
Synchrony-inducing temperature treatment, 182
Syncytia, 15 ff
Syncytial contacts, 55

Temperature
 effect on generation time, 329
 relation between multiplication rate and, 100 ff
Temperature cycle, 522 ff
Temperature-induced division synchrony of *Tetrahymena*, **99,** 144, 151
Temperature program, improvement of, 527 ff
Temperature shock, 442 ff, 451
 induction of division synchrony by, 104 ff
 responses of population to, 101
 sensitivity of cells to, 118 ff
 synchronization of cell division by, 601
Tetrahymena, 54, 72 ff, 75, 78, 80 ff, 88 ff
 centrifugation of cells, 556
 growth media for, 544
 inorganic medium for, 552
 nephelometer counting of cells, 513
 response to synchronizing treatment compared with *Acanthamoeba*, 240
 synchronized, biochemical studies on, **177**
 temperature-induced division synchrony of, **99**
 use of synchronizer, 548
Tetrahymena pyriformis, 374
 chromatography of NTP fraction, 207 ff
 components of NTP fraction, 207 ff
 morphology, 159 ff
 nucleotide metabolism in, **197**
 structural development in, **99**
Thermostat, 572
Thioglycolate, 331
 uptake rate and localization, 337 ff
Thymidine, 454 ff, 458, 460, 599
Thymidine triphosphate, 451
Thymine deprivation, 435
Time error of observation, 198, 557
Time scales, definitions, 198
Tissue cells, special methods relating to, **601**
Tissue differentiation, 60
Tmesipteris, 22
Tradescantia, 30
Transaminases, 399
Triclades, 43
Trigger event, 434
Trillium, 12, 24, 27, 28 ff, 31
 erectum, 12
Turbidostat, 599

Ultraviolet irradiation, 285
Ultraviolet shocks, 167
Unbalanced growth, 3
Unforced synchrony, 4

Vicia faba, 14, 460

Washed cells, 127, 131 ff
Welwitschia, 16

X-rays inducing synchronous division, 364 ff

Yeasts
 division synchrony in, **351**
 enzymes in synchronized populations of, **381**
 techniques for synchronizing cells of, **589**

Zamia floridana, 16
Zea, 20